IMPROVING EFFICIENCY BY SHRINKAGE

STATISTICS: Textbooks and Monographs

A Series Edited by

D. B. Owen, Founding Editor, 1972–1991

W. R. Schucany, Coordinating Editor
Department of Statistics
Southern Methodist University
Dallas, Texas

1. The Generalized Jackknife Statistic, *H. L. Gray and W. R. Schucany*
2. Multivariate Analysis, *Anant M. Kshirsagar*
3. Statistics and Society, *Walter T. Federer*
4. Multivariate Analysis: A Selected and Abstracted Bibliography, 1957–1972, *Kocherlakota Subrahmaniam and Kathleen Subrahmaniam*
5. Design of Experiments: A Realistic Approach, *Virgil L. Anderson and Robert A. McLean*
6. Statistical and Mathematical Aspects of Pollution Problems, *John W. Pratt*
7. Introduction to Probability and Statistics (in two parts), Part I: Probability; Part II: Statistics, *Narayan C. Giri*
8. Statistical Theory of the Analysis of Experimental Designs, *J. Ogawa*
9. Statistical Techniques in Simulation (in two parts), *Jack P. C. Kleijnen*
10. Data Quality Control and Editing, *Joseph I. Naus*
11. Cost of Living Index Numbers: Practice, Precision, and Theory, *Kali S. Banerjee*
12. Weighing Designs: For Chemistry, Medicine, Economics, Operations Research, Statistics, *Kali S. Banerjee*
13. The Search for Oil: Some Statistical Methods and Techniques, *edited by D. B. Owen*
14. Sample Size Choice: Charts for Experiments with Linear Models, *Robert E. Odeh and Martin Fox*
15. Statistical Methods for Engineers and Scientists, *Robert M. Bethea, Benjamin S. Duran, and Thomas L. Boullion*
16. Statistical Quality Control Methods, *Irving W. Burr*
17. On the History of Statistics and Probability, *edited by D. B. Owen*
18. Econometrics, *Peter Schmidt*
19. Sufficient Statistics: Selected Contributions, *Vasant S. Huzurbazar (edited by Anant M. Kshirsagar)*
20. Handbook of Statistical Distributions, *Jagdish K. Patel, C. H. Kapadia, and D. B. Owen*
21. Case Studies in Sample Design, *A. C. Rosander*

22. Pocket Book of Statistical Tables, compiled by R. E. Odeh, D. B. Owen, Z. W. Birnbaum, and L. Fisher
23. The Information in Contingency Tables, D. V. Gokhale and Solomon Kullback
24. Statistical Analysis of Reliability and Life-Testing Models: Theory and Methods, Lee J. Bain
25. Elementary Statistical Quality Control, Irving W. Burr
26. An Introduction to Probability and Statistics Using BASIC, Richard A. Groeneveld
27. Basic Applied Statistics, B. L. Raktoe and J. J. Hubert
28. A Primer in Probability, Kathleen Subrahmaniam
29. Random Processes: A First Look, R. Syski
30. Regression Methods: A Tool for Data Analysis, Rudolf J. Freund and Paul D. Minton
31. Randomization Tests, Eugene S. Edgington
32. Tables for Normal Tolerance Limits, Sampling Plans and Screening, Robert E. Odeh and D. B. Owen
33. Statistical Computing, William J. Kennedy, Jr., and James E. Gentle
34. Regression Analysis and Its Application: A Data-Oriented Approach, Richard F. Gunst and Robert L. Mason
35. Scientific Strategies to Save Your Life, I. D. J. Bross
36. Statistics in the Pharmaceutical Industry, edited by C. Ralph Buncher and Jia-Yeong Tsay
37. Sampling from a Finite Population, J. Hajek
38. Statistical Modeling Techniques, S. S. Shapiro and A. J. Gross
39. Statistical Theory and Inference in Research, T. A. Bancroft and C.-P. Han
40. Handbook of the Normal Distribution, Jagdish K. Patel and Campbell B. Read
41. Recent Advances in Regression Methods, Hrishikesh D. Vinod and Aman Ullah
42. Acceptance Sampling in Quality Control, Edward G. Schilling
43. The Randomized Clinical Trial and Therapeutic Decisions, edited by Niels Tyg strup, John M Lachin, and Erik Juhl
44. Regression Analysis of Survival Data in Cancer Chemotherapy, Walter H. Carter, Jr., Galen L. Wampler, and Donald M. Stablein
45. A Course in Linear Models, Anant M. Kshirsagar
46. Clinical Trials: Issues and Approaches, edited by Stanley H. Shapiro and Thomas H. Louis
47. Statistical Analysis of DNA Sequence Data, edited by B. S. Weir
48. Nonlinear Regression Modeling: A Unified Practical Approach, David A. Rat kowsky
49. Attribute Sampling Plans, Tables of Tests and Confidence Limits for Proportions, Robert E. Odeh and D. B. Owen
50. Experimental Design, Statistical Models, and Genetic Statistics, edited by Klaus Hinkelmann
51. Statistical Methods for Cancer Studies, edited by Richard G. Cornell
52. Practical Statistical Sampling for Auditors, Arthur J. Wilburn
53. Statistical Methods for Cancer Studies, edited by Edward J. Wegman and James G. Smith
54. Self-Organizing Methods in Modeling: GMDH Type Algorithms, edited by Stanley J. Farlow
55. Applied Factorial and Fractional Designs, Robert A. McLean and Virgil L. Anderson
56. Design of Experiments: Ranking and Selection, edited by Thomas J. Santner and Ajit C. Tamhane
57. Statistical Methods for Engineers and Scientists: Second Edition, Revised and Expanded, Robert M. Bethea, Benjamin S. Duran, and Thomas L. Boullion
58. Ensemble Modeling: Inference from Small-Scale Properties to Large-Scale Sys tems, Alan E. Gelfand and Crayton C. Walker
59. Computer Modeling for Business and Industry, Bruce L. Bowerman and Richard T. O'Connell
60. Bayesian Analysis of Linear Models, Lyle D. Broemeling

61. Methodological Issues for Health Care Surveys, *Brenda Cox and Steven Cohen*
62. Applied Regression Analysis and Experimental Design, *Richard J. Brook and Gregory C. Arnold*
63. Statpal: A Statistical Package for Microcomputers—PC-DOS Version for the IBM PC and Compatibles, *Bruce J. Chalmer and David G. Whitmore*
64. Statpal: A Statistical Package for Microcomputers—Apple Version for the II, II+, and IIe, *David G. Whitmore and Bruce J. Chalmer*
65. Nonparametric Statistical Inference: Second Edition, Revised and Expanded, *Jean Dickinson Gibbons*
66. Design and Analysis of Experiments, *Roger G. Petersen*
67. Statistical Methods for Pharmaceutical Research Planning, *Sten W. Bergman and John C. Gittins*
68. Goodness-of-Fit Techniques, *edited by Ralph B. D'Agostino and Michael A. Stephens*
69. Statistical Methods in Discrimination Litigation, *edited by D. H. Kaye and Mikel Aickin*
70. Truncated and Censored Samples from Normal Populations, *Helmut Schneider*
71. Robust Inference, *M. L. Tiku, W. Y. Tan, and N. Balakrishnan*
72. Statistical Image Processing and Graphics, *edited by Edward J. Wegman and Douglas J. DePriest*
73. Assignment Methods in Combinatorial Data Analysis, *Lawrence J. Hubert*
74. Econometrics and Structural Change, *Lyle D. Broemeling and Hiroki Tsurumi*
75. Multivariate Interpretation of Clinical Laboratory Data, *Adelin Albert and Eugene K. Harris*
76. Statistical Tools for Simulation Practitioners, *Jack P. C. Kleijnen*
77. Randomization Tests: Second Edition, *Eugene S. Edgington*
78. A Folio of Distributions: A Collection of Theoretical Quantile-Quantile Plots, *Edward B. Fowlkes*
79. Applied Categorical Data Analysis, *Daniel H. Freeman, Jr.*
80. Seemingly Unrelated Regression Equations Models: Estimation and Inference, *Virendra K. Srivastava and David E. A. Giles*
81. Response Surfaces: Designs and Analyses, *Andre I. Khuri and John A. Cornell*
82. Nonlinear Parameter Estimation: An Integrated System in BASIC, *John C. Nash and Mary Walker-Smith*
83. Cancer Modeling, *edited by James R. Thompson and Barry W. Brown*
84. Mixture Models: Inference and Applications to Clustering, *Geoffrey J. McLachlan and Kaye E. Basford*
85. Randomized Response: Theory and Techniques, *Arijit Chaudhuri and Rahul Mukerjee*
86. Biopharmaceutical Statistics for Drug Development, *edited by Karl E. Peace*
87. Parts per Million Values for Estimating Quality Levels, *Robert E. Odeh and D. B. Owen*
88. Lognormal Distributions: Theory and Applications, *edited by Edwin L. Crow and Kunio Shimizu*
89. Properties of Estimators for the Gamma Distribution, *K. O. Bowman and L. R. Shenton*
90. Spline Smoothing and Nonparametric Regression, *Randall L. Eubank*
91. Linear Least Squares Computations, *R. W. Farebrother*
92. Exploring Statistics, *Damaraju Raghavarao*
93. Applied Time Series Analysis for Business and Economic Forecasting, *Sufi M. Nazem*
94. Bayesian Analysis of Time Series and Dynamic Models, *edited by James C. Spall*
95. The Inverse Gaussian Distribution: Theory, Methodology, and Applications, *Raj S. Chhikara and J. Leroy Folks*
96. Parameter Estimation in Reliability and Life Span Models, *A. Clifford Cohen and Betty Jones Whitten*
97. Pooled Cross-Sectional and Time Series Data Analysis, *Terry E. Dielman*
98. Random Processes: A First Look, Second Edition, Revised and Expanded, *R. Syski*
99. Generalized Poisson Distributions: Properties and Applications, *P. C. Consul*
100. Nonlinear L_p-Norm Estimation, *Rene Gonin and Arthur H. Money*
101. Model Discrimination for Nonlinear Regression Models, *Dale S. Borowiak*

102. Applied Regression Analysis in Econometrics, *Howard E. Doran*
103. Continued Fractions in Statistical Applications, *K. O. Bowman and L. R. Shenton*
104. Statistical Methodology in the Pharmaceutical Sciences, *Donald A. Berry*
105. Experimental Design in Biotechnology, *Perry D. Haaland*
106. Statistical Issues in Drug Research and Development, *edited by Karl E. Peace*
107. Handbook of Nonlinear Regression Models, *David A. Ratkowsky*
108. Robust Regression: Analysis and Applications, *edited by Kenneth D. Lawrence and Jeffrey L. Arthur*
109. Statistical Design and Analysis of Industrial Experiments, *edited by Subir Ghosh*
110. U-Statistics: Theory and Practice, *A. J. Lee*
111. A Primer in Probability: Second Edition, Revised and Expanded, *Kathleen Subrah maniam*
112. Data Quality Control: Theory and Pragmatics, *edited by Gunar E. Liepins and V. R. R. Uppuluri*
113. Engineering Quality by Design: Interpreting the Taguchi Approach, *Thomas B. Barker*
114. Survivorship Analysis for Clinical Studies, *Eugene K. Harris and Adelin Albert*
115. Statistical Analysis of Reliability and Life-Testing Models: Second Edition, *Lee J. Bain and Max Engelhardt*
116. Stochastic Models of Carcinogenesis, *Wai-Yuan Tan*
117. Statistics and Society: Data Collection and Interpretation, Second Edition, Revised and Expanded, *Walter T. Federer*
118. Handbook of Sequential Analysis, *B. K. Ghosh and P. K. Sen*
119. Truncated and Censored Samples: Theory and Applications, *A. Clifford Cohen*
120. Survey Sampling Principles, *E. K. Foreman*
121. Applied Engineering Statistics, *Robert M. Bethea and R. Russell Rhinehart*
122. Sample Size Choice: Charts for Experiments with Linear Models: Second Edition, *Robert E. Odeh and Martin Fox*
123. Handbook of the Logistic Distribution, *edited by N. Balakrishnan*
124. Fundamentals of Biostatistical Inference, *Chap T. Le*
125. Correspondence Analysis Handbook, *J.-P. Benzécri*
126. Quadratic Forms in Random Variables: Theory and Applications, *A. M. Mathai and Serge B. Provost*
127. Confidence Intervals on Variance Components, *Richard K. Burdick and Franklin A. Graybill*
128. Biopharmaceutical Sequential Statistical Applications, *edited by Karl E. Peace*
129. Item Response Theory: Parameter Estimation Techniques, *Frank B. Baker*
130. Survey Sampling: Theory and Methods, *Arijit Chaudhuri and Horst Stenger*
131. Nonparametric Statistical Inference: Third Edition, Revised and Expanded, *Jean Dickinson Gibbons and Subhabrata Chakraborti*
132. Bivariate Discrete Distribution, *Subrahmaniam Kocherlakota and Kathleen Kocherlakota*
133. Design and Analysis of Bioavailability and Bioequivalence Studies, *Shein-Chung Chow and Jen-pei Liu*
134. Multiple Comparisons, Selection, and Applications in Biometry, *edited by Fred M. Hoppe*
135. Cross-Over Experiments: Design, Analysis, and Application, *David A. Ratkowsky, Marc A. Evans, and J. Richard Alldredge*
136. Introduction to Probability and Statistics: Second Edition, Revised and Expanded, *Narayan C. Giri*
137. Applied Analysis of Variance in Behavioral Science, *edited by Lynne K. Edwards*
138. Drug Safety Assessment in Clinical Trials, *edited by Gene S. Gilbert*
139. Design of Experiments: A No-Name Approach, *Thomas J. Lorenzen and Virgil L. Anderson*
140. Statistics in the Pharmaceutical Industry: Second Edition, Revised and Expanded, *edited by C. Ralph Buncher and Jia-Yeong Tsay*

141. Advanced Linear Models: Theory and Applications, *Song-Gui Wang and Shein-Chung Chow*
142. Multistage Selection and Ranking Procedures: Second-Order Asymptotics, *Nitis Mukhopadhyay and Tumulesh K. S. Solanky*
143. Statistical Design and Analysis in Pharmaceutical Science: Validation, Process Controls, and Stability, *Shein-Chung Chow and Jen-pei Liu*
144. Statistical Methods for Engineers and Scientists: Third Edition, Revised and Expanded, *Robert M. Bethea, Benjamin S. Duran, and Thomas L. Boullion*
145. Growth Curves, *Anant M. Kshirsagar and William Boyce Smith*
146. Statistical Bases of Reference Values in Laboratory Medicine, *Eugene K. Harris and James C. Boyd*
147. Randomization Tests: Third Edition, Revised and Expanded, *Eugene S. Edgington*
148. Practical Sampling Techniques: Second Edition, Revised and Expanded, *Ranjan K. Som*
149. Multivariate Statistical Analysis, *Narayan C. Giri*
150. Handbook of the Normal Distribution: Second Edition, Revised and Expanded, *Jagdish K. Patel and Campbell B. Read*
151. Bayesian Biostatistics, *edited by Donald A. Berry and Dalene K. Stangl*
152. Response Surfaces: Designs and Analyses, Second Edition, Revised and Expanded, *André I. Khuri and John A. Cornell*
153. Statistics of Quality, *edited by Subir Ghosh, William R. Schucany, and William B. Smith*
154. Linear and Nonlinear Models for the Analysis of Repeated Measurements, *Edward F. Vonesh and Vernon M. Chinchilli*
155. Handbook of Applied Economic Statistics, *Aman Ullah and David E. A. Giles*
156. Improving Efficiency by Shrinkage: The James–Stein and Ridge Regression Estimators, *Marvin H. J. Gruber*

Additional Volumes in Preparation

IMPROVING EFFICIENCY BY SHRINKAGE

The James–Stein and Ridge Regression Estimators

MARVIN H. J. GRUBER

Rochester Institute of Technology
Rochester, New York

CRC Press
Taylor & Francis Group
Boca Raton London New York

CRC Press is an imprint of the
Taylor & Francis Group, an **informa** business

First published in 1998 by Marcel Dekker

Published in 2020 by CRC Press
Taylor & Francis Group
6000 Broken Sound Parkway NW, Suite 300
Boca Raton, FL 3487-2742

First issued in paperback 2020

ISBN-13: 978-0-367-57936-4 (pbk)
ISBN-13: 978-0-8247-0156-7 (hbk)

Visit the Taylor & Francis Web site at
http://www.taylorandfrancis.com

and the CRC Press Web site at
http://www.crcpress.com

Library of Congress Cataloging-in-Publication Data

Gruber, Marvin H. J.
    Improving efficiency by shrinkage: the James-Stein and ridge regression esti-
mators / Marvin H. J. Gruber.
        p. cm.—(Statistics, textbooks and monographs; v. 156)
    Includes bibliographical references and indexes.
    ISBN 0-8247-0156-9 (alk. paper)
    1. Regression analysis. 2. Estimation theory. I. Title. II. Series
QA278.2.G77   1998
    519.5'36--dc21   98-5183
                                                                      CIP

In memory of my parents
Joseph and Adelaide Gruber

Preface

In 1956 Charles Stein published the surprising result that the maximum likelihood estimator for the mean of a multivariate normal distribution is inadmissible. In 1970 Hoerl and Kennard published a paper describing the properties of the ridge regression estimator and explaining how it could be used to obtain better estimates of parameters for multicollinear data. Since the publication of these two landmark papers a huge amount of research has been done on these two kinds of estimators and their variations. Both the ridge and the James-Stein estimator are special cases of a more general class of estimators called shrinkage estimators. The purpose of this book is to give a unified treatment of these two kinds of estimators from both a Bayesian and a non-Bayesian (frequentist) point of view.

The book is divided into three parts. The first part (Chapters I to III) explores how ridge and James-Stein estimators are better than least square or maximum likelihood estimators in the context of their original formulation. A historical summary in Chapter I provides the interested reader with many entry points to the vast literature on the subject. It should be of particular interest to graduate students and others contemplating research in the area. Chapter II includes a proof of the inadmissibility of the James-Stein estimator. Chapter III deals with the various practical issues that are important in the study of ridge regression estimators. Part II (Chapters IV to VI) deals with different Bayesian and non-Bayesian formulations of the ridge and James-Stein estimators. The estimators efficiencies are compared with the least square estimator. It is also shown how the efficiency of the JS is improved by using its positive parts. Part III (Chapters VII-X) considers similar problems for other linear model setups including simultaneous estimation of parameters for r linear models, the multivariate model, the seemingly unrelated regression equation model and the Kalman filter.

In addition to the theoretical development there are a number of numerical examples. Some of the computer programs used to generate these examples are included in both Mathematica and Maple. There is an abundance of exercises. These help the reader check his or her understanding, furnish more examples to illustrate the theory developed in the text, and extend the text.

Hopefully this book might be used as a textbook for a graduate course on Alternatives to Least Square Methods. It will be useful to graduate students and professionals in statistics, engineering and econometrics who would like to know more about the analytical properties and the applicability of James-Stein and ridge regression-type estimators.

A project like this can never be completed without the help of many other people. The results for a large portion of the material in Chapters IV , V, VII and IX were obtained in my Ph.D. thesis, written under the direction of Dr. Poduri S.R.S. Rao at the University of Rochester. Some of the results in Chapter V and VII were due to joint work done after the thesis was completed. The problems considered in Chapter VI and VIII were suggested to me by Dr. Rao when I was his student or later. I would like to express my gratitude to Dr. Rao for my long association with him. However, this volume is completely my own work and I take full responsibility for any errors.

This book was typed by the author on a Macintosh computer using Word 6 with an equation editor. An earlier version of the book was typed in TEX or AMS TEX by several people. These included students majoring in one of the three undergraduate programs offered by the Department of Mathematics and Statistics at Rochester Institute of Technology: Applied Mathematics, Computational Mathematics or Applied Statistics. Two students who did a great deal of work on the earlier version of the manuscript were Christopher Bean and Christine Jodoin. The original version of Chapter VII was typed by Joan Robinson, secretary of the Department of Mathematics at the University of Rochester. The contract to write the book was awarded based on this preliminary version. Certainly the excellent typing these people did helped make a good impression. I would like to express my appreciation for their fine work.

I would like to thank Dr. Patricia A. Clark for the many hours she spent with a highly computer-illiterate person providing instruction in Mathematica and the use of the Macintosh computer. Getting me to the level where I could write the programs in this book provided her with a major challenge. Many thanks to Professor Rebecca Hill, who provided a great deal of help with the graphics and other kinds of support and encouragement. Dr. Jack Weiss spent a considerable amount of time showing me how to use Word 6 to help compile the index. I am very grateful for this help.

During the Spring quarter of the 1992 academic year I was granted a sabbatical leave from my teaching duties. It was during this period that what eventually became part of Chapters I, II, IV, VII and IX was written. I was very grateful to have the time off to work on this project. The research reported in this book was supported by Dean's fellowships in the College of Science during summers from 1980 through 1996. The fellowships for the summers of 1995 and 1996 were specifically for work on the book. I would like to express my appreciation for these opportunities.

Other people who should be thanked for help and support in various ways include Dr. Mary Beth Krough Jesperesen, Dr. John Paliouras, Dr. Robert A. Clark, Dr. George Georgantas, Professor Richard Orr, Kelley Youngblood, Elinora Ayers and Suzanne Flynn- Parsons.

The staff of Marcel Dekker was very helpful in the production of the book. The book was originally to cover only James-Stein estimators. Russell Dekker suggested that I expand the scope of the book to include shrinkage estimators. The result was the present volume on both James-Stein and ridge regression estimators. The editorial staff was very helpful in guiding my preparation of the book in camera ready format and improve the typesetting of the equations. Maria Allegra, the acquisitions editor, was very helpful and patient and provided various kinds of support.

The content of the book was improved through thoughtful comments of two referees. I would like to thank these people for taking the time to read the work.

I am grateful for the friendship of Frances Johnson and for the help and support she has given me over the years.

This book is dedicated to the memory of my parents, Joseph and Adelaide Gruber, who always supported my personal and professional efforts during my growing-up years and my younger adulthood. Shortly before she died my mother informed me in a letter that this book would sell. She was often right about things.

Marvin H. J. Gruber

Contents

Preface

v

Part I Introduction to Shrinkage Estimators

Chapter I. Introduction   1
    1.0 Motivation for Writing This Book   1
    1.1 Purpose of This Book   4
    1.2 Introduction to the James-Stein Estimator (JS)   5
    1.3 Introduction to the Ridge Regression Estimator   7
    1.4 Historical Survey   15
    1.5 Empirical Bayes Methodology   61
    1.6 A Minimum Mean Square Error Estimator,
         Its Approximation as an Empirical Bayes
         Estimator   66
    1.7 The Structure of This Book   68

Chapter II. The Stein Paradox   71
    2.0 Estimation of the Mean   71
    2.1 The Multivariate Normal Distribution   73
    2.2 Maximum Likelihood Estimation   76
    2.3 The Decision Theory Framework   81
    2.4 Why the MLE May Not Be the Best   84
    2.5 Another Case Against the MLE   91

2.6   The Inadmissibility of the MLE   93
2.7   The JS as an Empirical Bayes Estimator (EBE)   99
2.8. How the JS is a Special Case of the Operational Ridge
      Regression Estimator   104
2.9   Introduction to the Positive Part of the JS   107
2.10 Summary   110

Chapter III. The Ridge Estimators of Hoerl and Kennard   111
3.0   The Need for Alternatives to the
      Least Square Estimators   111
3.1   Derivation of the Ridge Estimator   112
3.2   The Efficiency of the Ridge Regression Estimator   117
3.3   Estimating the Ridge Perturbation Factor k-
      The Ridge Trace   133
3.4   Estimating the Ridge Perturbation Factor k-
      Objective Methods   141
3.5   Evaluation of the Efficiency of Ridge Estimators
      by Computer Simulation   158
3.6   Summary   166

Part II Estimation for a Single Linear Model

Chapter IV. James-Stein Estimators for a Single Linear Model   167
4.0   Introduction   167
4.1   Review of Basic Linear Model Concepts   169
4.2   The James-Stein Estimator from a
      Bayesian Point of View   173
4.3   The Non-Bayesian Formulation of the JS and the JSL   186
4.4   The Average Mean Square Error (MSE)   190
4.5   The Conditional MSE   195
4.6   The MSE of the JS for Different Quadratic
      Loss Functions   202
4.7   The Efficiency of the JS as Compared with
      Optimum Estimators   206
4.8   Summary   223
4.9   Appendix   223

Chapter V. Ridge Estimators from Different Points of View   227
   5.0 Introduction   227
   5.1 Some More Review of Linear Models   229
   5.2 The Generalized Ridge Regression Estimator   240
   5.3 The Mixed Estimators   242
   5.4 The Linear Minimax Estimator   248
   5.5 The Linear Bayes Estimator   250
   5.6 Comparing the Efficiency of the Ridge and
       the Least Square Estimator from the
       Frequentist Point of View   266
   5.7 Comparing the MSE of Ridge-Type
       Estimators from the Bayesian Point of View   284
   5.8 The Jack-knifed Ridge Estimator   298
   5.9 Summary   305

Chapter VI. Improving the James-Stein Estimator: The Positive Parts   307

   6.0 Introduction   307
   6.1 The Positive Parts of the James-Stein
       Estimator (PP_0 - PP_4)   309
   6.2 The Average Mean Square Error   312
   6.3 The Conditional Mean Square Error   337
   6.4 The Estimators of Shao and Strawderman   364
   6.5 Summary   370

Part III Other Linear Model Setups
  Chapter VII.   The Simultaneous Estimation Problem   371
   7.0 Introduction   371
   7.1 The EBE of C. R. Rao   373
   7.2 The EBE of Wind   387
   7.3 The Estimator of Dempster   403
   7.4 A Generalization of the EBE   413
   7.5 Comparing the Efficiency of the Estimators   418
   7.6 The Positive Parts   432
   7.7 Summary   439

  Chapter VIII.   The Precision of Individual Estimators   441
   8.0 Introduction   441
   8.1 The Mean of a Multivariate Normal Distribution   442
   8.2 The Single Linear Model   448

8.3   The Case of r Linear Models   458
8.4   The Limited Translation Estimators   480
8.5   Summary   490

Chapter IX. The Multivariate Linear Model   491
9.0   Introduction   491
9.1   The Multivariate Linear Model   493
9.2   The BE   496
9.3   The EBE of Wind for One Linear Model   500
9.4   The EBE for r Linear Models   512
9.5   The Estimator of Dempster
      (An Example of an Approximate MSE)   525
9.6   Summary   529

Chapter X.   Other Linear Model Setups   531
10.0 Introduction   531
10.1 The Seemingly Unrelated Regression Model (SURE)   533
10.2 Some Aspects of the Simultaneous Estimation Problem   542
10.3 The Kalman Filter   556
10.4 Summary   578

Chapter XI. Summary and Conclusion   579
11.0 Introduction   579
11.1 Overview of the Chapters   580
11.2 Conclusion   590

References   591
Author Index   619
Subject Index   625

IMPROVING
EFFICIENCY
BY SHRINKAGE

Chapter I

Introduction

1.0 Motivation for Writing This Book

In 1956 a result due to Charles Stein was published that represented an important breakthrough in statistical estimation theory. Stein showed that the maximum likelihood estimator (MLE) for the mean of a multivariate normal distribution is inadmissible. This means that it is possible to construct an estimator with smaller risk than the MLE for the entire parameter space. James and Stein (1961) exhibited an estimator with risk uniformly smaller than that of the MLE. This estimator is now commonly referred to in the literature as the James-Stein estimator.

One of the reasons the James-Stein estimator (JS) is so revolutionary is that many different aspects of something being measured are used to describe a single measurement. For example, many batting averages of different baseball players can be used to estimate the batting average of a single player. The incidence of a disease in many different regions can help to estimate the incidence of the disease in a small area. Surprisingly one can do this and produce a better estimator.

1

The statistical community was indeed surprised by Stein's result. It is frequently called the Stein paradox. As a result, beginning with the early sixties many different papers and books have been written about the JS and its variants.

Often in the applications of Statistics when a linear regression model is fitted some of the independent variables are highly correlated and thus might have a linear relationship between them. When this happens the variance covariance matrix of the least square estimator will be almost singular and the least square estimators will be very imprecise. A possible solution to this problem suggested by Hoerl and Kennard (1970 a, b) was the formulation of the ridge regression estimator (RR). The ridge regression estimator has been shown to give more precise estimates of the regression parameters because its variance and its mean square error are generally smaller than that of the least square estimators. Ridge regression estimators have found diverse applications in dealing with multicollinear data in many fields, for example Chemistry, Econometrics and Engineering. Since 1970 many different papers and books have been written about the theoretical development and the application of ridge regression estimators.

It will be shown in the sequel that both the JS and the RR can be obtained by multiplication of the LS by a shrinkage factor. Furthermore shrinkage estimators are generally more efficient. That is why this book is called Improving Efficiency by Shrinkage with the subtitle The James-Stein and the Ridge Regression Estimator.

There are many ways to motivate or derive the JS. It can be derived from both the Bayesian and the frequentist point of view. The two methods that will be extensively considered in this book are obtaining the JS:

1. an approximation to the Bayes estimator, i.e., an empirical Bayes estimator (EBE);
2. as an approximation to a minimum mean square error estimator (MMSE).

The first method shows how the JS is almost optimal from the Bayesian point of view; the second method shows how the JS is almost optimal from the frequentist point of view. An interesting observation is the fact that the same estimator can be derived from two philosophically different points of view.

The ridge estimator can also be obtained in many different ways and from both the frequentist and the Bayesian point of view. It can be shown that the ridge estimator is a special case of the following:

1. the Bayes estimator (BE);
2. the mixed estimator;
3. the minimax estimator.

Originally Hoerl and Kennard (1970) derived the ridge regression estimator by finding the point on an ellipsoid that is closest to the point in the parameter space. When the ridge parameters are unknown or unspecified they may be estimated from the data. The operational ridge estimator that results may be thought of as an EBE, or as an approximation to the mixed or the minimax estimator. Certain kinds of ridge estimators, e.g., the generalized ridge regression estimator of Hoerl and Kennard are also approximate MMSEs. Moreover the JS can be obtained as a special case of the operational ridge estimator.

The JS, the RR and their variants may be considered in many different contexts and in many different practical settings. This volume considers different linear model setups. These include a single linear model, r linear models, the multivariate model, the Kalman filter, the seemingly unrelated regression model and the simultaneous equation model. These different kinds of linear models are useful for the solution to practical problems in many different diverse areas. Some areas of application include genetics, econometrics, aerospace tracking and quality control, only to mention a few.

The Stein type estimators have proved to have greater precision in many practical situations. A few of these include election night forecasting, sports statistics and fire alarm probabilities. The ridge estimators estimate the regression parameters more precisely. Chapter I of Gruber (1990) contains an example that illustrates this using data taken from the Economic Report of the President (1988). There the dependence of Personal Consumption Expenditures on Gross National Product, Personal Income and the total number of employed people in the civilian labor force is studied. Examples 1.3.1 and 1.3.2 to be given in Section 1.3 of this chapter will explain by example how the ridge estimators have greater precision than the least square estimators using data on the percentage of the Gross National Product spent on research and development by the United States and four foreign countries. The dependence of the percentage spent by the United States on the percentage spent by France, Japan, West Germany and the former Soviet Union will be studied.

The contents of this work overlaps with but is in many ways different from Regression Estimators: A Comparative Study written by the author and published by Academic Press in April 1990. Gruber (1990) considered ridge type estimators from both the Bayesian and the frequentist point of view. There prior information was completely known or available. The setups considered there included only one linear model or one Kalman filter. The present volume is broader in scope than Gruber (1990) because it:

1. treats both the cases where prior information is known and when the prior information is incomplete or unknown;
2. takes up different linear model setups, e.g., r linear models, multivariate models, more than one Kalman filter, etc.
3. summarizes many known results in the literature about the JS and the RR of interest to the practitioner.

1.1 Purpose of This Book

The intent of this book is to study the comparative properties of James-Stein, ridge type estimators and least square estimators for different linear model setups from both the Bayesian and the frequentist point of view.

The goals of this study include:

1. the formulation of ridge estimators from both the Bayesian and the frequentist point of view;
2. the formulation of JS and operational ridge estimators for different linear model setups from both the empirical Bayes and the pure frequentist point of view;
3. the evaluation of the efficiency of these estimators in both the Bayes and the frequentist sense;
4. comparison of the efficiencies of different JS and ridge type estimators with each other and the least square(LS) estimators;
5. illustrations of situations in both a theoretical and an applied framework where the JS and the ridge estimators are superior to the LS.

The rest of this chapter will contain:
1. a brief introduction to JS estimation (Section 1.2);
2. a brief introduction to ridge estimation (Section 1.3);
3. a historical summary of work done by this and other authors (Section 1.4);
4. two examples that illustrate empirical Bayes estimation (Section 1.5);
5. an example of an approximate MMSE (Section 1.6);
6. an outline of the structure of the rest of this book (Section 1.7).

The introduction to Stein estimation and ridge estimation explains why these two kinds of estimators might be used instead of the MLE or the LS for certain kinds of data sets. More details are given in Chapters II and III including the derivation of results stated without proof in Sections 1.2 and 1.3. An example using weather data in different cities is given that illustrates why the JS might be better than the MLE. An illustration of the usefulness of the ridge estimator is also given using econometric data.

The historical survey is intended to give the interested reader many entry points into the vast literature on this subject. It is not intended to be exhaustive. The author apologizes to anyone whose favorite paper or book is omitted. Chances are, however, that by consulting the Bibliographies given in the references that are cited in the historical survey the reader will be able to find those that are not listed. Much of the material that is taken up in the references cited in the historical survey will not be covered in this volume. These references are included to give an idea of the different kinds of research that has been done on shrinkage estimators and to provide suggestions for further reading.

The examples of empirical Bayes estimators are intended to give the reader a general idea of what empirical Bayes methodology is. The first example due to Robbins is very well known. The second one illustrates how the methodology can be applied to Analysis of Variance. The empirical Bayes methodology is applied extensively in the later chapters.

Likewise the example of an approximate MMSE will give an idea of the formulation of an almost optimal estimator from the frequentist point of view. The technique will also be extensively applied. It will also be shown how this same estimator is also an EBE.

1.2 Introduction to the James-Stein Estimator (JS)

This section introduces the JS and illustrates some of its important advantages. Motivating it, explaining why it works and deriving its properties is left to Chapter II and the succeeding chapters.

Consider the problem of estimating the mean of the multivariate normal distribution (MN) for the simple case where the individual variables are independent and the variance is known, i.e. let $X_i \sim N(\theta_i, 1)$, $1 \leq i \leq p$ be independent normal random variables. In Section 2.2 it will be shown that given a single multivariate observation the MLE of the vector $\theta = (\theta_1, \theta_2, \dots, \theta_p)$ is the p dimensional vector $X = (x_1, x_2, \dots, x_p)$ and that the risk

$$R = \frac{1}{p} \sum_{i=1}^{p} E(x_i - \theta_i)^2 = 1. \tag{1.2.1}$$

The estimator of θ proposed by James and Stein is a contraction of the MLE, i.e.,

$$\hat{\theta}_i = \left(1 - \frac{(p-2)}{\sum_{i=1}^{p} x_i^2}\right) x_i, \ 1 \leq i \leq p. \tag{1.2.2}$$

The estimator (1.2.2) has risk

$$R = 1 - \frac{(p-2)^2}{p} E\left[\frac{1}{p-2K+2}\right] \tag{1.2.3}$$

where K is a Poisson random variable with parameter $\sum_{i=1}^{p} \theta_i^2 / 2$ (See Section 2.6 for the derivation.). Since the risk of (1.2.3) is clearly smaller than that of the MLE (1.2.1) the MLE is inadmissible.

A noteworthy aspect of the estimator (1.2.2) is that the observations of all of the other coordinates are included in the estimator of θ_i. For example suppose the batting average of each member of a nine man baseball team is being estimated. Player 7's batting average is estimated using the observations of the batting averages of his other eight teammates. It is somewhat puzzling that an estimator produced this way actually should be better; for this reason the inadmissibility of the MLE is often referred to as the Stein paradox.

Another reason that one might suspect that the JS might be a better estimator than the MLE is that it is often ``closer'' to the true mean. This is illustrated by Example 1.2.1 below.

Example 1.2.1. Deviation of average temperatures in a given year from all time average. The data in Table 1.2.1 give average temperatures for July 1990 and the long time average July temperatures for 10 United States cities. The source is the Weather Almanac (1992). The data was used with the permission of Gale Research Inc.

Assuming that the deviations are normally distributed the MLEs are given by the last column. The true mean deviation should be zero. One measure of goodness is the sum of the squares of the deviations (SSD) divided by $p = 10$.

For the example under consideration

$$\text{SSD} = 20.63/10 = 2.063.$$

The JS contraction factor is

$$c = 1 - 8/20.63 = .612.$$

Thus the JS of the deviations is

$$\hat{\theta} = (-.428, -.551, 1.102, 1.163, 1.224, 1.408, .306, .857, -.428, .428).$$

For the JS

$$\text{SSD} = .773$$

less than half that of the MLE. Notice again that information about all of the cities is used to find the JS for a particular city. Thus, for example what happened in Lexington, Kentucky is part of the estimate for Rochester, New York.

Exercise 1.2.1. Find the JS of the deviation of the average 1990 July temperature from the all time average using the first five cities in the table.

Exercise 1.2.2. Consider the modified JS

$$\hat{\theta}_i = \bar{x} + \left(1 - \frac{(p-3)}{\sum_{i=1}^{p}(x_i - \bar{x})^2} \right)(x_i - \bar{x}).$$

Find the modified JS of the 1990 temperatures in each of the ten cities. How does the sum of the squares of the differences of this JS from that of the long time average compare with the sum of the squares of the differences between the MLE for 1990 and the long time average?

1.3 Introduction to the Ridge Regression Estimator

Statisticians often use the method of least squares to estimate the parameters of a linear regression model. The method of least squares consists of minimizing the sum of the squares of the differences between the observed and the predicted values of the dependent variable. Frequently, in applications there is a linear dependence between the independent variables. When this happens the precision of the least square estimators may be very poor, i.e., their variance may be very large. Ridge regression estimators often tend to be more precise for this type of data. To explain what is meant by this more fully the least square estimator and the well known ridge regression estimator of Hoerl and Kennard will now be presented.

Consider the general linear model

$$Y = X\beta + \varepsilon. \tag{1.3.1}$$

The matrix X is a known $n \times m$ matrix of rank $s \leq m$, Y is the n dimensional observation vector and ε is the n dimensional error vector. The β parameter is an m dimensional vector. The error vector ε is an n dimensional random variable assumed to have mean 0 and dispersion $\sigma^2 I$. Consequently, the random variable Y has mean $X\beta$ and dispersion $\sigma^2 I$. The components of Y represent independent observations.

The least square estimator is obtained by minimizing the sum of the squares of the error terms or equivalently the sum of the squares of the differences between the observed and the fitted values. Thus,

$$
\begin{aligned}
F(\beta) &= (Y - X\beta)'(Y - X\beta) \\
&= Y'Y + \beta'X'X\beta - 2\beta'X'Y
\end{aligned}
\tag{1.3.2}
$$

is minimized by differentiation and setting the result equal to zero. Thus,

$$
\frac{\partial F(\beta)}{\partial \beta} = X'X\beta - X'Y = 0.
\tag{1.3.3}
$$

The right hand equality in (1.3.3) may be rewritten

$$
X'X\beta = X'Y.
\tag{1.3.4}
$$

Equation (1.3.4) is called the normal equation. The solutions to the normal equation are called the least square estimators.

In real world data frequently there is a linear dependence between two or more of the m variables represented by the X matrix. When this happens the data is said to be multicollinear. There are two kinds of multicollinearity that will be studied in this book:

1. exact multicollinearity, i.e., the matrix X is of non-full rank;
2. near multicollinearity, i.e., the X'X matrix has at least one very small eigenvalue; consequently det X'X is very close to zero.

The solution to (1.3.4) for the full rank case is given by

$$
b = (X'X)^{-1}X'Y.
\tag{1.3.5}
$$

For the non-full rank case solutions to the normal equation may be obtained by finding a matrix G where

$$
b = GX'Y.
\tag{1.3.6}
$$

The matrix G is called a generalized inverse. It is not unique; for the non-full rank case the normal equation (1.3.4) has infinitely many solutions. Generalized inverses and their properties will be considered in Chapter IV and V and extensive use of them will be made throughout the book. The properties of the least square estimators for the non-full rank model will be dealt with in Chapters IV and V. The non-full rank case occurs more frequently in experimental design models where the X matrix is specified in advance and independent observations of the dependent or response variable Y are taken. For data where the X matrix is not specified near multicollinearity is often present.

The dispersion of b is

$$D(b) = \sigma^2 (X'X)^{-1}. \tag{1.3.7}$$

The trace of the dispersion matrix is the total variance. Thus, the total variance is

$$\text{Tr } D(b) = \sigma^2 \text{Tr}(X'X)^{-1} = \sigma^2 \sum_{i=1}^{s} \frac{1}{\lambda_i}, \tag{1.3.8}$$

where the λ_i are the non-zero eignevalues of $X'X$. From (1.3.8) it can be seen that the total variance would be severely inflated if one or more of the non-zero eigenvalues are very small.

The ordinary ridge regression estimator, suggested by Hoerl and Kennard as a possible remedy for this problem, is obtained by simply adding the scalar matrix kI to $X'X$ in the least squares estimator. Thus the ridge estimator takes the form

$$\hat{\beta} = (X'X + kI)^{-1} X'Y, \tag{1.3.9}$$

with k a positive constant. The total variance of the ridge estimator of the β parameter is

$$TV(\hat{\beta}) = \sigma^2 \sum_{i=1}^{s} \frac{\lambda_i}{(\lambda_i + k)^2} \tag{1.3.10}$$

It is clear that (1.3.10) is less than (1.3.8). Thus, the ordinary RR has a smaller total variance than the LS.

The ridge estimator (1.3.9) can be derived in many ways as will be explained in Chapters III and V. Originally Hoerl and Kennard (1970) obtained the ridge estimator by finding the point on the ellipsoid centered at the LS estimator b that is closest to the origin. This consists of minimizing the distance B'B of a vector of parameters B from the origin subject to the side condition

$$(B - b)'X'X(B - b) = \Phi_0. \tag{1.3.11}$$

The optimization problem is solved using Lagrange multipliers. Differentiate

$$L = B'B + \frac{1}{k}[(B - b)'X'X(B - b) - \Phi_0]$$ (1.3.12)

to obtain

$$B + \frac{1}{k}X'X(B - b) = 0.$$ (1.3.13)

The ridge estimator (1.3.9) is obtained by solving (1.3.13) for B.

Exercise 1.3.1. Minimize $B'X'XB$ subject to the constraint (1.3.11). What form of estimator is obtained?

Example 1.3.1. How Ridge Estimators are More Precise
Data found in economics are often multicollinear. Table 1.3.1 gives Total National Research and Development Expenditures -- as a Percent of Gross National Product by Country: 1972-1986. The data is highly multicollinear. It represents the relationship between the dependent variable Y the percentage spent by the United States and four other independent variables X_1, X_2, X_3 and X_4. The variable X_1 represents the percent spent by France, X_2 that spent by West Germany, X_3 that spent by Japan and X_4 that spent by the former Soviet Union. The data was reproduced with the position of Time-Life inc.
The telltale sign of multicollinear data is a high correlation between the independent variables. The variables X_1 and X_3 representing France and Japan have a correlation of 0.925. The variables X_2 and X_3 representing West Germany and Japan have an even higher correlation of 0.962. The matrix

$$X'X = \begin{bmatrix} 10 & 20.3 & 25.0 & 23.7 & 37.7 \\ 20.3 & 41.57 & 51.07 & 48.65 & 76.59 \\ 25.0 & 51.07 & 62.86 & 59.81 & 94.28 \\ 23.7 & 48.65 & 59.81 & 57.11 & 89.45 \\ 37.7 & 76.59 & 94.28 & 89.45 & 142.23 \end{bmatrix}$$

has eigenvalues

$$\lambda_1 = 312.932$$
$$\lambda_2 = 0.75337$$
$$\lambda_3 = 0.04531$$
$$\lambda_4 = 0.03717$$
$$\lambda_5 = 0.00186$$

Observe that:

1. The last three eigenvalues are very small.
2. The ratio of the largest eigenvalue to the smallest, a measure of the degree of multicollinearity is very large 168,238.
3. The matrix $X'X$ is almost singular, that is, det $(X'X) = 0.000739$.
4. The total variance

$$\sigma^2 \text{Tr}(X'X)^{-1} = \sigma^2 \sum_{i=1}^{5} \frac{1}{\lambda_i} = 587.569\sigma^2.$$

The values of the total variance for a few small values of k are given in Table 1.3.2 .

Table 1.3.1

OBS	YEAR	Y	X_1	X_2	X_3	X_4
1	1972	2.3	1.9	2.2	1.9	3.7
2	1975	2.2	1.8	2.2	2.0	3.8
3	1979	2.2	1.8	2.4	2.1	3.6
4	1980	2.3	1.8	2.4	2.2	3.8
5	1981	2.4	2.0	2.5	2.3	3.8
6	1982	2.5	2.1	2.6	2.4	3.7
7	1983	2.6	2.1	2.6	2.6	3.8
8	1984	2.6	2.2	2.6	2.6	4.0
9	1985	2.7	2.3	2.8	2.8	3.7
10	1986	2.7	2.3	2.7	2.8	3.8

Observe that even for small values of k the improvement in the total variance is quite substantial. When k=.01 the total variance is reduced by a factor of more than 12. When k=1 the reduction factor is greater than 1800.

Table 1.3.2

k	0.01	0.1	0.5	1
Total Variance	$46.03\sigma^2$	$5.332\sigma^2$	$0.769\sigma^2$	$0.325\sigma^2$

Example 1.2.2. Least Square and Ridge Regression Estimators for Data on Percentage of Gross National Product Spent on Research and Development in the United States Compared with Selected Foreign Countries. Recall that the solution to the normal equation

$$X'X\beta = X'Y.$$

is

$$b = (X'X)^{-1}X'Y.$$

For the data in Example 1.2.1

$$(X'X)^{-1}$$
$$= \begin{bmatrix} 459.083 & -13.0079 & -135.966 & 94.8876 & -84.2298 \\ -13.0079 & 19.1423 & 0.438158 & -11.2149 & 0.0973683 \\ -135.966 & 0.438158 & 59.9464 & -38.0737 & 20.0119 \\ 94.8876 & -11.2149 & -38.0737 & 31.6217 & -13.7614 \\ -84.2298 & -0.0973683 & 20.0119 & -13.7614 & 17.7751 \end{bmatrix}.$$

and

$$X'Y = \begin{bmatrix} 24.50 \\ 50.08 \\ 61.57 \\ 58.61 \\ 92.43 \end{bmatrix}.$$

Thus

$$\begin{bmatrix} b_0 \\ b_1 \\ b_2 \\ b_3 \\ b_4 \end{bmatrix} = \begin{bmatrix} 0.6921 \\ 0.6258 \\ -0.1154 \\ 0.2866 \\ 0.0256 \end{bmatrix}.$$

When k = 1 the ridge regression estimator

$$\begin{bmatrix} \hat{\beta}_0 \\ \hat{\beta}_1 \\ \hat{\beta}_2 \\ \hat{\beta}_3 \\ \hat{\beta}_4 \end{bmatrix} = \begin{bmatrix} 0.0643 \\ 0.1986 \\ 0.2058 \\ 0.2322 \\ 0.2417 \end{bmatrix}.$$

A very important issue in the study of ridge regression is how to find an appropriate parameter k. When k is estimated from the data the ridge estimator is called the operational ridge estimator. One data based estimate of k suggested by Hoerl and Kennard (1970) is $\hat{k} = s\hat{\sigma}^2 / b'b$. Here b is the least square estimator and $\hat{\sigma}^2 = (Y - Xb)'(Y - Xb)/(n - s)$ is the estimate of the error variance, i.e., the within sum of squares of the Analysis of Variance. The analysis of variance table obtained using the Statistical Software package Minitab was

Table 1.3.3
Analysis of Variance

SOURCE	DF	SS	MS	F	p
Regression	4	0.336860	0.084215	51.73	0.000
Error	5	0.008140	0.001628		
Total	9	0.345000			

Thus $\hat{k} = 0.00816645$. The operational ridge estimator is

$$\begin{bmatrix} \hat{\beta}_0 \\ \hat{\beta}_1 \\ \hat{\beta}_2 \\ \hat{\beta}_3 \\ \hat{\beta}_4 \end{bmatrix} = \begin{bmatrix} 0.1608 \\ 0.5716 \\ 0.0683 \\ 0.1988 \\ 0.1291 \end{bmatrix}.$$

The problem of choosing an appropriate value of k and the rationale for the value chosen above will be taken up in greater detail in Chapter III.

Exercise 1.3.2. The data in the Table 1.3.4 represents the percentage of the gross national product spent by four countries on Non-Defense Research and Development. It is also taken from the Times Almanac.
A. Find the linear regression equation where the United States is the dependent variable and France, West Germany and Japan are independent variables.
B. Find the correlation coefficients between each of three pairs of independent variables. Is there any reason to suspect that the data is multicollinear?
C. Find the eigenvalues, the total variance and the determinant of the X′X matrix? Does it appear that the matrix is almost singular?
D. Find the ridge estimator for k=.05, 0.5 and 5. Compare the total variance of the ridge estimator for each of these values to that of the LS.

Table 1.3.4

Year	United States	France	West Germany	Japan
1972	1.6	1.5	2.1	1.9
1975	1.6	1.5	2.1	2.0
1979	1.6	1.4	2.3	2.1
1980	1.7	1.4	2.3	2.2
1981	1.8	1.5	2.4	2.3
1982	1.8	1.6	2.5	2.4
1983	1.8	1.7	2.5	2.5
1984	1.8	1.7	2.5	2.6
1985	1.8	1.8	2.6	2.8
1986	1.8	1.8	2.6	2.8
1987	1.9	1.9	2.6	2.8

E. Find the operational ridge estimator using the estimator if k given above. In the light of what has been presented so far is it possible to find the total variance of the operational ridge estimator and compare it with that of the LS? Explain.

1.4 Historical Survey

The seminal paper was that of Stein (1956). There, Stein proved his celebrated result that the MLE for the mean of an MN with more than two variables is inadmissible. This means that there exists an estimator with uniformly smaller risk than the MLE. An explicit estimator was constructed by James and Stein (1961) for the following cases:

1. independent random variables with common variance one;
2. independent random variables with unknown common variance;
3. dependent variables where the dispersion is estimated by a Wishart matrix.

The work of James and Stein represents an important starting point in the study of alternatives to the least square estimator. More specific details will be given in Chapter II.

One of the important goals of this book is to study the JS as a special case of an empirical Bayes estimator (EBE). The original formulation of the general concept of empirical Bayes estimation is due to Robbins (1955). EBE's have the same basic form as BE's with estimates from the data replacing unknown parameters. EBE's are especially valuable when prior information is unknown or incomplete. Robbins (1955) gives an example of an EBE that is frequently cited in the literature, for the case of an unknown prior distribution and a Poisson population. Neymann (1963) in summarizing Robbins's work notes that it represents a very important breakthrough in the science of statistical decision making. Rutherford and Krutchoff (1967, 1969 a,b) and Griffin and Krutchoff (1971) describe EBE's as a consistent sequence of approximations to the BE.

Sclove (1968) considers estimation of the coefficients of an orthogonal linear model with at least three regression variables. He:

1. gives the form for the JS;
2. observes that its mean square error of prediction (MSEP) is uniformly smaller than that of the LS;
3. observes that when the regression coefficients are significantly different from zero the JS is very close to the LS.

The observation that the JS and the LS are very close when the regression coefficients differ significantly from zero motivates the idea of a preliminary test estimator. This kind of estimator is formulated as follows:

1. A test of hypothesis is performed to see which regression coefficients are significantly different from zero.
2. Regression coefficients that are not significantly different from zero are estimated by the JS.
3. Regression coefficients that are significantly different from zero are estimated by the LS.

Sclove also:

1. mentions a positive part estimator;
2. gives an indication of how to extend his results to the non-orthogonal case.

Baranchik (1964) observed that the JS positive part estimator had a smaller MSE than the ordinary JS. Stein (1966):

1. obtains the JS for the mean of an MN with independent variables and common unknown variance together with a formula for its risk;
2. shows how the JS may be obtained as an EBE;
3. shows how the JS estimators may be used in the recovery of interblock information in an incomplete block design.

Although when discussing the empirical Bayes approach to the JS most authors refer to Efron and Morris (1973) it appears that the original idea was first reported by Stein himself.

A large portion of this book will be devoted to the study of ridge regression estimators. They have proved to be particularly useful for the study of multicollinear data. Silvey (1969) gives a general discussion of the multicollinearity problem using the eigenvalues of the design matrix and the singular value decomposition.

Hoerl and Kennard (1970a) is the seminal paper on ridge regression most often cited in the literature. They explain how to formulate their ridge type estimators from both a Bayesian and a frequentist point of view. They also show the MSE optimality over least squares estimators and the connection with Stein estimators. They give examples of real world applications to Chemical data in Hoerl and Kennard (1970b).The contents of these papers will be summarized in Chapter III.

Although Hoerl and Kennard (1970a) is generally regarded as the seminal paper on the subject an earlier paper , Hoerl (1962), explains what ridge estimators are and how they can be used to solve problems involving ill conditioned data that arise in the measurement of a chemical process. Even earlier Hoerl (1959) explains ridge analysis in terms of finding the maximum values of a quadratic function (response surface) on an n dimensional sphere. A number of examples are given to illustrate the technique in two or three dimensions.

Marquardt (1970) gives some properties of the LS for the non-full rank model where the Moore-Penrose generalized inverse is employed. He also shows how the ridge regression estimator of Hoerl and Kennard is a mixed estimator, i.e., a special kind of LS estimator where additional "fictitious observations" are taken. The mixed estimator was described earlier by Theil and Goldberger (1961). Another similar characterization that is also an unbiased estimator of the regression parameters is given by Banerjee and Karr (1971).

Many journal articles consider families of shrinkage estimators that contain the JS as a special case. Like the JS they also have MSE's that are smaller than the MLE. One such article is that of Baranchik (1970).

The JS is formulated for the parameters of a multivariate regression model by Sclove (1971). It is shown there that the MSE averaging over a quadratic loss function with matrix equal to the dispersion matrix of the variables is smaller than that of the LS.

Efron and Morris (1971,1972b) observe that although the JS estimators have a smaller risk than the MLE when averaged over the parameters being estimated, i.e., the compound loss function, the individual estimators frequently do not have this property. They suggest a compromise; use the JS if it is close to zero or the sample mean; otherwise use the MLE. Estimators formulated this way are called limited translation estimators.

A desirable property of EBE's is asymptotic optimality (AO). This means that the limiting value of the risk as the sample size tends to infinity is equal to that of the BE. Meeden (1972) gives examples of empirical Bayes procedures of the type considered by Robbins (1955) that are AO but inadmissible.

Efron and Morris (1972) consider simultaneous estimation of the means of p independent multivariate normal distributions. The EBE obtained by them uses the information available from all p distributions for simultaneous estimation of the mean of each individual distribution. They obtain the risks of the EBEs with and without averaging over the prior distribution, i.e., the Bayes and the frequentist risks. Furthermore, they define a measure of efficiency called the relative savings loss (RSL), i.e., the ratio of the difference in the Bayes risks between the EBE and the BE to that between the MLE and the BE. Simultaneous estimation of p means is a special case of simultaneous estimation of the parameters of p regression models considered by C.R. Rao (1975). The work of C.R. Rao will be described in greater detail later.

Wind (1972) considers that properties of the risk of a family of EBE's one of whose members is the JS of a multivariate location parameter. Unlike James and Stein (1961), Wind does not assume any specific parametric distribution for the population or the prior distribution. For these more general conditions he obtains approximate formulae for the Bayes risks and uses them to establish asymptotic optimality of his EBE's.

Efron and Morris (1973) explain how the JS for the mean of an MN is an EBE. The BE is derived for the case of a normal prior. The EBE is obtained by substitution of sample estimates for unknown parameters into the BE. They also obtain the average MSE, the conditional MSE and the RSL of the JS.

Although the compound risk of the JS is uniformly smaller than that of the MLE the risk of the individual components can be quite a bit larger. Efron and Morris (1971,1972b) propose limited translation estimators. The Bayes case is considered in the 1971 article; the empirical Bayes case is considered in the 1972 paper. The compound risk of these estimators is smaller than that of the MLE but not quite as small as that of the JS or the BE. However, the maximum individual component risk of the limited translation estimators is considerably smaller for a large portion of the parameter space than the JS or BE while the sacrifice relative to the compound risk is not that great.

Wind (1973) considers r independent linear models for the case where the design matrix is orthogonal. Assuming a prior distribution with independent marginals, i.e., the individual coordinates of the regression parameters are independent, he obtains the BE and the EBE. Then he considers the case of a non-orthogonal design matrix. By means of a rather complex series of transformations he shows that r regression problems are equivalent to r problems of estimating the mean of an MN. The asymptotic optimality of the estimators is also examined.

Dempster (1973) explained how various ridge and contraction estimators are special cases of the BE. The JS for the parameters of a single linear regression model is derived using a prior distribution with dispersion matrix a constant multiple of that of the LS. This prior distribution will be referred to as Dempster's prior in the sequel. A prior of the same form is mentioned by Zellner and Vandaele (1974) and is often called a "g prior" in the literature.

An extension of the Gauss Markov theorem that includes Bayes and Ridge type estimators is discussed by Duncan and Horn (1972) and Harville (1976). Their methods and those of other authors are critiqued by Peffermann (1984).

By minimizing the length of the vector of predicted values in a linear model constrained to lie on an ellipsoid, Mayer and Willke (1973) obtain the contraction estimator. Their argument is analogous to that used by Hoerl and Kennard (1970) to obtain the ordinary ridge estimator. The contraction estimator can be shown to be the BE with respect to Dempster's prior and can thus be used to obtain the JS.

Baranchik (1973) discusses the JS in the context of a linear regression model and shows that the MLE is inadmissible.

Stein type estimators have proved useful in many applied settings. Carter and Rolph (1974) used Stein type estimators to estimate the probability that an alarm from a street firebox signals a structural fire. They derive a Stein type estimator using empirical Bayes methodology for the case where the components have unequal variances. The JS is used with fire alarm data from 1967-1969 in the Bronx, New York to develop estimates for 1970 box repeated alarms. The predictions were then compared with the actual 1970 data. The EBEs generally predicted fewer structural fires than the traditional MLEs with a lower incidence of not reporting a structural fire when there was one. The information gained from this study proved useful for the formulation of initial dispatch policies, i.e., more equipment would be dispatched in event of a structural fire.

The JS for different linear model setups is considered by Zellner and Vandaele (1974). They consider:

1. the k means problem;
2. the univariate regression problem;
3. the seemingly unrelated regression equation model (SURE);
4. the simultaneous equations model.

The estimators are shown to be special cases of:
1. BE with respect to a specific prior that is introduced in the paper;
2. approximate MMSE estimators.

Except for duplication of the risk formula of the JS they do not tell how to obtain the MSE of the approximately optimal estimators that they produce.

A class of shrinkage estimators is defined by Goldstein and Smith (1974). For a reparametized linear model with a diagonal design matrix they obtain conditions on the ridge parameters where the individual MSEs of the shrinkage estimators are smaller than that of the LS. Optimum ridge parameters are obtained for the generalized ridge regression estimator. A Bayesian formulation of ridge estimators is presented. They show how for the non-full rank case how the ridge estimator approximates the LS obtained using the Moore Penrose inverse.

Conditions for the MSE of ridge type estimators to be smaller than that of the LS estimators are given by Lowerre (1974). Both the full and the non-full rank case are considered.

Tukey (1975) gives a critical evaluation of the LS estimators in light of the hypothesis of the Gauss Markov Theorem. He concludes that often one would want to do more than or something different from least square estimation.

Bock (1975) gives general conditions for a family of estimators that include the JS to have a smaller MSE than the MLE for the cases of known and unknown population variance. The results will be given in Chapter III in the context of a single linear regression model.

Berger (1975) gives conditions for a family of estimators that include the JS to have a smaller MSE than the MLE for:

1. mixtures of normal densities;

2. a wider class of densities with finite second moments and discontinuity points of measure zero.

There have been many simulation studies done to compare ridge, LS and JS type estimators. Two such studies were those of Hoerl, Kennard and Baldwin (1975) and Dempster Schatzoff and Wermuth (1977). The study by Hoerl et al. was confined to ridge and LS estimators; that of Dempster et al. made comparisons of ridge, JS and other estimators. The findings of these and other simulation studies will be summarized in Chapter III.

Marquardt and Snee (1975) review the theory of ridge regression and its relation to generalized inverse regression. They present the results of a simulation experiment together with three examples of ridge regression in practice. In these examples they demonstrate the use of the ridge trace for finding a suitable estimate of the parameter k. The three examples studied illustrate how when the predictor variables are highly correlated ridge regression produces estimators and predictors that are better than least squares. They also demonstrate that ridge regression is a safe procedure for variable selection.

Farebrother (1975) notes that the ordinary ridge regression estimator of Hoerl and Kennard and the MMSE have similar mathematical forms. He explains how to obtain operational forms of the ridge estimator by substitution of sample estimates for the unknown parameters into the MMSE. These sample estimates may also be used to estimate optimal k.

The material in C.R.Rao (1975) is basic to the research to be reported in this book. The linear Bayes estimator is derived there for the full rank linear model and several algebraically equivalent alternative forms are obtained. These alternative forms are used to show that the ridge and contraction estimators are special cases of the BE and to obtain EBEs. Another important problem considered there is the simultaneous estimation of the parameters of a single linear regression model that is part of a setup involving r regression models. The parameters of each regression model is estimated by an EBE that uses information gained from all of the models. The average and conditional MSEs of the EBE are obtained. They are smaller than that of the LS. This problem will be considered at great length in Chapter VII.

Admissible linear estimators are presented in C.R. Rao (1976a). These optimal estimators are shown to be either Bayes linear estimators or limits of Bayes linear estimators. A brief explanation of how the ridge and the minimax estimators are a both special case of the Bayes linear estimators is given. The results of this paper are extended by Zontek (1987).

C.R. Rao (1976b) considers a superpopulation model. He proposes an alternative to the JS with shrinkage toward the grand mean of the observations instead of the origin. This type of shrinkage is often referred to as the Lindley mean correction in the literature. These estimators of the regression parameters are biased and have average and conditional MSE that is less than that of the LS. The JS with

Lindley mean correction (JSL), unlike the JS, enjoys the property of translation invariance.

Vinod (1976) proposes a modification of Farebrother's (1975) approximate MMSE based on a fixed point solution of an iterative process. The mathematical form of the estimator obtained is similar to but not the same as the JS. The results of a simulation study indicated that the JS had a smaller MSE than the approximate MMSE of Farebrother for intermediate values of the number of parameters. However the modified Farebrother estimator did better for three or four regression parameters.

The representation of ridge estimators as LS estimators multiplied by a shrinkage factor in matrix form will be described in Chapter V. The shrinkage may be toward the origin or a non-zero prior mean vector. The properties of ridge estimators that shrink toward a mean vector different from zero is considered by Swindel (1976). Included in his article are necessary conditions for the mean square error of parametric functions of ridge estimators to be less than that of the LS.

Necessary and sufficient conditions for ridge or general types of shrinkage estimators to have smaller MSE than the LS for parametric functions is given in two brief notes by Farebrother (1976,1978). Their results and some related results obtained by the author will be presented in Chapters III and V.

Efron and Morris (1977) give a nice intuitive presentation of the JS for the general scientific community in a Scientific American article. They gave two applied examples to illustrate the superiority of the JS estimators. They were:

1. batting averages of eighteen baseball players during the 1970 season;
2. incidence of Toxoplasmosis in thirty six cities of the Central American country El Salvador (Toxoplasmosis is a disease of the blood caused by a protozoan. Although it not serious or fatal, it can cause damage to unborn fetuses. It is one of the few diseases human beings can get from house cats.).

Comparison of batting averages from the first forty five at bats with the full season averages indicate the JS have smaller MSE than the actual batting averages. For the Toxoplasmosis data, it can be shown that the JS have a smaller total error of estimation. Furthermore, they also provide a more accurate ranking of the incidence of the disease in different cities.

The simultaneous estimation of vector parameters is considered by C.R. Rao (1977). For r separate problems of estimating the mean of a MN he formulates the BE and EBE and obtains the average and conditional MSE.

Classical F-tests and confidence regions for non-stochastically shrunken ridge regression estimators are discussed in Obenchain (1977). It is shown there that the central F-ratios and t statistics are the same for non-stochastically shrunken ridge regression estimators as the LS. Thus, although the biased point estimators produced by ridge regression differ from that of the LS they do not produce shifted confidence intervals. A concept called "associated probability" is defined and it is argued that ridge estimates are of little interest when they lie outside a LS region of 90% confidence.

C.R. Rao and Shinozaki (1978) consider the MSE properties of the individual components of a JS as compared and contrasted with its properties averaging over a compound loss function. Recall that Efron and Morris (1972) observed that although the JS has smaller MSE than the LS averaging over an appropriate quadratic loss function, the MSE of the individual coordinates may not have this property. Additional light on the problem is shed by Rao and Shinozaki. They:

1. compare the MSE of the JS with and without the Lindley mean correction;
2. obtain formulae for the individual MSEs of the JS and the LJS;
3. obtain some numerical values of the MSEs for individual and compound loss functions.

Further work on this problem is described in Chapter VIII.

The book of Judge and Bock (1978) deals with a number of issues related to those considered in this volume and is at a similar mathematical level. A presentation of the general decision theoretic model, Bayes estimation and mixed estimators is followed by an examination of the formulation, properties and performance of the preliminary test estimators. The Stein family of estimators is then presented. They propose a number of different positive parts and present the results of a Monte Carlo study performed to compare their performance. More work on the formulation and efficiency of these positive parts will be presented in Chapter VIII.

Ridge estimators for the non-full rank model is studied by Brown (1978). The LS is characterized as the limit of ridge regression estimators.

Ridge and James-Stein estimators are presented from the Bayesian point of view in a review article by Vinod (1978). An extensive bibliography is given with this article.

Ullah and Ullah (1978) present the double k class estimators. as a generalization of the minimum mean square error matrix. Two important special case of the double k class estimator are the least square estimator and the James-Stein estimator. An exact formula for the risk of the double k class estimator is given and the risk of the James-Stein estimator is obtained from making appropriate restrictions. Asymptotic expansions of the risk function are obtained . It is shown that in an approximate sense the double k class estimators dominate the LS estimators.

Haff (1978) observes that that the analysis of the fire alarm data in Carter and Rolph (1974) does not consider the fact that there may be a correlation between the estimated probabilities in adjacent neighborhoods. He deals with this problem by formulating Stein type EBEs for the case where the prior distribution of the mean vector has intraclass correlation structure. The estimators obtained outperformed those of Carter and Rolph in about 70% of the incidences investigated.

Draper and Van Nostrand (1979) provide an excellent review of the literature available and the state of the art at that time for ridge and Stein type estimators. It includes:

1. a review of the work of James and Stein (1961);
2. an explanation of the empirical Bayes approach given in Stein (1966) as an intuitive justification for the use of the JS;
3. a review of the work reported in Sclove (1966), a paper dealing with the JS in the regression context;
4. mention of the work reported by Efron and Morris (1972a,1973), Berger (1976a,b) and Bock (1975);
5. seventy-seven references to books, articles, etc., on ridge and JS estimators.

The references in the Draper and Van Nostrand article provide many excellent entry points to the literature on ridge and Stein estimators for papers written in the 1960s and 1970s.

Some of the material in the present volume is also contained in the author's 1979 Ph.D. thesis (Gruber (1979)). The thesis:

1. contains a formulation of Bayes and empirical Bayes estimators for setups with one and $r > 1$ univariate and multivariate non-full rank linear models;
2. shows that ridge and contraction estimators are special cases of the BE;
3. obtains the MSE's of the BE's and the EBE's with and without averaging over a prior distribution and compares them with each other where possible and with the LS;
4. shows how the EBE's are also approximate MMSE's.

An interesting and important application of the JS is small area estimation. Often more accurate estimators may be obtained for characteristics of small populations by combining information gained from several such groups and using the information to simultaneously estimate characteristics of each individual small population. This is done by Fay and Herriot (1979). They use the JS to estimate income from places with population less than 1000 by using information obtained from sample estimates of several such places as a result of the 1970 census. A linear regression model with unequal variances is employed. The estimates obtained using the JS have smaller average error than either the sample estimates or the county averages. The estimates for small places form the basis for the Census Bureau's updated estimates of per capita income for the general revenue sharing program.

Another application of the JS is in the use of standardized test scores for the prediction of academic achievement. Rubin (1980) deals with the prediction of the first year average in law school based on:

1. the Law School Aptitude Test (LSAT);
2. the undergraduate grade point average (GPA).

The JS EBE's prove to be better predictors than the LS for the following reasons:
1. they are more stable with relation to time, i.e., less likely to fluctuate from one year to the next;
2. they are more accurate, i.e., have a smaller MSEP.

· A general criterion for comparison of biased estimators is presented by D.Trenkler and G.Trenkler (1980). Their results generalize those of Farebrother (1976).

The results and methods of Hoerl and Kennard are generalized to a larger class of ridge estimators by Wax and Haitovsky (1980). An explanation is furnished by them of how a generalized ridge estimator is a BE or a mixed estimator.

The dispersion matrices of restricted regression and mixed regression estimators are compared in Srivastava and Agnihotri (1980). The restricted regression estimator is obtained when the restrictions are exact; the mixed estimator is obtained when the restrictions are stochastic. This will be taken up in Chapter V. The mixed estimators turn out to be less efficient.

Judge et al. (1980) explain in Chapter 3 how a James-Stein estimator may be viewed as a pretest estimator that combines the restricted and the unrestricted least square estimator. They explain how a Stein like pretest estimator originally formulated by Scolve, Morris and Radhakrishnan (1972) dominates the usual pretest estimator. In Chapter 12 they give an extensive discussion of the multicollinearity problem, its statistical implications and different ways of dealing with it. The suggested solutions to the multicollinearity problem include using additional sample information, imposing exact linear constraints stochastic or linear inequality restrictions, the use of prior information in the Bayesian sense, principal components regression and different kinds of James-Stein and ridge regression estimators.

Terasvirta (1981) compares the MSE of minimax estimators and mixed estimators for a quadratic loss function. He obtains necessary and sufficient conditions for minimax estimators to have a smaller MSE than the mixed estimators. The performance of the mixed and the minimax estimators is compared with that of the LS.

Some nice material on Shrinkage Estimators is contained in Vinod and Ullah (1981). The JS is discussed from both the Bayesian and the frequentist point of view. They observe that unlike the JS, the JSL is invariant with respect to location and scale transformations. They also present the example about batting averages of baseball players that was considered by Efron and Morris (1977). This book also contains introductory material on ridge regression from both the Bayesian and the frequentist point of view. The problem of choosing the biasing parameter k is dealt with and an example is given of how to select the correct ridge solution.

Menjoge (1981) in his Ph.D. thesis completed at the University of Rochester under the direction of Poduri S.R.S. Rao :

1. obtained conditions for a more general family of estimators than those considered by earlier authors, e.g. Shinozaki (1974), Bock (1975) and Strawderman (1978), to have a smaller compound weighted risk than the MLE;

2. developed a procedure to improve estimators having a smaller risk under a compound loss function so that estimators having a smaller risk under a compound weighted loss function could be obtained;

3. used the procedure in 2 to obtain improved estimators with smaller compound weighted risk than the MLE for all weight functions averaging over a quadratic loss function;

4. developed bounds on the parameters for the estimators to have a smaller individual component risk than the MLE;

5. showed that the positive part estimator has a smaller individual risk than the JS;

6. showed how his estimators have smaller compound weighted risk than the LS in the context of linear regression;

7. obtained translation invariant empirical Bayes estimators of C.R.Rao (1975) in the context of r linear models as approximate minimum mean square error estimators; this technique was originally used by Zellner and Vandeale (1974).

Some of the results of the thesis are also available in Menjoge and Rao (1983).

Vinod Ullah and Kadiyala (1981) present the ridge estimators of Hoerl and Kennard (1970) as special cases of double f class estimators. Using confluent hypergeometric functions they show how to obtain the exact bias and MSEs of these estimators. They also present asymptotic expansions of the bias and the MSE. They do a limited simulation study that compares the bias and MSE of the double f class estimators with the LS.

An explicit formula for the risk of the JS using an integral representation for the sum of an infinite series was obtained by Egerton and Laylock(1982) . Similar formulae are derived in Chapter III and IV for the purpose of obtaining numerical values for the risk of the JS for different parameter values.

Biased estimators for the non-full rank linear model are presented by Wang (1982). He also explains how ridge regression estimators are special cases of the BE. Some of the results of Wang's paper were obtained independently by Gruber and P.S.R.S.Rao(1982).

C.R. Rao (1975) presented different alternative forms for the BE of the parameters of a full rank regression model. Generalizations to the non-full rank model may be found in Gruber and P.S.R.S. Rao (1982) and Gruber (1990). The alternative forms of the BE are useful and important because they:

1. give an indication of the relative importance of sample and prior information;
2. may be used to show how ridge and contraction estimators are special cases of the BE;
3. may be used to formulate EBE's for situations where the prior distribution is unknown or only partially known.

Peele and Ryan (1982) explain how ridge estimators are special cases of minimax estimators and show that the minimax estimators have a smaller MSE than LS.

Comparisons of the MSE of parametric functions of different ridge estimators with each other and with parametric functions of the LS are taken up in Price (1982).He observes that it is possible to compare the ridge estimators with the LS but it is not always possible to compare parametric functions of ridge estimators with each other.

Toutenburg (1982) explains the relationship between the ridge, minimax and mixed estimators. He considers the role of prior information that is different from what is used by Bayesians. The behavior of the minimax and the mixed estimators are considered relative to correct prior information when the prior information used to derive them is incorrect. Unlike this volume their work does not consider the Bayesian point of view. The book contains a lengthy bibliography.

Robbins (1983) gives several different examples of applications of EBE's. These include:

1. simultaneous estimation of r independent means of a population with known unequal variances;
2. estimation of unknown variances for normal populations;
3. EBE's for the parameters of binomial and geometric distributions;
4. an empirical Bayes test of hypothesis and an empirical Bayes confidence interval.

The Kalman Filter (KF) was originally of primary interest to control engineers. Meinhold and Singpurwalla (1983) show how the KF may be formulated as a recursive BE using well known results from multivariate statistics. This formulation makes the KF accessible to statisticians. The KF has found application to many diverse fields, e.g., aeorospace tracking and quality control. A discussion of the KF together with entry points into the literature is available in Gruber (1990), Chapter IX. The simultaneous estimation of parameters for r KF's will be taken up in Chapter X of the present volume.

Ullah, Srivastava and Chandra (1983) present a general class of shrinkage estimators that include a Stein and a ridge type estimator as a special case. Assuming that the first four moments of the error terms are finite and nothing else they derive conditions for these estimators to dominate the LS. They observe that these conditions are quite different from the case where the error terms are normally distributed.

Farebrother (1984) extends the results of Farebrother(1976,1978) in order to compare the MSE of a generalized ridge estimator with that of a restricted least square estimator.

Reinsel (1984) obtains results similar to those of C.R. Rao (1975) for growth models. The results include approximate MSE formulae of EBEs for the parameters of r linear models with different design matrices.

Saxena (1984) shows how the Theil-Goldberger mixed estimator is a special case of the BE. The same result is obtained somewhat differently in Gruber (1985,1990). Gruber also considers the problem of incorrect prior information for non-full rank linear models.

Many of the topics covered in this book are presented at a basic level in Fomby, Hill and Johnson (1984). In particular they give good brief presentations at a basic level of:

1. ordinary and generalized least squares;
2. restricted least squares and mixed estimators;
3. Bayes estimators;
4. pretest estimators and Stein rule estimators;
5. the multicollinearity problem and ridge regression;
6. the seemingly unrelated ridge regression model.

A reader who is interested in econometrics might like to study these topics in Fomby et. al, in conjunction with the present work.

The small σ and large T asymptotic expansion of the distribution function of a general class of shrinkage estimators and their F ratios is obtained by Ullah, Carter and Srivistava (1984). It is concluded from an numerical study that the small σ approximation is more accurate.

Ullah and Vinod (1984) observe that many shrinkage estimators shrink toward a non stochastic target., usually the origin. They generalize results concerning the range of stochastic values of the ridge biasing parameter k where the MSE of a shrinkage estimator is better than that of the LS to estimators that shrink toward stochastic targets.

Swamy et al. (1985) present a technique for the measurement of multicollinearity. They observe that the mixed estimator and the BE are not always equivalent when the prior distribution is of non-full rank. Gruber (1985,1990) gives conditions for the equivalence of the BE and the mixed estimator for non full rank models

Casella (1985) observes that for a range of parameter values the conditional MSE of an EBE is less than that of the corresponding BE. Some results about the crossing values of the MSE of the BE and the EBE are contained in Chapter IV and VII. A new measure of efficiency, the relative loss, is proposed there, i.e., the ratio of the difference between the MSE of the JS and the MMSE to that of the LS and the MMSE.

Berger (1985) is a textbook on Bayesian analysis that includes a discussion of Stein type estimators and a good source of references. This book contains a good treatment of Bayes, empirical Bayes and minimax estimators. Bayesian robustness is touched upon. The subject of Bayesian robustness is also taken up in Chapter VIII of Gruber (1990).

A simulation study the MSE of different kinds of JS and ridge type estimators in comparison with that of the LS is performed by Precht and P.S.S.N.V.P. Rao (1985). The biased estimators generally have a smaller MSE than the LS and in some cases the improvement is quite substantial.

Press and Rolph (1986) formulate BE's and EBE's for the mean of a MN for different kinds of normal prior distributions. The special cases considered include a prior with diagonal dispersion matrix and one whose dispersion matrix has equal off diagonal elements.

George (1986a) describes a multiple shrinkage estimator. This is a convex weighted linear combination of JS estimators that each shrink toward a different given prior subspace. George (1986b) shows how the multiple shrinkage estimator my be applied to a given class of partitioned shrinkage estimators that include individual JS estimators and gives a Bayesian motivation for the estimators. He shows how for the prediction of baseball team batting averages the multiple shrinkage estimator was a better predictor than the individual JS estimators.

The LINEX loss function was introduced by Varian (1975) to study real estate assessment. In that context squared error loss is not appropriate because of its symmetry properties; the LINEX loss function is aysmmetric. Zellner (1986a) obtains the risk functions and derives the BE relative to the LINEX loss function. For a univariate or bivariate population the sample mean is admissible relative to the squared error loss function; relative to the LINEX loss function the sample mean is inadmissible.

Berger and Berliner (1986) consider robustness of prior information. Their discussion includes ε contaminated prior distributions.

When the MSE of ridge and least square estimators are compared the regions where the ridge estimator dominates the LS is an ellipsoid. The geometric form of the region obtained where one BE dominates another BE, for example, can be much more complicated. In order to deal with this problem necessary and sufficient conditions for the comparison of two heterogeneous linear estimators are given by Terasvirta (1986a).The results are applicable to the comparison of Bayes, minimax or mixed estimators. Mixed estimators are compared in Terasvirta (1986b).

Toutenburg (1986) takes up weighted mixed regression. The auxiliary information is weighted by an appropriate scalar weight with value between zero and one. The MSE of the weighted mixed regression estimator is compared with the

LS. He shows how the methods of mixed estimation can be applied to the missing value problem.

A very mathematically sophisticated discussion of ridge, minimax and James-Stein estimators is given in Bunke and Bunke (1986). This book is an English version of Humak (1977) written in German. Good bibliographic references are available at the end of Chapters 1-4.

Copas (1983) advocates the use of the JS to anticipate the shrinkage associated with prediction in a linear multiple regression model. If a sample of observations is taken and it is assumed that the first two population moments associated with individuals about whom predictions are going to be made coincide with the original sample moments Copas claims that Stein type estimators give a uniformly lower prediction mean square error (PMSE) then the LS. When this paper was presented to the Royal Statistical Society a number of discussants questioned the validity of the assumption. Jones and Copas (1986) attempts to answer these objections by finding regions where the future mean vector and variance matrix must lie so that the JS still has a smaller PMSE than the LS. They call these regions "robustness regions." Numerical observations suggest that the JS is robust to small and large differences between the future moments and the sample moments because its MSEP is less than that of the LS.

Some basic information about pre-test estimators, different kinds of James-Stein estimators and empirical Bayes confidence intervals is available in Judge and Yancy (1986).

Swindel (1976) proposes a modified ridge estimator based on a prior mean. Pliskin (1987) gives a necessary and sufficient condition for the modified ridge estimator to have a smaller risk than the ordinary ridge estimator of Hoerl and Kennard when the same value of the parameter k is used. Trenkler (1988) further studies this problem for the case of different ridge parameters.

An ordinary linear regression model with parameters constrained to a given ellipsoid is considered by Lauterbach and Stahlecker (1987,1988). A sequence of approximate minimax estimators is obtained by them with an exact mimimax estimator as a cluster point. The exact minimax estimator is compared to the LS using a simulation study. They consider a setup that is more general than that to be used in the present volume.

Minimax estimators are developed by Stahlecker and Trenkler (1988) for two kinds of constrained linear regression models. In one case the parameters are simultaneously constrained to an ellipsoid; in the other they are constrained by exact linear restrictions.

Toutenburg (1988) compares the MSE of restricted LS, mixed and weighted LS estimators. In addition, he considers nested restrictions.

The case where the parameters of a linear model are constrained both by linear equations and lying in a given ellipsoid is considered by Schipp et al. (1988).They establish the existence of the restricted minimax estimator containing both kinds of prior information for the case when the weight matrix of the risk function is of rank one. They propose other estimators for the case when the weight matrix is just non-negative definite and compare them to the LS using a simulation study.

Quenouille (1956) introduced the Jack-knife estimator . When the jack-knife procedure is applied to applied to a biased estimator the resulting estimator has a smaller bias. The use of "jack-knifing" to reduce the bias of ridge regression estimators and the properties of jack-knifed ridge estimators is studied by Kadiyala (1984), Nyquist (1988), Singh, Chaubey and Dwivedi (1986) , Nomura (1988) and Gruber (1991). Gruber (1991) observes that although the use of the Jack-knife procedure reduces bias considerably the jack-knifed estimators have a larger variance and may have a larger MSE than the usual ridge estimators.

Necessary and sufficient conditions for the matrix MSE of a generalized ridge regression estimator to be smaller than that of a LS are obtained by Chawla (1988,1990). Similar results obtained by the author are taken up in Gruber (1990).

Terasvirta (1988) applies the results of Terasvirta (1986) for comparison of two heterogeneous estimators to the comparison of two mixed estimators.

Comparison criteria for linear admissible estimators including ridge estimators similar to those developed by other authors are developed by Liski (1988). He also develops a test procedure for choosing a linear estimator instead of a LS.

Alternatives to EBE's are suggested by Kempthorne (1988). These include the solution to Bayes compromise problems and minimax Bayes compromise problems. For the Bayes compromise problem Bayes risks under different prior distributions or loss functions are minimized simultaneously. For a minimax Bayes compromise problem a Bayes risk under some loss function for a given prior distribution and a maximum risk under a possibly different loss function are minimized simultaneously. Kempthorne gives existence proofs for solutions to the minimax Bayes and the Bayes compromise problem. However, these solutions cannot be obtained in closed form and consequently numerical approximations must be resorted to. The claimed advantage is that the optimal solutions are admissible. Nevertheless they cannot be displayed explicitly. Although Stein estimators are inadmissible their advantages include:

1. a closed form representation;
2. an MSE considerably smaller than that of the MLE especially when the number of variables is large.

As has already been observed, although the JS is an improvement over the MLE it, too, is inadmissible. An estimator with uniformly smaller conditional and average MSE than the JS is its positive part. The positive part corresponds to the JS when the contraction factor is positive; otherwise it is zero. Two papers about JS positive parts are those of Nickerson (1988) and Robert (1988). Nickerson (1988) shows that the MSE of the traditional positive part of a family of estimators that include the JS is smaller than that of the estimators themselves. Robert (1988) obtains a formula for the risk of the traditional positive part.

Liang and Zeger (1988) propose a Stein type estimator of the common odds ratio in one-to-one case-control studies based on estimating functions. They also give a different justification for the use of Stein type estimators in terms of the estimating functions.

Ghosh, Saleh and Sen (1989) use empirical Bayes methodology to obtain a JS that shrinks toward a restricted LS. They obtain its Bayes risk. They show how their JS could be used to estimate the main effects in a factorial design model that does not have significant higher order interaction terms.

One method of comparison of estimators that has become popular in recent years is Pitman nearness. One estimator is Pitman closer to the estimated parameter than another estimator if the probability that its loss is less than that of the other estimator is greater than one half. Keating and Czitrom (1989) show that in the sense of Pitman closeness the JS does better than the LS but nevertheless can be improved upon. A reader who wishes to find more out about Pitman Nearness may consult Keating, Mason and Sen (1993). The Bibliography contains a few references concerning the comparison of ridge and JS estimators using Pitman nearness as the criterion. The seminal paper on Pitman's measure of closeness is Pitman (1937).

Peddada et al. (1989) consider three criteria for comparison of ridge and LS estimators Pitman nearness, the MSE averaging over a quadratic loss function. and the MSE averaging over a Mahalanobis loss function. They obtain necessary and sufficient conditions for the ridge estimator to have a smaller MSE averaging over a quadratic loss function than the LS. They also observe that these conditions are the same as those for the ridge estimator to do better than the LS in the sense of Pitman nearness. Under the Mahalanobis loss function they find that the ridge estimator never has a smaller MSE than the LS.

Ghosh et al. (1989) estimate a subset of the parameters using a JS type EBE that shrinks the unrestricted LS estimator toward a restricted LS estimator given the null hypothesis that the parameters outside the estimated subset are zero. These estimators with slight modification have a smaller frequentist and Bayes risk than the preliminary test estimators. The following application of this idea is considered. In a factorial experiment in order to estimate the main effects the unrestricted LS estimator of the main effects is shrunk using the JS toward the restricted LS under the hypothesis that higher order interactions are not significant.

Edlund (1989) investigates the preliminary estimation of Box-Jenkins transfer function weights. They generate time series using statistical simulation. Fourteen estimators including the LS, different ridge and principal component estimators are compared with respect to the MSE and the standard error of the weight estimators. The estimators are investigated for :

1. different levels of multicollinearity;
2. signal to noise ratio;
3. number of independent variables ;
4. length of the time series;
5. the number of lags included in the estimation.

It is found that ridge estimators nearly always have a lower MSE than the LS. When the level of multicollinearity is high and the signal to noise ratio is low the ridge estimators give much lower MSE than the LS. For some cases the principal components estimators give lower values of the MS than the LS but they can also give higher values. Nearly always all unbiased estimators have a much lower estimated standard error than the LS.

Cellier et al. (1989a) consider a class of shrinkage estimators of the mean whose shrinkage factors are not necessarily differentiable. They establish a sufficient condition for the shrinkage estimator to have a smaller MSE than the MLE with respect to an arbitrary quadratic loss function.

Srivastava and Bilodeau (1989) formulate a JS estimator that is essentially the same as the one formulated in Berger (1976) for the parameters of a linear regression model. This estimator and its MSE properties for the normal case will be taken up in Chapter IV. They show that the necessary and sufficient conditions for this estimator to have a smaller MSE than the LS for the case where the error terms are normally distributed are also valid for the more general problem where the error terms have a spherical distribution. The robustness of the JS of the location parameter of certain kinds of elliptical distributions is also demonstrated.

Four different ridge type shrinkage estimators are studied by Noor and Mehta (1989). The estimators considered are proposed by Hoerl and Kennard (1970), Hemmerle (1975), Obenchain (1975) and the authors. The estimator proposed by the authors is similar to one first suggested by Mehta and Srinivasan (1971). The ratio of the MSE of these estimators to that of the LS times 100 is compared for the four kinds of estimators. The estimator proposed by the author has the largest range of values as compared with the other three where the shrinkage estimator has smaller MSE than the LS.

Flack(1989) considers the problem of identifying dissimilar data when doing a ridge regression. Two methods described include the leverage for the regression hat matrix and the predictive MSE bound. She applies these methods to asphalt pavement rut depth data of Daniel and Wood (1971) discussed in Willan and Watts (1978).

Mackinnon and Puterman (1989) define collinearity of generalized linear models (see Macullagh and Nelder (1983)). They note that data that is not multicollinear for a standard linear model may be multicollinear for a generalized linear model and vice versa. They give bounds that relate the degree of collinearity in these two kinds of models. Estimation based on ridge methods is then discussed.

Firinguetti (1989) does simulation studies on several different kinds of ridge regression estimators taking into account autocorrelated disturbances. Two quantities are compared the ratio of the MSE of the ridge estimator to that of the LS (NMSE) and the ratio of the absolute bias of the ridge estimator to that of the operational generalized least square estimator. Some important findings are:

1. Ridge estimators performance improves as the degree of multicollinearity increases.
2. As the autocorrelation coefficient increases the generalized ridge estimator that takes the autocorrelated disturbances into account does better than the ordinary ridge estimator.

Liski (1989) compares restricted least square estimators. His results include:

1. comparison of two restricted LS estimators when the restriction are different;
2. comparison of a restricted estimator with a stochastically constrained estimator (mixed estimator).

This extends the work of Baksalary (1984) who compares restricted LS estimators when the restrictions are the same and Srivastava and Agnihotri (1980) who compare a restricted estimator and a mixed estimator.

Sen et.al (1989) considers the Stein paradox in the context of the Pitman measure of closeness. They consider a general form of shrinkage estimator that includes the original JS and a modified JS considered by Keating and Mason (1988). For this general shrinkage estimator they obtains a condition for it to be closer than the MLE in the Pitman sense. The JS and the estimator of Keating and Mason satisfy this condition. They note that the Stein estimator can be improved on using its positive part for Pitman closeness. They also observe that for Pitman closeness the conditions for the shrinkage estimator to dominate the MLE extends to estimators that shrink toward a subspace instead of the origin.

Hosmane (1988) obtained a generalized JS by minimizing the MSE of a linear estimator and then estimating the unknown parameters using sample values. Ali (1989) modifies Hosmane's procedure by minimizing the sum of the bias and a constant multiple of the variance. Ali develops approximate MSE formulae for these estimators using Kadane's (1970,1971) small disturbance asymptotics. The estimators are compared and conditions are given for Ali's estimator to dominate Hosamane's estimator in small sample asymptotics up to order σ^4.

An important application of the Kalman filter to be described in Section 10.3 is to tracking targets. Bar-Shalom and Fortmann (1988) consider the problem of tracking a single target and multiple targets. The discrete and the extended discrete Kalman filter is developed and used for this purpose.

Bilodeau and Kariya (1989) estimate the coefficient matrix in a normal multivariate regression model. Shrinkage estimators with smaller risk than the MLE are obtained similar those of Baranchik (1970), Stein (1981), Efron and Morris (1976a,b) and Strawderman (1973).

Shrinkage estimators in the form of a weighting of the sample mean and a prior value of the mean are considered by Hawkins and Han (1989). Certain weightings lead to the commonly studied preliminary test and shrinkage estimators. They consider a choice of weightings that minimizes the average risk relative to a specific prior distribution. They also consider how the MSE can be controlled by varying the form of the prior and thus observe a robustness property of the family of posterior mean estimators that correspond to the conjugate normal priors.

Laird and Louis (1989) compare the performance of three kinds of confidence intervals on a random sample of bioassays from the National Cancer Institute data base on potential chemical carcinogens. The three kinds of confidence intervals compared are:

1. classical confidence intervals;
2. empirical Bayes confidence intervals based on a Bayes hyperprior distribution (see Morris (1983) ;
3. empirical Bayes confidence intervals based on using bootstrap sample to adjust for uncertainty in the estimate of the prior distribution (see Laird and Louis (1987)).

Shinozaki (1989) shows how to construct a confidence interval for the mean of a multivariate normal distribution by shrinking the usual one for the mean that is based on the MLE toward the origin. These confidence sets have the same coverage probability as the usual one but a smaller volume.

Steece (1989) uses the concept of an extended data set where additional observations represent the information contained in a prior distribution (See Zellner (1986b), Gruber (1985)) and the generalized singular value decomposition (see Van Loan (1976)) to derive the hat matrix for Bayes estimators of regression coefficients. The hat matrix used in regression diagnostics shows how much influence or leverage the observed responses and the prior means have on each of the posterior fitted values. They apply their results:

1. to a model relating the real gross investment of the Westinghouse firm to the value of its outstanding shares, a proxy measure for expected profitability and the value of its real capital stock;
2. to a model relating the gross national product of forty nine countries to infant death rate, inhabitants per physician, population density, population density computed on the basis of agricultural land, literacy rate and the enrollment rate in higher education.

Mason et al (1990) presented a method for determining the exact closeness probabilities when comparing two linear forms under Pitman's measure of closeness. The technique is illustrated by comparing two ridge regression estimators.

Mason and Blaylock (1991) apply the methodology of Mason et al. (1990) to two kinds of regression data sets. One involves a strong mullticollinearity. The other data set is almost orthogonal. The regions where one estimator is preferable to another is clarified for these two example. They point out that the technique can be applied to other kinds of shrinkage estimators.

Two papers that explore motivations for the use of JS estimators are those of Brandwein and Strawderman (1990) and Stigler (1990). Brandwein and Strawderman give a nice presentation of a geometrical motivation of Stein. Stigler gives a motivation based on the least square principle. These rationales for the use of the JS will be presented in Chapter II.

In Gruber (1990) and in the present volume prior parameters are assumed known for the derivation of Bayes and mixed estimators. However the derivations of the EBE's and the approximate MMSE's in the present volume also are done assuming no or incomplete knowledge about the prior distributions or parameters. Gruber (1990) will be frequently referred to. The reader might like to consult its Bibliography for more entry points into the literature.

Hoerl and Kennard (1990) claim that the common practice of obtaining the sum of squares decomposition and assigning degrees of freedom for ridge regression estimators the same way as it is done for least squares is wrong. Their objections stem from:

1. the non-uniqueness of the decomposition because the residual and the regression vectors are not orthogonal;
2. the degrees of freedom do not apply to either the distributional theory or the expected mean squares.

They propose a sum of squares decomposition that takes into account the non-orthogonality of the residual and regression vectors and explain what they believe is the correct way to define the degrees of freedom.

Wang et al. (1990) consider two measures of multicollinearity the eigenvalues and the condition number, that is the ratio of the largest to the smallest non-zero eigenvalue. One suggested remedy to the multicollinearity problem is to add more data. As will be seen in Chapter V this leads to ridge type shrinkage estimators. Wang et al. suggest based on some theoretical results what they believe is a proper way to select additional data to eliminate the multicollinearity . They also apply their results to regression diagnostics.

For the most part this book will be confined to the estimation of the parameters of a multivariate normal distribution . George (1986 a,b) showed how data weighted shrinkage estimators of the mean of a multivariate normal distribution that shrink to different target subspaces lead to substantial risk reduction on a large portion of the parameter space. Ki and Tsui (1990) extend the multiple shrinkage estimators to the case of simultaneous estimation of the means of several one-parameter exponential families. Their results are developed using an identity similar to that of Haff and Johnson (1986). A computer simulation is done to indicate how much the risk is reduced. One application of their problem is the correct choice of component random variables to combine with the goal of obtaining a suitable shrinkage estimator.

The confidence set for a multivariate mean vector can be improved upon by centering it at a JS estimator instead of the sample mean. Work in this area was done by Hwang and Chen (1986). Robert and Casella (1990) note that at the time their paper was written analytical work was not done for the case of an unknown variance. They adapt the techniques of Hwang and Chen (1986) to deal with the unknown variance case.

Withers (1990) observes that one way to reduce the risk of the JS is to decompose the sample space into orthogonal components and apply the JS method within each. They propose an adaptive method of choosing how to decompose the sample space. They study the effect of large n and compare their estimators with other shrinkage estimators.

Md et al. (1990) give four estimators :

1. the restricted least square estimator (RLSE);
2. the unrestricted least square estimator (URLSE);
3. the preliminary test least square estimator (PTLSE);
4. the shrinkage least square estimator (SLSE) (a JS type estimator).

They derive and compare their risk functions. They find that:

1. The RLSE has the smallest risk if the regression coefficients satisfies the restriction but is unbounded when the parameter moves away from the subspace of the restriction.

2. The SLSE generally has the smallest risk but not when the parameter is in or near the restriction subspace; in that case the PTSLSE is better.

The authors recommend the use of the PTLSE when the restriction subspace has dimension less than three; otherwise they advocate the PRLSE and SLSE with the SLSE preferred.

Nebebe and Sim (1990) study the relative performances of Jackknifed ridge estimators and an EBE using Monte Carlo simulation. It is found that the EB method performs better with respect to the criteria of smaller MSE and more accurate empirical coverage. A bootstrap procedure is proposed and some theorems are given. While this improves on the empirical coverage of the ridge estimators the EBE still performs better.

Firinguetti (1990) obtains some results on the expected values of the product of ratios of random variables. He uses these results to obtain expressions for the MSE of the operational generalized ridge regression estimator of Hoerl and Kennard (1970) and the Jackknifed ridge estimator originally proposed by Kadiyala (1984).

Miller (1990) discusses ridge and James-Stein estimators from the standpoint of choosing an appropriate subset of the independent variables in fitting a linear regression model. He points out that often the best fitting model using LS estimators give estimates of regression coefficients biased in the direction of being too large. Ways to deal with this problem include the use of ridge and James-Stein type estimators , the jack-knife and conditional maximum likelihood estimation.

Brownstone (1990) observes that nonparametric bootstrapping provides good estimates of the sampling distributions for Mundlak's(1981) restricted principle components estimator and a Stein rule that shrinks this estimator toward the LS. For a squared error loss function the Stein estimator performed better than the LS most of the time. The principal components estimator generally does better than the study of Hill and Judge (1987).

Judge, Hill and Bock (1990) note that for shrinkage estimators to achieve significant risk improvement as compared with MLEs it is necessary to identify the region or subspace where the location vector being estimated is either known or thought to lie as a result of prior information. The best shrinkage estimators are those that shrink toward the correct subspace or region. Frequently vague or conflicting priors suggest that a broad class of estimators may be effective for risks using a squared error loss function. In these cases an adaptive JS identified from the use of Stein's unbiased estimator of the risk proves useful.

A well known measure of information is the condition number, that is the ratio of the largest to the smallest eigenvalue of the regression matrix. Soofi (1990) shows how to use the information measures proposed by Shannon (1948), Lindley(1956) and Kullback and Liebler (1951) to quantify collinearity effects[1]. Shannon's entropy for the posterior distribution is shown to be related to the condition number. The information measures provide additional diagnostics to help the ridge analyst choose an appropriate value of the ridge parameter. This is illustrated for data on air pollution that Mc Donald and Schwing (1973) proposed ridge estimates for.

An unbiased estimator of the MSE matrix of the Stein-rule estimator of the vector of parameters of a normal linear regression model is presented by Carter et.al. (1990). Three kinds of confidence ellipsoids are derived. They are denoted by C_a, C_s and C_{ls}. The confidence ellipsoid C_a uses the Stein-rule estimator and the LS estimated MSE matrix. The confidence ellipsoid C_s employs the Stein rule point estimator together with the unbiased estimator of its MSE proposed in the paper. The region C_{ls} is the standard LS confidence set. It is found that both C_a and C_s are preferable to C_{ls} because C_a has approximately the same expected squared volume as C_{ls} but a larger coverage probability while C_s has a smaller expected squared volume that C_{ls} with the same coverage probability. It is impossible to compare C_a and C_s with respect to coverage probability. An F test is developed based on the Stein rule

---

[1] A basic reference on the use of Information Theory in Statistics is Kullback (1959).

estimator has smaller power than one based on the LS. They conclude that the LS is most useful for inferences about individual regression coefficients while the JS is more useful for inferences about the whole vector when the number of regression coefficients is greater than three.

George (1986 a,b) considered shrinkage estimators where the shrinkage was to parameter subspaces. Sengupta (1991) considers the problem of shrinkage toward an arbitrary estimator. He develops a concept of optimal shrinkage by considering the choice of an estimator that is Bayes with respect to some prior.

Ghosh and Shieh (1991) propose a class of EBE's with smaller frequentist risk than the MLE's for many quadratic loss functions. The EBE's are compared in terms of their simulated risks and concrete recommendations are made about the choice of a particular EBE. The construction of these EBE's is the result of solving certain types of differential inequalities.

Brandwein and Strawderman (1991) shows how to extend Stein's results to spherically symmetric distributions. They investigate conditions where Stein type estimators have a smaller risk than the MLE for quadratic and concave loss functions.

A method of truncating a JS estimator to improve its risk is given by Kubokawa (1991). It consists of constructing a sequence of estimators that converge to the generalized BE of Strawderman (1971) and Berger (1976).The resulting estimator is admissible and has smaller risk than the usual JS.

Srivastava and Giles (1991) obtains an exact unbiased estimator for the MSE of an operational ridge estimator. Exact expressions for the MSE of the operational ridge estimator are very cumbersome and difficult to obtain. Their expression is relatively simple; the problem is that it is an estimator and as a result the estimation error has to be dealt with.

It is noted by Silvapulle (1991) that the ordinary ridge estimator is sensitive to outliers in the y variable. For this reason he proposes a class of ridge-type M estimators which are obtained by shrinking an M estimator. An M estimator is one where the sum of arbitrary functions of the observed and the value to be predicted are minimized. Special cases include the LS estimator or the estimator obtained by minimizing the sum of the absolute values. Sivapulle suggests a procedure for selecting k. For a particular choice of M estimator he found that if the M estimator is more efficient than LS the ridge-type M estimator is more efficient than the ordinary ridge estimator for a suitable choice of k. He illustrates how the estimators proposed in his article are less sensitive to outliers in the y variable than the ordinary ridge estimators.

The properties of mixed regression estimators when data contain outliers are studied by Ali (1991). He notes that the mixed regression approach reduces the effects of two kinds of outliers those due to mean-shift and those due to variance inflation. It will be shown in Chapter V that the ridge estimator is a special kind of mixed estimator.

Two popular techniques for finding optimal k according to Girard (1991) are generalized cross validation and Mallows C_L. These techniques choose k as the solution to the problem of minimizing some function of the residual and the trace of a matrix function of k called the influence matrix. Generally in applications the influence matrix is neither explicitly known or easy to compute. Girard (1989) proposed a randomized procedure. It consist of generating a few standard normal random vectors and replacing the trace function by the average of estimates obtained from these simulated vectors in the function to be optimized. The value of k that is obtained by the optimization procedure is selected. The main goal of the paper is to show that consistency and asymptotic optimality properties of the non- randomized procedures also hold for the randomized procedures.

The estimation of the variance of Stein estimators by the bootstrap method is considered by Yi (1991). The bootstrap estimates are found to be reasonably close to the true estimates but they tend to be biased downward. A basic reference for bootstrapping regression models Freedman (1981) is mentioned.

Fomby and Samanta (1991) offer two combination forecasting methods based on Stein shrinkage techniques. Monte Carlo studies showed that these methods performed well in comparison with other available methods.

It has been suggested to the author by a referee that some of the ideas in this book may be formulated using estimating functions. Estimating functions are expressions involving the sample and the parameter values. Estimates of the parameters are obtained by setting the estimating function equal to zero and solving for the parameter values algebraically in terms of the sample values. Godambe (1991) is a collection of papers on the use of estimating functions in different branches of theoretical and applied statistics. Topics covered include an Overview, applications to Biostatistics, Stochastic Processes and Survey Sampling and discussions of their role in both the Foundations of Statistics and General Statistical Methodology. Two papers of particular importance to the material covered in this book are those of Godambe and Kale (1991) and Liang and Liu (1991).

Godambe and Kale (1991):

1. explain what estimating functions are;
2. give an alternative version of the Gauss Markov theorem in terms of the theory of estimating functions;
3. develop an optimality criterion for estimating functions.

Liang and Liu (1991) study the estimation of regression coefficients in a generalized linear model when some of the study variables are measured with error. They examine an approach suggested by Whittemore (1989) that involves replacing the study variables with Stein estimators and then estimating the regression parameters the usual way. Consistent estimators are thus produced for the regression parameters.

Brandwein, Ralescu and Strawderman (1992) consider estimation of a location vector in the presence of a known or unknown scale parameter in three dimensions. They obtain conditions for a Stein type estimator to have a smaller MSE than the MLE for spherically symmetric unimodal distributions.

Guo and Pal (1992) define a sequence of estimators inductively that have smaller risks than the JS. Each succeeding member of the sequence has smaller risk than its predecessor. These estimators, unlike the JS positive parts are smooth. They note that the problem of finding a limiting estimator in closed form is still an open question. They then apply the technique of Kubokawa (1991) to obtain a non smooth estimator with smaller risk than the improved smooth estimators that they constructed.

Kleffe and J.N.K. Rao (1992) consider a random effects linear model. They derive the linear BE or the best linear unbiased predictor (BLUP) for the regression parameters. The unknown parameters in the model are estimated and substituted into the BLUP to obtain the EBE or the empirical BLUP (EBLUP). They then derive:

1. a second order approximation to the MSE of the BLUP;
2. an approximately unbiased estimator of the MSE.

The accuracy of these approximations is confirmed by the results of a simulation study.

As has already been mentioned George (1986) introduced multiple shrinkage estimators. The classical JS shrinks toward the origin; the modified JS with the Lindley correction shrinks toward the sample mean; the multiple shrinkage estimators shrink toward several prior subspaces simultaneously. Ki (1992) performs a simulation to show that in most cases multiple shrinkage estimators are preferable to estimators obtained by the stepwise regression procedure. Application of the multiple shrinkage technique to real data analysis examples give evidence that the these estimators have better prediction power than those obtained by stepwise regression. Ki also gives conditions for these multiple shrinkage estimators to have a smaller MSE than the LS.

In recent years a lot of work has been done on the use of different regression diagnostics to identify outliers and influential observations. A basic reference on this subject is Belsey et al. (1980). A number of diagnostic measures and graphical methods have been proposed for this purpose. Chalton and Troskie (1992) carried out a simulation study with ridge regression and fractional rank estimators (see Marquardt (1970) for a discussion of the fractional rank estimators) to explore the use of standard cutoffs for the diagnostic measures. They found that no cutoff could be adopted as standard and that the diagnostic values should be interpreted with caution. In recent years a lot of work has been done on the use of different regression diagnostics to identify outliers and influential observations. A basic reference on this subject is Belsey et al. (1980). A number of diagnostic measures and graphical methods have been proposed for this purpose. Chalton and Troskie (1992) carried out a simulation study with ridge regression and fractional rank estimators (see Marquardt (1970) for a discussion of the fractional rank estimators) to explore the use of standard cutoffs for the diagnostic measures. They found that no cutoff could be adopted as standard and that the diagnostic values should be interpreted with caution.

As has already been pointed out an important aspect of ridge regression is the choice of the perturbation factor k. Hadzivukovic et al. (1992) compare the performance empirically of methods of choosing k based on the ridge trace (Hoerl and Kennard (1970) and variance inflation factors (Neter et al. (1985)) with objective methods based on functions of the sample estimates. They use empirical data on different factors in the agriculture of Yugoslavia and its regions. They found that in general with respect to the mean square error of estimation and the mean square error of prediction the estimators based on the ridge trace and variance inflation factors did worse than those obtained using sample estimates from the data.

Le Cessie and Van Houwenlinging (1992) used ridge regression estimators to estimate the parameters of a logistic regression model. They consider different ways to estimate k with the primary focus on cross validation. Three different kinds of prediction error are considered, classification error , squared error and minus the log likelihood. Ridge estimators are applied to the problem of developing a prognostic index for the two year survival of patients with ovarian cancer as a function of their DNA histogram. Good, clinically interpretable ridge estimators are obtained by defining a restriction so that neighboring intervals in the DNA histogram only have a slight influence on survival. The model based on ridge estimation turns out to be a more accurate predictor of survival.

Harewood (1992) presents a Stein rule estimator that shrinks toward the mixed estimator. It will be seen in Chapter V that the mixed estimator may be formulated as a special case of the ridge estimator. This fact was also noted for example by Gruber (1990), Bacon and Hausman (1974) and Fomby and Johnson (1977). Harewood observes from a simulation study that his estimator appears to offer slightly greater risk reduction over the LS when compared with the usual JS.

The finite sample moments of the bootstrap estimator of the James-Stein rule are derived and shown to be biased in Adkins (1992). Adkins notes that approximate confidence intervals were constructed by Chi and Judge (1985) and Adkins and Hill (1990). These studies indicated that bootstrap confidence intervals and ellipsoids centered at the JS rule have a tendency to be larger than is necessary for coverage near the origin; this is not as much of a problem for larger values of the regression parameters. Thus, bootstrap standard errors for the JS may be more accurate estimators of the actual standard errors in some parts of the parameter space than others. This fact appears to be true from the results of a small simulation study.

Tracy and Srivastava (1992) consider a family of adaptive ridge estimators of Strawderman (1978). Large sample approximations of the bias vector and the MSE matrix are made and the different members of the family are compared with respect to them.

He'bel et al. (1993) show that Stein type shrinkage estimators are better predictors than the LS estimators of the regional yield of winter wheat, the most widespread crop in France. The reason noted by them is that the shrinkage estimators have a smaller mean square error averaged over a predictive loss function (MSEP) than the LS. The relative merits of three different Stein type shrinkage EBEs are compared. An attempt is made using cross validation and simulation to see if the use of the Stein type estimator reduces the forecasting error.

The book Alternative Methods of Regression by Birkes and Dodge (1993) reviews least squares regression in the first three chapters. Each single chapter that follows considers a different alternative to least squares. Included are chapters on Least Absolute-Deviations Regression, M-Regression, Nonparametric Regression, Bayesian Regression and Ridge Regression. The chapter on ridge regression reviews the basic theory and shows using practical examples how to implement ridge regression for the solution of practical problems using collinear data.

In Chapter IV it will be shown that the risk dominance of JS type estimators over the MLE that holds true for squared error loss does not always hold true under arbitrary quadratic loss. A decision rule is said to be Γ- minimax if for the class of priors Γ its Bayes risk is not larger than that of any other decision rule with finite Bayes risk for this class of priors. Ghosh (1993) proves that a class of shrinkage estimators introduced by Baranchik (1970) and some related estimators are Γ-minimax.

In this volume the comparison of Stein , ridge and LS estimators will generally be done with respect to symmetric quadratic loss functions. Giles and Giles(1993) consider the comparison of preliminary test estimators risk averaging over an aysmmetric loss function the LINEX loss function. The behavior of some commonly used estimators and pre-test estimators of the scale parameter are robust to small departures from symmetry. However, this is not likely to be the case for larger departures.

Brandwein, Ralescu and Strawderman (1993) consider estimation of the location vector for certain subclasses of spherically symmetric distributions. The case of a known and an unknown scale parameter is considered. The class of estimators considered include the JS as a special case and are more general than those considered by Brandwein and Strawderman (1991). Conditions are obtained for the shrinkage estimators considered to dominate the maximum likelihood estimator for quadratic loss functions, concave functions of quadratic loss functions and general quadratic loss functions.

Baksalary and Pordzik (1993) extend the theory of preliminary test estimation based on the matrix risk of two competing estimators of the regression parameters or the predictors for the general Gauss-Markov model (GGM) to the case of a given vector of estimable parametric functions. As an illustration they discuss preliminary test estimation of a vector of treatment contrasts in a fixed effect model corresponding to a block design. The estimators considered involve restricted least square estimators. One might be able to develop the theory using shrinkage estimators instead. A reference to Bock et al. (1973) where preliminary test estimators involving JS estimators is given.

Marchand (1993) considers estimation of the mean of a spherical distribution. Prior information is available about the norm of the parameter that represents the mean. The best estimator invariant under an orthogonal group of transformations for a given constant value of the parameter norm (best equivariant estimator BEE) is obtained. A lower bound is obtained for the risk functions. The BEE and the best linear estimator (BLE) are compared under departures from the constant norm assumption.

Marchand and Giri (1993) are interested in estimating the mean of a multivariate normal distribution whose variance is a positive random variable with a specific known distribution function. He considers a class of JS type estimators and finds the best one in this class assuming:

1. the estimated parameter has a known norm;
2. the norm of the estimated parameter is restricted to an interval.

Lu (1993) compares the performance of JS type EBEs and hierarchical BEs (HBE). He presents an HBE that dominates the EBE. He illustrates the results by an application to the two way classification analysis of variance model.

Kejian (1993) considers a modification of the ridge estimator where a constant multiple d between zero and one of the LS is added to $X'Y$. He obtains an estimate for optimal d by approximating the value of d in terms of the parameter that minimizes the MSE. He refines his estimates by iteration. He gives conditions for the estimator obtained in this way to have smaller MSE than the LS. He shows how his estimator represents an improvement over the LS for a numerical example based on aggregate about import activity in the French economy. It is not clear to the author how this estimator is an improvement over the ridge or the JS that Kejian criticizes.

Perron (1993) considers two independently conducted related experiments. He is interested in estimating the mean of the multivariate normal distribution from the first experiment using the information obtained from both experiments. He compares three shrinkage estimators by means of a simulation. The three estimators are the JS positive part , an estimator proposed by Ghosh and Sinha (1988) and one that he proposes. For smaller parameter values the estimator of Perron appears to have a smaller risk; it is not a significant improvement over the other two for larger values; in fact for larger parameter values the JS positive part appears to be the dominant estimator.

A rather nice discussion of ridge regression estimators is available in Brown (1993). It includes:

1. the motivation for their use;
2. the derivation of the Hoerl-Kennard ordinary ridge estimator from both the frequentist and the Bayesian point of view;
3. methods for finding k;
4. applications.

Brown is careful to point out that when used correctly ridge estimators can be very helpful in the study of multicollinear data but it is not a panacea for the problem He also includes a discussion of partial least squares and principal component estimation.

A book that studies the Kalman Filter primarily from an engineering viewpoint is Grewal and Andrews (1993). In addition to technical information about the Kalman filter and the extended Kalman filter the book contains:

1. a nice account of how the Kalman filter came to be including a brief biography of Kalman;
2. an application to inertial navigation;
3. a diskette containing computer software that demonstrates how the Kalman filter algorithms work.

The effects of misspecifying the disturbances in a linear regression model as spherical on the efficiency of Stein-rule estimators is investigated by Chaturvedi, Van Hoa and Shulka (1993).   Edgeworth type asymptotic expansions of the distribution of Stein-rule estimators based on an erroneous assumption of spherical disturbances are derived. The risk of the Stein-rule estimators averaged over a quadratic loss function are compared with that of the usual LS estimators. For a locally non-scalar disturbance covariance matrix two kinds of Stein estimators are compared. One is based on the ordinary LS estimator. The other is based on a weighted LS where the weight matrix is estimated from the data. This weighted LS is referred to as a feasible generalized least square estimator (FGLS) in the literature. The results of the paper are applied to two kind of linear models. The first one deals with n observations classified into two groups. The observations are independent. The within group variance is the same, but observations belonging to different groups have different variances. The second model is a Box Jenkins AR(1) model.

The James-Stein (1961) estimator was used to prove the inadmissibility of the MLE. It too is inadmissible. Baranchik (1964,1970) showed that the JS positive part estimator has uniformly smaller risk than the JS. Brown (1971) observed that the positive part estimator was also inadmissible. An explicit estimator that dominates the JS positive part estimator was given by Shao and Strawderman (1994). They assumed that the dispersion of the multivariate normal distribution whose mean was being estimated was known.

A critical evaluation of the JS is given by Hill (1994) in the light of the philosophy of mathematics. He criticizes the use of Stein type estimators and the concept of admissibility for infinite spaces.

A class of shrinkage estimators that includes the JS and ridge regression estimators as special cases is proposed by Singh et al. (1994). The risk associated with these estimators is approximated using the small σ asymptotic approach due to Kadane (1971). Some new estimators in the class they propose are compared with existing ones.

The Bayesian Choice (1994) by Christian Robert is an English version of a French graduate level textbook on Bayesian methods entitled Analyse statistique bayésienne. A number of sections are devoted to the "Stein effect." In Section 2.4 there is a nice general historical overview of Stein estimators with many references to key works in the literature. Some of these are also cited in this volume. A proof of the inadmissibility of the MLE of the mean of a multivariate normal distribution is given that holds true for all spherically symmetric distribution.[2] Section 8.5 gives the empirical Bayes formulation of JS estimators. A discussion of the use of Stein estimators for the construction of empirical Bayes confidence intervals. Results of Morris (1983), Hwang and Casella (1982), Casella and Hwang (1983) and George and Casella (1994) are described and used.

Three different kinds of pre-test estimators are considered by Venter and Steel (1994). They are compared to one another and other pre-test estimators with respect to a risk function that combines mean square error with cost complexity. They find that the three kinds of estimators perform similarly and do better than estimators considered previously by other authors, e.g., combinations of Stein and pre-test estimators considered by Ghosh and Dey (1986).

Zellner (1994) describes quadratic balanced loss functions. These loss functions are weighted averages of loss due to goodness of fit and the variability of the estimators. The optimal estimators are obtained for estimating a scalar mean, a vector mean and a vector of regression coefficients. Their MSEs are then compared to the MLE or the LS. The optimal estimators dominate the MLE or the LS for a portion of the parameter space.

---

[2] The multivariate normal distribution is spherically symmetric.

C.R.Rao (1994) gives a survey of the Kalman Filter (KF) methodology used for single and multitarget tracking of satellites. For multitarget tracking separate observations on each target are generally unavailable. Estimation of the positional coordinates of the individual targets using the information provided by the dynamical equations and the previous estimates is considered. Several different methods of associating observations with individual targets are discussed and critically evaluated. A number of references are given to Technical Reports co-authored by C.R. Rao and other authors.

Hwang and Ullah (1994) compare the confidence set for the parameters of a linear regression model centered at a James-Stein estimator with that centered at a least square estimator. The confidence intervals centered at the JS are better because they generally have a smaller volume and a higher coverage probability than that of the LS. They note that previous studies, for example Hwang and Casella (1982,1984), Chen and Hwang(1988) and Carter et al. (1990) claim that for the unknown variance case the bound on the shrinkage factor where the JS is better is similar to that of the known variance case Much to their surprise Hwang and Ullah(1994) found that for the unknown variance case the bound can be as much as thirteen times higher.

Godambe (1994) provides the connection between estimating functions and the linear Bayes estimator as studied by Hartigan (1969). Hartigan's linear BE is the same as that of C.R. Rao (1975).

Rukhin (1995) surveys the study of admissibility in statistical decision theory. The problems reviewed include:

1. Stein's necessary and sufficient admissibility condition and its extensions;
2. Brown's heuristic method of determining admissibility;
3. a differential inequality useful to the statistical estimation problem;
4. admissible estimators of functions of normal parameters;
5. complete class theorems in hypothesis testing;
6. problems with finite sample spaces.

The role of the JS in this development is emphasized. The connection between admissibility and variational problems is also stressed.

Nieto and Guerrero (1995) derive the Kalman filter equations when the error terms in the observation equation and the system equation are correlated. The cases where the state space models are singular and the disturbance probability distributions are conditional on some information related to the observation process is considered. An application to time series is presented. The JS and other EBEs for the parameters of a Kalman filter will be taken up on Chapter X.

Kuks and Olman (1971,1972) proposed a minimax linear estimator for linear parametric functions in a regression model assuming that the parameter vector belongs to a known ellipsoid. However their estimator sometimes takes on values outside of the ellipsoid. To deal with this problem Hoffmann (1995) proposes a modified estimator the does not have this flaw. The modified estimator turns out to be a least square estimator for the restricted model. The ridge regression estimator of Hoerl and Kennard is a special case of Hoffmann's estimator. The minimax linear estimators were studied in Gruber (1990) and will be taken up in the present work in Chapter V.

Baksalary, Markiewicz and C.R.Rao (1995) characterize the class of admissible estimators for the parametric functions of the parameters of a linear model of less than full rank with error terms whose dispersion need not be of full rank. The criterion for comparison is a weighted quadratic loss function. They determine the extent to which this characterization is independent of the weight matrix. They show that except for some special cases admissibility is independent of the choice of weight matrices that have the same range.

Fourdrinier and Roberts (1995) consider the problem of estimating the mean of a normal and a Poisson distribution using an entropy loss function. For the normal case the entropy loss corresponds to the MSE. A class of EBEs is derived that includes among its special cases a class of estimator derived by Alam (1973) and the JS. Under entropy loss the EBE for the Poisson distribution is a shrinkage of x+1 instead of the MLE x.

Sun (1995) considers a class of EBE's that include the JS with the Lindley correction as a special case. He obtains the risk of these estimators and gives conditions for them to dominate the MLE. A lower bound is given for the ratio of the risks of the EBE and the MLE. The lower bound is shown to be the limit of the risk ratio as the number of parameters tend to infinity. The hierarchical BE is obtained. A modal estimator is obtained by replacing a function of the unknown parameters in the BE by the mode of the posterior distribution. Comparisons of the efficiencies of the JS, its positive part, the hierarchical BE and the modal estimator.

Although Stein proved that the MLE is inadmissible when the number of variables $p > 2$ it is well known that when $p \leq 2$ the MLE is admissible. In a linear model with normally distributed error terms the LS is the MLE. Srivastava and Ullah (1995) show that for a dynamic regression model with a lagged dependent variable:

1. if $p > 2$ the JS has a smaller MSE than the LS;
2. if $p \leq 2$ the LS does not dominate the JS.

They do not show that if $p \leq 2$ that the LS is admissible. in the linear model context.

It is well known that for a multivariate normal distribution the positive part of the JS has a smaller MSE than the MLE. Cellier et. al. (1995a) show that this result holds for spherically symmetric unimodal distributions. The spherically symmetric distributions include the normal distribution as a special case. They explain that

Berger and Bock (1976) obtain the same kind of result with a wider class of distributions but with stronger conditions on the shrinkage factor. They give conditions for dropping the unimodality assumption. They also consider two kinds of extensions:

1. a general positive part rule of the type defined by Judge and Bock (1978);
2. estimators that shrink differently in different directions.

A family of counterexamples is given to show that the positive rules do not always have smaller risk than the JS for distributions in general.

Hasegawa (1995) considers the performance of the risk for estimators of location and scale parameters of a linear model. He proves that the pre-test estimator for the location parameter is dominated by the JS under Ignaki's loss function when the distribution of the error term of the linear model is expressible as a scale mixture of normal distributions with unknown variances. Ignaki's (1977) loss function considers two kinds of models for statistical model fitting, one due to modeling and the other due to estimation. These errors are measured by the Kullbach-Leibler information measure. The paper evaluates the risks numerically and analytically for the unrestricted estimator, restricted estimator , JS estimator and the pretest estimator. It is noted that the results of the paper are an extension of those of Nagata (1983).

Recall that Kejian (1993)[3] described a modification of the ridge estimator where a constant multiple of the LS was added to X'Y. He also suggested using a diagonal matrix D in place of the constant multiple in place of the constant multiple thus obtained a generalized Liu estimator. Akdeniz and Kaciranlar (1995) derive an almost unbiased generalized Liu estimator by using the Jackknifing procedure suggested by Ohtani (1986). They obtain:

1. an optimal choice for the elements of the matrix D that minimizes the MSE of the almost unbiased generalized Liu estimator;
2. an unbiased estimator for the bias of the generalized Liu estimator;
3. conditions for the MSE of the almost unbiased generalized Liu estimator to have a smaller or larger MSE than the generalized Liu estimator and the LS.

Gnot, Trenkler and Zmyslony (1995) are interested in estimating the sum of a quadratic form and a non-negative scalar because it has the same structure as the TMSE of a linear estimator. They note that the naive unbiased estimator is unacceptable because it can have negative values. They derive and give the explicit form of a non-negative minimum quadratic unbiased estimator. In one of their

---

[3] The article is referenced by Akdeniz and Kacinralar as Kejian (1993). The full name is Liu Kejian. The estimator proposed in the article is called the Liu estimator by Akdeniz and Kacinralar. It will be referred to as the Liu estimator in this book. The name Liu is in the header of Kejian (1993).

examples they show how to use this kind of estimator to estimate the TMSE of the ridge estimator of Hoerl and Kennard.

Crivelli et al (1995) note that because the distribution of the operational ridge regression estimators are unknown only asymptotic confidence intervals can be obtained. They report on a technique of Rayner (1989) that combines the Edgeworth expansion and the Bootstrap technique to obtain an approximation to the distribution of some ridge regression estimators. They do simulation experiments to compare the asymptotic confidence intervals with those obtained using Rayner's technique.

Fourdrinier and Wells (1995) are interested in estimating the loss of a point estimator for the case of sampling from a spherically symmetric distribution. They consider two location estimators the least square estimator and a shrinkage type estimator. The class of shrinkage estimators considered are similar to but do not contain the JS. Their unbiased loss estimator is compared with an improved loss estimator. Since the domination results are true for a large class of spherically symmetric distributions the loss estimators have good robustness properties.

Chang (1995) applies JS type estimators to a two stage estimation procedure for the mean of a multivariate normal distribution that was suggested by Waikar and Katti (1971). They propose five different JS type estimators and prove that they all have smaller MSE than the two stage estimators that were proposed by Waikar and Katti (1971). They illustrate the relative performance of these JS estimators graphically for a few special cases.

Crouse, Jin and Hanumara (1995) develop an unbiased ridge regression estimator using a random vector of empirical prior information. They develop a robust user oriented non-iterative method for estimating k. The operational ridge estimator is unbiased for the β parameters and takes a form similar to that of the JS. A simulation study comparing the proposed estimator with that of Hoerl, Kennard and Baldwin shows that its MSE is substantially smaller especially when the population variance is very large.

The use of ridge regression in the linear probability model where the dependent variable takes on two values zero and one is taken up by Gana (1995). One reason that ordinary least squares is problematic is that often the estimated values of the dependent variable do not lie between zero and one. The paper illustrates using an example how by the use of ridge regression estimators the estimated values of the dependent variable can be made to lie between zero and one.

A different way to use a ridge parameter is suggested by Takada, Ullah and Chen (1995). They consider the seemingly unrelated ridge regression model where the estimated error variance covariance matrix is singular. They obtain two step generalized least square estimators after adding a ridge parameter to the diagonal elements of the covariance matrix. An important difference between their approach and the usual ridge estimators is that the usual ridge estimators are obtained by adding the ridge parameter to the diagonal elements of the $X'X$ matrix; they add the ridge parameter to the diagonal elements of the weight matrix. They discuss the choices of the ridge parameter, implement them and compare them through a simulation study. They use their estimators to do an empirical analysis of the geographic variation of diffusion processes of the videocassette recorder market.

Breiman (1995) proposes a new method called the nonnegative garrote as a technique for doing subset-selection regression. It chooses nonnegative shrinkage coefficients that minimize the difference between the observed values and a weighted sum of products of LS estimators and values of the independent variable where the sum of the weights(shrinkage factors) are constrained to be less than some upper bound, the garrote.[4] A shrinkage estimator is produced for some of the subsets of the regression parameter and it others are zeroed. The technique is compared to ridge regression and subset selection. Subset selection is generally very unstable. Ridge regression is very stable. Nonnegative garrote is intermediate between these two methods in stability. Unstable estimators generally cause a larger prediction error. Breiman shows that nonnegative garrote unlike ridge regression is scale invariant. He also shows that its accuracy is better than subset selection and comparable to that of the ridge estimator.

A number of topics also considered in the present volume are presented in the recent work of C.R. Rao and Toutenburg. Rao and Toutenburg (1995) discuss many different topics of interest to readers of the present volume. Some of these are

1. methods for dealing with multicollinearity including principle component regression, ridge estimation and partial least squares;
2. linear minimax estimation;
3. estimators derived for exact and stochastic linear restrictions including the restricted LS and the mixed estimator;
4. comparison of the mean dispersion error (MDE) of biased estimators.

Barry (1995) reports the results of a simulation study that compares the performance of the Bayes estimate for the parameters of a linear model obtained using Jeffrey's prior with ridge and smoothing spline estimators with parameters estimated using generalized cross validation methods. The generalized cross validation methods used were described in Golub, Heath and Wahba (1979) and Craven and Wahba (1979). The performance of the MSEs of the ridge regression estimator and the Bayes estimator obtained using Jeffrey's prior were comparable. The simulation studies suggested that the Bayes estimator using Jeffrey's prior does

---

[4]According to the American Heritage Dictionary a garrote is an iron collar that was used in Spain to execute prisoners by tightening the collar to either strangle or break the neck of the condemned person. In the regression context the garrote is tightened by decreasing the upper bound in the regression equation.

better when the MSEs are compared than the smoothing spline estimator when the error variance is large and the sample size is small.

Ohtani (1995) obtains a sufficient condition for the risk of the generalized ridge regression estimator to be smaller than that of the LS using the LINEX loss function. He also compares the risk functions of the feasible generalized ridge estimator, that is, the generalized ridge estimator where a sample estimator is substituted for the ridge parameters, to that of the LS numerically.

Shiaishi and Konno (1995) compare the performance of several different kinds of shrinkage estimators and a feasible least square(FLSE) estimator for the SURE model of Zellner (1962). The SURE model will be taken up in Chapter IX. The estimators compared include a preliminary test estimator (PTE), a shrinkage estimator(SE), a positive rule shrinkage(PSE) estimator and a Kubowaka (1994) type shrinkage estimator(KSE). The PTE, SE, PSE and KSE all have a smaller asymptotic distributional quadratic risk than the FLSE and thus in this sense perform better. The asymptotic performance of the estimators are also compared using the Mahalinobis loss function.

Noor and Mehta (1988) propose a class of ridge type shrinkage estimators. For their estimators they:

1. investigate relative-bias, relative mean square error and relative efficiency;
2. show that for small values of the non-centrality parameter they are more efficient than the LS;
3. observe that for larger values of the noncentrality parameter they are as efficient as the LS.

Noor and Ahmad (1995) obtain and evaluate the exact and the approximate relative efficiency of the estimator of Noor and Mehta (1988). Noor and Ahmad compare the values that they obtain with those obtained by Noor and Mehta. They observe that for small values of the non-centrality parameter the approximate and the exact values of the relative efficiency are very close.

Wei and Trenkler (1995) formulate Bayes and empirical Bayes estimators in model that is misspecified in the sense that some of the variables are missing. They show that the EBE has a smaller MDE than the LS . They also observe that the EBE has a smaller MSE than the LS when viewed as a predictor instead of as an estimator.

As noted by Ravishanker, Wu and Dey (1995) concurrent time series with strong inter-series dependence occur in different applications. Each time series may be modeled by the same Box-Jenkins ARIMA model. They show that the inter-series dependence may be incorporated in the construction of JS type shrinkage estimators of the model parameters by bootstrapping the covariance matrix of the marginal parameter estimates. Simulation studies are presented. that indicate that the MSE and the Pitman nearness of the estimators are substantially improved

Vinod (1995) considers applications of the double bootstrap to ridge regression. The motivation to this work is inference problems studied by Vinod and Raj (1988) associated with the 1984 breakup of Bell system telephone companies. Two solutions to the lack of a pivot problem for ridge and related shrinkage estimators are provided using an adjustment of Efron's (1987) bias-corrected percentile and Beran's (1987) double bootstrap.

The Stein rule and ridge type estimators studied in this book will be for linear models. However there has been work done on these estimators for nonlinear regression models. Kim and Hill (1995) present a positive part Stein rule estimator with MSE that is smaller than the MLE and pretest estimators. For a Monte Carlo experiment the shrinkage estimator dominates the MLE under unweighted quadratic loss. Under other unweighted quadratic loss functions the MLE may have lower risk. Kim and Hill mention some earlier studies for nonlinear models. Sheafer, Roi and Wolfe (1984) proposed a ridge type estimator for the logistic regression model. They showed that its MSE was smaller than that of the MLE for a sufficiently large sample size and severe multicollinearity. Sheafer (1986) showed that for a logistic model a Stein rule estimator had a smaller risk than the MLE when the data was multicollinear. Adkins and Hill (1989) studied the risk properties of a Stein rule estimator in comparison with the constrained MLE and pretest estimators. The positive part outperformed the MLE and the Stein estimator for small to moderate degrees of hypothesis error.

Fule (1995) applies ridge regression to a problem of ecological inference, the use of aggregate data to describe behavior at the individual level. The problem is inference from the results of two consecutive elections in different electoral districts the proportion of voters who voted for the same party or who changed their party choice between elections. A multivariate regression using the seemingly unrelated regression equation model to be studied in Chapter X is performed. The data about the parties voted for in the second election is the response or dependent variable and the data about the parties voted for in the first election is the independent variable. Ridge regression estimates are obtained. The estimation procedure is applied to two consecutive parliamentary elections during 1988 and 1991 in Malmo, Sweden.

Vinod and Srivastava (1995) propose double k class estimators that are consistent and asymptotically normal, a property possessed by the LS estimators. The error terms in the linear regression model need not be normally distributed. These double k class estimators are called rejuvenated double k class estimators. The large sample approximation of their MSE is less than that of the LS to the order of approximation that is considered. Conditions are given for the rejuvenated k class estimator to do better than the James-Stein estimator. A simulation study indicates that double k class estimators generally dominate JS estimators for the MSE and MSE averaging over the predictive loss function. References are made to the study of double K class estimators by Carter(1981, 1984), Menjoge (1984), Srivastava and Chatuvedi (1986), Ullah and Ullah (1978) and Vinod (1980).

Properties of Stein confidence sets for a moderate to a large number of variables are studied by Beran (1995). The main results are:

1. As the number of variables tends to infinity the classical confidence spheres are dominated by Stein type spheres; that means a sphere with a smaller radius will have the same or a higher probability of containing the true parameter value;
2. Bootstrap critical values require resampling from a normal distribution using a good estimator of the mean;
3. The rate of convergence for simple asymptotic or bootstrap constructions are $O(p^{-1/2})$; for more sophisticated bootstrap methods $O(p^{-1})$.

Srivastava (1996) shows that the power generalization of the mixed regression estimator and the mixed regression estimator are the same. Thus, using this kind of generalized mixed estimator is fruitless. Hoerl and Kennard (1975) gives the same kind of generalization and discussion for the ordinary ridge estimator. The paper also notes the similarity of the mixed estimator and a shrinkage estimator proposed by Schmidt (1976).

The general Gauss Markov model where the design matrix and the weight matrix can both be of non-full rank is considered by Gross (1996). He considers estimators that:

1. are admissible for the regression parameters among the class of linear estimators with respect to the usual mean square error risk;
2. are linearly sufficient.

He refers the reader to Baksalary and Markiewicz(1990) for the properties of linearly admissible estimators and to Baksalary and Kala (1981) and Drygas (1983) for the properties of linearly sufficient statistics. In the light of the work of Markiewicz (1991) he notes that the estimators considered in his paper belong to the class of general ridge estimators.

Markiewicz (1996) characterizes the general ridge estimators as linear sufficient and admissible among the set of all linear estimators for a nonsingular linear model. The counterparts of the general ridge estimator are presented for a weakly singular linear model and a nonsingular model with linear restrictions. For the weakly singular model :

1. The design matrix is of full or non-full rank.
2. The dispersion of the error term is non-singular.
3. The range of the design matrix belongs to the range of the dispersion of the error vector.

Heiligers and Markiewicz (1996) identify the linearly sufficient and admissible linear estimators with bounded mean square error in linear models with parameter restrictions as special general ridge estimators. They base their result on the characterization of linearly sufficient and admissible estimators as ridge estimators found in Markiewicz (1996).

Empirical Bayes estimators of a multivariate normal mean for the case where the components are independent and the variances are unequal are considered by Shinozaki and Chang (1996). They obtain conditions for the EBE to dominate the MLE when the prior distribution is normal with unknown mean and variance for shrinkage toward the grand mean and a regression estimate. They note that earlier results of Shinozaki and Chang (1993) gave conditions for EB estimators that shrunk toward the origin to dominate the MLE. The 1996 paper treats empirical Bayes estimators suggested by Carter and Rolph (1974) and Fay and Herriot (1979). Earlier in this historical summary it was pointed out that these estimators proved useful in estimating fire alarm probabilities and the estimation of income for small places.

A great many of the results in this work particularly those that involve a non-full rank linear model make use of the singular value decomposition of a matrix. It is taken up and used to derive many of the results in Chapter IV and V. A discussion of the singular value decomposition (SVD) for a general mathematical audience is given by Kalman (1996). Kalman shows how the SVD may be used to derive and study the least square estimator and for reduced rank approximations of matrices. He mentions a number of good sources of information about the SVD including Golub and Van Loan (1983) and Strang (1980, 1993).

Recall that Bilodeau and Kariya (1989) gave sufficient conditions for an estimator of the mean coefficient matrix in the context of the MANOVA model to have a smaller risk than the MLE and gave several examples of Stein type shrinkage estimators with this property. In the context of a GMANOVA model or equivalent growth curve model Kariya, Konno and Strawderman(1996) give a number of examples of double shrinkage estimators that dominate the MLE with respect to an invariant risk matrix. Earlier Tan (1991) and Kubokawa et al.(1992) proposed some estimators that had uniformly smaller scalar risk than the MLE. The GMANOVA model contains the MANOVA model as a special case and was originally formulated as a growth curve model by Pothoff and Roy (1964).

Adkins and Hill (1996) study the risks of Bayes, empirical Bayes and Stein estimators for the probit model, i.e., a linear model whose dependent variable takes on the values zero or one. They derive an approximate Bayes estimator (BE) identical to that of Zellner and Rossi (1984), an empirical Bayes estimator and a James-Stein estimator. They then do a simulation study of the MLE, the restricted MLE (RMLE), pretest, Bayes, empirical Bayes and Stein estimators. Four economic examples are used. These include:

1. voting decisions of individuals in a local school election (see Rubenfield (1977);
2. data in labor force participation by women (see Berndt (1991);

3. financial and personal characteristics that influence home buyers to select either a fixed rate or variable mortgage (see Dhillon et.al. (1987);
4. an exploration of factors which affect the probability that the Democratic candidate wins a state in the 1976 election (see Greene (1990)).

They find that :

1. The use of prior information via the Bayes, empirical Bayes or Stein estimator for the estimation of parameters in the probit model can reduce risk significantly as compared with the MLE and pretest estimators.
2. The risk performance of the Stein probit estimator is similar to that of the Stein estimator for the parameters in the classical regression model.

A rather nicely written paper of Srivastava and Shulka (1996) compares three estimators the LS , the quasi minimax estimator and the mock minimax estimator with respect to bias, minimax risk and the mean square error matrix. The least square estimator is unbiased and both the quasi minimax estimator and the mock minimax estimator are shown to have smaller minimax risk and matrix mean square error. The mock minimax estimator and the quasi minimax estimator are compared and conditions for one estimator to have a larger bias and dominate the other with respect to minimax risk and matrix mean square error are given.

Although, as has already been mentioned, Pitman admissibility and Pitman closeness are not central to the ideas mentioned in this book a great deal of work has been done on this topic in recent years. Ghosh, Mukerjee and Sen (1996) study second order Pitman admissibility and Pitman closeness properties for first order efficient estimators in multiparameter estimation problems. As a case in point they give conditions for Stein rule estimators to dominate the MLE with respect to second order Pitman efficiency and thus give conditions for the MLE to be second order Pitman inadmissible.

Schipp and Toutenburg (1996) derive approximate minimax estimators based on partial constraints on the structural parameters in a single equation of a simultaneous equation model. The form of these estimators are similar to ridge type estimators. The feasible estimators obtained by substituting sample estimators of the unknown ridge parameters are biased but consistent.

The fact that the JS has a smaller risk than the MLE when estimating the mean of a multivariate normal distribution motivates the formulation of shrinkage estimators for the solution of other problems. In the context of estimation of reliability Tse and Tso (1996) propose three shrinkage estimators for the mean of an exponential distribution. They do a simulation study and observe that two of the estimators have a substantially smaller MSE than the MLE.

Ohtani (1996a) considers exact small sample properties of an operational version of the minimum MSE estimator in a linear regression model where the unknown population variance is replaced by the within sum of squares and the unknown regression coefficients are replaced by the LS estimator. Expressions are derived for the MSE. It is then shown that the predictive MSE is always smaller than that of the LS when the number of regression parameters is at least three. For the case where the number of regression parameters is less than two a sufficient condition is obtained for the minimum MSE estimator to dominate the LS.

The risks of empirical Bayes estimators converge to the risks of the corresponding Bayes estimators as the number of observations become very large. This property is called asymptotic optimality. Tong (1996) investigates the convergence rates of the risks of the empirical Bayes joint estimators of the regression coefficients and the error variance in a linear regression model. In particular, he does this for the parameters of the variance component model. He mentions the investigation of the empirical Bayes estimators by Wind (1972), the rates of convergence of the EBE obtained by Singh (1977,1979) for one parameter exponential families and the corresponding results of Tong (1996) for multiparameter exponential families.

Giri (1996) has a brief discussion of the James-Stein estimator and its positive part in Chapter 5 of the book Multivariate Statistical Analysis. A proof of the inadmissibility of the MLE, the JS and its positive part is given. The JS is defined for the case where the covariance matrix of the multivariate normal distribution being estimated is unknown. There is a valuable list of references at the end of the chapter.

An interesting application of the Kalman Filter is presented by Huang and Cressie (1996). They point out that the National Weather Service use current snow water equivalent (SWE) data and a purely spatial model for prediction at sites where no data is available. To improve these predictions a spatio-temporal model is introduced that uses past data. This results in the use of a Kalman filter prediction algorithm. A procedure for estimation of the parameters of the model is presented and an example is presented for the Animas River basin in southwest Colorado. The Kalman Filter is described in Gruber (1990) and will be taken up in Chapter X of the present volume.

Sajjan and Basawa (1996) consider a mixed model where the error terms follow a first order autoregressive process. Such mixed linear models have been used for the study of animal breeding, genetics and agricultural experiments. They derive and compare the prediction error of :

1. the Bayes predictor;
2. an empirical Bayes predictor that is an approximation to the best linear unbiased predictor (BLUP).
They establish the consistency and the asymptotic normality of the empirical Bayes predictor.

A class of empirical Bayes estimators is proposed for a two way multivariate normal model by Sun (1996). The mean square errors are derived, their lower bounds are obtained and their asymptotic behavior is studied. The JS estimator is one of the empirical Bayes estimators in the class considered. The JS is compared with a modal estimator that is derived in the using a hierarchical prior distribution for the unknown parameters. It appears that some of the results and ideas described in this paper might also be available in Sun (1992).

Some interesting generalizations of the inadmissibility results for the JS are given in Evans and Stark (1996). Using stochastic analysis they establish that when estimating the mean of a multivariate distribution certain shrinkage estimators dominate the MLE for a large class of error distributions including the spherically symmetric distributions and the normal distribution. The proof is accomplished by representing the error distribution as that of a stopped Brownian motion and using stochastic analysis to obtain a generalization of Stein's (1981) integration by parts lemma for the normal distribution.

Ravishanker, Dey and Wu (1996) observe that often industrial firms are organized into smaller units. Consequently data series are available for each unit thus leading to a set of repeated time series that covers the same time period. As a result the estimated model parameters from the different time series are often correlated. Often each of these time series contain a block of irregular observations called contemporaneous outliers. This paper explains how to use Stein type shrinkage estimators to improve the accuracy of the estimates of the model parameters and the contemporaneous outliers. It shows an application of the techniques to IBM regional revenue data. The simulation study shows a high percentage improvement for both the MSE and the Pitman nearness.

The shrinkage estimators that will be considered in this book will be primarily of the ridge and James-Stein type. However another kind of shrinkage estimator can be obtained using partial least squares. Partial least squares consists of choosing subspaces of the column space of the design matrix and projecting the observation vector into these subspaces and finding the least square fit. The resulting least square estimators have a smaller norm. Goutis (1996) gives a geometric proof that the estimates of the parameters of a regression model obtained by partial least squares shrinks the ordinary least square estimates. This result was also proved algebraically by de Jong (1995).

There are a number of different ways to improve the efficiency of a JS estimator. For example Berry (1994) shows that a modified Stein type estimator that incorporates an improved variance estimator proposed by Stein (1964) has a smaller MSE than the traditional JS estimator. Kubokawa (1991) obtains an estimator that has a smaller MSE than the JS. Sometimes a regression model is misspecified because relevant regressors are omitted from the model. Mittelhanmmer(1984) and Ohtani (1993) showed that for this situation the JS does not dominate the LS. Ohtani (1996b) considers a broader class of pre test estimators than Berry (1994). He shows that if a certain condition on the number of regressors and degrees of freedom is satisfied and an appropriate critical value of the test statistic is used that the Stein rule estimators studied by Berry (1994) can be further improved. However his numerical studies show that for a misspecified model the pre-test estimators no longer dominate the OLS.

Kibria (1996) estimate the coefficients of a general linear regression model when the error term follows a Student t distribution. Three estimators are considered the unrestricted ridge regression estimator (URRRE) the restricted ridge regression estimator (RRRE) and the preliminary test ridge regression estimator (PTRRE). The URRRE is the ordinary ridge regression estimator of Hoerl and Kennard. The RRRE takes the form of the ridge estimator of Hoerl and Kennard with a restricted LS associated with a linear hypothesis replacing the ordinary LS. For a given linear hypothesis H_0 the PTRRE is the ridge estimator when the test rejects H_0 for a given level of significance and the RRRE when H_0 is not rejected. The RRRE performs better than the URRRE if the linear restriction holds true; otherwise the URRRE is still an efficient estimator but the RRRE is not. A better performance is obtained from the PTRRRE. The paper compares the three kinds of estimators and gives conditions for one type of estimator to dominate another.

Baye and Parker (1984) proposed a general $r - k$ class estimator that includes the ordinary least square estimator, the ridge regression estimator and the principal component estimator as a special case. They showed that the $r - k$ class estimator had a smaller MSE than the principal components estimator . These estimators were also studied by Nomura and Ohkubo (1985) and Sarkar (1989). Sarkar (1996) compares the performance of the LS , ridge estimator and the principal components estimator with respect their mean dispersion error (matrix MSE)(MDE). He obtains necessary and sufficient conditions for the $r - k$ class estimator to have a smaller MDE than the LS, ordinary ridge estimator and the principal components estimator and suggests tests of hypothesis to see whether those conditions are satisfied.

The balanced loss function considered by Zellner (1994) is a convex combination of goodness of fit and estimation precision. Giles, Giles and Ohatani (1996) show that under Zellner's loss function that the efficiency of a restricted MLE based on a pre-test estimator is better than an unrestricted MLE if less than half of the weight is given to goodness of fit. Otherwise the unrestricted MLE is more efficient. Unless all of the weight is given to goodness of fit it is shown using the JS and its positive part that the unrestricted MLE is inadmissible. When at least half of the weight is given to goodness of fit the optimal estimator amongst those considered in the paper is the positive part.

Cellier, Fourdrinier and Robert (1989b) and Cellier and Fourdrinier (1995b) present a robust JS that unlike the usual JS dominates the MLE for a wide class of non-normal as well as normal distributions. Fourdrinier and Strawderman (1996) show that under certain conditions robust JS that use the residual vector to estimate the shrinkage constant actually have a smaller risk than the usual JS where the known variance is used. This result is somewhat paradoxical because an estimator that uses an estimated quantity does better than one that uses a known quantity.

George and Oman (1996) propose a multiple shrinkage JS type estimator of the kind considered by George (1986a) that shrinks toward the principal components regression estimator. The predictive performance of this estimator, the principal components estimator and the LS are compared for three data sets relating to

1. air pollution (see Mac Donald and Schwing (1973));
2. car price date (see Becker et al. (1988));
3. highway accident data (see Weisberg (1980)).

It was found that most of the time the principal components estimator and the multiple shrinkage estimator had a smaller MSE than the LS. The multiple shrinkage estimator usually   performed better when the principal component estimator did poorly.

Gross (1996) explains how the estimator of Kuks and Olman(1972), an optimal estimator when the parameters are constrained to lie on an ellipsoid, may be extended from the full to the non-full rank case. The full rank case was taken up by Kuks and Olman (1972), Bunke (1975) , Lauter (1975) and Toutenburg (1982). Its form is very similar to the Bayes estimator in C.R. Rao (1975). Like the estimator of C.R. Rao the Kuks and Olman estimator contains the ridge estimator as a special case.

A policy capturing method that combines human judgment in the form of prior information with ridge regression is presented by Holzworth (1996). The estimators employed are ridge estimators derived from priors with non-zero mean. They are called "smart ridge estimators"; the term was coined by Crouse and Holzworth (1988).   Relative to two cross validation indices, cross- validated multiple correlation and mean square error of prediction of new judgments "smart ridge regression" was found to outperform ordinary LS regression and the usual kind of ridge regression. The applications studied include judgments about:

1. whether a student wants to take a course from a particular professor;
2. the likelihood of teenagers getting into trouble;
3. the desirability of jobs;
4. about hypothetical job related situations by police officers;
5. concerning the attractiveness of roommates by students.

General formulae for the moments of the parametric functions of Stein-rule and the classic positive part Stein-rule are derived in Ohtani and Kozumi (1996). The authors note that their methodology is different from that of Phillips (1984). It is shown that the positive part rule has a smaller MSE than the usual Stein rule. A sufficient condition is obtained for the risk of the Stein rule for parametric functions to dominate the LS. The first four moments of the Stein rule estimator and its positive part are evaluated numerically. The bias, variance MSE, skewness and kurtosis are examined. In Chapter VIII of this book it will be shown that parametric functions of Stein estimators do not always have smaller risk than parametric functions of the LS. Formulae for the risk of different positive parts of the JS will be obtained in Chapter VI.

Pal and Chano (1996) investigate the risks of four shrinkage estimators first assuming that the random vector follows a multivariate normal distribution and then for the case of a multivariate t distribution. The four shrinkage estimators considered are :

1. the JS.
2. the classical positive part of the JS;
3. an estimator of Kubokawa(1991) that is uniformly better than the JS and admissible under a quadratic loss function;
4. a sequence of smooth estimators due to Guo and Pal(1992) that give significant improvement over the JS.

While the JS positive part, the estimators of Kubokawa and the estimators of Guo all have uniformly smaller risk than the conventional JS there is no clear-cut winner among the three estimators.

Recall that Shao and Strawderman (1994) gave the form of an explicit estimator that had a smaller risk than the JS positive part when the dispersion of the multivariate normal distribution whose mean was being estimated was known. Sugiura and Takagi (1996) extend the result of Shao and Strawderman to the case where the dispersion is unknown using a similar type of argument.

The MLE of a normal mean that is known to be positive is the positive part of the observation of the normal mean. The parameters of the normal mean estimated by the Stein type estimators considered in this book may be positive or negative. Shao and Strawderman (1996) produce a class of explicit estimators that dominate the MLE of a positive normal mean parameter with respect to squared error loss. They believe but have not proved that some members of their class are admissible. The methodology is similar to that of Shao and Strawderman (1994). Methodology similar to that of this paper is used by Shao and Strawderman (1995) to improve the positive part of the UMVUE of the noncentrality parameter in a chi-square distribution.

One kind of improved estimator for the parameters of a linear regression model that was introduced by Theil (1971) was the minimum mean square error estimator. This estimator includes known parameters. A variant of this that replaces the known parameters with functions of sample statistics was proposed by Farebrother (1975). Ohtani (1996c) presents an estimator similar to that of Farebrother with an adjustment to the degrees of freedom. The exact MSE is derived and compared with that of the JS and its positive part and the minimum mean square error estimator by numerical evaluation. The adjustment of the degrees of freedom is particularly effective when the noncentrality parameter is close to zero. Ohtani also shows that his estimator can have a smaller mean square error than the positive part of the JS for a wide range of values of the noncentrality parameter.

A general discussion of the multicollinearity problem, principal component and ridge regression estimators is given in Chapter 8 of Hocking (1996).

Nagata (1997) constructs a Stein type confidence interval for the error variance under the normality assumption. The coverage probability is obtained numerically under the multivariate Student t distribution. Sufficient conditions for the Stein type confidence interval to improve on the usual confidence interval are given.

The literature on ridge and Stein estimators is so vast that many more articles could have been reviewed in this survey. However given the entry points to the literature described here one should be able to find almost any available reference in his/her area of interest.

1.5 Empirical Bayes Methodology

The correctness of the inferences made by Bayesian methods often depend very strongly on the accuracy of the knowledge of the prior distribution. Often when solving a practical problem such knowledge is unavailable or incomplete. When prior information is incomplete or unavailable the statistician can obtain estimators for the prior parameters from the data and use them in their place. This technique is called empirical Bayes.

Empirical Bayes (EB) methodology as it will be used in this book entails:

1. for the given form of the prior and the sampling distribution, e.g., normal, Poisson, Binomial, etc., finding the posterior distribution and the resulting Bayes estimator for a squared error loss function;
2. substituting estimators for the unknown parameters (usually unbiased) obtained from past data.

Two illustrations of EB methodology will now be given; the first deals with a Poisson distribution; the second deals with estimation of the variance components in a random effects Analysis of Variance model.

1.5.1 The Poisson Distribution

The following example is due to Robbins (1955), the originator of empirical Bayes methodology. It considers a Poisson population with a completely unknown prior distribution.

Let X be an observation of the Poisson random variable X with unknown mean . Let $F(\mu)$ represent the cumulative distribution function of the prior distribution of μ. By Bayes theorem the posterior distribution is

$$\Pi(\mu \mid x) = \frac{(e^{-\mu}\mu^x / x!)dF(\mu)}{G(x)} \tag{1.5.1}$$

where

$$G(x) = \int \frac{e^{-\mu}\mu^x}{x!} dF(\mu) \tag{1.5.2}$$

is the marginal distribution of X. For squared error loss the BE is the mean of the posterior distribution. Thus

$$\begin{aligned}
\hat{\mu}^b = E(\mu \mid x) &= \int \frac{\mu e^{-\mu}\mu^x}{x! G(x)} dF(\mu) \\
&= \int \frac{(x+1)\mu^{x+1}e^{-\mu}}{(x+1)! G(x)} dF(\mu) \\
&= \frac{(x+1)G(x+1)}{G(x)}
\end{aligned} \tag{1.5.3}$$

Since $G(x)$ and $G(x+1)$ are unknown they must be estimated from the data. Suppose past observations $x_1, x_2, \cdots, x_n$ are available. One way to estimate $G(x)$ is

$$\hat{G}(x) = \frac{\text{\# of observations} = x}{N} \tag{1.5.4}$$

and

$$\hat{G}(x+1) = \frac{\text{\# of past observations} = x+1}{N}. \tag{1.5.5}$$

The EBE of μ is

$$\hat{\mu} = \frac{(x+1)\hat{G}(x+1)}{\hat{G}(x)} \tag{1.5.6}$$

Example 1.5.1. The number of accidents in a week on a superhighway has a Poisson distribution. Suppose the number of accidents for each of ten weeks was

$$5 \quad 8 \quad 7 \quad 4 \quad 4 \quad 1 \quad 4 \quad 4 \quad 2 \quad 5$$

Sometime later in a particular week there are 4 accidents. To estimate μ, observe from (1.5.4) and (1.5.5)

$$\hat{G}(4) = \frac{4}{10}$$

$$\hat{G}(5) = \frac{5}{10}$$

Then the EBE is

$$\hat{\mu} = \frac{5(\frac{4}{10})}{\frac{2}{10}} = 10.$$

1.5.2 The Analysis of Variance

The following example, due to C.R. Rao (1973) illustrates EBE's in the context of Analysis of variance.

Let $x_1, x_2, .., x_n \sim N(\theta, \sigma_2^2)$. Assume that θ has a prior distribution, i.e., $\theta \sim N(\mu, \sigma_1^2)$. The objective is to construct an empirical Bayes estimate of θ. Before the EBE can be formulated the BE must first be obtained. A squared error loss function will be assumed. Thus, the BE is the mean of the posterior distribution. The posterior distribution is obtained by combining the joint distribution of θ and $x_1, x_2, ..,x_n$,i.e., the likelihood, with the prior distribution of θ employing Bayes theorem. The likelihood is given by

$$f(\underline{\theta},\underline{x}) = \frac{1}{\sqrt{2\pi}\sigma_1} \exp\left[\frac{-(\theta-\mu)^2}{2\sigma_1^2}\right] \frac{1}{(2\pi\sigma_2^2)^{y_2}} \exp\left[-\sum_{i=1}^{n} \frac{(x_i-\theta)^2}{2\sigma_2^2}\right]$$

(1.5.7)

The marginal density $g(\underline{x})$ is obtained by integrating (1.4.7) with respect to θ. Then

$$g(\underline{x}) \propto \exp\left[-\frac{n(\bar{x}-\mu)^2}{2(n\sigma_1^2+\sigma_2^2)} - \frac{s^2}{2\sigma_2^2}\right]$$

(1.5.8)

where $\bar{x} = \dfrac{x_1+x_2+...x_n}{n}$ and $s^2 = \sum_{i=1}^{n}(x_i-\bar{x})^2$. Hence by Bayes theorem and division of (1.5.7) by (1.5.8) the posterior distribution

$$h(\theta \mid x) \propto e^{-(\theta-\alpha)^2/2\beta^2}$$

(1.5.9)

where

$$E(\theta \mid x) = \frac{\mu/\sigma_1^2 + n\bar{x}/\sigma_2^2}{1/\sigma_1^2 + 1/\sigma_2^2} \text{ and } V(\theta \mid x) = \beta^2 = \frac{\sigma_1^2\sigma_2^2}{n\sigma_1^2+\sigma_2^2}$$

(1.5.10)

Suppose past data consists of p independent sets of n observations

$$x_{11}, \ x_{21}, \cdots, x_{n1}$$
$$\vdots$$
$$x_{1p}, \ x_{21}, \cdots, x_{np}$$

For a random effects model the ANOVA table is

	SS	Expectation
Between Sets	B	$(p-1)(\sigma_2^2+n\sigma_1^2)$
Within Sets	W	$p(n-1)\sigma_2^2$

By the classic ANOVA method

$$\hat{\sigma}_2^2 = \frac{W}{p(n-1)} \text{ and } \sigma_1^2 = \frac{B}{n(p-1)} - \frac{W}{np(n-1)}$$

(1.5.11)

An estimate of μ is provided by the grand mean of the observations

$$\hat{\mu} = \frac{\sum_{i=1}^{p} \sum_{j=1}^{n} x_{ij}}{np} \qquad (1.5.12)$$

and the EBE is

$$\hat{\theta} = \frac{(\hat{\mu}/\sigma_1^2) + (n\bar{x}/\sigma_2^2)}{(1/\sigma_1^2) + (1/\sigma_2^2)}. \qquad (1.5.13)$$

Example 1.5.2. Three groups of eight light bulbs is selected from a large population of light bulbs. The number of hours until burnout is given below. A later observation was also obtained for another group of light bulbs that is given in the last column.

Table 1.5.1

Group / Bulb	1	2	3	Later observation
1	1013.07	981.15	993.92	1006.98
2	992.80	999.88	1010.31	995.81
3	992.04	992.28	1014.65	987.01
4	995.73	1002.03	1007.87	996.03
5	1001.80	998.51	1005.49	1000.10
6	989.12	997.73	1011.11	989.12
7	1007.61	993.81	1007.63	1007.28
8	995.60	1013.18	1028.10	997.47

The ANOVA table for the past data, i.e., the first three sets is

Source	Degrees of Freedom	Sum of Squares	Mean Square	F	p
Factor	2	848.6	424.3	4.68	0.021
Error	21	1905.7	90.7		
Total	23	2754.3			

Now $\hat{\mu} = 1001.47$, $\sigma_2^2 = 90.7$, $\bar{x} = 997.5$. Hence the future value of θ is

$$\hat{\theta} = \frac{\dfrac{1001.47}{41.7} + \dfrac{8(997.5)}{90.7}}{\dfrac{1}{41.7} + \dfrac{8}{90.7}} = 998.34$$

1.6 A Minimum Mean Square Error Estimator, Its Approximation as an Empirical Bayes Estimator

The objectives of this section are:

1. to give an example of how a MMSE may be found and approximated;
2. to show how the same estimator may be formulated as an EBE.

The techniques illustrated here will be used extensively in the rest of the book. Consider a model of the form

$$Y_i = \mu + \varepsilon_i, \ 1 \le i \le n \tag{1.6.1}$$

where Y_i, $1 \le i \le n$ are independent observations μ represents the parameter for the mean to be estimated an the ε_i are the errors. Assume that $\varepsilon_i \sim N(0, \sigma^2)$. The least square estimator of μ is

$$\hat{\mu} = \frac{\sum_{i=1}^{n} Y_i}{n} = \overline{Y}, \tag{1.6.2}$$

i.e. the usual sample mean. Consider the problem of minimizing

$$\begin{aligned} H(c) &= E(c\overline{Y} - \mu)^2 \\ &= c^2(\frac{\sigma^2}{n} + \mu^2) - 2c\mu^2 + \mu^2 \end{aligned} \tag{1.6.3}$$

By the usual process of differentiating and setting the result equal to zero

$$H'(c) = 2c(\frac{\sigma^2}{n} + \mu^2) - 2\mu^2 = 0 \tag{1.6.4}$$

and

$$c = \frac{\mu^2}{\frac{\sigma^2}{n} + \mu^2} \ . \tag{1.6.5}$$

The MMSE is

$$\hat{\mu} = \frac{\mu}{\frac{\sigma^2}{n} + \mu^2} \overline{Y} = \frac{n\overline{Y}}{\frac{\sigma^2}{\mu^2} + n} \ . \tag{1.6.6}$$

There is an obvious difficulty with the estimator in (1.6.6). It contains the parameter μ which is being estimated as well as σ^2 which is usually not known in practice. This problem may be circumvented by substituting the usual unbiased sample estimates for μ and σ^2, namely $\overline{Y}$ and $s^2 = \sum_{i=1}^{n-1}(Y_i - \overline{Y})^2/(n-1)$ Thus, the approximate MMSE is

$$\hat{\mu} = \frac{n\overline{Y}}{s^2/\overline{Y}^2 + n} \ . \tag{1.6.7}$$

The estimator (1.6.7) can be obtained from a Bayesian approach. Assume that the prior distribution of μ is $N(0, \tau^2)$. Minimize the expectations in (1.6.3) to obtain estimator (1.6.6) with τ replacing μ. It will be shown later that when the prior distribution is normal minimizing the mean square error averaging over the prior produces the Bayes estimator; i.e. the mean of the posterior distribution. Then $s^2/\overline{Y}^2$ may be substituted for σ^2/τ^2 thus yielding (1.6.7). Thus, for the assumed form of prior with unknown variance (1.6.7) may be looked upon as an EBE.

Exercise 1.6.1. Obtain the estimator in 1.6.7 for the percentage of the Gross National product spent on Non-Defense Research and Development by the United States.

Exercise 1.6.2. Let x_i, $1 \le i \le p$ be independent random variables with mean θ_i and variance 1.

A. Find c so that $\sum_{i=1}^{p}(cx_i - \theta_i)^2$ is minimized.

B. Formulate the MMSE.

C. Explain how to obtain the JS as an approximation to the MMSE.

D. Was a normality assumption needed for your proof? Explain.

E. Suppose that θ_i has a prior distribution with mean 0 and variance A. Formulate the MMSE for this assumption.

F. How can the JS be obtained as an approximation to the estimator in E.?

1.7 The Structure of This Book

Different motivations for and formulations of the JS for the mean of a MN are taken up in Chapter II. These include heuristic formulations due to Stein, the EBE formulation, the approximate MMSE formulation. The MLE is shown to be inadmissible by showing that the JS has uniformly smaller risk than the MLE. A brief summary of the needed basic decision theory concepts is included. The positive part estimator is introduced.

Chapter III presents different RRs by generalizing the original argument of Hoerl and Kennard (1970). The variance and the MSE of Hoerl Kennard RR are compared to the LS. The results of different Monte Carlo studies for operational ridge estimators are summarized. Some of the controversies concerning the RR are explored. Examples of practical applications of the RR are given.

The ideas and concepts considered in Chapter II are extended to a single linear model in Chapter IV. For the sake of generality the non-full rank model is considered. The discussion of the non-full rank case is facilitated by a brief discussion of estimability and generalized inverses. The JS is formulated both as an EBE and an approximate MMSE. The average and conditional MSEs are obtained. Both the MSE averaging over the prior distribution and the frequentist MSEP is smaller than that of the LS. This enables two measures of relative efficiency to be defined, the relative savings loss (RSL) and the relative loss (RL). Their mathematical properties are presented together with graphics illustrating the results for special cases. It is explained how the JS is a special case of the operational RR.

Chapter V is a condensation of material in Gruber (1990). The formulation of ridge type estimators for the non-full rank model is taken up from different points of view. Ridge type estimators are formulated:

1. from a generalization of the derivation of the ridge estimators in Chapter III;
2. as a mixed estimator;
3. as a linear minimax estimator;
4. as a special case of the BE.

The efficiencies of the RR and the LS are compared when the estimators are looked at from both the Bayesian and the frequentist viewpoint. A Jackknifed ridge estimator is described and a critical evaluation of various measures of its efficiency is given.

Originally James and Stein (1961) used the MSE properties of their estimator to prove the inadmissibility of the MLE. However the JS is inadmissible. Consequently, estimators with smaller frequentist risk can be constructed. There are various ways to truncate the JS to do this. Chapter VI presents five different positive parts and evaluates their MSEs, RSL and RL. These positive parts of the JS do better than the JS with respect to the measures of efficiency. Graphical illustrations of the properties of these different measures of efficiency are given. These positive parts are inadmissible. However it is very difficult to find an estimator with uniformly smaller MSE than the classical positive part. Shao and Strawderman (1994) produce such an estimator. Their work is summarized.

The problem of simultaneous estimation of the parameters for r linear regression models is presented in Chapter VII. This is a very rich problem because:

1. It is applicable to genetic experiments and replicated industrial experiments.
2. Many different kinds of EBEs may be formulated and compared.

Three basic kinds of EBEs are derived, those of C.R. Rao, Wind and Dempster. The forms of these estimators are quite different because they are derived assuming different amounts of available information about the prior parameters. These EBE's are also shown to be approximate MMSE's. Like Chapter III the average and conditional MSE's, the RSL and the RL are evaluated.

As has already been mentioned the properties of the MSE of the individual components of the JS are quite different from that of the MSE averaging over a compound loss function. The comparative properties of these two kinds of MSE are taken up in Chapter VIII. Often the individual components of the MSE of the JS are not smaller than that if the LS. However, for certain compound loss functions, e.g., the predictive loss function, the JS does have a smaller MSE than the LS.

Questions similar to those considered in Chapters II, IV and VII are dealt with in Chapter IX for the multivariate analysis of variance model (MANOVA). It is found that in many of the cases the MSE cannot be computed exactly. To illustrate how this may be dealt with a method of obtaining a large sample approximation is presented and an illustrative example is given.

Chapter X extends the results of Chapters III-V and VII and IX to other linear model setups. These include the Seemingly Unrelated Regression Equation model (SURE), r Kalman filters and the Simultaneous Equations model.

Chapter XI is a summary of the entire book and conclusions reached by the author as a result of his work.

Chapter II

The Stein Paradox

2.0 Estimation of the Mean

Many statistical analyses are done assuming that the underlying population is normal. When the dimension of the parameter space is at least two the multivariate normal distribution and the distributions that can be derived from it, e.g., the chi square or Wishart distributions, prove to be very useful. Frequently, a linear regression model is assumed to have normally distributed error terms. The least square estimators of the parameters then follow a multivariate normal distribution. A focal point of this book is different estimation procedures for the parameters of a multivariate normal distribution. To help make this book reasonably self contained and for ready reference the main properties of the multivariate normal distribution are summarized in Section 2.1.

The method of maximum likelihood of R.A. Fisher is a popular estimation procedure because:

1. Often the estimators are quite efficient, for example in the regression problem the maximum likelihood estimators (MLE) are minimum variance unbiased estimators (MVUE).
2. The MLEs are functions of sufficient statistics.
3. The MLEs are often asymptotically unbiased.

The MLE of a parameter is a function of the observations that maximizes the likelihood function. The likelihood function is the joint probability distribution of the observations of a random sample. Section 2.2 gives examples of the application of the maximum likelihood estimation technique to univariate and multivariate random samples.

This book is about the comparison of the efficiency of different estimators from both the Bayesian and the Frequentist point of view. The chief measure of goodness that will be used to compare the efficiencies of the different kinds of estimators is the mean square error (MSE), i.e., the expectation of the sum of the squares of the differences between the estimated and the observed values of a parameter. If the parameters are known to have a prior distribution, as is the case in the Bayesian context, the MSE is evaluated averaging over the prior distribution. The MSE can also be evaluated without averaging over the prior distribution. The efficiency of an estimator can be quite different relative to these two different kinds of MSE. Section 2.3 explains these and other relevant decision theoretic concepts.

The statistical community was indeed surprised when, in 1956, Charles Stein established the inadmissibility of the MLE. However there are a number of different mathematical and heuristic arguments to suggest that this should happen. Some of these are summarized in Sections 2.4 and 2.5.

One way of establishing the inadmissibility of the MLE is to demonstrate the existence of an estimator with uniformly smaller risk. This was done by James and Stein (1961). The risk of the James-Stein estimator (JS) for the mean of a multivariate normal distribution with more than two variables is computed in Section 2.6 and shown to have uniformly smaller risk than the MLE thus demonstrating its inadmissibility.

The empirical Bayes approach is one of the main methods for the formulation of JS type estimators. This is the subject of Section 2.7. The MSE of the JS averaging over a prior distribution is obtained and shown to be less than that of the MLE.

Section 2.8 explains how the JS is a special case of the operational ridge regression estimator.

The study of the risk of the JS is one way to establish the inadmissibility of the MLE. The JS is also inadmissible. One way of showing this is to consider a truncated form of the estimator. One such modification of the JS is the classical positive part. The classical positive part will be formulated in Section 2.9. It will be shown there that its risk is less than that of the JS. Other different positive parts of the JS will be studied in Chapter VI.

This chapter lays the groundwork for the consideration of JS type estimators for different model setups that will be taken up in the later chapters.

2.1 The Multivariate Normal Distribution

The mathematical form of the multivariate normal distribution and its important properties will be given for the cases where the dispersion matrix is of full rank and of non-full rank. The non-full rank case is important when considering the distributional properties of estimators of parameters in a non-full rank linear model.

2.1.1 The Full Rank Case

Let $X' = [X_1, X_2, X_3, \ldots, X_p]$ be a p dimensional random variable, $\theta' = [\theta_1, \theta_2, \cdots, \theta_p]$ be a p dimensional vector of parameters and Σ be a non-singular symmetric matrix. The multivariate normal density (MN) is

$$f(x) = \frac{1}{(2\pi)^{p/2} (\det \Sigma)^{1/2}} \exp[-\frac{1}{2}(x - \theta)'\Sigma(x - \theta)]. \qquad (2.1.1)$$

Theorem 2.1.1 gives the interpretation of the θ vector and the Σ matrix.

Theorem 2.1.1. For the MN (2.1.1) the vector θ is the mean vector and the matrix Σ is the dispersion (variance covariance matrix), i.e.,

$$E(X) = \theta \text{ and } D(X) = E(X - \theta)'(X - \theta) = \Sigma. \qquad (2.1.2)$$

Proof. The main idea of the proof is to rewrite the X vector as a linear transformation of a vector whose components are independent, univariate, standard, normal random variables. To do this make the transformation

$$Z = (X - \theta)\Sigma^{-1/2} \qquad (2.1.3)$$

with Jacobian $|\det \Sigma|^{1/2}$. Substitution of Z into (2.1.1) yields

$$g(z) = \frac{1}{(2\pi)^{\frac{p}{2}}} \exp[-\frac{1}{2}z'z] = \prod_{i=1}^{p} \frac{1}{\sqrt{2\pi}} \exp(-\frac{z_i^2}{2}). \qquad (2.1.4)$$

From (2.1.4)

$$E(z_i) = 0 \text{ and } D(z_i) = 1, \tag{2.1.5a}$$

hence

$$E(Z) = 0 \text{ and } D(Z) = I. \tag{2.1.5b}$$

Solving the matrix equation (2.1.3) for the vector X yields

$$X = \theta + \Sigma^{1/2} Z \tag{2.1.6}$$

and thus the result in (2.1.2).

One of the important goals in this study is to compare the different estimators of θ.

Exercise 2.1.1. Let $\theta = \begin{bmatrix} -1 \\ 1 \end{bmatrix}$ and $\Sigma = \begin{bmatrix} 2 & -1 \\ -1 & 2 \end{bmatrix}$. Find the components of the

vector Z in Equation (2.1.3) in terms of the components of X. [Hint: Find the eigenvalues and the orthogonal eigenvectors of Σ. Express Σ in the form $\Sigma = P\Lambda P'$ where P is an orthogonal matrix. This is the singular value decomposition of Σ. Then $\Sigma^{-\frac{1}{2}} = P\Lambda^{-\frac{1}{2}}P'$.

2.1.2 The Non-Full Rank Case

When Σ is not nonsingular, i.e., of rank q<p the distribution (2.1.1) cannot be written down because Σ^{-1} does not exist. However, from Cramer(1946), there is a q dimensional vector Y of uncorrelated hence independent random variables, such that $X-\mu = L'Y$ (almost surely, i.e., with probability one). Clearly

$$D(Y) = \Delta \tag{2.1.7}$$

is a p x p positive definite matrix and Y has a normal distribution. Furthermore

$$\Sigma = D(X) = LD(Y)L' = L\Sigma L'. \tag{2.1.8}$$

The random variable X is said to have a singular normal distribution.

Exercise 2.1.1. Given

$$f(x_1, x_2) = \frac{1}{4\pi} \exp\{-\frac{1}{2}[2(x_1 - 1)^2 + 2\sqrt{2}(x_1 - 1)(x_2 + 1) + 3(x_2 + 1)^2]\}.$$

A. What is θ and Σ?

B. Find an orthogonal linear affine transformation such that the transformed variables Y_1 and Y_2 are standard normal independent random variables. (Hint: Find the singular value decomposition of Σ.)

Exercise 2.1.2. Consider a singular normal distribution with $\theta = 0$ and

$$\Sigma = \begin{bmatrix} \dfrac{7}{6} & -\dfrac{5}{6} & -\dfrac{1}{3} \\ -\dfrac{5}{6} & \dfrac{7}{6} & -\dfrac{1}{3} \\ -\dfrac{1}{3} & -\dfrac{1}{3} & \dfrac{1}{3} \end{bmatrix}.$$

Find a 3x 2 matrix A such that

$$Y = \begin{bmatrix} Y_1 \\ Y_2 \\ Y_3 \end{bmatrix} = A \begin{bmatrix} X_1 \\ X_2 \end{bmatrix}$$

and D(Y) is nonsingular. (Hint: Find the eigenvalues and eigenvectors of Σ.)

2.2 Maximum Likelihood Estimation

2.2.1 The Basic Idea and Its Importance

Let X_1, $X_2.,...,X_n$ be a random sample from a univariate population with distribution $f(X_i|\theta)$ where θ is a parameter. The joint distribution of the x_i,

$$L = f(x_1, x_2, ... x_n, \theta) = \prod_{i=1}^{n} f(x_i, \theta) \tag{2.2.1}$$

is called the likelihood. The estimator $\hat{\theta} = \theta(x_1, x_2, ..., x_n)$ that maximizes L is called the maximum likelihood estimator (MLE). For distributions belonging to the exponential family it is easier to solve the equivalent problem of maximizing the logarithm of the likelihood.

The MLEs are popular estimators because they:

1. are often simple functions of sample statistics, e.g., the sample mean, median or variance;
2. are functions of complete and sufficient statistics;
3. are often but not always admissible estimators.

When p<3 the MLEs are admissible estimators for the mean of an MN but as has been already mentioned are inadmissible when p≥3.

2.2.2. Some Standard Examples of MLE's

Example 2.2.1. Let X_1, $X_2....X_n$ be a random sample from an exponential distribution

$$f(x_i | \theta) = \frac{1}{\theta} e^{-x_i/\theta}, \ x_i > 0. \tag{2.2.2}$$

The likelihood

$$L = \frac{1}{\theta^n} e^{-\sum_{i=1}^{n} x_i / \theta} \tag{2.2.3}$$

and

$$\log L = -n \log \theta - \frac{\sum_{i=1}^{n} x_i}{\theta}. \tag{2.2.4}$$

Differentiate log L with respect to θ. Set the result equal to zero. Thus,

$$\frac{d \log L}{d\theta} = -\frac{n}{\theta} + \frac{\sum_{i=1}^{n} x_i}{\theta^2} = 0 \tag{2.2.5}$$

and the MLE is

$$\hat{\theta} = \frac{1}{n} \sum_{i=1}^{n} x_i. \tag{2.2.6}$$

Example 2.2.2. Let $X_1, X_2....X_n$ be a random sample from a normal distribution, $N(\mu, \sigma^2)$. Then

$$L = \frac{1}{(2\pi)^{n/2} \sigma^n} \exp[-\frac{1}{2\sigma^2} \sum_{i=1}^{n} (x_i - \mu)^2]$$

and

$$\log L = -\frac{n}{2} \log 2\pi - n \log \sigma - \frac{1}{2\sigma^2} \sum_{i=1}^{n} (x_i - \mu)^2.$$

Obtain the partial derivatives of log L with respect to σ and μ. These are

$$\frac{\partial \log L}{\partial \mu} = \frac{1}{\sigma^2} \sum_{i=1}^{n} (x_i - \mu) = 0$$

and

$$\frac{\partial \log L}{\partial \sigma} = -\frac{n}{\sigma} + \frac{1}{\sigma^3} \sum_{i=1}^{n} (x_i - \mu) = 0.$$

The MLE for μ and σ^2 are

$$\hat{\mu} = \bar{x}$$

and

$$\hat{\sigma}^2 = \frac{1}{n}\sum_{i=1}^{n}(x_i - \bar{x})^2$$

The following exercises furnish standard examples of MLEs.

Exercise 2.2.1. Let $X_1, X_2....X_n$ be a random sample from a normal distribution, $N(\mu, \sigma^2)$. Find the MLE of μ and σ^2 for the cases where
A. $\mu = 0$.
B. $\sigma^2 = 1$.

Exercise 2.2.2. Let $X_1, X_2....X_n$ be a random sample from an Erlang distribution

$$f(x \mid \theta) = \frac{1}{(m-1)!}\left(\frac{1}{\theta}\right)^m x^{m-1}e^{-x/\theta}, \ x > 0.$$

Find the MLE of θ.

Exercise 2.2.3. Let $X_1, X_2....X_n$ be a random sample from an exponential distribution with $X_{n-r+1},...,X_n = t$, i.e., n items are being tested to observe their failure times and are taken off if they do not fail by time t.
A. Show that the likelihood function is given by

$$L = \frac{1}{\theta^r}\exp\left[-\sum_{i=1}^{r}\frac{x_i}{\theta}\right]\exp\left[-\frac{(n-r)t}{\theta}\right]$$

(This is an example of a problem involving censored data.)

B. Find the MLE of θ.

Exercise 2.2.4. Let $X_1, X_2....X_n$ be a random sample from a uniform distribution

$$f(x_i, \theta) = \frac{1}{\theta}, \ 0 \le x_i \le \theta.$$

Show that $\hat{\theta} = \max_{1 \le i \le n} x_i$ is the MLE of θ.

2.2.3 Estimating the Mean of a Multivariate Normal Distribution

James and Stein (1961) showed that the MLE for the mean vector of a MN with more than two variables is not the best by constructing an estimator with uniformly smaller risk. The MLE for the mean of a multivariate normal distribution will be derived. Some numerical comparisons of its risk with that of the JS will be given.

Let $x = (x_1, x_2, ..., x_p)$ be a single multivariate observation where

$$x \sim N(\theta, I).$$

The likelihood function is

$$L = \frac{1}{(2\pi)^{p/2}} \exp\left[-\frac{1}{2} \sum_{i=1}^{p} (x_i - \theta_i)^2 \right] \tag{2.2.7}$$

Thus,

$$\log L = -\frac{p}{2} \log 2\pi - \frac{1}{2} \sum_{i=1}^{p} (x_i - \theta_i)^2. \tag{2.2.8}$$

Now

$$\frac{\partial \log L}{\partial \theta_i} = -(x_i - \theta_i) = 0. \tag{2.2.9}$$

The MLE is

$$\theta_i = x_i. \tag{2.2.10}$$

Exercise 2.2.4. Find the MLE of θ for n>1 multivariate observations $\underline{x}_j$. Denote individual components of $\underline{x}_j$, $1 \le j \le m$ by x_{ij}, $1 \le i \le p$.

Example 2.2.3. The MLE and the JS. Consider a multivariate normal distribution with $\theta' = (.1, -.1, .05, -.03, .2)$. The MLE are the single simulated multivariate observations of θ, $\hat{\theta}$ = (.780993,1.27306,-1.52148,2.2847,1.93347). The risk is estimated by calculating

$$R = \frac{1}{p}\sum_{i=1}^{p}(\hat{\theta}_i - \theta_i)^2$$

For the MLE

R=2.53136.

In the next section it will be shown that the MLE does not have the smallest risk. An estimator with smaller risk is the JS with components

$$\hat{\theta}_i = \left(1 - \frac{(p-2)}{\sum_{i=1}^{p} x_i^2}\right) x_i, \ 1 \le i \le p.$$

For the above observations

$$\hat{\theta} = (0.607487,.990235,-1.18347,-1.77713,1.50393)$$

and

R=1.58671

a reduction of 37.1% over the risk of the MLE.

Exercise 2.2.5. Given a MN with mean vector θ and dispersion Σ find the MLEs
A. For one observation.
B. For n>1 observations.

Exercise 2.2.6. Given a MN with θ =(.02,-.02,.1,.08,-.09,-.1) with MLE
$$\hat{\theta} = (-1.68,-.559,-1.638,-.489,.706,-1.843).$$
A. Estimate the risk of the MLE.
B. Estimate the risk of the JS and compare your result to that of A.

Exercise 2.2.7. Given a MN with θ=(5,4,7,9,10), Σ= I, and
MLE = (6.327,3.535,5.357,8.449,10.642).
A. Estimate the risk of the MLE.
B. Estimate the risk of the JS and compare your answer to A.
C. Consider the JS with Lindley mean correction (JSL)

$$\hat{\theta}_i = \bar{x} + \left(1 - \frac{(p-3)}{\sum_{i=1}^{p}(x_i - \bar{x})^2} \right)(x_i - \bar{x}),\ 1 \le i \le p.$$

where $\bar{x} = \sum_{i=1}^{p} x_i / p$. Estimate the risk of this estimator and compare your result to that of A and B.

D. Repeat part C using the parameter values and the MLE of Exercise 2.2.6 and compare your results to 2.2.6A and B.

Remark: For the solution to Exercise 2.2.6, the estimated

$$\text{Risk(JS)} < \text{Risk(JSL)} < \text{Risk(MLE)}$$

while in Exercise 2.2.7, the estimated

$$\text{Risk(JSL)} < \text{Risk(JS)} < \text{Risk(MLE)}.$$

In Exercise 2.2.7B the JS does not offer much of an improvement over the MLE. For parameters close to the origin it is best to shrink toward the origin. Otherwise it is best to shrink toward the sample mean. This important idea will be dealt with in greater detail in Chapter VIII.

2.3 The Decision Theory Framework

The basic concepts and notations of decision theory that forms the framework for the study of the comparative properties of estimators will be formulated.

First some notation. Let $Z = (Z_1, Z_2, \dots, Z_m)$ be an m dimensional random variable and let $X = (X_1, X_2, \dots, X_n)$ be an n dimensional random variable. Then:

1. the conditional density

$$a(z|x) = a(z_1, z_2, \dots, z_m | x_1, x_2, \dots, x_n) ;$$

2. the density

$$c(x) = c(x_1, x_2, \dots, x_n).$$

3. $\int a(z|x) c(x) dx = \int ... \int a(z|x) c(x) dx_1 ... dx_n.$

The concept of a decision space is contained in Definition 2.3.1.

Definition 2.3.1. A Decision space consists of a triple (Θ, A, L) where:

1. θ represents the set of m dimensional parameters;
2. A is the m dimensional set of actions;
3. L is a function with domain the Cartesian product set $\Theta \times A$ and range the real numbers. (The function L is usually called the loss function.)

Let χ be a sample space of n dimensional vectors. Let X be an n dimensional random variable defined on χ and let $d(x)$ be a mapping from χ to A. The mapping $d(x)$ is called a decision rule.

The risk is the expected value of the loss function. Let θ be a true value of the parameter. The risk function

$$R(\theta, d) = E_\theta L(\theta, d(x)) = \int L(\theta, d(x)) f(x|\theta) dx. \qquad (2.3.1)$$

A number of different loss functions are described in the literature (see, for example, Feurguson (1969)). The squared error loss function

$$L(\theta, d(x)) = (\theta - d(x))'(\theta - d(x)) = \| \theta - d(x) \|^2 \qquad (2.3.2)$$

will be used throughout this book. The expected value of the loss function (2.3.2), i.e., the risk is called the mean square error (MSE). An estimator is admissible if no other estimators exist with equal risk or smaller risk for all parameter values and smaller risk for at least one parameter value. Definitions 2.3.2 to 2.3.4 define the concept of admissibility more formally.

Definition 2.3.2. Let $(X_1, X_2, ..., X_m)$ be an m dimensional random variable. An estimator $\hat{\theta}$ of θ is a decision rule $d(x)$.

Definition 2.3.3. Let $\hat{\theta}_1$ and $\hat{\theta}_2$ be estimators of θ. If

$$R(\hat{\theta}_1, \theta) \le R(\hat{\theta}_2, \theta)$$

where

$$R(\hat{\theta}_1, \theta) < R(\hat{\theta}_2, \theta)$$

for at least one value of θ then $\hat{\theta}_1$ is better than $\hat{\theta}_2$.

Definition 2.3.4. The estimator $\hat{\theta}$ of θ is said to be admissible if there does not exist a better estimator.

In the sense of Definitions 2.3.2 to 2.3.4 the best estimators are the admissible ones. The inadmissibility of an estimator may be demonstrated by finding an estimator with uniformly smaller risk. This technique will be used in Section 2.6 to establish the inadmissibility of the MLE for the mean of a MN for more than two variables, i.e., $p>2$.

Two other general concepts important to statistical inference that will be used mostly in the later chapters in the book for MSE computations are sufficiency and completeness. A brief informal discussion of these ideas is given here for the readers convenience. More detailed presentations are available in Lindley (1965), Hogg and Craig (1995), Mood Graybill and Boes (1974) and Lehmann (1959).

A sufficient statistic is one that contains all the information about the parameter being estimated. A statistic $t(x)$ is sufficient for a parameter θ if the conditional distribution of X given $t(x)$ is independent of θ. A necessary and sufficient condition for a statistic $t(x)$ to be sufficient for the parameter θ is that there exists a factorization where

$$f(x, \theta) = g(t(x), \theta)h(x). \qquad (2.3.3)$$

The sufficiency principle states that if $t(x)$ is sufficient for θ then for any prior distribution the posterior distribution given x and $t(x)$ are the same. This fact will be used in some of the MSE computations in Chapters IV, VII and IX. Many commonly used estimators are sufficient statistics, for example in a normal population the sample mean is a sufficient statistic for the population mean; likewise the sample variance is sufficient for the population variance. The MLE of the parameter of an exponential distribution is a sufficient statistic for that parameter.

A statistic $t(x)$ is complete if for all values of a parameter from a particular family of distributions $E[f(t(x)]=0$ implies that $f(t(x))=0$ on a set with probability one. In other words $t(x)$ is complete if and only if the only unbiased estimator of 0 is the statistic that is zero with probability one. Functions of complete sufficient statistics that are unbiased estimators of parameters are minimum variance unbiased estimators.

2.4 Why the MLE May Not Be the Best

The statistical community was indeed surprised to learn that the MLE of the mean of a MN was inadmissible. This fact is frequently referred to in the literature as the Stein paradox. However, there are a number of intuitive arguments which would lead one to believe that an estimator of the Stein type would be a more precise estimator than the MLE. Three of these arguments will now be given. They include:

1. Stein's original (1956) argument;
2. a geometric argument given later by Stein (1962);
3. a motivation based on the least square principle due to Stigler (1990).

2.4.1 Stein's Original Argument

One way to measure the goodness of an estimator is by how close it is to the true value of the parameter being estimated. Thus, if the MLE

$$\hat{\theta} = (x_1, x_2, \ldots, x_p) \tag{2.4.1}$$

is a good estimator of $\theta = (\theta_1, \theta_2, \ldots, \theta_p)$, then each of the x_i coordinates should be close to the corresponding θ_i coordinate. Furthermore $\sum_{i=1}^{p} \theta_i^2$ should be close to $\sum_{i=1}^{p} x_i^2$. What Stein (1956) actually observed that appears to be contrary to intuition is that $\sum_{i=1}^{p} \theta_i^2$ is close to $\sum_{i=1}^{p} x_i^2 - p$ with a high probability. (See Theorem 2.4.2). If c is slightly less than 1 and the MLE is used θ would be estimated to lie outside the convex set $C = \{\theta : \sum_{i=1}^{p} \theta_i^2 \leq \sum_{i=1}^{p} x_i^2 - cp \}$ although one would expect that θ could lie inside C. To deal with this problem   Stein suggests multiplication of the MLE by the contraction factor

$$\hat{c} = 1 - \frac{p}{\sum\limits_{i=1}^{p} x_i^2}. \tag{2.4.2}$$

This seems reasonable because it can be shown (Exercise 2.4.5) that $\sum_{i=1}^{p} (1 - p/\sum_{i=1}^{p} x_i^2) x_i^2$ is close to $\sum_{i=1}^{p} \theta_i^2$ with a high probability.

The above discussion will now be made a bit more precise. First a well known result will be stated without proof.

Theorem 2.4.1. (Chebychev) Let Y be a random variable with mean μ_Y and standard deviation σ_Y. Then

$$P[|\,Y - \mu_Y \,| \geq k\sigma] \leq \frac{1}{k^2}. \tag{2.4.3}$$

Corollary 2.4.1. Suppose $\hat{\theta}$ is an unbiased estimator of θ. Then

$$P[|\,\hat{\theta} - \theta\,| \geq \varepsilon] \leq \frac{\text{Var}(\hat{\theta})}{\varepsilon^2}. \tag{2.4.4}$$

Inequality (2.4.4) is a simple consequence of (2.4.3), i.e., let $\varepsilon = k\sqrt{\text{Var}(\hat{\theta})}$. Theorem 2.4.1 and its corollary will be used in the proof of Theorem 2.4.2 to show that $\sum_{i=1}^{p} \theta_i^2$ and $\sum_{i=1}^{p} x_i^2 - p$ are close. The proof of Theorem 2.4.1 is available in almost any Calculus based Probability or Mathematical Statistics text (see, for example, Hogg and Craig (1995)).

Theorem 2.4.2. If X_i, $1 \leq i \leq p$ are independent $N(\theta_i, 1)$ random variables then as $p \to \infty$, $\sum_{i=1}^{p} x_i^2 / p$ converges in probability to $1 + \sum_{i=1}^{p} \theta_i^2 / p$.

Proof. Since $x_i \sim N(\theta_i, 1)$, $1 \leq i \leq p$,

$$E\left[\frac{1}{p}\sum_{i=1}^{p} x_i^2\right] = \frac{1}{p}\sum_{i=1}^{p} \theta_i^2 + 1 \tag{2.4.5}$$

and

$$\text{Var}\left(\frac{1}{p}\sum_{i=1}^{p} x_i^2\right) = \frac{1}{p}. \tag{2.4.6}$$

Thus from (2.4.4)

$$P\left[\left|\frac{1}{p}\sum_{i=1}^{p}x_i^2 - 1 - \frac{1}{p}\sum_{i=1}^{p}\theta_i^2\right| > \varepsilon\right] < \frac{1}{p\varepsilon}. \tag{2.4.7}$$

As $p \to \infty$, $1/p\varepsilon^2 \to 0$, and the result follows.

Exercise 2.4.1. Let $X_1, X_2, \ldots, X_m$ be a random sample from $N(\mu, \sigma^2)$. Show that $\hat{\mu} = \bar{x}$ is admissible among all estimators of the form $\hat{\mu} = c\bar{x}$, $0 \le c \le 1$, c is a constant.

Exercise 2.4.2. Let X_i, $1 \le i \le p$ be independent $N(\theta_i, 1)$ random variables. Show that among all estimators of the form $\hat{\theta} = cx_i$, with c a constant, $\hat{\theta} = x_i$ is admissible. Observe that the JS does not belong to this class of estimators.

Exercise 2.4.3. Show that for the estimators of the form in Exercises 2.4.1 and 2.4.2 there exists a range of values of c where $c\bar{x}$ has a uniformly smaller MSE than $\bar{x}$ and cx_i has a uniformly smaller risk than x_i. How do you explain the fact that $\bar{x}$ is admissible.

Exercise 2.4.4. Show that if x_i are independent $N(\theta_i, 1)$ random variables then $\sum_{i=1}^{p}(1 - p/\sum_{i=1}^{p}x_i^2)$ converges in probability to $\sum_{i=1}^{p}\theta_i^2/p$.

2.4.2 A Geometrical Argument

Stein (1962) gave a geometric motivation for the JS. A nice presentation of his ideas is available in Brandwein and Strawderman (1990). This presentation is summarized below.

Let X be an observation vector in p dimensions with mean vector θ and uncorrelated components with equal variance σ^2. For the sake of discussion, let $\sigma^2 = 1$. Let

$$\|X\|^2 = \sum_{i=1}^{p}x_i^2 \text{ and } \|\theta\|^2 = \sum_{i=1}^{p}\theta_i^2. \tag{2.4.8}$$

Since $E(X-\theta) = 0$, it is expected that $X-\theta$ and X are nearly orthogonal. Also

$$E\|X\|^2 = p + \|\theta\|^2 \qquad\qquad (2.4.9)$$

In equation (2.4.9) the p that is added to $\|\theta\|^2$ suggests that X as an estimator of θ is "too long". Perhaps, instead of X, the projection of θ on X is a better estimator of a. Denote this projection by (1-a)X. The problem now reduces to the estimation of a.

Assume:

1. the angle between and is exactly a right angle;
2. the quantity

$$\|X\|^2 = \theta'\theta + p \qquad\qquad (2.4.10a)$$

and

$$\|X - \theta\|^2 = p. \qquad\qquad (2.4.10b)$$

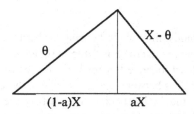

Figure 2.4.1

From Figure 2.4.1, the Pythagorean Theorem and (2.4.9)

$$\begin{aligned}
\|Y\|^2 &= \|\theta\|^2 - (1-\hat{a})^2\|X\|^2 \\
&= \|X\|^2 - p - (1-\hat{a})^2\|X\|^2 \qquad\qquad (2.4.11) \\
&= (2\hat{a} - \hat{a}^2)\|X\|^2 - p
\end{aligned}$$

and

$$\|Y\|^2 = \|X - \theta\|^2 - \hat{a}^2\|X\|^2 = p - \hat{a}^2\|X\|^2. \qquad\qquad (2.4.12)$$

Equating the expressions in (2.4.11) and (2.4.12)

$$2\hat{a}\|X\|^2 = 2p$$

and

$$\hat{a} = \frac{p}{\|X\|^2}. \tag{2.4.13}$$

Thus, the suggested estimator is

$$(1-\hat{a})X = (1 - \frac{p}{\|X\|^2})X \ . \tag{2.4.14}$$

2.4.3. An Argument Based on the Least Square Principle

Stigler (1990) gives a nice motivation for the JS using the least square principle. For a multivariate normal distribution the MLE is the least square estimator (LS) for the regression of X on θ. The least square estimator is the estimator such that the sum of the squares of the differences between the observed and the predicted values is minimized. Stigler obtains the JS by finding the least square estimator for the regression of θ on X, i.e., doing the inverse of the regression used to find the MLE. Since the BE is the mean of the posterior distribution of $\theta|X$ (See Section 2.7.), in a certain sense Stigler's derivation is from the Bayesian point of view. His argument will now be summarized.

Consider the class of estimators of θ_i that are linear in X with zero intercept, i.e.,

$$\hat{\theta}_i = bx_i \tag{2.4.15}$$

The $\hat{\theta}_i$ in (2.4.15) would be the predicted values from regressing on θ_i. From this standpoint, the LS of b is

$$\hat{\beta} = \frac{\sum_{i=1}^{p} \theta_i x_i}{\sum_{i=1}^{p} x_i^2} \qquad (2.4.16)$$

Equation (2.4.16) is the result of minimizing the loss function

$$L(\theta, \hat{\theta}) = \sum_{i=1}^{p} (\hat{\theta}_i - \theta_i)^2 \qquad (2.4.17)$$

At the outset it was assumed that $X_i \sim N(\theta_i, 1)$. Thus,

$$E[\sum_{i=1}^{p} \theta_i x_i] = \sum_{i=1}^{p} \theta_i E(x_i) = \sum_{i=1}^{p} \theta_i^2 \qquad (2.4.18)$$

and

$$E[\sum_{i=1}^{p} x_i^2] = p + \sum_{i=1}^{p} \theta_i^2 \qquad (2.4.19)$$

Hence,

$$E[\sum_{i=1}^{p} \theta_i x_i] = E[\sum_{i=1}^{p} x_i^2] - p. \qquad (2.4.20)$$

Equation (2.4.20) suggests replacement of $\sum_{i=1}^{p} \theta_i x_i$ with $\sum_{i=1}^{p} x_i^2 - p$ resulting in

$$\hat{\theta}_i = \left(1 - \frac{p}{\sum_{i=1}^{p} x_i^2}\right) x_i, \qquad (2.4.21)$$

a form of the JS.

The problem solved above assumed that the intercept term is zero. More generally, it is desirable to find the best linear estimator of the θ_i of the form

$$\hat{\theta}_i = a + bX_i, \quad 1 \le i \le p \qquad (2.4.22)$$

Then

$$\hat{\theta}_i = \bar{\theta} + \hat{\beta}(x_i - \bar{x}) \tag{2.4.23}$$

where

$$\hat{\beta} = \frac{\sum_{i=1}^{p}(x_i - \bar{x})(\theta_i - \bar{\theta})}{\sum_{i=1}^{p}(x_i - \bar{x})^2} \tag{2.4.24}$$

Observe that

$$A_1 = \sum_{i=1}^{p}(x_i - \bar{x})(\theta_i - \bar{\theta}) \tag{2.4.25}$$

and

$$A_2 = \sum_{i=1}^{p}(x_i - \bar{x})^2 - (k-1) \tag{2.4.26}$$

have the same expectation (Exercise 2.4.3). Now estimate $\hat{\beta}$ by

$$\hat{\beta} = \frac{\sum_{i=1}^{p}(x_i - \bar{x})^2 - (p-1)}{\sum_{i=1}^{p}(x_i - \bar{x})^2} = 1 - \frac{(p-1)}{\sum_{i=1}^{p}(x_i - \bar{x})^2}. \tag{2.4.27}$$

The resulting least square line is

$$\hat{\theta}_i = \bar{x} + \left(1 - \frac{(k-1)}{\sum_{i=1}^{p}(x_i - \bar{x})^2}\right)(x_i - \bar{x}),\ 1 \le i \le p. \tag{2.4.28}$$

This is a form of the JSL considered in Exercise 2.27, i.e., a JS type estimator with a correction for the mean. Estimators with smaller risks than (2.4.21) and (2.4.28) may be obtained by replacing k by $k-1$ and $k-2$ by $k-3$. This fact will become clear in Section 2.6.

Exercise 2.4.3. Show that the random variables A_1 and A_2 have the same expectation.

2.5 Another Case Against the MLE

In Section 2.4 it was shown how an estimator of the JS type appeared to be much closer to the parameter being estimated that the MLE. Estimators that are contractions of the MLE will now be considered, i.e., the MLE multiplied by positive constants less than one. It will now be shown that the JS is an approximation to the contraction of the MLE that is optimum in the sense of having the smallest MSE when compared with other estimators of the same type.

Consider the class of estimators of the mean of a MN with components

$$\hat{\theta}_i = cx_i, \, 0 \le c \le 1, \, 1 \le i \le p \tag{2.5.1}$$

For this class of estimators:

1. There exists a range of values where the risk is less than that of the MLE.
2. There is an optimal estimator, i.e., one with smallest risk that is uniformly smaller than that of the LS.
3. The JS is an approximation to this optimal estimator.

The risk of estimators of the type in (2.5.1) is given by (Carry through the details in Exercise 2.5.1.)

$$R = \frac{1}{p}\sum_{i=1}^{p} E(\hat{\theta}_i - \theta_i)^2 = c^2 + \frac{(c-1)^2}{p} \theta'\theta. \tag{2.5.2}$$

The risk of the MLE is 1. Thus, the estimators (2.5.1) have smaller risk than the MLE if $c < 1$ and

$$\frac{\theta'\theta}{p} < \frac{1+c}{1-c}. \tag{2.5.3}$$

When θ lies in the region(2.5.3) and $c < 1$ the estimator (2.5.1) has a smaller risk than the MLE. However when θ lies outside the region (2.5.3) even when $c<1$ the risk of (2.5.1) is larger than that of the MLE. Thus, for the class of estimators (2.5.1) there is no estimator better than the MLE. The MLE is admissible for the class of estimators (2.5.1) but not for estimators in general.

Since the MLE is admissible for a certain class of estimators but not in general the following definition is appropriate.

Definition 2.5.1. Let C be a class of estimators of a parameter θ. Then $\hat\theta$ is admissible for C if there is no better estimator contained in C.

The JS does not belong to the class of estimators C given in Equation (2.5.1) because it is not a constant multiple of the MLE, i.e., the contraction factor is a random variable.

The contraction estimator with the smallest MSE will now be obtained. To do this differentiate (2.5.2) and set the result equal to zero. Thus,

$$\frac{dR}{dc} = 2c + \frac{2(c-1)}{p}\theta'\theta = 0. \tag{2.5.4}$$

Solve (2.5.4) to obtain optimal c. Thus,

$$c = \frac{\theta'\theta/p}{1+\theta'\theta/p} = 1 - \frac{1}{1+\theta'\theta/p}. \tag{2.5.5}$$

Thus, the optimal estimator is

$$\hat\theta_i = \left(1 - \frac{1}{1+\theta'\theta/p}\right)x_i, \ 1 \le i \le p. \tag{2.5.6}$$

The risk of the estimator (2.5.6) is

$$R = 1 - \frac{1}{1+\theta'\theta/p} < 1. \tag{2.5.7}$$

There is a serious problem with the estimator (2.5.6), the MMSE.   The estimator depends on the parameters being estimated.   If the parameters being estimated were known there would be no point is estimating them. However, Inequality (2.5.7) suggests that substitution of sample estimates for functions of θ_i in (2.5.6) will produce an estimator with risk uniformly smaller than 1, the risk of the MLE. If, in (2.5.6), the contraction factor

$$f = \frac{1}{1+\theta'\theta/p} < 1 \tag{2.5.8}$$

is replaced by a scalar multiple of $p/x'x$, the JS is obtained in the form

$$\hat{\theta} = (1-\frac{hp}{x'x})x_i, \ 1\le i \le p. \tag{2.5.9}$$

In Section 2.6, it will be established that for a range of values of h , the risk of the estimator (2.5.9) is uniformly smaller than that of the MLE and thus the MLE is inadmissible.

Exercise 2.5.1. Show that the risk of the estimators in (2.5.1) is given by (2.5.2).

Exercise 2.5.2. Show that the risk of the MMSE (2.5.6) is given by (2.5.7).

Exercise 2.5.3. Find the minimum risk estimator of the form

$$\hat{\theta}_i = \bar{x} + c(x_i - \bar{x}).$$

Obtain the JSL by an argument similar to that used to obtain (2.5.9).

2.6 The Inadmissibility of the MLE

James and Stein (1961) show that if p>2 the MLE is inadmissible by observing that the risk of the JS estimator

$$\hat{\theta}_i = (1 - \frac{(p-2)}{x'x})x_i \, , \, 1 \le i \le p \tag{2.6.1}$$

is less than that of the MLE. In the proof of Theorem 2.6.1 the expression for the risk is obtained and shown to be less than the MLE.

Theorem 2.6.1. If p>2 the JS is inadmissible.

Proof. The proof consists of computing the expectations in

$$R = \frac{1}{p}E(\hat{\theta} - \theta)'(\hat{\theta} - \theta)$$

$$= \frac{1}{p}E(x - \theta)'(x - \theta) - \frac{2(p-2)}{p}E\left[\frac{(x - \theta)'x}{x'x}\right] + \frac{(p-2)^2}{p}E\left[\frac{1}{x'x}\right].$$

$$\tag{2.6.2}$$

and showing that R<1. The result will follow easily once it is shown that

$$E\left[\frac{(x - \theta)'x}{x'x}\right] = (p-2)E\left[\frac{1}{x'x}\right] \tag{2.6.3}$$

To establish (2.6.3) observe that

$$E\left[\frac{(x - \theta)'x}{x'x}\right] = \sum_{i=1}^{p} E\left[\frac{(x_i - \theta_i)x_i}{\sum_{i=1}^{p}x_i^2}\right]$$

$$= \int_{-\infty}^{\infty} \cdots \int_{-\infty}^{\infty} \frac{\sum_{i=1}^{p}(x_i - \theta_i)x_i e^{-\frac{1}{2}\sum_{i=1}^{p}(x_i - \theta_i)^2}}{\sum_{i=1}^{p}x_i^2}dx_1 dx_2 \ldots dx_p \, . \tag{2.6.4}$$

For each summand in (2.6.4) rearrange the order of integration so that for the i'th term, the first integration in evaluating the iterated integral is with respect to x_i. By integration by parts (Exercise 2.6.1)

$$\int_{-\infty}^{\infty} \frac{(x_i - \theta_i)x_i}{x'x} e^{-\frac{1}{2}(x_i - \theta_i)^2} dx_i = \int_{-\infty}^{\infty} (\frac{1}{x'x} - \frac{2x_i^2}{(x'x)^2}) e^{-\frac{1}{2}(x_i - \theta_i)^2} dx_i.$$

$$(2.6.5)$$

Integration of both sides of (2.6.5) with respect to the other $p - 1$ variables and summing from 1 to p yields (2.6.3). Substituting (2.6.3) into (2.6.2),

$$R = 1 - \frac{(p-2)^2}{p} E\left[\frac{1}{x'x}\right] < 1. \qquad (2.6.6)$$

How much of an improvement in the risk is (2.6.6)? To find out, it is necessary to find an expression for $E[\frac{1}{x'x}]$. Observe that $u = x'x$ has a Noncentral Chi Square distribution with noncentrality parameter $\lambda = \frac{1}{2}\theta'\theta$. This follows from the assumption that $x_i \sim N(\theta_i, 1)$.. From Searle (1971), the distribution of u is

$$f(u) = e^{-\lambda} \sum_{k=0}^{\infty} \frac{\lambda^k}{k!} \frac{u^{\frac{1}{2}n+k-1} e^{-\frac{1}{2}u}}{2^{\frac{1}{2}n+k} \Gamma(\frac{1}{2}n + k)}. \qquad (2.6.7)$$

Recall that

$$\Gamma(\alpha) = \int_0^{\infty} x^{\alpha-1} e^{-x} dx \qquad (2.6.8)$$

From (2.6.7),

$$E[\frac{1}{u}] = e^{-\lambda} \sum_{k=0}^{\infty} \frac{\lambda^k}{k!} \int_0^{\infty} \frac{u^{\frac{1}{2}n+k-2} e^{-\frac{1}{2}u}}{2^{\frac{1}{2}n+k} \Gamma(\frac{1}{2}n + k)} du \qquad (2.6.9)$$

Make the transformation $y = \frac{u}{2}$ so that $dy = \frac{1}{2} du$. With the help of (2.6.8),

$$\int_0^{\infty} u^{\frac{1}{2}n+k-2} e^{-\frac{1}{2}u} du = \int_0^{\infty} 2^{\frac{1}{2}n+k-1} y^{\frac{1}{2}n+k-2} e^{-y} dy = 2^{\frac{1}{2}n+k-1} \Gamma\left(\frac{1}{2}n + k - 1\right).$$

$$(2.6.10)$$

Substitute (2.6.10) into (2.6.9) and make use of the functional equation for the gamma function

$$\Gamma(\alpha+1) = \alpha\Gamma(\alpha). \tag{2.6.11}$$

to obtain

$$E\left[\frac{1}{u}\right] = \sum_{k=0}^{\infty} \frac{\lambda^k e^{-\lambda}}{k!(p+2k-2)}. \tag{2.6.12}$$

Thus, (2.6.6) may be written

$$R = 1 - \frac{(p-2)^2}{p} \sum_{k=0}^{\infty} \frac{\lambda^k e^{-\lambda}}{k!(p+2k-2)}. \tag{2.6.13}$$

To really get an idea of how much of an improvement the JS is when compared to the MLE, it is useful to obtain some numerical values of R for different values of λ and p. To facilitate the computations, there is a useful expression for the sum of the series (2.6.12) in terms of an integral. To obtain this result first observe that

$$\int_0^\lambda x^{\frac{p}{2}+k-2} dx = \frac{\lambda^{\frac{p}{2}+k-2}}{(\frac{p}{2}+k-1)}. \tag{2.6.14}$$

From (2.6.14)

$$\begin{aligned}
S_{1,p}(\lambda) &= \sum_{k=0}^{\infty} \frac{\lambda^k e^{-\lambda}}{k!(p+2k-2)} \\
&= \frac{1}{2}\sum_{k=0}^{\infty} \frac{e^{-\lambda}\lambda^k}{k!(\frac{p}{2}+k-1)} = \frac{1}{2\lambda^{\frac{p}{2}-1}}\sum_{k=0}^{\infty} \frac{e^{-\lambda}\lambda^{\frac{p}{2}+k+1}}{k!(\frac{p}{2}+k-1)} \\
&= \frac{e^{-\lambda}}{2\lambda^{\frac{p}{2}-1}}\sum_{k=0}^{\infty} \int_0^\lambda \frac{x^{\frac{p}{2}+k-1}}{k!} dx = \frac{e^{-\lambda}}{2\lambda^{\frac{p}{2}-1}}\int_0^\lambda e^x x^{\frac{p}{2}-2} dx.
\end{aligned}$$

$$\tag{2.6.15}$$

since $\sum_{k=0}^{\infty}\frac{x^k}{k!} = e^x$. The integral in (2.6.15) may be evaluated by repeated integration by parts for even p at least 4. For small even values of p the risk may be obtained using a hand held calculator that does exponential and logarithmic functions. For other cases, numerical evaluation of the risk may be done using Mathematica or some other similar software.

Table 2.6.1 contains some exact values of R. Figure 2.6.1 is a plot of R vs. λ for various values of p. The following observations are suggested by Figure 2.6.1:

1. The risk R is an increasing function of λ and a decreasing function of p.

2. The improvement of the risk by using the JS instead of the MLE is most dramatic for small values of the parameter λ.

3. The lower bound of the risk is 2/p; the upper bound is 1.

Table 2.6.1

p/λ	0	.1	.5	1	2	5	10
4	.500	.524	.607	.684	.784	.901	.950
8	.250	.268	.335	.406	.514	.694	.816
12	.167	.180	.231	.288	.381	.562	.709
16	.125	.136	.177	.223	.303	.472	.628

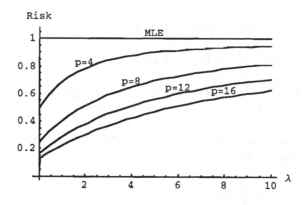

Figure 2.6.1
Risk of JS for Different p, λ

Exercise 2.6.1. Let $X_i \sim N(\theta, \sigma^2)$, $1 \le i \le p$.
A. Find the best estimator of the form

$$\hat{\theta}_i = cx_i, \ 0 \le c \le 1, \ 1 \le i \le p.$$

B. Write down the JS when σ^2 is unknown.
C. Evaluate the risk R.

Exercise 2.6.2. Let $X \sim N(\theta, \Sigma)$ where Σ is a px p positive definite matrix.
A. What would be the form of the JS? (Hint: Estimate Σ by a Wishart matrix.)
B. Evaluate the risk

$$R = \frac{1}{p} E(x - \theta)' \Sigma (x - \theta).$$

Exercise 2.6.3. Find an upper bound on the risk (2.6.13) by finding the sum of 10 terms of the series for $\lambda = .25, .75$ and 2 when p= 4, 6, 8, 10.

Exercise 2.6.4.
A. Using integration by parts and the result of (2.6.15) show that

$$S_{1,p} = \frac{1}{2\lambda} - (\frac{p}{2} - 2)\frac{S_{1,p-2}}{\lambda}, \ p > 4. \qquad (2.6.16)$$

B. From the result of A or otherwise show that

$$2\lambda S_{1,p} + (p - 4)S_{1,p-2} = 1, \ p > 4. \qquad (2.6.17)$$

Exercise 2.6.5. Find the exact value of the risk for $\lambda = .25, .75, 3$, for p=4, 6, 8, 10. (Hint: The results of Exercise 2.6.4 might prove useful.)

Exercise 2.6.6. Let R_m denote the risk of the MMSE, R_l the risk of the MLE and R_j the risk of the JS. Define

$$RL = \frac{R_j - R_m}{R_l - R_m} \qquad (2.6.18)$$

A. Obtain this efficiency for p=4, 6, 8, 10, λ= 01, .1, 1, 10, 100, 1000.
B. Find the limiting value of the efficiency:

(1) for each p as $\lambda \to 0$ and $\lambda \to \infty$?

(2) as $p \to \infty$

Exercise 2.6.7.

A. Show that the JSL has risk

$$R = 1 - \frac{(p-3)^2}{p} \sum_{k=0}^{\infty} \frac{e^{-\alpha}\alpha^k}{k!(p+2k-3)} \qquad (2.6.19)$$

with

$$\alpha = \frac{1}{2}\sum_{i=1}^{p}(\theta_i - \bar{\theta})^2, \ \bar{\theta} = \frac{1}{p}\sum_{i=1}^{p}\theta_i.$$

B. Obtain an integral representation of the sum of the series in (2.6.19) similar to that of (2.6.15). Compute some exact numerical values of R.

Exercise 2.6.8.

A. Show that the risk of the JS is an increasing function of with lower bound $\frac{2}{p}$ and upper bound 1.

B. State and establish an analogous result for the JSL.

2.7 The JS as an Empirical Bayes Estimator (EBE)

2.7.1 The Methodology

Bayesian Statistics is characterized by the assumption of a prior distribution for the parameters about which inferences are being made. Both the prior information and the data obtained by sampling together with Bayes theorem are used to make inferences about the parameters of interest. When the prior distribution is unknown or information about it is incomplete, its parameters are often replaced by sample estimates. This technique is called empirical Bayes.

Bayes estimators (BE) are obtained as functions of prior parameters, population parameters and sample estimates. The posterior distribution is obtained by using Bayes theorem to mix the sampling distribution with the prior distribution. The BE are optimal because they have the smallest MSE averaging over the prior distribution. For the squared error loss function the BE is the mean of the posterior distribution.

When the prior distribution is unknown or not completely known:

1. The BE is obtained as if the prior were known.
2. The unknown prior parameters are then replaced by functions of sample estimates.

The approximate BE obtained by this technique is called an empirical Bayes estimator(EBE).

2.7.2 How the JS is an EBE

Consider again the problem of estimating the mean of a MN. Assume that:

1. the population is MN but the individual coordinates are independently distributed, i.e., $X \sim N(\theta, I)$.
2. the prior distribution is multivariate normal and the individual coordinates of the p dimensional vector θ are independent, i.e., $\theta \sim N(0, a)$.

Thus, the distribution of X_i conditional on θ_i is

$$f(x_i \mid \theta_i) = \frac{1}{\sqrt{2\pi}} \exp[-\frac{1}{2}(x_i - \theta_i)^2], \ 1 \le i \le p. \qquad (2.7.1)$$

and the prior distribution of the θ_i is

$$\pi(\theta_i) = \frac{1}{\sqrt{2\pi a}} \exp[-\frac{\theta_i^2}{2a}], \ 1 \le i \le p. \qquad (2.7.2)$$

The posterior distribution (the Bayes transform) of $\pi(\theta)$ will be obtained by combining the distributions in (2.7.1) and (2.7.2) using Bayes Theorem. In the context of the estimation problem being considered Bayes Theorem may be expressed by the equation

$$f(\theta_i \mid x_i) = \frac{f(x_i \mid \theta_i)\pi(\theta_i)}{m(x_i)}, \ 1 \le i \le p \qquad (2.7.3)$$

with

$$m(x_i) = \int f(x_i \mid \theta_i)\pi(\theta_i)d\theta_i \qquad (2.7.4)$$

Calculation of the Bayes transform simply involves substitution of (2.7.1) and (2.7.2) into (2.7.3) and (2.7.4). Thus

$$f(x_i \mid \theta_i)\pi(\theta_i) = \frac{1}{2\pi\sqrt{a}} \exp\left[-\frac{1}{2}(x_i - \theta_i)^2 - \frac{\theta_i^2}{2a}\right]$$

$$= \frac{1}{2\pi\sqrt{a}} \exp\left[-\frac{(\theta - \frac{a}{a+1}x_i)^2}{\frac{2a}{a+1}}\right] \exp\left[-\frac{1}{2(a+1)}x_i^2\right]. \qquad (2.7.5)$$

and

$$m(x_i) = \frac{1}{\sqrt{2\pi(a+1)}} \exp[-\frac{1}{2(a+1)}x_i^2]. \qquad (2.7.6)$$

The posterior distribution of $\theta_i \mid x_i$ is

$$f(\theta_i \mid x_i) = \frac{1}{\sqrt{\frac{2\pi a}{a+1}}} \exp[-\frac{(\theta_i - \frac{ax_i}{a+1})^2}{\frac{2a}{a+1}}]. \qquad (2.7.7)$$

As was explained in Section 2.7.1, the BE for squared error loss is the mean of the posterior distribution (2.7.7). Thus, the BE is

$$\hat{\theta}_i = \frac{a}{a+1}x_i = (1 - \frac{1}{a+1})x_i, \ 1 \le i \le p. \qquad (2.7.8)$$

The derivation of the BE in Equation (2.7.8) assumed knowledge of the prior parameter a. Suppose a is unknown. One way to estimate a from the sample is to employ an unbiased estimator of $e = \frac{1}{(a+1)}$. Observe that

$$Y = \frac{X'X}{1+a} \sim \chi^2(p). \qquad (2.7.9)$$

Thus,

$$E[\frac{1}{x'x}] = \frac{1}{(a+1)} E[\frac{1}{y}] \qquad (2.7.10)$$

and

$$E[\frac{1}{y}] = \frac{1}{\Gamma(\frac{p}{2})2^{\frac{p}{2}}} \int_0^\infty y^{\frac{p}{2}-2} e^{-y} dy = \frac{\Gamma(\frac{p}{2}-1)}{2\Gamma(\frac{p}{2})}$$

$$= \frac{\Gamma(\frac{p}{2}-1)}{2(\frac{p}{2}-1)\Gamma(\frac{p}{2}-1)} = \frac{1}{p-2}. \qquad (2.7.11)$$

From (2.7.10) and (2.7.11), an unbiased estimator of e is

$$\hat{e} = \frac{(p-2)}{x'x} \qquad (2.7.12)$$

Thus, the JS may be viewed as an approximate BE or as an EBE.

   The methodology used in this section to obtain the BE and the EBE will be used repeatedly throughout the book as different EBEs are formulated and studied.

Exercise 2.7.1.
A. Carry through the details of the algebra needed to verify the equivalence of the two expressions in (2.7.5).
B. Fill in the details of the integration needed to verify (2.7.6).
C. Verify (2.7.7).

Exercise 2.7.2.
A. Show that $x'x/p$ is an unbiased estimator of $1+a$.
B. Show that for the EBE

$$\hat{\theta}_i = (1 - \frac{cp}{x'x}) x_i$$

for the average and conditional risk optimal $c = (p-2)/p$.

2.7.3 The Efficiency of the BE and the EBE

The Bayes risk of the BE and the EBE will be obtained. The relative savings loss (RSL) of Efron and Morris (1973) will be defined and its properties will be investigated. The RSL is the ratio of the difference between the Bayes risk of the EBE and the BE to that of the BE and the MLE. Thus it is a measure of efficiency of the EBE relative to the BE.

For the BE the Bayes risk is given by

$$
\begin{aligned}
r &= \frac{1}{p} E[((1-\frac{1}{a+1})x-\theta)'((1-\frac{1}{a+1})x-\theta)] \\
&= \frac{1}{p} E(x-\theta)'(x-\theta) - \frac{2}{p(a+1)} E[x'(x-\theta)] + \frac{1}{p(a+1)^2} E(x'x) \\
&= 1 - \frac{2p}{p(a+1)} + \frac{p(a+1)}{p(a+1)^2} = 1 - \frac{1}{a+1}.
\end{aligned}
$$

$$(2.7.13)$$

Let E_θ denote the expectation conditional on θ. For the JS employing (2.7.11), the Bayes risk is

$$
\begin{aligned}
r &= E[R] = \frac{1}{p} E_\theta (\hat{\theta}-\theta)'(\hat{\theta}-\theta) \\
&= 1 - \frac{(p-2)^2}{p} E[\frac{1}{x'x}] \\
&= 1 - \frac{(p-2)^2}{p} \frac{1}{1+a}.
\end{aligned}
$$

$$(2.7.14)$$

The defining equation for the RSL is

$$
RSL = \frac{r(EBE) - r(BE)}{r(MLE) - r(BE)}.
$$

$$(2.7.15)$$

For the JS

$$RSL = \frac{2}{p}.$$ (2.7.16)

As p→∞, RSL→ 0. The EBE is asymptotically optimal(AO). This means that as p→∞, the risk of the EBE approaches that of the BE.

Exercise 2.7.3.
A. Show that the JS is AO.
B. Show that for the JS, RL(0)=RSL and $RL(\infty) = \frac{4(p-1)}{p^2}$.

Exercise 2.7.4. Let $X_i \sim N(\theta_i, 1)$, $1 \leq i \leq p$, $\theta_i \sim N(\mu, a)$.
A. Obtain the BE of θ_i.
B. Show that the EBE of θ_i. is the JSL.
C. Find the average risk of the JSL.
D. Find the RSL of the JSL.

Exercise 2.7.5. Show that averaging over the prior distribution

$$E\sum_{k=0}^{\infty}\left[\frac{\lambda^k e^{-\lambda}}{k!(p + 2k - 2)}\right] = \frac{1}{p-2}.$$

Thus the average MSE can be obtained using the formula for the conditional MSE.

2.8. How the JS is a Special Case of the Operational Ridge Regression Estimator

The JS can be viewed as a special case of the operational ridge estimator by considering p linear models with design matrix [1] and using the information from all of the linear models to simultaneously estimate k. Consider the p linear models

$$x_i = \theta_i + \varepsilon_i \, , \, 1 \leq i \leq p \text{ where } \varepsilon_i \sim N(0,1).$$ (2.8.1)

The x_i, θ_i and ε_i are one dimensional vectors. They represent a single observation, the regression parameter for the i'th model and the error term. The least square estimator for the i'th model is

$$\hat{\theta}_i = x_i \tag{2.8.2}$$

which is also the MLE. The ridge regression estimator for the i'th model is

$$\hat{\theta}_i = \frac{1}{1+k} x_i. \tag{2.8.3}$$

By finding the risk using Equation (2.5.2) with $c = 1/(1+k)$ the optimal value of k may be found by differentiation. The expression for k will contain the parameters to be estimated. By substituting an appropriate sample estimate the JS will be obtained. The risk (2.5.2) as a function of k is

$$R = \frac{1}{(1+k)^2} + \frac{k^2}{(1+k)^2} \cdot \frac{\theta'\theta}{p}. \tag{2.8.4}$$

Finding the derivative of R with respect to k and setting the result equal to zero gives

$$\frac{dR}{dk} = \frac{2(k\theta'\theta - p)}{p(1+k)^3} = 0. \tag{2.8.5}$$

Optimal

$$k = \frac{p}{\theta'\theta}. \tag{2.8.6}$$

An unbiased estimator of $\theta'\theta$ is $x'x - p$. Thus, one way to estimate k is

$$\hat{k} = \frac{p}{x'x - p}. \tag{2.8.7}$$

Substituting k into (2.8.3) gives

$$\hat{\theta}_i = (1 - \frac{p}{x'x}) x_i, \tag{2.8.8}$$

a form of the JS. To obtain the form of the JS with the smallest MSE estimate (2.6.1) the biased estimator $px'x/(p-2)-p$ can be used in place of $\theta'\theta$ in (2.8.6) and subsequently (2.8.3).

Exercise 2.8.1. Obtain the ridge estimator (2.8.3) by minimizing θ_i^2 subject to the constraint $(\theta_i - x_i)^2 = \Phi_0$. Notice that this is a specialization of the argument given in Equations (1.3.11) to (1.3.13).

Exercise 2.8.2. Show how to formulate the JS as a special case of the ridge estimator when the standard deviation is unknown.

Exercise 2.8.3. Find the bias of the estimator $px'x/(p-2)-p$. Is it asymptotically unbiased for $\theta'\theta$?

Exercise 2.8.4. Find the variance and the bias of the ridge estimator (2.8.3). Your answer should be the same as the two terms of the risk (2.84).

Exercise 2.8.5. A. Show that of for a particular value of k

$$\theta'\theta < p(\frac{2}{k}+1)$$

then the risk of the ridge estimator is less than that of the LS.
B . Show that if $\theta'\theta > p$ there is a range of values of k where the risk of the ridge estimator (2.8.2) is less than that of the LS. What happens if $\theta'\theta < p$?

Exercise 2.8.6. A. Find the minimum risk of the ridge estimator (2.8.3).
B. Show that the minimum risk of the ridge estimator is uniformly smaller than the risk of the JS.

Exercise 2.8.7. Show that the risk of the ridge estimator (2.8.3) is not always less than that of the JS estimator (the operational ridge estimator) by producing an explicit value of λ and k where this holds true. This will be discussed in greater detail in Chapter IV.

Exercise 2.8.8. Show that if $\theta'\theta \gg p$, for optimum k,

$$\frac{1}{1+p/\theta'\theta} \cong 1 - \frac{p}{\theta'\theta}.$$

Could this be a reason the JS has smaller MSE than the MLE?

2.9 Introduction to the Positive Part of the JS

As has already been pointed out the JS is inadmissible. The MLE was shown to be inadmissible by producing an estimator with smaller risk the JS. The JS will be shown to be inadmissible by producing an estimator with smaller risk the classical positive part.

This section will:

1. introduce the classical positive part;
2. show that its risk is uniformly less than that of the JS;
3. give a numerical illustration of how much less the risk of the classical positive part is less than the JS.

2.9.1 The Positive Part Estimator

The positive part estimator is defined in such a way to insure that the contraction factor is positive. For the classical positive part estimator the contraction factor is set equal to zero for values of $x'x$ where it is negative for the JS. Thus, the classical positive part estimator PP_0 is defined to be

$$\hat{\delta}_i = \left(1 - \frac{(p-2)}{x'x}\right)x_i, \quad 1 \le i \le p \quad \text{if} \left(1 - \frac{(p-2)}{x'x}\right) \ge 0$$
$$0 \hspace{7cm} \text{otherwise} \tag{2.9.1}$$

2.9.2 The Inadmissibility of The JS

The inadmissibility of the JS is proved in Theorem 2.9.1 below.

Theorem 2.9.1. The JS is inadmissible.

Proof. It suffices to show that the estimator (2.9.1) has a smaller risk. To simplify the notation let $w = x'x$ and $\phi(w) = 1 - \frac{(p-2)}{x'x}$. Let $A_+ = \{x : \phi(w) \ge 0\}$ and $A_- = \{x : \phi(w) < 0\}$. For any set C define the indicator function

$$I_C(x) = 1 \text{ of } x \in C$$
$$\qquad\quad 0 \text{ otherwise.}$$

The result will follow once it is verified that

$$
\begin{aligned}
R(JS) \\
&= E[(\phi(w)x - \theta)'(\phi(w)x - \theta)] \\
&= E[\phi(w)^2 w] - 2\theta' E[\phi(w)x] + \theta'\theta \\
&= E[\phi(w)^2 I_{A_+}(w)w] + E[\phi(w)^2 I_{A_-}(w)w] \\
&\quad -2\theta' E[\phi(w) \, I_{A_+}(w)x] - 2\theta' E[\phi(w) \, I_{A_-}(w)x] + \theta'\theta \\
&\geq E[\phi(w)^2 I_{A_+}(w)w] - 2\theta' E[\phi(w) \, I_{A_+}(w)x] + \theta'\theta \\
&= R(PP_0).
\end{aligned}
$$

(2.9.2)

To establish the inequality in (2.9.2) it must be shown that $E[\phi(w)^2 I_{A_+}(w)w] \geq 0$ and $\theta' E[\phi(w)I_{A_+}(w)x] \leq 0$. The first inequality is obvious because w and $\phi(w)^2$ are positive quantities. To establish the second inequality notice that

$$E[\phi(w)I_{A_+}(w)x_i] = E[\phi(w)I_{A_+}(w)(x_i - \theta_i)] + \theta_i E[\phi(w)I_{A_+}(w)].$$

(2.9.3)

Now

$$
\begin{aligned}
E[\phi(w)I_{A_+}&(w)(x_i - \theta_i)] \\
&= \int_A (x_i - \theta_i)\phi(w)\exp[-\tfrac{1}{2}(x-\theta)'(x-\theta)]d\underline{x} \\
&= \frac{\partial}{\partial\theta_i} \int_A \phi(w)\exp[-\tfrac{1}{2}(x-\theta)'(x-\theta)]d\underline{x} \\
&= \frac{\partial}{\partial\theta_i} E[\phi(w)I_{A_+}(w)]
\end{aligned}
$$

(2.9.4)

where $d\underline{x} = dx_2 \cdots dx_p$.

Since w has a noncentral chi square distribution with noncentrality parameter $\lambda = \tfrac{1}{2}\theta'\theta$ and p degrees of freedom

$$E[\phi(w)I_{A_+}(w)] = \int_0^{p-2} \phi(w)f(w)dw$$

(2.9.5)

where f is defined in Equation (2.6.7). The reader can show in Exercise 2.9.1 that

$$\frac{\partial}{\partial\theta_i}E[\phi(w)I_{A_-}(w)] = E[\phi(\chi^2_{p+2})I_{A_-}(\chi^2_{p+2})]\,\theta_i - E[\phi(w)I_{A_-}(w)]\,\theta_i.$$

$$(2.9.6)$$

Substituting (2.9.6) into (2.9.3) gives

$$\theta'E[\phi(w)I_{A_-}(w)] = \theta'\theta E[\phi(\chi^2_{p+2})I_{A_-}(\chi^2_{p+2})] \le 0 \qquad (2.9.7)$$

establishing (2.9.2) and thus establishing the result.

As has already been pointed out the positive part estimator is also inadmissible. Examples of estimators that dominate PP_0 will be given in Chapter VI.

Exercise 2.9.1. A. Establish (2.9.6).
B. Convince yourself that the expectation in (2.9.7) is negative.

Exercise 2.9.2. Show that the Bayes risk of PP_0 is smaller than that of the JS.

Expressions for the frequentist and the Bayes risk of PP_0 and other positive part estimators will be given in Chapter VI. However to get some idea of how much of an improvement PP_0 is over the JS some values of the risk are given in Table 2.9.1 below and the risks of the two estimators is plotted in Figure 2.9.1.

Table 2.9.1

λ	0	0.1	0.5	0.9	1	3	5
JS	0.333	0.355	0.432	0.495	0.509	0.696	0.786
PP_0	0.226	0.251	0.342	0.420	0.437	0.671	0.779

p=6

λ	0	0.1	0.5	0.9	1	3	5
JS	0.200	0.216	0.274	0.325	0.337	0.514	0.621
PP_0	0.126	0.143	0.208	0.266	0.280	0.488	0.611

p=10

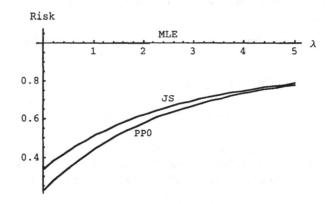

Figure 2.9.1
Risk of the MLE , the JS and PP$_0$ p = 6

2.10 Summary

In this chapter:

1. Properties of the multivariate normal distribution (MN) were reviewed.
2. The maximum likelihood estimator (MLE), its derivation and its properties were reviewed.
3. The MLE was obtained for the mean of a MN.
4. Important decision theory concepts were reviewed.
5. Admissibility of estimators was defined.
6. Several different arguments were given to show how the MLE is suspect as the best estimator for the mean of a MN.
7. The James-Stein estimator (JS) was formulated by both the Bayesian and the frequentist approach.
8. The inadmissibility of the MLE was established.
9. The JS is shown to be an operational ridge estimator.
10. The positive part was introduced and the inadmissibility of the JS was established.

Chapter III

The Ridge Estimators of Hoerl and Kennard

3.0 The Need for Alternatives to the Least Square Estimators

In Chapter I it was shown by example how the variability of least square estimators can be seriously inflated for collinear data. It was also shown how the ridge estimator of Hoerl and Kennard can be used to reduce this variability. The ridge estimator was originally proposed by Hoerl (1962) because data in chemical processes was often multicollinear. Since then it has found applications in a wide variety of diverse fields.

Actually the term ridge estimator can be applied to a larger class of estimators than that of Hoerl and Kennard. In place of kI any positive semidefinite matrix could be added to the X'X matrix. This is the generalized ridge regression estimator of C.R.Rao (1975). The ridge estimators of Hoerl and Kennard and a number of other estimators to be presented in this chapter are special cases of this estimator.

This chapter is a study of different ridge estimators from both the theoretical and applied standpoint. In Section 3.1 several different kinds of ridge estimators will be derived by generalizing the method used to obtain the ridge estimator of Hoerl and Kennard in Chapter I. Three kinds of ridge estimators will be obtained a generalized ridge estimator of Hoerl and Kennard, the contraction estimator of

Mayer and Willke and the generalized ridge estimator of C.R. Rao. A generalized contraction estimator will be presented in the exercises.

Section 3.2 will show how to compare. ridge estimators analytically with each other and with the least square estimator for non stochastic k. Three kinds of mean square error (MSE) will be studied:

1. the MSE of the parametric functions of the regression parameters;
2. the total MSE (TMSE);
3. the individual MSE of regression parameters after a reparametization of the regression model where the design matrix is orthogonal.

The mean square error of prediction will be presented in the exercises.

One of the important problems is the appropriate choice of the value of k. Some of the methods of choosing k will be discussed in Section 3.3 and Section 3.4. Two general methods of choosing k that will be discussed are:

1. the ridge trace, i.e., plotting the ridge estimators as a function of k and seeing where they stabilize (Section 3.3);
2. estimating k as a function of the sample estimators (Section 3.4).

They will be illustrated using the data of Section 1.3.

In general it is very hard to construct interval estimators of regression parameters based on ridge regression methods. A viable alternative is provided by the double bootstrap. This method is discussed in a general framework by Vinod (1993) and Beran (1987) and for ridge regression in Vinod(1995). An outline of the double bootstrap in the context of ridge regression following Vinod (1995) will be given at the end of Section 3.4.

It is very difficult to obtain formulae for the variance, bias and the mean square error of a ridge regression estimator when k is estimated from the data. As a result there has been a great deal of Monte Carlo work done on the performance of ridge estimators for stochastic k. The results of these Monte Carlo studies wiil be summarized in Section 3.5.

Section 3.6 is a brief summary of the chapter.

3.1 Derivation of the Ridge Regression Estimator

The ordinary ridge estimator was obtained in Section 1.3. Examples were given to show how the ridge estimator was an improvement over the least square estimator for multicollinear data. The form of the ridge estimator was given in (1.3.9). It differed from the LS because of the addition of the perturbation matrix kI to $X'X$. Actually the perturbation matrix could have taken a number of different forms. For example it could have been a general positive definite matrix or a constant multiple of X′X. Different kinds of ridge estimators can thus be formulated.

These different kinds of ridge estimators may be derived if the argument of Hoerl and Kennard that was given in Section 1.3 is generalized slightly. The generalization of their derivation will take into account whether some of the variables deserve more weight than others.

Instead of minimizing B′B minimize the weighted distance

$$D = B'HB \qquad (3.1.1)$$

where H is a positive semidefinite matrix subject to the side condition that B lies on the ellipsoid

$$(B - b)'X'X(B - b) = 0. \qquad (3.1.2)$$

A slightly more general optimization problem than that of Section 1.3 may now be solved by differentiating

$$L = B'HB + \lambda[(B - b)'X'X(B - b) - \Phi_0]. \qquad (3.1.3)$$

The result is

$$H\beta + \lambda X'X(B - b) = 0. \qquad (3.1.4)$$

In this chapter it will be assumed that X′X is of full rank. The non-full rank case will be dealt with in Chapter IV and V. The optimum estimator is the solution to the matrix equation obtained from (3.1.4)

$$(H + \lambda X'X)B = \lambda X'Xb. \qquad (3.1.5)$$

The solution to (3.1.5) is

$$\hat{\beta}^r = (H + \lambda X'X)^{-1}\lambda X'Xb = (\tfrac{1}{\lambda}H + X'X)^{-1}X'Xb$$
$$= (\tfrac{1}{\lambda}H + X'X)^{-1}X'Y. \qquad (3.1.6)$$

Then if k = 1/λ the estimator (3.1.6) takes the form

$$\hat{\beta}^r = (kH + X'X)^{-1}X'Y. \tag{3.1.7}$$

There are four special cases of (3.1.7) that are of interest.

1. When H=I (3.1.7) reduces to the ordinary ridge regression estimator of Hoerl and Kennard that was given in (1.3.9).

2. When $H = (1/k)$ G where G is any positive semidefinite matrix

$$\hat{\beta}^r = (G + X'X)^{-1}X'Y. \tag{3.1.8}$$

This is a generalized ridge regression estimator that was proposed by C.R. Rao (1975). It will be discussed further in Chapter V and VII.

3. From linear algebra it is well known that there exists an orthogonal matrix Q such that $X'X = Q\Lambda Q'$ where Λ is the diagonal matrix of eigenvalues and Q is the orthogonal matrix of eigenvectors. Let $H = (1/k) QKQ'$ where K is a positive definite diagonal matrix. The ridge estimator takes the form

$$\hat{\beta}^r = (QKQ' + X'X)^{-1}X'Y. \tag{3.1.9}$$

The ridge estimator (3.1.9) was proposed by Hoerl and Kennard (1970a). It is known in the literature as the generalized ridge estimator of Hoerl and Kennard.

4. When H=X'X

$$\hat{\beta}^r = \frac{1}{1+k}b. \tag{3.1.10}$$

This estimator was originally proposed by Mayer and Willke (1973) and has been derived by those readers who did Exercise 1.3.1. The proper choice of k leads to the JS (See Section 2.8).

Exercise 3.1.1.   Instead of minimizing the distance of B'B form the origin in deriving the ridge estimator minimize the distance from an arbitrary point in m dimensional space. Minimize

$$(B - \theta)'H(B - \theta) = \Phi_0 \tag{3.1.11}$$

subject to the constraint in (3.1.2) and obtain the optimum estimator.

Exercise 3.1.2. The ordinary ridge estimator and the ridge estimators in (3.1.9) and (3.1.10) are special cases of (3.2.8) for the proper choice of the matrix G. What matrix G should be used in each case. Check your assertion.

Exercise 3.1.3. Consider a regression problem where $X'X = I_3$. Suppose the LS estimators are $b_0 = 1$, $b_1 = 2$ and $b_2 = 3$. What is the value of the ordinary ridge estimator ?

Exercise 3.1.4. Redo Exercise 3.1.3 with $X'X=$

A.
$$\begin{bmatrix} 16 & 0 & 0 \\ 0 & 9 & 0 \\ 0 & 0 & 4 \end{bmatrix}$$

B.
$$\begin{bmatrix} 2 & 1 & 1 \\ 1 & 2 & 1 \\ 1 & 1 & 2 \end{bmatrix}$$

Exercise 3.1.5. In the derivations of the ridge regression estimator how is specification of k related to Φ_0 ?

Exercise 3.1.6. Let $G = QK\Lambda Q'$ in the estimator (3.8).
A. Show that the ridge estimator of the parametric functions may be expressed as $p'\hat{\beta} = q'\hat{\gamma}$ where $\hat{\gamma} = (I + K)^{-1}g$, $q = Qp$ and $g = Q'b$.
B. How could the estimator in A have been derived using the argument at the beginning of the section? What form should H take?

Exercise 3.1.7. An attempt to generalize the ordinary ridge estimator of Hoerl and Kennard. (See Hoerl and Kennard 1975.) Sommers (1964) proposed the following generalization of the ridge regression estimator. Let q be a non-negative integer. The proposed estimator is

$$\hat{\beta} = [\ (X'X)^{q+1} + kI\ ]^{-1}(X'X)^{q+1}b. \tag{3.1.12}$$

Show that this is a special case of the generalized ridge estimator of Hoerl and Kennard.

Exercise 3.1.8. An estimator that is similar but not the same as the ordinary ridge regression estimator. (See Farebrother (1975)).
A. Show that the linear estimator $L'Y$ of the parametric functions $p'\beta$ that minimizes

$$m = E(L'Y - p'\beta)^2 \tag{3.1.12}$$

is given by

$$\hat{\beta} = \beta\beta'X'(X\beta\beta'X' + \sigma^2 I)^{-1} Y. \tag{3.1.13}$$

B. Show that an equivalent form of (3.1.13) is given by

$$\hat{\beta} = \frac{\beta'X'Y}{\sigma^2 + \beta'X'X\beta} \tag{3.1.14}$$

C. Show that the operational forms of the ordinary ridge estimator and (3.1.14) are

$$\hat{\beta} = \left(X'X + \frac{s^2}{b'b}\right)^{-1} X'Y \tag{3.1.15}$$

and

$$\hat{\beta} = \frac{b'X'Y}{s^2 + b'X'Xb} . b , \tag{3.1.16}$$

where b is the LS and s^2 is the usual unbiased estimator of σ^2, are consistent estimators of β.

Exercise 3.1.9. Consider the augmented linear model

$$\begin{bmatrix} Y \\ r \end{bmatrix} = \begin{bmatrix} X \\ R \end{bmatrix} \beta + \begin{bmatrix} \varepsilon \\ \eta \end{bmatrix}$$

where the error vector has mean and variance

$$V = \begin{bmatrix} \sigma^2 I & 0 \\ 0 & \tau^2 I \end{bmatrix}.$$

A. Let $Z = \begin{bmatrix} X \\ R \end{bmatrix}$. Find the weighted LS, i.e., $\hat{\beta} = (Z'V^{-1}Z)^{-1}Z'V^{-1}Y$.

B. Give matrices R and a vector r where this weighted LS takes the form each of the different ridge estimators derived in this section.

3.2 The Efficiency of the Ridge Regression Estimator

Various measures of the efficiency of the ridge regression estimator will be taken up in this section. The results of Hoerl and Kennard (1970) will be summarized and some special cases of a result obtained by Farebrother (1976,1978) will also be obtained. They include:

1. the comparison of the MSE of the ridge regression estimators with the least square estimators;
2. the study of the components of the Mean Square Error, the variance or dispersion and the bias.

The ridge regression estimators will be compared to the LS with respect to the following measures of efficiency:

1. The Total Mean Square Error (TMSE)[1]

$$m_1 = E(\hat{\beta} - \beta)'(\hat{\beta} - \beta) = \text{tr}M \tag{3.2.1}$$

where $M = E(\hat{\beta} - \beta)(\hat{\beta} - \beta)'$.

2. The Mean Square Error of the Parametric Functions $p'\beta$ (MSE)

$$m_2 = p'E(\hat{\beta} - \beta)(\hat{\beta} - \beta)p = p'Mp \tag{3.2.2}$$

3. The Mean Square Error of the Individual Components of $\gamma = Q'\beta$. (IMSE)[2]. Let Q_j be the j'th column of Q. Then $\gamma_j = Q_j'\beta$. The individual components of m_3 are

$$m_{3j} = E(\hat{\gamma}_j - \gamma)^2 = Q_j'E(\hat{\beta} - \beta)(\hat{\beta} - \beta)Q_j = Q_j'MQ_j. \tag{3.2.3}$$

---

[1] The TMSE is usually called just the MSE in the literature. The term is used here to distinguish it from the MSE of the parametric functions.
[2] In this book IMSE will mean individual mean square error. However other authors use this notation to denote integrated mean square error.

Exercise 3.2.1. Consider the linear model

$$Y_i = \mu + \varepsilon_i \quad 1 \leq i \leq n$$

Let $\hat{\mu}$ be any estimator of μ. Show that all three measures of efficiency defined above are the same.

An expression in terms of G and $X'X$ will now be derived for the matrix M for the ridge estimator of C.R. Rao. This expression will make it easier to write down the various expressions for the bias variance and MSE s for the special cases to be considered in the sequel.
Observe that

$$M = D + B \tag{3.2.4}$$

where $D = E[(\hat{\beta} - E(\hat{\beta}))'(\hat{\beta} - E(\hat{\beta}))]$ and $B = (\beta - E(\hat{\beta}))(\beta - E(\hat{\beta}))'$. The matrix D is the variance covariance matrix of the estimator and the matrix B is the squared bias. Thus, the variance of $p'\hat{\beta}$ is given by $p'Dp$ and the square of the bias of $p'\hat{\beta}$ is given by $p'Bp$. For any vector $\text{var}(L'Y) = L'D(Y)L$ where $D(Y)$ is the variance covariance matrix of Y. Thus, it follows that for the ridge estimator (3.1.8) of C.R.Rao that

$$D = (X'X + G)^{-1}\sigma^2(X'X)(X'X + G)^{-1}. \tag{3.2.5}$$

Since

$$E(\hat{\beta}) = (X'X + G)^{-1}X'X\beta \tag{3.2.6}$$

and

$$(X'X + G)^{-1}X'X - I = I - (X'X + G)^{-1}G, \tag{3.2.7}$$

$$B = (X'X + G)^{-1}G\beta\beta'G(X'X + G)^{-1}. \tag{3.2.8}$$

Substitution of (3.2.8) and (3.2.5) into (3.2.4) gives

$$M = (X'X + G)^{-1}(G\beta\beta'G + \sigma^2(X'X))(X'X + G)^{-1}. \tag{3.2.9}$$

The ordinary ridge estimator (1.3.9) will be studied with respect to m_1. It will be shown that:

1. There are values of k where the ridge estimator has a smaller MSE than the LS.
2. The TMSE has two components the total variance (TV) and the total bias (TB).
3. The TV is a decreasing function of k.
4. The TB is an increasing function of k and is bounded above by $\beta'\beta$.

The ridge estimator of C.R. Rao (3.1.8) will be studied with respect to m_2. An ellipsoid will be obtained where its MSE is always smaller than that of the LS. The results obtained for (3.1.8) will specialize to both the ordinary and generalized ridge estimator of Hoerl and Kennard because they are special cases.

The study of the generalized ridge estimator with respect of m_3 will suggest a technique for estimating an optimal value of the parameter k.

3.2.1 The TMSE of the Ordinary Ridge Regression Estimator

This subsection will deal with a number of important properties of the ordinary ridge regression estimator. Recall that the ordinary ridge estimator is a special case of the ridge estimator of C.R. Rao with G = kI. Thus, the total variance is

$$TV(k) = \sum_{i=1}^{m} \frac{\sigma^2 \lambda_i}{(\lambda_i + k)^2} \qquad (3.2.10)$$

and the total squared bias is

$$TB(k) = k^2 \beta'(X'X + kI)^{-2} \beta. \qquad (3.2.11)$$

Adding (3.2.10) and (3.2.11) gives the total MSE

$$TMSE(k) = TV(k) + TB(k). \qquad (3.2.12)$$

It is easy to see from (3.2.10) that the total variance is a continuous decreasing function of k. Since $Q'Q = I$ and $X'X = Q'\Lambda Q$ where Λ is the matrix of eigenvalues of $X'X$, from (3.2.11)

$$TB(k) = k^2 \sum_{i=1}^{m} \frac{\gamma_i^2}{(k^2 + \lambda_i)^2} = \sum_{i=1}^{m} \frac{\gamma_i^2}{(1 + \lambda_i/k)^2} \ . \tag{3.2.13}$$

From (3.28) it is readily apparent that TB(k) is a decreasing function of $1/k$, hence an increasing function of k. Moreover since $\sum_{i=1}^{m} \gamma_i^2 = \beta'\beta$ it is easy to show that $\lim_{k \to \infty} TB(k) = \beta'\beta$. Thus TB(k) is an increasing function of k that is bounded above by $\beta'\beta$. It would indeed be disappointing if the TMSE of the ridge estimator was never less than that of the LS. Fortunately this is not the case as shown by Theorem 3.2.1 below. Notice that the ordinary ridge estimator reduces to the LS when k = 0 and thus TMSELS = TMSE(0).

Theorem 3.2.1. There exists a k > 0 such that the TMSE of the ordinary ridge estimator is less than that of the LS.
Proof. The Theorem will be proved by showing that if $0 < k < \sigma^2/\gamma_{max}^2$ then $\frac{dTMSE(k)}{dk} < 0$. This will establish that for an interval with 0 as a lower bound TMSE(k) is a decreasing function. Hence there must exist a k where the MSE of the ordinary ridge regression estimator is less than that of the LS. Since

$$\frac{dTMSE(k)}{dk} = \sum_{i=1}^{m} \frac{-2\sigma^2\lambda_i + 2k\lambda_i\gamma_i^2}{(\lambda_i + k)^3}, \tag{3.2.14}$$

the result follows.

It has been shown that for the ordinary ridge regression estimator there is a value of k where the TMSE is smaller than the LS . However, in general, it is impossible to determine the value of k analytically where the TMSE of the ridge estimator is minimized. The reader may obtain the theoretically optimum k for two statistically uninteresting cases in Exercise 3.2.2.

Exercise 3.2.2. A. Obtain the TMSE for the model given in Exercise 3.2.1.
B. For the model in Exercise 3.2.1 find a theoretically optimum k, i.e., an estimator for k that minimizes the TMSE.
C. Assume all of the eigenvalues of the X′X matrix are equal. Find a theoretically optimum k. What kind of difficulty is encountered when the assumption of equal

eigenvalues is dropped or when the number of regression coefficients being estimated is greater than one?

3.2.2 The MSE of the Ridge Estimator of C.R.Rao

An ellipsoid will be derived inside of which the parametric functions of the ridge estimator of C.R. Rao will have a smaller MSE than that of the LS. When appropriate substitutions are made for G these ellipsoids will specialize to the ridge estimator of Hoerl and Kennard and the contraction estimator of Mayer and Willke. The shape of these ellipsoids will be related to the amount of multicollinearity in a data set.

Some notation must be fixed first. First the definition of positive semidefinite and positive definite matrices will be reviewed.

Definition 3.2.1. A symmetric matrix A is positive semidefinite of for each vector p, $p'Ap \geq 0$ and for at least one p, $p'Ap > 0$. A symmetric matrix A is positive definite if for all non-zero p, $p'Ap > 0$.

A useful and important property of positive semidefinite matrices is that all of its eigenvalues are greater than or equal to zero; in other words it has no negative eigenvalues. Likewise all of the eigenvalues of a positive definite matrix are positive.

Exercise 3.2.3. Which of the following matrices are positive semidefinite? positive definite?

$$
\text{A. } \begin{bmatrix} 10 & -6 \\ -6 & 10 \end{bmatrix} \quad
\text{B. } \begin{bmatrix} 3 & 4 \\ 4 & 4 \end{bmatrix} \quad
\text{C. } \begin{bmatrix} 3 & 1 & 0 \\ 1 & 3 & 0 \\ 0 & 0 & 0 \end{bmatrix}
$$

A ridge estimator is considered better than a LS if $M_{LS} - M_R$ is positive semidefinite. This would imply that the MSE of the parametric functions of the ridge estimator is not greater than the MSE of the parametric functions of the LS. If $M_{LS} - M_R$ is positive definite then the MSE of all of the parametric of the ridge estimator are less than the corresponding MSE of the LS.

If the MSE of the parametric functions of a ridge estimator is less than the corresponding MSE of a LS it is easy to show that the same is true of the TMSE and the IMSE. Exercise 3.2.4 calls for the proof of a slightly more general result.

Exercise 3.2.4. Show that if $p'\hat{\beta}_1$ and $p'\hat{\beta}_2$ are two regression estimators of the parametric functions of the regression parameter p' and if $MSE(p'\hat{\beta}_1) \leq MSE(p'\hat{\beta}_2)$ the same is true of the TMSE and the IMSE.

Exercise 3.2.5. Show that the difference between the variances of the LS and the ridge estimator of C.R. Rao is positive semidefinite.

Exercise 3.2.6. It was shown that for the ridge estimator the total variance is an decreasing function of k and the total squared bias is an increasing function of k. Let D_i and B_i I = 1, 2 be the dispersion and the squared bias matrices of two estimators where for the ridge matrices G_1 and G_2 $G_2 - G_1$ is positive semidefinite. Show that:

A. It does not always follow that $D_1 - D_2$ is positive semidefinite. Hint: The matrix $A - B$ is positive semidefinite does not always imply that $A^2 - B^2$ is positive semidefinite.

B. Show that if α and θ are linearly independent vectors then $\alpha\alpha' - \beta\beta'$ is not a definite matrix. It has two eigenvalues one positive and one negative. (See Gruber (1990) or Toutenburg (1982)).

C. Using the result of B show that if m > 1 the difference of the bias of two estimators is never a positive semidefinite matrix. Thus the result that the total bias is an increasing function of k for the ordinary ridge estimator does not generalize to the parametric functions for the ridge estimator of C.R.Rao.

D. Show that if $G_i = Q\Delta_i Q'$ i= 1,2 where the Δ_i are positive semidefinite diagonal matrices $G_2 - G_1$ is positive semidefinite implies $p'D_2p \leq p'D_1p$.

E. Show that when the hypothesis of part D are true then tr $B_1 \leq trB_2$. This means that for the class of estimators defined in D the total bias is an increasing function of G.

The case considered in part D and E of Exercise 3.2.5 above includes the ordinary and the generalized ridge estimator , the contraction estimator and the generalized contraction estimator.

The main result of this subsection gives a necessary and sufficient condition for the ridge estimator to have a smaller MSE than the LS. It is contained in Theorem 3.2.4 and its Corollary below. Before this result can be established some inequalities must be proved. These will be used to establish Theorem 3.2.5 and in the sequel, especially in Chapter V. The first one is very well known.

Theorem 3.2.2. Cauchy-Schwarz Inequality (CS). Let b and c be non-zero n x 1 column vectors. Then

$$[b'c]^2 \leq (b'b)(c'c). \tag{3.2.15}$$

Equality holds only when b and c are scalar multiples of each other.

Proof. The proof is based on the following fact. If the discriminant of a quadratic equation is negative it has no real roots. Let λ be a scalar. Then

$$0 \le (c - \lambda b)'(c - \lambda b) = c'c - 2\lambda b'c + \lambda^2 b'b. \qquad (3.2.16)$$

The relationship in (3.2.16) is a quadratic inequality in λ. From the properties of the discriminant stated above inequality (3.2.16) holds true if and only if(iff)

$$4(b'c)^2 - 4(b'b)(c'c) \le 0. \qquad (3.2.17)$$

Inequality (3.2.17) and (3.2.15) are equivalent. When equality holds $c = \lambda b$.

The following simple Corollary of the CS will be needed.

Corollary 3.2.1. Let b and c be as defined in the CS. Let A be an nxn positive definite matrix. Then

$$[b'c]^2 \le b'A^{-1}b \cdot c'Ac \qquad (3.2.18)$$

Proof. The matrix A may be written in the form $A=P'\Lambda P$ where P is an orthogonal matrix and Λ is the diagonal matrix of eigenvalues of A. Let $\Lambda^{1/2}$ be the diagonal matrix whose elements are the positive square roots of the eigenvalues of A. Then the square root of A, $A^{1/2} = P'\Lambda^{1/2} P$ and $A = A^{1/2} A^{1/2}$. Application of the CS gives

$$[b'c]^2 = [b'A^{-\frac{1}{2}}A^{\frac{1}{2}}c]^2 \le b'A^{-1}b \cdot c'Ac. \qquad (3.2.19)$$

The following result , due to Farebrother (1976), will be used to establish the main result.

Theorem 3.2.4. (Farebrother) . Let d be a positive scalar. Let A be a positive definite matrix. Then dA - bb' is positive semidefinite iff $b'A^{-1}b \le d$.
Proof. Observe that dA - bb' is positive definite iff, for all $c \ne 0$,

$$dc'Ac > c'bb'c = (c'b)^2. \qquad (3.2.20)$$

Dividing both sides of (3.2.20) by $c'Ac$ it follows that (3.2.20) is equivalent to

$$d > \frac{(c'b)^2}{c'Ac}. \tag{3.2.21}$$

For all non-zero c (3.2.21) is equivalent to

$$\frac{d}{b'A^{-1}b} > \frac{(c'b)^2}{c'Ac \cdot b'A^{-1}b}. \tag{3.2.22}$$

By the CS inequality the most the right hand side of (3.2.22) can be is 1. Furthermore the inequality in (3.2.22) is strict and (3.2.21) holds for all non-zero c. Thus (3.2.22) holds iff

$$\frac{d}{b'A^{-1}b} \geq 1. \tag{3.2.23}$$

Inequality (3.2.23) is equivalent to the result to be proved.

In the proof of Theorem 3.2.4 the matrix $A \geq 0$ means A is positive semidefinite. The strict inequality means A is positive definite. Likewise for matrices A and B $A \geq B$ means that $A - B$ is positive semidefinite. Again the strict inequality implies that $A - B$ is positive definite. Moreover the if A is positive semidefinite and M is a non- singular matrix MAM' is positive semidefinite. Positive definiteness is also preserved. Note that the comparison of matrices is only a partial ordering. Given two arbitrary definite matrices their difference is not always positive semidefinite.

Theorem 3.2.5. All of the parametric functions of the ridge estimator of C.R.Rao (3.1.8) have a mean square error that is less than or equal to that of the LS , iff

$$\beta'(2G^{-1} + (X'X)^{-1})^{-1}\beta \leq \sigma^2 \tag{3.2.24}$$

Proof. The MSE of the parametric functions of the ridge estimator is less than that of the LS iff

$$(X'X + G)^{-1}(G\beta\beta'G + \sigma^2(X'X))(X'X + G)^{-1} \leq \sigma^2(X'X)^{-1}. \tag{3.2.25}$$

Multiplication of both sides of inequality (3.2.25) by $X'X + G$ preserves non-negative definiteness of the difference. Thus, (3.2.25) is equivalent to

$$G\beta\beta'G + \sigma^2(X'X)^{-1} \le (X'X + G)(\sigma^2(X'X)^{-1})(X'X + G) \tag{3.2.26}$$

But

$$(X'X + G)(\sigma^2(X'X)^{-1})(X'X + G) = \sigma^2((X'X) + 2G + G(X'X)^{-1}G)$$

$$\tag{3.2.27}$$

Thus after substitution of (3.2.27) into (3.2.26) the equivalent inequality

$$G\beta\beta'G \le \sigma^2(2G^{-1} + (X'X)^{-1}) \tag{3.2.28}$$

results. Multiplication of (3.2.28) by G^{-1} yields

$$\beta\beta' \le \sigma^2(2G^{-1} + (X'X)^{-1}). \tag{3.2.29}$$

Inequality (3.2.29) is equivalent to (3.2.24) by virtue of Theorem 3.2.3. The result then follows.

Corollary 3.2.2 specializes the ellipsoid to the ridge estimators of Hoerl and Kennard and the contraction estimator of Mayer and Willke.

Corollary 3.2.2. The parametric functions $p'\beta$ of :

1. the ordinary ridge regression estimator (1.3.10) has a smaller MSE than the LS iff

$$\beta'(\tfrac{2}{k}I + (X'X)^{-1})^{-1}\beta \le \sigma^2 ; \tag{3.2.30}$$

2. the generalized ridge estimator (3.1.9) has a smaller MSE than the LS if

$$\beta'(2K^{-1} + (X'X)^{-1})^{-1}\beta \le \sigma^2 ; \tag{3.2.31}$$

3. the contraction estimator of Mayer and Willke (3.1.10) has a smaller MSE than the LS if

$$\beta'(X'X)\beta \le \frac{k}{k+2}\sigma^2 . \tag{3.2.32}$$

These ellipsoids can give useful visual information concerning the amount of multicollinearity in the data. The following reformulation of Corollary 3.2.2 is helpful in this regard.

Corollary 3.2.3. The ellipsoids (3.2.29)-(3.3.32) may be rewritten in the form

$$\sum_{i=1}^{m} \frac{\gamma_i^2}{c_i^2} \leq 1 \tag{3.2.33}$$

where:

1. for the ordinary ridge estimator

$$c_i = \sigma \sqrt{\frac{2}{k} + \frac{1}{\lambda_i}} \, ; \tag{3.2.34a}$$

2. for the generalized ridge estimator of Hoerl and Kennard

$$c_i = \sigma \sqrt{\frac{2}{k_i} + \frac{1}{\lambda_i}} \, ; \tag{3.2.34b}$$

3. for the contraction estimator of Mayer and Willke

$$c_i = \sigma \sqrt{\frac{1}{\lambda_i} \left(\frac{2}{k} + 1 \right)} \, . \tag{3.2.34c}$$

When λ_i is very small the corresponding axis of the ellipsoid is large. Likewise when λ_i is large the axis is short. Thus for data with a great deal of collinearity the ellipsoid will be very flat. For ordinary ridge estimator when $X'X=I$, the orthogonal case the ellipsoid reduces to a sphere. This is clarified in Example 3.2.1.

The above results give an idea of the shape of the regions where the ridge type estimators have a smaller MSE than LS. However the results are limited in usefulness because they are about unknown parameters.

Example 3.2.1.   An Illustration of Corollary 3.2.1.   Consider the Data in Example 1.3.1. When $k = 1$ the equation of the five dimensional ellipsoid will be written down. For the purposes of this illustration the relative not the actual size of the ellipsoid is what is important. With this in mind let $a = \gamma/\sigma$. Then the equation of the five dimensional ellipsoid is

$$\frac{a_1^2}{(1.4534)^2} + \frac{a_2^2}{(1.82411)^2} + \frac{a_3^2}{(4.90614)^2} + \frac{a_4^2}{(16.4631)^2} + \frac{a_5^2}{(73.33)^2} \leq 1 .$$

Consider the projection into the two dimensional plane that contains variables one and five.

$$\frac{a_1^2}{(1.4534)^2} + \frac{a_5^2}{(73.33)^2} \leq 1 .$$

The graph of this ellipse is given in Figure 3.2.1

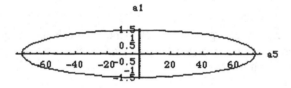

Figure 3.2.1

This projection is the one derived from the largest and the smallest eigenvalue of X′X. It is really much flatter than the picture would indicate. If the picture were really drawn to scale the ellipse would practically look like a straight line.

Exercise 3.2.7. What is the form of the ellipsoid for the generalized contraction estimator analogous to those obtained in Corollaries 3.2.2 and 3.2.3?

When the MSE of the parametric functions of the ridge estimator have a smaller MSE than the LS the same holds true for the IMSE and the TMSE. However it does not always follow that of the IMSE or the TMSE of the ridge estimator is smaller than that of the LS the MSE of the parametric functions behaves the same way. The

MSE of the parametric functions for one estimator is smaller than that of another estimator iff the same is true for all quadratic loss functions. That is the content of Theorem 3.2.5 below due to Theobald (1974).

Theorem 3.2.1. Theobald's Result (1974). A symmetric mxm matrix A is non-negative definite iff tr $AB \geq 0$ for all non-negative definite B.

Proof. Suppose tr $AB \geq 0$ for all non-negative definite B. Let $B = p'p$ where p is a vector. Then

$$tr\ Ap'p = p'Ap \geq 0. \qquad (3.2.35)$$

Thus A is non-negative definite Suppose A is non-negative definite. Let Λ be the diagonal matrix of eigenvalues of B. Let [Q R] be an orthogonal matrix where the rows of Q are the eigenvectors corresponding to the non-zero eigenvalues and the rows of R correspond to the eigenspace of zero. Then

$$B = Q\Lambda Q' = \sum_{i=1}^{m} \lambda_i Q_i Q_i', \qquad (3.2.36)$$

where $\lambda_1, \lambda_2, ..., \lambda_m$ are the eigenvalues of B. Denote the columns of Q by Q_i. The column vectors Q_i are the orthogonal eigenvectors of B. Since A is non-negative definite and $\lambda_i \geq 0$ it follows that

$$tr(AB) = \sum_{i=1}^{m} \lambda_i Q_i' A Q_i \geq 0. \qquad (3.2.37)$$

It would seem that Theobald's result proved above is just an abstract result in Linear Algebra. However, it was established to facilitate the comparison of the estimators. This comparison will be done in Corollary 3.2.3 below.

Corollary 3.2.3. Let $\hat{\theta}_1$ and $\hat{\theta}_2$ be estimators of a parameter θ. Then

$$M_1 = E(\hat{\theta}_1 - \theta)(\hat{\theta}_1 - \theta)' \leq E(\hat{\theta}_2 - \theta)(\hat{\theta}_2 - \theta)' = M_2 \qquad (3.2.38)$$

iff

$$ms_1 = E(\hat{\theta}_1 - \theta_1)'A(\hat{\theta}_1 - \theta_1) \leq E(\hat{\theta}_2 - \theta_2)'A(\hat{\theta}_2 - \theta_2) = ms_2 . \qquad (3.2.39)$$

Thus, the MSE of the parametric functions of a ridge estimator is less than that of a LS iff the same is true for the MSE averaging over any quadratic loss function with a non-negative definite matrix.

Exercise 3.2.8. Let θ be an m dimensional vector . Let D be a positive semidefinite diagonal matrix. Let $\hat{\theta}_1$ and $\hat{\theta}_2$ be estimators of θ. Then

$$E(\hat{\theta}_1 - \theta)'D(\hat{\theta}_1 - \theta) \leq E(\hat{\theta}_2 - \theta)'D(\hat{\theta}_2 - \theta)$$

iff for the individual MSEs of the coordinates

$$E(\hat{\theta}_{1i} - \theta_{1i})^2 \leq E(\hat{\theta}_{2i} - \theta_{2i})^2 .$$

3.2.3 The Individual MSE of the Ridge Estimator Using Orthogonal Coordinates

The generalized ridge estimator of Hoerl and Kennard will now be obtained using canonical coordinates. The linear model will be reparametized so that the analogue of the $X'X$ matrix will be diagonal and the design matrix will be orthogonal. The individual MSE of the regression coefficients will be obtained . A value of k_i will be obtained that minimizes the MSE. This value will suggest the form of an estimate of k for an operational ridge regression estimator.

To reparametize the linear model recall that $\gamma = Q'\beta$. Then the linear model

$$Y = X\beta + \varepsilon \tag{3.2.40}$$

may be written

$$Y = XQQ'\beta + \varepsilon = XQ\gamma + \varepsilon. \tag{3.2.41}$$

The reparametized model is (3.2.41). The least square estimator for γ will be denoted by g. Now $Q'X'XQ = \Lambda$. Thus, the LS of γ is given by

$$g = \Lambda^{-1} Q' X'Y. \tag{3.2.42}$$

Let S be the matrix of orthogonal eigenvectors of XX'. Then

$$X = S'\Lambda^{1/2}Q' \tag{3.2.43}$$

and

$$g = \Lambda^{-1} \Lambda^{1/2} SY = \Lambda^{-1/2}SY, \qquad (3.2.44a)$$

since $QQ' = I$. Let $Z = \Lambda^{1/2} SY$. Then the individual coordinates of g are

$$g_i = \frac{z_i}{\lambda_i}, \ 1 \le i \le m. \qquad (3.2.44b)$$

Observe that $g = Q'b$. The estimator (3.1.9) may be written

$$\begin{aligned} \hat{\beta} &= (QKQ' + Q\Lambda Q')^{-1}X'Xb = Q(K + \Lambda)^{-1}Q'Q\Lambda Q'b \\ &= Q(K + \Lambda)^{-1}\Lambda g = Q\hat{\gamma} \end{aligned} \qquad (3.2.45)$$

where the ridge estimator in the reparametized model is

$$\hat{\gamma} = (K + \Lambda)^{-1}Z. \qquad (3.2.46)$$

The individual coordinates of (3.2.46) are

$$\hat{\gamma}_i = \frac{z_i}{\lambda_i + k_i} = \frac{\lambda_i g_i}{\lambda_i + k_i}, \ 1 \le i \le m. \qquad (3.2.47)$$

The estimators in (3.2.47) have variance

$$v_i = \frac{\lambda_i \sigma^2}{(\lambda_i + k_i)^2} \qquad (3.2.48)$$

and squared bias

$$sb_i = \left(\frac{k_i}{\lambda_i + k_i}\right)^2 \gamma_i^2 \qquad (3.2.49)$$

and IMSE

$$(IMSE)_i = \frac{\lambda_i \sigma^2 + k_i^2 \gamma_i^2}{(\lambda_i + k_i)^2}. \qquad (3.2.50)$$

The IMSE of the generalized ridge estimator may be compared to that of the LS. The IMSE of the LS is σ^2/λ_i. Solving the inequality

$$\frac{\lambda_i\sigma^2 + k_i^2\gamma_i^2}{(\lambda_i + k_i)^2} \leq \frac{\sigma^2}{\lambda_i} \qquad (3.2.51)$$

for k_i when $\gamma_i > \sigma$ the $(IMSE)_i$ of the ridge estimator is less than that of the LS provided that

$$k_i \leq \frac{2\lambda_i}{\gamma_i^2/\sigma^2 - 1}. \qquad (3.2.52)$$

If $\gamma_i \leq \sigma$ the $(IMSE)_i$ of the ridge estimator is less than that of the LS for all values of k_i. Solving inequality (3.2.52) for γ_i^2, it can be seen that if

$$\gamma_i^2 \leq \left(\frac{2}{k_i} + \frac{1}{\lambda_i}\right)\sigma^2 \qquad (3.2.53)$$

then the $(IMSE)_i$ of the ridge estimator is less than that of the LS.

Exercise 3.2.9. Show that if Inequality (3.233) with the c_i given by (3.2.34b) holds then (3.2.53) holds. However the reverse implication may not hold. Why? This is another way of showing that if the MSE of the parametric functions of the ridge estimator is less than that of the LS the IMSE also has that property and that the converse statement may not be true.

Theoretically optimum k_i is found by differentiating the expression in (3.2.50), setting the derivative equal to zero and solving the resulting equation for k_i. Thus,

$$\frac{d(IMSE)_i}{dk_i} = \frac{2\lambda_i(k_i\gamma_i^2 - \sigma^2)}{(\lambda_i + k_i)^3} = 0. \qquad (3.2.54)$$

Solving Equation (3.2.54) for k_i its optimum value is

$$k_i = \frac{\sigma^2}{\gamma_i^2} \qquad (3.2.55)$$

The minimum MSE is after substitution of (3.2.54) into (3.2.50) and some algebra

$$m_{3\,i\,min} = \frac{\sigma^2}{\lambda_i} - \frac{\sigma^2/\lambda_i}{\gamma_i + \sigma^2/\lambda_i}. \tag{3.2.55}$$

The first term is the $(IMSE)_i$ of the LS; the second term represents the maximum possible improvement.

The sample estimator of k_i suggested by Hoerl and Kennard is

$$\hat{k}_i = \frac{\hat{\sigma}^2}{g_i^2}. \tag{3.2.57}$$

A closed form analytical optimum k is not available for the ordinary ridge estimator. However if the g_i^2 are averaged

$$\frac{1}{m}\sum_{i=1}^{m} g_i^2 = \frac{1}{m} g'g = \frac{1}{m} b'QQ'b = \frac{1}{m} b'b, \tag{3.2.58}$$

noting that Q is an orthogonal matrix so $QQ'=I$. This motivates the choice of optimum k for the ordinary ridge estimator. Recall that this choice was

$$\hat{k} = \frac{m\hat{\sigma}^2}{b'b}. \tag{3.2.59}$$

Different analytical and ad hoc ways to choose the ridge parameter is the subject of the next section.

Exercise 3.2.10. Carry through the details of the algebra necessary to obtain (3.2.55).

Exercise 3.2.11. Find optimum k_i and the values of k_i and γ_i where the IMSE of the generalized contraction estimator is smaller than that of the LS.

Exercise 3.2.12. For the contraction estimator of Mayer and Willke:
A. Find k so that the MSE of prediction

$$m = E(\hat{\beta} - \beta)'X'X(\hat{\beta} - \beta)$$

is minimized.

B. For what values of k and for what condition on β is the MSE of the contraction estimator less than that of the least square estimator?
C. Suggest an estimator of k that will lead to a generalization of the JS to the linear model. (This will be taken up in Chapter IV.)

3.3 Estimating The Ridge Perturbation Factor k - The Ridge Trace

3.3.1 Overview

Section 3.2 dealt with the properties of ridge regression estimators for non-stochastic k. This and the next section deals with different methods for choosing k . The two principle methods for choosing k are:

1. the ridge trace, a subjective method;
2. as a function of sample estimates from the data, an objective method.

The ridge trace is a simultaneous plot of the ridge regression estimators for each of the parameters of a linear model vs. k. When data is collinear the regression coefficients appear to change very rapidly as k varies for small values. At some value of k they stabilize; the rate of change slows down to almost zero. This value of k is picked from examining the plot. Of course this method of choosing k is somewhat subjective because two people examining the plot might have different opinions as to where the regression coefficients stabilize. This technique will be dealt with in this section.

Using a function of sample estimates from the data is an objective method because it is based on a mathematical formula. The choice of k proposed by Hoerl and Kennard mentioned in Chapter I and Section 3.2 is an example of an objective method of choosing k. Section 3.4 will deal with different objective methods of choosing k.

Obtaining the ridge trace will be illustrated using the data from Example 1.3.1. A number of different objective methods for estimating k will be presented and illustrated using Example 1.3.1.

3.3.2 The Ridge Trace

The use of the ridge trace for selecting an appropriate value of the perturbation parameter k will be illustrated in this subsection using the data from Example 1.3.1.

Most authors recommend standardizing the data so that the X'X matrix is in the form of a correlation matrix. An advantage of standardization of the data is that the regression coefficients will then be expressed in comparable numerical units. Vinod (1981) points out that the appearance of a ridge trace that does not plot standardized regression coefficients may be dramatically changed by a simple translation of the origin and scale transformation of the variables. On the other hand standardization may not be advisable for most theoretical results and when the investigator prefers to do the centering , scaling and the MSE computations in the original units.

The standardization is accomplished by transforming the linear model

$$Y = X\beta + \varepsilon \qquad\qquad\qquad (3.3.1a)$$

to

$$Y_s = X_s\beta_s + \varepsilon \ . \qquad\qquad\qquad (3.3.1b)$$

The matrix X_s is n x (m − 1) with elements

$$x_{sij} = \frac{x_{ij} - \overline{x}_j}{\sqrt{n-1}\ s_j} \ . \qquad\qquad\qquad (3.3.2a)$$

The vector Y_s has components

$$y_{si} = \frac{y_i - \overline{y}}{\sqrt{n-1}s_y} \qquad\qquad\qquad (3.3.2b)$$

where $s_j = \dfrac{1}{n-1}\sum_{i=1}^{n}(x_{ij} - \overline{x}_i)^2$ and $s_y = \dfrac{1}{n-1}\sum_{i=1}^{n}(y_i - \overline{y})^2$.

Example 3.3.1 shows how to plot the ridge trace for the standardized data in Example 1.3.1.

Example 3.3.1. The Ridge Trace

The standardized matrix

$$X_s = \begin{bmatrix}
-0.2164 & -0.5000 & -0.4845 & -.02203 \\
-0.3828 & -0.5000 & -0.3814 & 0.0944 \\
-0.3828 & -0.1667 & -0.2783 & -0.5350 \\
-0.3828 & -0.1667 & -0.1752 & 0.0944 \\
-0.0499 & 0.0000 & -0.0722 & 0.0944 \\
0.1165 & 0.1667 & 0.0309 & -0.2203 \\
0.1165 & 0.1667 & 0.2371 & 0.0944 \\
0.2829 & 0.1667 & 0.2371 & 0.7237 \\
0.4494 & 0.5000 & 0.4433 & -0.2203 \\
0.4494 & 0.3333 & 0.4433 & 0.0944
\end{bmatrix}$$

and

$$Y_s = \begin{bmatrix}
-0.2554 \\
-0.4256 \\
-0.4256 \\
-0.2554 \\
-0.0851 \\
0.0851 \\
0.2554 \\
0.2554 \\
0.4256 \\
0.4256
\end{bmatrix}.$$

Theoretically the sums of the columns of both the X_s and Y_s matrices should add up to zero and the sums of the squares should add up to one. This may be checked by the reader for the X_s and the Y_s matrices above if he or she allows for some slight deviation due to rounding off. Notice that to three decimal places

$$X_x' X_s = \begin{bmatrix}
1 & 0.888 & 0.925 & 0.309 \\
0.888 & 1 & 0.962 & 0.157 \\
0.925 & 0.962 & 1 & 0.328 \\
0.309 & 0.157 & 0.328 & 1
\end{bmatrix}.$$

Another advantage of standardizing the X matrix is that it can be seen which variables are highly correlated. From the standardized X'X matrix. it is easily seen that variables 2 and 3 (West Germany and Japan) are highly correlated, variables 1 and 3 (France and Japan) are also highly correlated, variables 1 and 2 (France and West Germany) are moderately highly correlated and Japan and the former Soviet Union have little in common. The ridge regression estimator

$$\hat{\beta}_s(k) = (X_s'X_s + kI)^{-1}X'Y$$

is tabulated for k = 0 to k = 0.5 in Table 3.3.1 below. Note that the case where k=0 corresponds to the LS. i.e. $\hat{\beta}_s(0) = b_s$

Table 3.3.1

b_s	$\hat{\beta}_s(0.1)$	$\hat{\beta}(0.2)$	$\hat{\beta}(0.3)$	$\hat{\beta}(0.4)$	$\hat{\beta}(0.5)$
0.640	0.492	0.424	0.384	0.357	0.336
−0.118	0.160	0.203	0.219	0.225	0.226
0.473	0.296	0.287	0.280	0.274	0.268
0.014	0.067	0.076	0.079	0.081	0.081

The corresponding ridge trace is given in Figure 3.3.1 below.

As has already been pointed out the choice of k is a judgment call. However looking at Figure 3.3.1 the parameters appear to change rapidly for small values of k but begin to stabilize between k = 0.2 and k= 0.3. A reasonable choice of k might be k = 0.3. Are there any other ways to see if this choice is reasonable? Two other things that might be investigated are:

1. the behavior of the variance inflation factors(VIFs);
2. how much of the variation is accounted for by the ridge estimator as compared to the least square estimator.

The correlation coefficients given in the standardized X'X matrix only indicate how a given variable is influenced by one other variable. The VIF is found by:

1. finding the least square estimators or the ridge estimator of the regression of each of the exploratory variables on the remaining exploratory variables;
2. obtaining

$$R^2 = \frac{\text{Regression Sum of Squares}}{\text{Total Sum of Squares}} = \frac{\text{SSR}}{\text{SST}};$$

3. substituting the values found into

$$VIF = \frac{1}{1 - R^2}.$$

For the standardized data

$$SST = 1,$$

$$\begin{aligned} SSR &= \hat{\beta}_s[k]'X'X\hat{\beta}_s[k] \\ &= 1 - (Y - X\hat{\beta}_s[k])'(Y - X\hat{\beta}_s[k]) \\ &= 1 - SSE. \end{aligned}$$

Notice that when $R^2 = 0$ the VIF is 1 and as $R^2 \to 1$ the VIF tends to infinity. One way to write the standard error of a regression estimator is

$$SD(\hat{\beta}_j) = \frac{(\sigma/s_j)\sqrt{VIF_j}}{\sqrt{n-1}}.$$

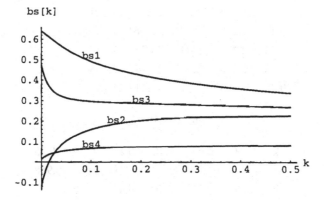

Figure 3.3.1

Thus, a larger VIF means a less precise estimate of the regression coefficient. Some values of the VIFs are tabulated in Table 3.3.2 below. A plot of the VIFs is given in Figure 3.3.2. Table 3.3.3 and Figure 3.3 show how much of the variation is accounted for by the regression.

From Figures 3.3.1 it can be seen that at k = 0.3:

1. the bs[k] appear to be stabilizing;
2. there is a substantial reduction in the two largest of the VIFs;
3. more than 95% of the variation is accounted for by the regression model.

For this reason k = 0.3 might be a good choice. On the other hand maybe some observers might feel that $bs_3[0.3]$ is not stable enough and might want to choose a larger value of k. The sacrifice that would be made is that R^2 could be substantially smaller and consequently the overall fit might not be as good.

Table 3.3.2

k	VIF_1	VIF_2	VIF_3	VIF_4
0	12.062	21.581	29.756	1.795
0.05	10.951	18.604	27.982	1.479
0.10	10.432	15.887	25.436	1.341
0.15	10.045	14.018	23.037	1.275
0.20	9.679	12.612	20.863	1.235
0.25	9.313	11.483	18.915	1.209
0.30	8.946	10.543	17.186	1.191
0.35	8.582	9.741	15.661	1.177
0.40	8.225	9.048	14.322	1.166
0.45	7.880	8.443	13.147	1.156
0.50	7.550	7.912	12.117	1.149

Table 3.3.3[3]

k	0	0.05	0.10	0.15	0.20	0.25
R^2	0.976	0.973	0.969	0.969	0.961	0.958
k	0.30	0.35	0.40	0.45	0.50	
R^2	0.954	0.951	0.947	0.943	0.939	

---

[3] Note that $100R^2$ is the percentage of the variation accounted for by the regression estimator.

Another graph yields completely different results. Looking at a plot for 0 to 0.1 one might conclude that stabilization occurs around k = .06. This is quite different from the conclusion drawn above. However the smaller value of k is closer to the values of k obtained by the objective methods studied in the next section. Perhaps the chosen value of k should be based on both the ridge trace and the value suggested by objective methods.

Exercise 3.3.1.   Consider a regression model with two exploratory variables.
A. Show that the standardized X'X matrix is

$$X_s'X_s = \begin{bmatrix} 1 & \rho \\ \rho & 1 \end{bmatrix}.$$

B. Find the dispersion matrix of $X_s'X_s$. How is the variability and the covariance of the regression coefficients affected by the value of ρ?
C. Find the two eigenvalues of $X_s'X_s$ .Describe their dependence on ρ. Also describe the behavior of the ratio of the larger to the smaller eigenvalue as ρ varies between -1 and 1.

Exercise 3.3.2.   For the data in Example 3.3.1 find the eigenvalues   of the standardized X'X matrix and compare them to the corresponding non-standardized matrix.

Exercise 3.3.3.   Repeat all of the calculations and plot all the graphs for the data in Example 3.3.1 again using the non-standardized X and Y matrix by employing Mathematica or some other software package. Suggest a choice of k and compare it to that of Example 3.3.1.

Exercise 3.3.4.   Go through the analogous calculations of this section and plot the corresponding graphs for the data in Exercise 1.3.2 for with and without standardization of the data.

Exercise 3.3.5.   A test of significance that could be one indicator of multicollinearity is the Farrar Glauber test[4] based on the chi-square statistic

---

[4] A referee pointed out that the consensus of recent econometrics texts is that the Ferrar Glauber test is not meaningful because in a regression model of the form E(Y|X)=Xβ, X is not stochastic. The multicollinearity in the available X must be dealt with.

$$\chi^2 = \left[-n - 1 - \frac{1}{6}(2p + 5)\right] \ln |R|.$$

Based on comparison with the 0.05 and 0.01 significance levels does the above statistic indicate mullticollinearity for the data of Exercise 1.3.2 and Example 1.3.1?

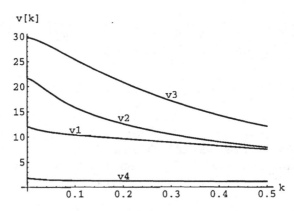

Figure 3.3.2[5]

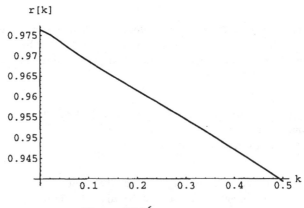

Figure 3.3.3[6]

---

[5] The notation v[k] means VIF, the variance inflation factor. Likewise V_i is (VIF)$_i$
[6] The r[k] is R^2 for the given value of k.

Exercise 3.3.6. For values if k at intervals of 0.005 up to 0.1 find
A. The variance inflation factors.
B. The percentage $100R^2$ accounted for by the regression equation.
C. Make a guess at a reasonable value of k if possible.

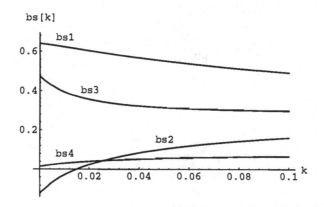

Figure 3.3.4

Exercise 3.3.7. Find the derivatives of the components of bs[k] for k=0.01,0.02,..0.1,0.2...1. What estimate of k would you pick on this basis?

3.4 Estimating the Perturbation Factor k–Objective Methods

3.4.1 Estimation Procedures Based on Using Sample Statistics

A number of different formulae will be presented and illustrated for estimating the perturbation factor k for the ordinary ridge estimator and the elements of the perturbation matrix K for the generalized ridge regression estimator. These formulae are listed below. For the ordinary ridge regression estimator four estimators of k are:

1.
$$\hat{k} = \frac{m\hat{\sigma}^2}{g'g} = \frac{m\hat{\sigma}^2}{b'b}$$
(3.4.1)

2.
$$\hat{k} = \frac{\hat{\sigma}^2}{\max_{1 \le i \le m} g_i^2}$$
(3.4.2)

3.
$$\hat{k} = \frac{m\hat{\sigma}^2}{\sum_{i=1}^{m} \lambda_i g_i^2} = \frac{m\hat{\sigma}^2}{b'X'Xb}$$
(3.4.3)

4. Let Q be defined by

$$Q = g'g - \hat{\sigma}^2 \left(\sum_{i=1}^{m} \frac{1}{\lambda_i} \right).$$
(3.4.4 a)

Then $\hat{k}$ is the solution to the equation

$$\hat{\gamma}(k)'\hat{\gamma}(k) = Q$$
(3.4.4 b)

if $Q > 0$. Otherwise $k = 0$. The λ_i are the eigenvalues of $X'X$ and

$$\hat{\gamma}_i(k) = \frac{\lambda_i}{\lambda_i + k} g_i.$$
(3.4.5)

Some other procedures for estimating k will be mentioned in Section 3.5.

Some estimation procedures for k_i for the generalized ridge estimator will now be given. As has already been mentioned Hoerl and Kennard obtained the theoretically optimum value of k_i to be

$$k_i = \frac{\sigma^2}{\gamma_i^2}$$
(3.4.6a)

with estimator

$$\hat{k}_i = \frac{\hat{\sigma}^2}{g_i^2}.$$ (3.4.6b)

Notice that the estimators of k and k_i for the ridge and the generalized ridge estimator are functions of the within sum of squares variance and the LS estimators. One may take the operational ridge estimators obtained using $\hat{k}$ or $\hat{k}_i$ and in the expression for the perturbation factor substitute this operational ridge estimator instead of the LS to obtain a new estimate of k or k_i. Then a new operational ridge estimator may be formulated using the new estimate of the ridge perturbation factor. This procedure may be repeated ad infinitum or until the difference between successive values of the estimators of k or k_i are less than some desired value. Hemmerle (1975) gave conditions for convergence of the estimates of k_i and found explicit expressions for their limiting values. These will be discussed in Section 3.4.3.

There are some interesting mathematical relationships between the estimators of k_i for the generalized ridge estimator and the estimators of k for the ordinary ridge estimator. Notice that:

1. for the estimator in (3.4.1)

$$\frac{1}{\hat{k}} = \frac{1}{m}\sum_{i=1}^{m}\frac{1}{\hat{k}_i},$$ (3.4.7)

i.e., the reciprocal of the estimator of k is the mean of the reciprocals of the estimators of k_i ;
2. for the estimator in (3.4.2)

$$\hat{k} = \min_{1\le i\le m}\hat{k}_i,$$ (3.4.8)

i.e., the estimator of k is chosen to be the minimum of the estimators of k_i ;
3. for the estimator in (3.4.3)

$$\frac{1}{\hat{k}} = \frac{1}{m}\sum_{i=1}^{m}\frac{\lambda_i}{\hat{k}_i},$$ (3.4.9)

i.e., the reciprocal of the estimator of k is a weighted average of the estimators of k_i where the weights are the eigenvalues of the $X'X$ matrix. Thus, using Hemmerle's method to be described in Section 3.4.3 it may be possible sometimes to estimate k using the estimators of k_i.

The subsections that follow will give the rationale for the various estimators of k and will show how to calculate them numerically.

3.4.2 The Rationale for the Sample Estimators of k in the Ordinary Ridge Estimator

The rationale for the formulae for k presented above will be given. These justifications were suggested by Wichern and Churchill (1978). For (3.4.1) they note that if $X'X = I$ that $k = m\sigma^2/g'g$ minimizes the mean square error. Recall Exercise 3.2.2.c. Also see Hoerl, Kennard and Baldwin (1975). Of course one might question the validity of this rationale on the grounds that ridge estimators are used for data where the X matrix is far from being orthogonal. Another rationale might be the relationship derived in Equation (3.4.9) connecting the estimator of k with that of the generalized ridge estimator.

For (3.4.2) notice that when $k = \sigma^2/\max_{1 \le i \le m} \gamma_i^2$ from the proof of Theorem 3.2.1 the TMSE is smaller than that of the LS. The LS in orthogonal coordinates with the largest absolute value uses the smallest eigenvalue of $X'X$. It also has the largest variability of all of the LS estimators of individual coordinates.

The estimator of k in (3.4.3) was suggested by Lawless and Wang (1976). It may be justified by a Bayesian argument. Assume that the error terms of the linear model (3.2.41) are normally distributed with mean zero and dispersion $\sigma^2 I$. Assume also that the γ_i are independently normally distributed with

$$E(\gamma_i) = 0 \text{ and } D(\gamma_i) = \sigma_\gamma^2. \qquad (3.4.10)$$

The linear model in orthogonal coordinates may be written in the form

$$z_i = \lambda_i^{1/2}\gamma_i + \varepsilon_i, \ 1 \le i \le m. \qquad (3.4.11)$$

Then the prior distribution of γ_i is

$$\pi(\gamma_i) = \frac{1}{\sqrt{2\pi\sigma_\gamma^2}} \exp\left(-\frac{\gamma_i^2}{2\sigma_\gamma^2}\right) . \qquad (3.4.12)$$

Since g_i is a sufficient statistic for γ_i the posterior distribution of γ_i conditional on g_i is the same as that of γ_i conditional on z_i. Now, for a single coordinate the likelihood is

$$f(z_i \mid g_i) = \frac{1}{\sqrt{2\pi\sigma^2/\lambda_j}} \exp\left(-\frac{(g_i-\gamma_i)^2}{2\sigma^2/\lambda_j}\right) \tag{3.4.13}$$

Application of Bayes Theorem and the sufficiency principle yields the posterior distribution

$$g(\gamma_i \mid z_i) = \frac{1}{\sqrt{2\pi}\sigma_b} \exp\left(-\frac{(\gamma_i-\hat{\gamma}_i)^2}{2\sigma_b^2}\right) \tag{3.4.14}$$

where

$$\hat{\gamma}_i = \frac{\sigma_\gamma^2 g_i}{\sigma_\gamma^2 + \sigma^2/\lambda_i} = \frac{\lambda_i g_i}{\lambda_i + \sigma^2/\sigma_\gamma^2} \tag{3.4.15}$$

and

$$1/\sigma_b^2 = \lambda_i/\sigma^2 + 1/\sigma_\gamma^2 \ . \tag{3.4.16}$$

The Bayes estimator (3.4.15) is a ridge regression estimator with $k = \sigma^2/\sigma_\gamma^2$. Thus, the ridge regression estimator is a special case of the Bayes estimator. This will be demonstrated again in a more general setting in Chapter V.

For the fourth technique of estimating k observe that

$$E(g_i) = \gamma_i \text{ and } Var(g_i) = \frac{\sigma^2}{\lambda_i} \ . \tag{3.4.17}$$

Thus

$$E(g_i^2) = \gamma_i^2 + \frac{\sigma^2}{\lambda_i} \tag{3.4.18}$$

and

$$E(b'b) = E(g'g) = \sigma^2 \sum_{i=1}^{m} \frac{1}{\lambda_i} + \gamma'\gamma. \qquad (3.4.19)$$

Since $E(\hat{\sigma}^2) = \sigma^2$, Q is an unbiased estimator of $\gamma'\gamma = \beta'\beta$. Thus, if the estimate of the length of the coefficient vector is positive k is chosen so that the expected length of the ridge estimator of γ is the same as that of Q. Otherwise the least square estimator is used.

Exercise 3.4.1. Show by direct comparison of the expressions for the MSE that when $k = \sigma^2 / \max_{1 \le i \le m} \gamma_i^2$ the ridge estimator has a smaller MSE than the LS. Is the value of k optimum? Explain by providing a proof or a counterexample.

Exercise 3.4.2. Carry through the details of the algebra needed to obtain the posterior distribution (3.4.14) and its mean and variance (3.4.14) and (3.4.15).

Exercise 3.4.3. What prior assumptions should be made about β to show that the ridge estimator of β is a special case of the Bayes estimator? This will be taken up in Chapter V.

Exercise 3.4.4. Suppose that in the prior assumptions (3.4.10) are modified so that the γ_i have prior variance $\sigma_{\lambda_i}^2 = \omega^2 / \lambda_i$. Show that the contraction estimator is a special case of the Bayes estimator with $k = \sigma^2/\omega^2$. What prior assumptions should be made about β ?

Exercise 3.4.5. How should the prior assumptions be changed to obtain the generalized ridge estimator and the generalized contraction estimator as special cases of the Bayes estimator?

Exercise 3.4.6. A. Find optimum k_i for the generalized contraction estimator.
B. Show that if k is given using Lawless and Wang's equation with the parameter values replacing the estimators then for optimum k_i found in A

$$\frac{1}{k} = \frac{1}{m} \sum_{i=1}^{m} \frac{1}{k_i}$$

3.4.3 Hemmerle's Method of Estimating k_i for the Generalized Ridge Estimator

Recall that Hoerl and Kennard (1970a) obtained the optimum value of k_i (Equation 3.4.5) as a function of the population parameters by minimizing the MSE of the generalized ridge estimator. To estimate this optimum value they proposed an estimator by replacing the population parameters by unbiased estimators. They then suggested taking the LS as a starting value finding the ridge estimator and computing $\hat{k}_i$ by the formula $\hat{k}_i = \hat{\sigma}^2 / \hat{\gamma}_i^2$ where $\hat{\gamma}_i$ is the generalized ridge estimator. This value is used to form a new ridge estimator and the process is repeated until a satisfactory level of convergence is obtained. Hemmerle (1975) showed that under certain conditions it is possible to obtain an explicit formula . This sub-section obtains Hemmerle's result . The method employed is slightly different from his but uses the same basic ideas.

   If the sequence of ridge estimators described above converges to a given value, loosely speaking for a large enough value of j the j'th and j+1'st value of k_i should be almost equal. The limiting value, if it exists, can be thought of as a stationary value. Thus, it may be obtained by solving the equations for $1 \leq i \leq m$

$$\hat{k}_{ij} = \frac{\hat{\sigma}^2}{\hat{\gamma}_{ij}^2} = \frac{\hat{\sigma}^2}{\left(\dfrac{\lambda_i}{\lambda_i + k_{ij}} g_i\right)^2}. \tag{3.4.20}$$

Some algebra yields the equivalent quadratic equations

$$\hat{\sigma}^2 \hat{k}_{ij}^2 + (2\lambda_i \hat{\sigma}^2 - \lambda_i^2 g_j^2)\hat{k}_{ij} + \hat{\sigma}^2 \lambda_i^2 = 0. \tag{3.4.21}$$

Using the formula for solving a quadratic equation and some algebra to simplify the resulting expression the solutions to (3.4.21) and (3.4.20) are

$$\hat{k}_{ij} = \frac{\lambda_i}{2\hat{\sigma}^2}\left[(\lambda_i g_i^2 - 2\hat{\sigma}^2) \pm \sqrt{\lambda_i^2 g_i^4 - 4\lambda_i \hat{\sigma}^2 g_i^2}\right]. \tag{3.4.22}$$

Notice that $\hat{k}_{ij}$ is a real number iff

$$\lambda_i \geq \frac{4\hat{\sigma}^2}{g_i^2} = 4k_{i0}. \tag{3.4.23}$$

Hemmerle shows that the condition for the iterations of the ridge estimators to converge is given by the condition in (3.4.21). When the condition is not satisfied the values of $\hat{k}_{ij} \to \infty$ and $\hat{\gamma}_{ij} \to 0$. Which of the two solutions in (3.4.22) is valid? A basic condition that must be satisfied is that $\hat{k}_{ij} \geq 0$. It is not hard to show that the root with the negative sign in front of the radical in (3.4.22) is positive and that the other root is negative. Thus,

$$\hat{k}_{ij} = \frac{\lambda_i}{2\hat{\sigma}^2}\left[(\lambda_i g_i^2 - 2\hat{\sigma}^2) - \sqrt{\lambda_i^2 g_i^4 - 4\lambda_i \hat{\sigma}^2 g_i^2}\right] \tag{3.4.24}$$

The generalized ridge estimator may now be written after substitution of (3.4.24) for k_i into Equation (3.2.47) and some algebraic simplification.

$$\hat{\gamma}_i = \left(\frac{1}{2} + \sqrt{\frac{1}{4} - \frac{\hat{\sigma}^2}{\lambda_i g_i^2}}\right)g_i = \left(\frac{1}{2} + \sqrt{\frac{1}{4} - \frac{\hat{k}_{i0}}{\lambda_i}}\right)g_i$$

$$= \left(\frac{1}{2} + \sqrt{\frac{1}{4} - \frac{1}{F_i}}\right)g_i \tag{3.4.25}$$

where $\hat{k}_{i0}$ is the initial estimate of k_i and F_i is the usual F statistic for testing the linear hypothesis $\gamma_i = 0$.

Exercise 3.4.7. A. Obtain the roots in (3.4.24) as the solution to the algebraic equation (3.4.22).
B. Verify that the root with the positive sign in front of the radical is negative and that the other root is positive.
C. Verify Equation (3.4.25).
D. When $F_i < 4$ the LS of the component γ_i is recommended. For different values of n and s approximate the probability that this is the case.

3.4.4 Numerical Example of Calculating Objective Values of k

The examples below illustrate the computation of the objective values of k, i.e., values of k obtained by a mathematical formula or algorithm, using the data in the example of Section 1.3.

Example 3.4.1 Objective values of k for the ordinary ridge estimator

For the standardized data since there are ten observations and four parameters

$$\hat{\sigma}(k)^2 = \frac{1}{6}(Y_s - X_s b_s(k))'(Y_s - X_s b_s(k)) = \frac{1}{6} e(k)' e(k). \qquad (3.4.26)$$

The e(k) is the difference between the observed and the fitted value by the ridge regression estimator. The reader may show that

$$\hat{\sigma}^2(0) = .003932.$$

The LS is given by

$$b_s = \begin{bmatrix} 0.6402 \\ -0.1179 \\ 0.4733 \\ 0.0139 \end{bmatrix}.$$

In orthogonal coordinates

$$g_s = \begin{bmatrix} -0.5670 \\ -0.0342 \\ -0.4449 \\ 0.3569 \end{bmatrix}.$$

Using these values the objective estimates of k and the corresponding ridge estimators are given in Table 3.4.1 below.

Table 3.4.1

Method	k	$\hat{\gamma}_1$	$\hat{\gamma}_2$	$\hat{\gamma}_3$	$\hat{\gamma}_4$
1	0.0243	−0.5624	−0.0333	−0.3643	0.1620
2	0.0122	−0.5647	−0.0337	−0.4004	0.2225
3	0.0161	−0.5640	−0.0336	−0.3880	0.1986
4	0.0526	−0.5571	−0.0323	−0.3008	0.0990

Example 3.4.2 Objective values of k_i for the generalized ridge estimator (Hemmerle's method)

The initial value of the K matrix is

$$K = \begin{bmatrix} 0.0122 & 0 & 0 & 0 \\ 0 & 3.3709 & 0 & 0 \\ 0 & 0 & 0.0199 & 0 \\ 0 & 0 & 0 & 0.0309 \end{bmatrix}.$$

The corresponding generalized ridge estimator is

$$\hat{\gamma} = \begin{bmatrix} -0.5647 \\ -0.0073 \\ -0.3762 \\ 0.1411 \end{bmatrix}.$$

Using (3.4.22) choosing the formula with the negative sign Hemmerle's estimate of the K matrix is

$$K = \begin{bmatrix} 0.0123 & 0 & 0 & 0 \\ 0 & \infty & 0 & 0 \\ 0 & 0 & 0.0341 & 0 \\ 0 & 0 & 0 & \infty \end{bmatrix}.$$

Formula (3.4.22) gives complex values for k_2 and k_4. To illustrate the convergence Table 3.4.2a contains the first ten k values . Table 3.4.2 contains the first ten generalized ridge estimators. The generalized ridge estimator of Hemmerle is

$$\hat{\gamma} = \begin{bmatrix} -0.5647 \\ 0.0000 \\ -0.3393 \\ 0.0000 \end{bmatrix}.$$

For this particular example k_1, and k_3, converge. However k_2 and k_4 diverge to infinity; that is why the corresponding ridge estimate is zero.

Using the values obtained by Hemmerle's method and Equations (3.4.7) and (3.4.9) estimators of k can be obtained for the ordinary ridge regression estimator. Using (3.4.7) $\hat{k} = 0.0113$. Using (3.4.9) $\hat{k} = 0.0164$.

Exercise 3.4.8. Find the generalized ridge estimators of β using the estimators of Hoerl and Kennard and of Hemmerle for the data in Examples 3.4.1 and 4.4.2 above.

Table 3.4.2a[7]

	k_1	k_2	k_3	k_4
1	0.0122	3.371	0.0199	0.0309
2	0.0123.	74.32	0.0277	0.1974
3	0.0123	22929	0.0312	3.5834
4	0.0123	∞	0.0327	982.65
5	0.0123	∞	0.0335	∞
6	0.0123	∞	0.0338	∞
7	0.0123	∞	0.0340	∞
8	0.0123	∞	0.0340	∞
9	0.0123	∞	0.0341	∞
10	0.0123	∞	0.0341	∞

[7] The value of k_1 and k_3 become very large after a few iterations so loosely speaking they can be thought of as infinite for all practical purposes.

Table 3.4.2b[8]

	$\hat{\gamma}_1$	$\hat{\gamma}_2$	$\hat{\gamma}_3$	$\hat{\gamma}_4$
1	−0.5647	−0.0073	−0.3767	0.1411
2	−0.5647	−0.0004	−0.3552	0.0331
3	−0.5647	0	−0.3465	0.0020
4	−0.5647	0	−0.3426	0
5	−0.5647	0	−0.3409	0
6	−0.5647	0	−0.3401	0
7	−0.5647	0	−0.3397	0
8	−0.5647	0	−0.3395	0
9	−0.5647	0	−0.3394	0
10	−0.5647	0	−0.3394	0

Exercise 3.4.9. Find the ordinary ridge estimators of γ and β using the estimators of k based on (3.4.7) and (3.4.9).

Exercise 3.4.10. In the iteration procedure above $\hat{\sigma}(0)$ was used throughout the procedure. Compute the iterated estimates of k_i and γ_i using

$$\hat{\sigma}^2(K) = \frac{1}{6}(Y - XQ\hat{\gamma})'(Y - XQ\hat{\gamma})$$

where $\hat{\gamma}$ is the generalized ridge estimator using the value of K obtained from the last iteration.

Exercise 3.4.11. For the data of Exercise 1.3.2 find
A. The four estimators of k for ordinary ridge regression and the corresponding ridge estimators.
B. The estimators of k_i by the method of Hoerl and Kennard and Hemmerle and the corresponding generalized ridge regression estimators.
C. The estimators of k based on (3.4.7) and (3.4..9) and the corresponding ridge regression estimators.

---

[8] The values of γ_1 and γ_3 are denoted by zero once the first four digits after the decimal point are zero. They converge to zero very rapidly.

3.4.5 Computer Intensive Methods of Choosing k

There are several ways to choose k that require a fair amount of computation. They include:

1. Allen's Press (see Allen (1971,1974);
2. the generalized cross validation of Golub, Heath and Wahba (1979);
3. the Bootstrap (see Efron and Gong (1983)).

This section will describe these methods. A related technique the Jackknife will be discussed in Chapter V. The focus will be on the estimation of k; the inference problem for such estimators will not be considered.

Allen's Press and generalized cross validation of Golub, Heath and Wahba are based on cross validation. This is an old idea whose time has come again due to the advent of modern high speed computers. Suppose a regression model is to be fit to data but the researcher is unsure of the form of the model. The data set is randomly divided into two halves. The first half is used for model fitting. This may be done using hunches, preliminary testing, looking for patterns, trying large numbers of different models and rejecting outliers. The cross validation consists of using the fitted regression model together with the first half of the data to predict the second half. A modification of this used in Allen's Press and Golub et al. is to leave out one data point at a time, fit the model to the remaining points and see how well it predicts the excluded point each time.

PRESS stands for predicted residual sum of squares for selection of k. It is implemented by:

1. dropping one observation at a time, estimating the model with the missing observation and predicting the left out observation using the ridge estimator for each choice of k;
2. obtaining a mean square error of prediction (MSEP) for each choice of k;
3. choosing the value of k that minimizes the MSEP.

For this purpose let $e_1(k)$, $e_2(k)$,..., $e_n(k)$ be the residuals from ridge regression for a given k. These residuals are the difference between the observed value and the predicted value of the left out observation. Let

$$H(k) = X(X'X + kI)^{-1}X'.$$

(3.4.26)

The Press criterion is to find k that minimizes

$$CV(k) = \frac{1}{n}\sum_{i-1}^{n}\frac{e_i(k)}{[1-h_{ii}(k)]^2} \qquad (3.4.27)$$

where the $h_{ii}(k)$ are the diagonal elements of $H(k)$.

The generalized cross validation method of Golub, Heath and Wahba (1979) minimizes a weighted MSEP in a manner similar to that of Allen. Thus, the criterion is to find k that minimizes

$$GCV(k) = \frac{\frac{1}{n}\sum_{i=1}^{n}e_i(k)^2}{\left[\frac{1}{n}\sum_{i=1}^{n}(1-h_{ii}(k))\right]^2}. \qquad (3.4.28)$$

Golub, Heath and Wahba (1979) show that the expression in (3.4.28) is:

1. invariant under rotations;
2. is a weighted MSEP , i.e., a weighted version of Allen's Press;
3. does not require an estimate of σ^2.

The bootstrap was originally proposed by Efron (1979,1980). It is a way of generating more data points by a resampling scheme. The goal is to estimate some population parameter, e.g., the mean or the variance. Suppose a random sample of size n is taken for a particular population. From this sample choose n observations with replacement. This sample will contain observations that are duplicated and will not contain all of the observations. The sample obtained in this fashion is called the bootstrap sample. The bootstrap sample is then used to compute the value of the statistic that is being used to estimate the parameter in question. Suppose a large number of bootstrap samples are taken. The bootstrap estimate is the arithmetic average, the standard deviation or whatever aspect of the population desired computed from the values of the estimates obtained from each bootstrap sample.

The use of the bootstrap together with cross validation to obtain an optimal choice of k will now be explained following the work of Delaney and Chatterjee (1986). Suppose n observations on p independent variables and one dependent variable are available. Choose a bootstrap sample of size n from this random sample. Select G values of k. For each bootstrap sample and each of the G values of k selected obtain the ridge estimator. In choosing a bootstrap sample not all of the n observations are chosen. Use the ridge estimators obtained for the bootstrap samples to predict the observations that were not chosen. The prediction vector for the unchosen observations is $\hat{Y}_{K_j}(k_g)$. The subscript K_j is the number of unchosen observations when the j'th sample was selected. The subscript on k represents the g'th value of k that was used. Let Y_{K_j} be the vector of observed values of the response variable for the set of unchosen observations. When k $=k_g$ the MSEP for the j'th bootstrap sample is

$$MSEP_j(\theta_g) = (\hat{Y}_{K_j}(k_g) - Y_{K_j})'(\hat{Y}_{K_j}(k_g) - Y_{K_j})/K_j,$$
$$g = 1,2,\cdots,G. \tag{3.4.29}$$

Repeat the above procedure for a large number B of bootstrap samples. For each of the g selected values of k obtain a final weighted average for the MSEP where the weights are the number of unchosen observations in the bootstrap samples. This weighted average may be obtained from the formula

$$MSEP(k_g) = \frac{\sum_{j=1}^{B}(MSEP_j(k_g)K_j}{\sum_{j=1}^{B}K_j}, \quad g = 1,2,\cdots,G. \tag{3.4.30}$$

The bootstrap choice of the ridge parameter is k* where

$$k^* = \min\{MSEP(k_g), \quad g = 1,2,\cdots,G. \tag{3.4.31}$$

Golub, Heath and Wahba (1979) point out that a conservative estimate of the number of published estimates of k would be several dozen. The interested reader can consult any of the following references and also many others for further information. Some of the references contain estimators of k already discussed. Some references include Dempster (1973), Dempster, Schatzoff and Wermuth (1977), Farebrother (1975), Goldstein and Smith (1974), Hemmerle (1975), Hoerl and Kennard (1976), Hoerl, Kennard and Baldwin (1975), Hudson (1974), Lawless and Wang (1976), Lindley and Smith (1972), Marquardt and Snee (1975) , McDonald and Galarneau (1975), Obenchain (1975), Rolph (1976), Swindel (1976) , Thisted (1976) and many others. More recent articles are by Hadzivukovic et.al. (1992) and Vinod (1995).

Exercise 3.4.12. Divide the data of Example 1.3.1 into two subsets of five observations. Fit a linear regression model to one of the subsets and obtain the LS. Find the predicted Y values for the members of the other subset. Compute the MSEP and comment.

Exercise 3.4.13. For the data in Example 1.3.1 and Exercise 1.3.2 find the estimate of k
A. Using Allen's Press.
B. Using generalized cross-validation.
C. Using the bootstrap with 50 bootstrap samples.

Exercise 3.4.14. Show that (3.4.28) is invariant under orthogonal transformations.

3.4.6 The Double Bootstrap

The double bootstrap consists of:

1. obtaining a bootstrapped ridge regression estimator;
2. 'studentizing' this ridge estimator;
3. constructing the empirical distribution function of the studentized ridge estimator; thus obtaining a first stage cumulative distribution function;
4. compute pseudo y values using sampling with replacement from the ridge residuals;
5. obtaining a new ridge regression estimator and its empirical distribution function from these pseudo y values;
6. using the inverse at 1–α as the upper limit of a (1–α)% confidence interval.

Using appropriate mathematical notation the steps go as follows.

1. Obtain the ridge regression estimator

$$\hat{\beta} = (X'X + (m\hat{\sigma}^3 / b'b))^{-1}X'Y \qquad (3.4.32)$$

where the biasing parameter k is estimated using the estimate suggested by Hoerl, Kennard and Baldwin (1975). Define the ridge residuals by $\tilde{e} = Y - X\hat{\beta}$ and the Recentered ridge residuals by $e = \tilde{e} - \bar{e}$. Let J be a large number, for example J=1000. Draw J sets of bootstrapped samples of size T from the Recentered residuals. The elements of these samples are denoted by $e_{\bullet jt}$ (j = 1, 2,..., J and t =1, 2, ...,T). Thus, generate J sets of Tx1 vectors $e_{\bullet j}$ with elements $e_{\bullet jt}$ (t =1, 2, ...,T) and obtain the pseudo y data

$$y_{\bullet j} = X\hat{\beta} + e_{\bullet j}, \, j = 1,2,...J. \qquad (3.4.33)$$

Obtain ridge regression estimators $\hat{\beta}_{\bullet j}$ using the y values in (3.4.33) with the variance estimated by

$$\sigma_e^2 = \hat{\sigma}^2 \frac{(T-m)}{T}.$$ (3.4.34)

2. Define the 'studentized' ridge estimator by

$$\tau(\hat{\beta}_{\bullet j}) = (WX)^{-1}(\hat{\beta}_{\bullet j} - \hat{\beta})(SD_{\bullet}^{-1})$$ (3.4.35a)

where $W = (X'X + (\hat{\sigma}^2/b'b)I)^{-1}X'$ and $SD_{\bullet}$ is the diagonal matrix of standard deviations using the estimators of the variance

$$var(\hat{\beta}_{\bullet i}) = \frac{1}{J-1}\sum_{j=1}^{J}(\hat{\beta}_{\bullet ji} - \bar{\beta}_{\bullet i})^2$$ (3.4.35b)

with

$$\bar{\beta}_{\bullet i} = \frac{1}{J}\sum_{j=1}^{J}\hat{\beta}_{\bullet ji}.$$ (3.4.35c)

Appending an i means use the i'th component of the vector.

3. The first stage cumulative distribution function is constructed by defining $R_{\bullet j} = \tau(\hat{\beta}_{\bullet j})$ and $H_{\bullet}(w) = P(R_{\bullet j} < w)$ using the empirical distribution function.
4. Compute a T x 1 vector of pseudo y values using the equation

$$y_{\bullet\bullet k} = X\hat{\beta}_{\bullet j} + e_{\bullet jk} \quad k = 1,2,\ldots, K$$ (3.4.36)

where the $e_{\bullet jk}$ are obtained from the T x 1 vector of $e_{\bullet j}$ values by sampling with replacement.
5. Obtain the ridge regression estimators $\hat{\beta}_{\bullet\bullet jk}$ and $R_{\bullet\bullet jk} = \tau(b_{\bullet\bullet jk})$ for each j and k. This amounts to 1,000,000 estimates. Define

$$Z_j = \frac{\#(R_{\bullet\bullet jk} < R_{\bullet j})}{K}.$$

6. Use the empirical distribution function to estimate the CDF H_{**} of Z_j . The inverse evaluated at 1-α is the upper limit of a confidence interval.

3.5 Evaluation of the Efficiency of Ridge Estimators by Computer Simulation

So far a number of different ridge regression estimators have been presented, their mean square errors have been evaluated and compared with that of the LS. However the comparisons of the estimators efficiencies has always been for non-stochastic k. A number of different estimators of k have been mentioned. Up until now nothing has been said about the comparative merits of these estimators. It is very difficult to make analytical comparisons of the efficiencies of ridge estimators for non-stochastic k because the forms of the MSE are very cumbersome and difficult to derive. For this reason there have been a number of simulation studies done on the behavior and the comparative properties of the MSE of the ridge and the LS estimators. The purpose of this section is to summarize some of the results of some of the different simulation studies of ridge type estimators. Only the results of the studies will be presented. The interested reader may consult the referenced articles to see how the simulations were actually done.

In general a simulation study consists of:

1. specifying a large variety of values of the parameters to be studied, e.g., parameters of a regression model;
2. using some kind of random number generator to generate observations of a response variable for each value of the parameters , e.g., the mean of a distribution or the observation of the Y values in a linear regression model;
3. using these observations to calculate the estimator being studied, e.g., the sample mean or variance, the LS or a particular form of ridge estimator;
4. comparison of values of the estimators with values of the parameters that were used in generating the simulated observations often using some estimator of the measure of efficiency being evaluated, e.g., the MSE or MSEP.

Simulation studies are often useful in the study of ridge estimators for stochastic k because the probability distribution of these estimators for small samples is often very complicated or mathematically intractable. The disadvantage however is that they still are empirical studies whose conclusions may not be valid for parameter values that are not chosen. Many including the author would argue : "Example is not proof!"

Two kinds of comparisons will be considered. They are:

1. comparison of the MSE of ridge estimators and LS estimators;

2. comparison of different ridge type estimators with each other.

The second kind of comparison will give a handle on the worthiness of different estimators of k. The main results of a few of the many simulation studies will be presented here.

Hoerl, Kennard and Baldwin (1975) do a simulation study of the ridge estimator with biasing parameter $\hat{k} = m\hat{\sigma}^2 / b'b$,already considered in Section 3.4. (See Equation 3.4.1.) They made the following conclusions about this particular ridge regression estimator.

1. The ridge estimator has a probability greater than 0.5 of producing estimates whose MSE is smaller than LS.
2. As the dimension m of the factor space increases the probability that the ridge estimator has a smaller MSE than the LS increases.
3. For a given number of regression parameters m the probability that the ridge estimator dominates the LS increases as the spread in the eigenvalues of X′X increases; that is as the level of multicollinearity increases.
4. For a given X matrix and a given set of regression parameters β , the probability of the ridge estimator having a smaller MSE than the LS increases as the magnitude of the noise, that is measured by σ^2 increases.

They suggest that every linear regression program compute a set of regression coefficients and the associated information using the ridge estimator studied in their paper. Furthermore they recommend that every ridge trace should compute the point corresponding to the estimator of k that they study explicitly.

Hoerl and Kennard (1976) consider the iterative method of choosing k described in Section 3.4. together with a rule proposed by them for terminating the iteration. They recommend that the iteration process be terminated when

$$\frac{\hat{k}_{i+1} - k_i}{k_i} \le 20T^{-1.30} \tag{3.5.1}$$

where $\hat{k}_i$ is the i'th iterate and $T = \text{Trace } (X'X)^{-1}/p$. They do a simulation similar to that of Hoerl Kennard and Baldwin (1975). They make the following observations and conclusions.

1. A significant reduction can be made by using the criterion in (3.5.1) and making more than one iteration when the criterion calls for it.
2. The improvement in MSE increases as X′X becomes less well conditioned, i.e., as T increases.

3. More than one iteration is needed when there is a significant but relatively low signal to noise ratio[9] and when $X'X$ is moving toward poor conditioning.

4. The error distribution has a smaller standard deviation.

5. The criterion (3.5.1) is insensitive to an exact value.

6. The ridge estimator based on the criterion (3.5.1) has a probability greater than 0.5 of having a smaller MSE than the LS.

7. The fraction of the time the iterated estimator has a smaller MSE than the LS is smaller than the single iteration estimator, but this is offset by its MSE being smaller.

Lawless and Wang (1976) do a simulation study similar to that of Hoerl Kennard and Baldwin (1975) on six estimators. They include using orthogonal coordinates:

1. the ordinary ridge estimators with k estimated as in objective methods 2 and 3 in Section 3.4;

2. two principal component estimators in the general form

$$\hat{\gamma}_i = g_i \ \text{if} \ \left|\sqrt{\lambda_i}g_i/\hat{\sigma}\right| \geq C$$
$$0 \ \text{if} \ \left|\sqrt{\lambda_i}g_i/\hat{\sigma}\right| < C \qquad i = 1,2,...,p \quad ; \qquad (3.5.2)$$

where C=1 and C=$\sqrt{2}$.

3. the generalized ridge regression estimator of Hoerl and Kennard with the estimator of optimum k_i;

4. the first iterate of the generalized ridge estimator of Hoerl and Kennard.

They find that:

1. the ordinary ridge estimator with k recommended by Hoerl and Kennard performs very well but is outperformed by the ordinary ridge estimator with k recommended by Lawless and Wang;

2. although the principal component estimators can give an improvement over the LS they are relatively ineffective;

3. the generalized ridge estimators are better than the principal component estimators but not as good as the ordinary ridge estimators.

They conclude that:

1. ordinary ridge estimators do best with respect to MSE and MSEP;

2. it may not be worthwhile to consider generalized ridge estimators;

3. principle component estimators are not recommended as estimators with good MSE properties.

---

[9] The signal to noise ratio is (Bias)^2/Variance.

Mc Donald and Galarneau compare three estimators, the LS, the estimator given in (3.4.4) and (3.4.5) and a modified estimator where if $Q \leq 0$ $k = \infty$. They denote these estimators by R_1, R_2 and R_3 respectively. They also consider the "best possible ridge estimator", i.e., the one with k chosen so that the MSE is minimum to get a handle on the maximum, possible improvement. They consider the measure of efficiency

$$M = \text{MSE(Ridge Estimator)/MSE (LS)}. \qquad (3.5.3)$$

The simulation study shows that:

1. The evaluated and the potential performance of the ridge estimator depends on the variance of the random error, the correlations among the exploratory variables and the unknown coefficient vector.
2. Neither of the two ridge estimators considered are always better than the LS.
3. For the cases investigated for R_2 and R_3

$$0.33 \leq M \leq 1.34.$$

4. The potential reduction measured by the rule R_0 is between 0.001 and 0.875.

They believe that the potential reduction warrants further investigation of ridge type estimators.

Dempster, Schatzoff and Wermuth (1977) perform a simulation study of fifty seven kinds of estimators including the LS, various ridge and JS type estimators, principal component estimators and estimators obtained by different variable selection methods. They compare these estimators with respect to two criteria, the sum of squared errors of betas

$$SEB = (\hat{\beta} - \beta)'(\hat{\beta} - \beta)/\sigma^2 \qquad (3.5.4a)$$

and the sum of prediction errors

$$SPE = (\hat{\beta} - \beta)'X'X(\hat{\beta} - \beta)/\sigma^2. \qquad (3.5.4b)$$

They find that :

1. With respect to SEB the LS is worse than all other regression methods considered, in one case worse than just setting all the coefficients equal to zero.
2. The reductions in SEB, over the observed data sets achieved by some of the methods under study are on average as large as 90%.

3. The average reductions in SPE are at most 20-30% and the LS performs better than some of the other estimators studied with respect to this criterion.

4. The best performers are the various ridge type estimators and another type of estimator called RGEF not considered in this book. (see Dempster, Schatzoff and Wermuth (1977)).

Wichern and Churchill (1978) perform a simulation study of the ridge estimators for k defined by objective methods in (3.4.1)-(3.4.4). He attempts to compare these with estimators that one might obtain using the ridge trace. The estimates of k produced by the ridge trace are often inconsistent with those produced by objective methods. The estimates produced by the ridge trace are frequently larger. Recall that for the data of Example 1.3 the first estimate of k using the ridge trace was much larger than that obtained by objective methods. The four objective methods perform about the same and generally do better than the LS. The estimators obtained using the ridge trace do not do as well and frequently are poorer than the LS.

Gibbons (1981) does a simulation study of twelve estimators. These include the already considered LS, the ridge estimator of Mc Donald and Galarneau(1975), the ridge estimator as originally proposed by Hoerl (1962), the estimator of Hoerl, Kennard and Baldwin (1975), the iterative ridge estimator of Hoerl and Kennard (1976), the cross validation estimator of Golub, Heath and Wahba (1979) and the best possible ridge estimator. They also study the estimator of Lindley and Smith (1972), Obenchain (1975) and two estimators considered by Dempster , Shatzoff and Wermuth (1977).

The Lindley and Smith estimator is formulated as follows. They assume that the regression parameters have a prior distribution and use the BE , i.e., the mode of the posterior distribution. The estimator for β is

$$\hat{\beta} = (X'X + (s^2/s_\beta^2)I)^{-1}X'Y \qquad (3.5.5a)$$

with

$$s^2 = \frac{(Y - X\hat{\beta})'(Y - X\hat{\beta})}{n+2} \qquad (3.5.5b)$$

and

$$s_\beta^2 = \frac{\hat{\beta}'\hat{\beta}}{p+2}. \qquad (3.5.5c)$$

The initial estimate for $\hat{\beta}$ is b. The quantities s^2 and $s_\beta^{\ 2}$ can be determined in terms of b. Use an iteration procedure similar to that used on the Hoerl and Kennard estimator until the absolute value of the difference between successive values of $k = s^2 / s_\beta^2$ is less than 10^{-4}. If convergence is not obtained after a specified number of iterations the procedure defaults to least squares.

The estimator suggested by Obenchain (1975) takes the form

$$\hat{\beta} = [(X'X)^{-q+1} + kI]^{-1}(X'X)^{-q} X'Y \qquad (3.5.6a)$$

where q is chosen to minimize

$$C(q) = \frac{\sum_{i=1}^{m} |g_i| \lambda_i^{(q-q)/2}}{[(\sum_{i=1}^{m} g_i^2) \sum_{i=1}^{m} \lambda_i^{(1-q)}]^{1/2}}. \qquad (3.5.6b)$$

The parameter k is chosen to minimize

$$L = n \ln(2\pi e \sigma^{**}) + \xi'\xi - g'\xi/\sigma^{**} \qquad (3.5.6c)$$

where the parameters are defined as follows,

$$\xi_i = \text{sign}(g_i)[\delta_i /(1 - \delta_i)]^{1/2}, \ \delta_i = \lambda_i /(\lambda_i + k\lambda_i^q)$$
$$\text{and } \sigma^{**} = 2\{[(g'\xi)^2 + 4n]^{1/2} + g'\xi\}.$$

The two methods of Dempster, Shatzoff and Wermuth (1977) involve:

1. differentiating the MSE function for the ordinary ridge estimator, estimating σ^2 by s^2 and evaluating k by numerical methods (SRIG);
2. choose k so that

$$\sum_{i=1}^{m} \hat{\gamma}_i^2 /(\sigma_\beta^2 + s^2 /\lambda_i) = m \qquad (3.5.7)$$

where $\sigma_\beta^{\ 2} = s^2/k$.(RIDGM) (This was the best estimator in the Dempster et al. study.)

When comparing the performance of the estimators two values of the parameter β were considered. When the constraint $\|\beta\| = 1$ is applied the MSE is minimized when β is the normalized eigenvector corresponding to the maximum eigenvalue of

X'X. Likewise the MSE is maximized when β is the normalized eigenvector corresponding to the minimum eigenvalue of X'X. These eigenvectors, denoted by β_L and β_S respectively are the values considered. The main results that were obtained were:

1. All estimators are better than the LS if $\beta = \beta_L$.
2. No estimator is better than the LS if $\beta = \beta_S$.
3. Overall the estimators of Hoerl, Kennard and Baldwin, Golub, Heath and Wahba and RIDGM perform well.
4. No estimator is dominated by any other estimator throughout the study.
5. When the square of the correlation coefficient is high than 0.5 the iterative Hoerl and Kennard and Lindley and Smith estimators converge and have smaller MSE than the estimator of Hoerl, Kennard and Baldwin.
6. The estimator of Obenchain offers potential improvement over the one parameter ridge estimators.
7. None of the estimators track the optimum k.

Tamarkin (1982) examines the stochastic ridge K derived from the iteration procedure suggested by Hoerl and Kennard. They found that some of the k_i s were widely scattered and highly unstable contrary to what was expected. As a result of their simulations they do not believe that the iteration procedure is likely to lead to optimum k_i s.

The simulation study of Delaney and Chatterjee (1986) evaluates the bootstrap choice of the ridge parameter. The bootstrapped ridge estimator is compared with the ordinary least square estimator, Allen's PRESS, the Hoerl Kennard Baldwin estimator and the generalized cross validation ridge estimator. The findings were:

1. The results for the average MSE and MSEP were very similar.
2. The MSEs of all the estimators studied decreased as the signal to noise ratio increased.
3. The Hoerl Kennard Baldwin estimator was outperformed by all of the other estimators.
4. The condition number ,i.e., the ratio of the largest to the smallest eigenvalue of the X'X matrix, had a negligible impact on both the MSE and the MSEP for the bootstrap, the PRESS and the ordinary least square estimator.
5. The ordinary least square estimator did better than expected.

The following procedures were simulated by Hoerl, Schuenemeyer and Hoerl (1986):

1. least squares;
2. ridge regression using Lawless-Wang k;
3. the basic ridge estimator, i.e. , that with k of Hoerl, Kennard and Baldwin;
4. principal components;

5. the Efroymson stepwise procedure;

6. ridge selection using least squares with the final subset;

7. ridge selection using a basic ridge with the final subset;

8. ridge selection using a Lawless-Wang ridge using least squares with the final subset.

The principal component estimator is obtained by finding the LS in orthogonal coordinates and replacing the insignificant components with zero and transforming back to the original coordinates. A special case of a criterion described below may be used to evaluate whether a given component is significant or not.

Subset selection techniques attempt to find subsets that explain the response adequately. The Efroymson (1960) stepwise procedure is the most popular (See Draper and Smith (1981).). It starts with no variables in the equation and adds variables one at a time according to the partial F ratios until either all variables are included or no excluded variable's F ratio is significant. At each step the technique includes exclusion of variables whose F ratio in the full model has become insignificant.

The ridge selection method examines a full ridge solution and deletes the terms that are not significant. The significance test performed is a modified t test. The t statistic is given by

$$t = \frac{\hat{\beta}}{S_{ii}}$$

(3.5.8a)

where S_{ii} is the i'th diagonal element of the dispersion matrix of the ridge estimator

$$D = \sigma^2 (X'X + kI)^{-1} (X'X)(X'X + kI)^{-1}.$$

(3.5.8b)

When k=0 this reduces to a popular significance test to decide whether to eliminate components of principal component estimators.

The findings were:

1. For a large $r^2 = \gamma'\gamma$ the SEB and SPE is very large for the stepwise technique.

2. Principal components performed very poorly overall. It was rarely noticeably better than LS and uniformly poorer than the basic ridge estimator with respect to all criteria evaluated.

3. Despite poor performance of the principal components and the stepwise procedure it is not clear that biased estimation or subset selection is always better for either estimation or prediction.

4. Ridge selection using least squares with the final subset fared well when compared with the stepwise procedure indicating that this method shows promise.

5. All of the techniques performed better relative to LS when a larger number of regression parameters were being estimated.

6. Use of small α levels when performing the tests of hypothesis to decide whether to include or delete parameters in the model using the stepwise or ridge selection procedure proved to be dangerous. The ridge selection procedure performed poorly when compared to the LS.

7. Subset methods improved slightly when the sample size was increased.

8. For small r^2 the basic ridge estimator was dominated by that of Lawless and Wang. For large r^2 the ridge estimator of Lawless and Wang was sometimes inferior to that of the LS and the SEB of the basic ridge estimator tended to approach that of the LS from below.

9. When collinearity levels were increased all of the techniques improved relative to the LS.

Based on the results of their study the authors recommend ridge regression over principal components as a biased estimation procedure. They also believe that there is potential is using biased estimation to select subsets.

Exercise 3.5.1. For the data in Section 1.3:

A. Find the estimators of k and the ridge estimators of Lindley and Smith, and Obenchain.

B. Estimate k and find the corresponding estimators for SRIG and RIDGM.

3.6 Summary

In this chapter:

1. Different kinds of ridge estimators were derived. These included the ordinary and generalized ridge estimator of Hoerl and Kennard, the generalized ridge estimator of C.R. Rao and the contraction estimator of Mayer and Willke.

2. The MSE of these estimators was evaluated for non-stochastic k and compared with that of the LS. Necessary and sufficient conditions were derived for the various ridge estimators to have a smaller MSE than the LS.

3. Different estimates for the perturbation matrix K and scalar k were presented. These included those of Hoerl Kennard and Baldwin, Lawless and Wang, Mc Donald and Galarneau and Hemmerle.

4. Illustrations were given of the numerical computation of the ridge estimator, the objective estimates of K and k and the ridge trace.

5. The results of some of the important Monte Carlo simulations for the comparison of ridge type estimators were summarized.

Chapter IV

James-Stein Estimators for a Single Linear Model

4.0 Introduction

Many statistical models are concerned with data where a response (dependent) variable has a dependence on one or more control (independent) variables. The response variable may have only one aspect as is the case for data where a univariate model is appropriate or the response variable may have several aspects making the use of a multivariate model necessary. The data may be generated by an experiment that is repeated several times the same way necessitating the use of several models. This chapter will consider the case of a single univariate model. Chapter VII will consider r univariate models and the multivariate model will be considered in Chapter IX.

Chapter II considered the problem of estimating the mean of a multivariate normal distribution with known variance. This chapter considers the estimation of the parameters and their linear combinations for a single univariate regression model. The variability of the observations of the dependent variable will be unknown as is usually the case in practice.

Sometimes due to the real world situation being considered or the type of experimental design employed or study being conducted some of the independent variables are related to one another. Data where this occurs is said to be multicollinear. Recall that multicollinearity is present if either:

167

1. The X′X matrix is nearly singular with the smallest eigenvalue close to zero (near multicollinearity).

2. There is an exact linear relationship between several independent variables(exact multicollinearity).

In Chapter III the case of near mullticollinearity was considered. Often in data with near multicollinearity the is a high correlation with absolute value almost one between two independent variables. This Chapter will focus on the case of exact multicollinearity. For the case of exact multicollinearity the linear model is of less than full rank. One way of developing the least square estimator requires the use of generalized inverses. Section 4.1 develops least square estimation for a non-full rank linear model including a brief discussion of generalized inverses. Generalized inverses are discussed in greater detail in many other sources, e.g., Gruber (1990), Rao and Mitra (1971), Searle (1971) and Campbell and Meyer (1979).

The JS is developed as an empirical Bayes estimator (EBE) in Section 4.2 using the form of prior dispersion considered by Dempster (1973). The case of a nonzero prior mean is considered by proposing a JS with a Lindley mean correction (JSL). The JSL proposed here is slightly different than that of Vinod and Ullah (1981). Some numerical examples for the calculation of these estimators are given.

The development of the JS and the JSL as an approximate MMSE with respect to a predictive loss function is the subject of Section 4.3. The JS actually belongs to a larger class of generalized JS estimators proposed by Berger (1976a,b). The generalization is treated in the exercises.

Sections 4.4 and 4.5 consider the efficiencies of the JS estimators from the Bayesian and the non-Bayesian points of view. The measure of efficiency is the mean square error (MSE). In Chapter II two kinds of risk were evaluated:

1. the Bayes risk, i.e., the risk averaging over the prior;

2. the risk conditional on the estimated parameters, i.e., the risk without averaging over the prior.

In this chapter, by analogy, the MSE of the linear parametric functions of the parameters is evaluated with and without averaging over the prior. These two types of MSE are taken up in Sections 4.4 and 4.5 respectively.

The conditional MSE of the JS is smaller than that of the LS when averaged over the predictive loss function. This is not always true for other quadratic loss functions. The MSE properties of the JS averaging over different quadratic loss functions is dealt with in Section 4.6.

The JS is both an approximate BE and an approximate MMSE. Thus, it is an approximately optimal estimator from both the Bayesian and the non-Bayesian point of view. The relative savings loss (RSL) proposed by Efron and Morris (1973) is a measure of efficiency from the Bayesian point of view. The relative loss (RL) to be proposed is the corresponding measure of efficiency from the frequentist point of view. Both measures of the efficiency of the approximately optimal estimators as compared with the optimal estimators are studied in Section 4.7. Regions where the JS has a smaller MSE than the BE are obtained using the properties of the RL and an inequality contained in Rao and Shinozaki (1978).

4.1 Review of Basic Linear Model Concepts

Throughout this chapter a linear model of the form

$$Y = X\beta + \varepsilon \qquad (4.1.1)$$

will be considered. The n dimensional vector Y represents independent observations. The matrix X is n x m with rank $s \le m$, β is an m dimensional vector of parameters and ε is an n dimensional error vector. The least square estimator (LS) is the solution to the optimization problem of minimizing the sum of the squares between the observed values and the values that would be predicted by the least squares equation. The LS is obtained by minimizing

$$Q = (Y - X\beta)'(Y - X\beta) \qquad (4.1.2)$$

by setting the vector of partial derivatives $\partial Q/\partial \beta$ equal to zero and obtaining the normal equation

$$X'X\,b = X'Y \qquad (4.1.3)$$

The solution to the normal equation is

$$b = (X'X)^{-1}X'Y \qquad (4.1.4)$$

when X is of full rank and

$$b = GX'Y \qquad (4.1.5)$$

when X is of non-full rank with G a generalized inverse of X'X. The form of (4.1.5) provides the motivation for one way of defining a generalized inverse. Definitions 4.1.1 and 4.1.2 give two equivalent definitions of a generalized inverse.

Definition 4.1.1. A matrix G is a generalized inverse (GI) of a matrix A iff

$$x = Gy \qquad (4.1.6)$$

is a solution to

$$A x = y. \tag{4.1.7}$$

The more commonly used definition of a GI is:

Definition 4.1.2. A matrix G is a GI of a matrix A iff G satisfies the equation

(1) $AGA = A$ (4.1.8a)

The equivalence of Definitions 4.1.1 and 4.1.2 are proved in Gruber (1990) and also in Searle (1971). When A is of non-full rank Equation (4.1.7) can have infinitely many solutions and as a result A can have infinitely many generalized inverses. A special unique generalized inverse that will be used a lot in the sequel is the Moore Penrose inverse (MP).

Definition 4.1.3. A matrix M is a Moore Penrose inverse (MP) of A if A satisfies (4.1.8a) and in addition

(2) $MAM = M$ (4.1.8b)

(3) MA is a symmetric matrix;
(4) AM is a symmetric matrix.

A matrix can have infinitely many GIs but only one MP. (Exercise 4.1.3). Examples of GI and MP can be found for some specific linear models in Gruber (1990).

The singular value decomposition (SVD) (Definition 4.1.4) will be very important in the work that follows because it provides a helpful characterization of the MP and estimable parametric functions. It was used in Section 3..2 in the context of the full rank model. Now the formal definition will be given that works for both the full and the non-full rank model.

Definition 4.1.4. Let [S′ T′] and [U′ V′] be orthogonal matrices where S′ is of dimension n x s, T is of dimension nx (m-s), U is mx s and V is m x (m-s). Let Λ be the sxs matrix of eigenvalues of X'X where $\lambda_1 \geq \lambda_2, ..., \geq \lambda_s$. The SVD of

$$X = [S' \ T'] \begin{bmatrix} \Lambda^{\frac{1}{2}} & 0 \\ 0 & 0 \end{bmatrix} \begin{bmatrix} U' \\ V' \end{bmatrix} \tag{4.1.9}$$

When X has full rank V = 0. The reader may show that

$$X'X = U\Lambda U', \tag{4.1.10}$$

$$XX' = S'\Lambda S \tag{4.1.11}$$

and that

$$(X'X)^+ = U\Lambda^{-1}U'. \tag{4.1.12}$$

(See Exercises 4.1.1 and 4.1.2 below.)
From (4.1.6) the form of the LS is

$$b = (X'X)^+ X'Y \tag{4.1.13}$$

For the non-full rank case there are infinitely many solutions to the least squares problem because a nonsingular matrix has infinitely many generalized inverses. However, for linear parametric functions of the regression parameters that can be estimated unbiasedly by linear combinations of the observations the corresponding linear combinations of the LS are unique. Such linear parametric functions are said to be estimable. More formally:

Definition 4.1.5. A parametric function $p'\beta$ is estimable iff there exists a vector t where

$$p'\beta = E(t'Y) \tag{4.1.14}$$

or equivalently

$$p = Xt. \tag{4.1.15}$$

An equivalent characterization of estimability is to observe that p lies in the column space of X. This characterization can be formulated in terms of the eigenvectors of X'X (Definition 4.1.6).

Definition 4.1.6. A parametric function $p'\beta$ is estimable if there exists a vector d such that

$$p = Ud. \tag{4.1.16}$$

When dealing with the non-full rank model many results can be derived easily using a reparametization of the model to one of full rank. The following reparametization to a full rank orthogonal model will be used repeatedly in the sequel. It is similar to the reparametization of the full rank model to an orthogonal model that was used in Section 3.2. Since by (4.1.1) X=XUU' the non-full rank linear model (4.1.1) may be rewritten in the form

$$Y = XUU'\beta + \varepsilon. \tag{4.1.17}$$

Let $\gamma = U'\beta$. The reparametized model of full rank s is

$$Y = XU\gamma + \varepsilon. \tag{4.1.18}$$

Since $UX'XU' = \Lambda$ the LS of γ is

$$g = \Lambda^{-1}U'X'Y. \tag{4.1.19}$$

Let w_i, $1 \le i \le s$ be the components of $U'X'Y$. The components of g are

$$g_i = \frac{w_i}{\lambda_i} \tag{4.1.20}$$

More details and illustrative examples concerning the above ideas may be found in Gruber (1990). The derivation of the LS is similar to that of Section 3.2.

Exercise 4.1.1. Verify Equations (4.1.10), (4.1.11) and (4.1.12).

Exercise 4.1.2.
A. Show that $(X'X)^+$ as obtained in Equation (4.1.12) satisfies the axioms for a MP.
B. Show that the MP of $X'X$ is unique.

Exercise 4.1.3. Show that if G is a Gl of $X'X$ then

A. $U'GU = \Lambda^{-1}$

B. If $p'\beta$ is estimable then for the LS

$$p'GX'Y = p'(X'X)^+X'Y.$$

C. If $p'\beta$ is estimable then $p'b$ is independent of the choice of GI.

Exercise 4.1.4. Assume that for the reparametized model (4.1.18) $E(\varepsilon) = 0$ and $D(\varepsilon) = \sigma^2 I$. Find the variance covariance (dispersion) matrix of g in (3.1.20) and observe that the g_i are independent random variables for normally distributed error terms.

Some of the results of the above exercises are available in Gruber (1990).

4.2 The James-Stein Estimator from a Bayesian Point of View

The JS and the JSL will be formulated as a special case of the empirical Bayes estimator (EBE). Again the basic methodology will be to:

1. obtain the Bayes estimator BE;
2. substitute sample estimates of functions of unknown parameters into the expression for the BE to formulate the EBE.

A general formulation of the BE that will be used in this and subsequent sections of the text will now be given.

4.2.1. The Bayes Estimator (BE)

The linear BE was derived by C. R. Rao (1973) as the solution to the optimization problem of minimizing

$$v = \text{Var}(p'\beta - a - L'Y) \tag{4.2.1}$$

subject to the constraint that

$$E(p'\beta - a - L'Y) = 0 \tag{4.2.2}$$

Assuming the prior information that

$$E(\beta)=0 \text{ and } D(\beta)=F \tag{4.2.3}$$

and

$$E(\epsilon|\beta)=0 \quad \text{and } D(\epsilon|\beta) = \sigma^2 I \tag{4.2.4}$$

the linear BE is

$$p'\hat{\beta} = p'\theta + p'FX'(XFX' + \sigma^2 I)^{-1}(Y - X\theta) \tag{4.2.5}$$

An alternative form of (4.2.5) useful for the formulation of EBEs for estimable parametric functions is

$$p'\hat{\beta} = p'b - p'(X'X)^+ (UU'FUU' + \sigma^2 (X'X)^+)^+ (b - \theta)\sigma^2. \tag{4.2.6}$$

When the prior distribution and the sampling distribution is multivariate normal the BE and the linear BE are the same (see Kagan, Linnik and Rao (1973)). A detailed solution to the optimization problem and derivation of Equations (4.2.5) and (4.2.6) will be presented in Chapter V.

The derivation of the BE in this and succeeding chapters of the text will consist of using (4.2.5) or some other mathematically equivalent form and making the appropriate substitution for the prior mean and dispersion.

4.2.2 The JS

The JS is obtained by using the prior distribution suggested by Dempster (1973) It is also called a g prior see Zellner and Vanele(1974). The prior mean is assumed to be zero, i.e., $\theta = 0$, and the prior dispersion is assumed to be a constant multiple of the inverse or in the non-full rank case the MP of $X'X$, i.e., $F = \omega^2 (X'X)^+$. Thus, from (4.2.5) and (4.2.6)

$$p'\hat{\beta} = p'b - \frac{\sigma^2}{\sigma^2 + \omega^2} b. \tag{4.2.7}$$

The parameter σ^2 may be estimated by the unbiased estimator

$$\hat{\sigma}^2 = \frac{1}{n-s}(Y - Xb)'(Y - Xb), \qquad (4.2.8)$$

the error sum of squares divided by the number of degrees of freedom needed to estimate error. Also $d = \sigma^2 + \omega^2$ may be estimated by the uncorrected sum of squares due to regression divided by the number of regression parameters estimated, i.e.,

$$\hat{d} = \frac{1}{s} b'X'Xb \qquad (4.2.9)$$

The expression (4.2.9) is unbiased for d (Exercise 4.2.1). Thus, for some scalar c,

$$p'\hat{\beta} = \left(1 - \frac{cs\hat{\sigma}^2}{b'X'Xb}\right)p'b. \qquad (4.2.10a)$$

The scalar c will be chosen so that the MSE is minimized.

Exercise 4.2.1. Establish the unbiasedness of (4.2.9).

Exercise 4.2.2. Show that $\dfrac{(s-2)\hat{\sigma}^2}{b'X'Xb}$ is an unbiased estimator for $\dfrac{\sigma^2}{\sigma^2 + \omega^2}$. Thus, an alternative formulation of (4.2.10a) is

$$p'\hat{\beta} = \left(1 - \frac{c(s-2)}{b'X'Xb}\right)p'\beta \qquad (4.2.10b)$$

Exercise 4.2.3. Given the prior assumptions

$$E(\beta)=0$$

and

$$D(\beta) = (\sigma^2 + \omega^2)(X'X)^+\Sigma(X'X)^+ - \sigma^2(X'X)^+$$

where:

1. the dispersion F is positive semidefinite;

2. the matrix $U'\Sigma U$ is positive definite.

A. Show that the BE is

$$p'\hat{\beta} = p'b - \frac{p'M^+(X'X)b}{\sigma^2 + \omega^2}$$

where $M = UU'\Sigma UU'$.

B. Show that the EBE is

$$p'\hat{\beta} = p'b - \frac{c(s-2)\hat{\sigma}^2 p'M^+(X'X)b}{b'X'X\Sigma^+X'Xb}.$$
(4.2.11)

C. Show that if $\Sigma = X'X$ (4.2.11) reduces to the JS.

Estimator (4.2.11) was formulated by Berger (1976) in the context of estimating the mean of a MN. In the sequel it will be referred to as the generalized James-Stein estimator (GJS).

4.2.3 The JSL

As will be seen in Example 4.2.1 shrinkage toward the origin often does not produce an estimator which is "closer" to the true value of the parameter being estimated. A "closer" estimator is often produced by shrinkage toward a sample estimate of the prior mean. The JSL to be formulated in this section is one estimator that is a contraction of the LS and shrinks toward an estimate of the prior mean.

Consider the setup (4.1.1) together with the prior assumptions

$$E(\beta) = \theta(U\Lambda^{-\frac{1}{2}}1_s) \text{ and } D(\beta) = \omega^2(X'X)^+$$
(4.2.12a)

where θ is a constant. The BE of the estimable parametric functions is

$$p'\hat{\beta} = p'b - \frac{\sigma^2}{\sigma^2 + \omega^2}p'(b - \theta(U\Lambda^{-\frac{1}{2}}1_s)).$$
(4.2.12b)

Reparametize (4.1.1) to a full rank linear model (4.1.19). Now let Z=SY. Thus, the reparametized model (4.1.19) may be rewritten,

$$Z=SY= \alpha + S\varepsilon \qquad (4.2.13)$$

where $\alpha = \Lambda^{\frac{1}{2}}U'\beta$. Then $p' = d'U' = q'\Lambda^{\frac{1}{2}}U'$, i.e., $q' = d'\Lambda^{\frac{1}{2}}$. For the model (4.2.13) the LS of the parameters is

$$a = SY = \Lambda^{\frac{1}{2}}U'b. \qquad (4.2.14)$$

For the reparametized model (4.2.13) the prior assumptions (4.2.11) become

$$E(\alpha) = \theta 1_s \text{ and } D(\alpha) = \omega^2 I. \qquad (4.2.15)$$

Thus the BE of α is

$$q'\hat{\alpha} = q'a - \frac{\sigma^2}{\sigma^2 + \omega^2}q'(a - \theta 1_s) \qquad (4.2.16)$$

To formulate the EBE replace $e = \sigma^2/(\sigma^2 + \omega^2)$ by the unbiased estimator

$$\hat{e} = \frac{(s-3)\hat{\sigma}^2}{\sum_{i=1}^{s}(a_i - a_{\cdot})^2} \qquad (4.2.17)$$

where $a_{\cdot} = \sum_{i=1}^{s} a_i/s$ is an unbiased estimator of θ and the a_i are the coordinates of a. Thus, the JSL for α is

$$q'\hat{\alpha} = q'a - \frac{c(s-3)\hat{\sigma}^2}{\sum_{i=1}^{s}(a_i - a_{\cdot})^2}q'(a - a_{\cdot}1_s). \qquad (4.2.18)$$

Since $q' = d'\Lambda^{-\frac{1}{2}} = p'U'\Lambda^{-\frac{1}{2}}$ for estimable parametric functions let $\hat{\beta} = U\Lambda^{-\frac{1}{2}}\hat{\alpha}$. Now $U\Lambda^{-\frac{1}{2}}a = b$. Also $a_{\cdot} = \sum_{i=1}^{s}\lambda_i^{\frac{1}{2}}U_i'b/s$. Let $b_{\cdot} = (\sum_{i=1}^{s}\lambda_i^{\frac{1}{2}}(U_i'b)U\Lambda^{-\frac{1}{2}}1_s)/s$. Thus,

$$p'\hat{\beta} = p'b - \frac{c(s-3)\hat{\sigma}^2 p'(b - b_{\cdot})}{(b - b_{\cdot})'X'X(b - b_{\cdot})}. \qquad (4.2.19)$$

For the full rank case Vinod and Ullah (1981) derive an estimator that is similar to (4.2.19) using the prior assumptions

$$E(\beta) = \theta 1_m \quad \text{and} \quad D(\beta) = \omega^2 (X'X)^{-1} \tag{4.2.20}$$

Their estimator does not admit an exact MSE in closed form. The estimator in (4.2.19) corresponds to the JSL for the mean of a multivariate normal distribution with unknown variance σ^2. (See Equation 2.4.28.) The risk of the estimator may be obtained using Exercise 2.6.7.

Example 4.2.1.   Illustration of Transformations Used to Obtain the JSL.

Let

$$
\begin{bmatrix} Y_1 \\ Y_2 \\ Y_3 \\ Y_4 \end{bmatrix} =
\begin{bmatrix}
\frac{4}{\sqrt{2}} & -\frac{4}{\sqrt{2}} & 0 & 0 & 0 \\
\frac{3}{\sqrt{6}} & \frac{3}{\sqrt{6}} & -\frac{6}{\sqrt{6}} & 0 & 0 \\
\frac{2}{\sqrt{12}} & \frac{2}{\sqrt{12}} & \frac{2}{\sqrt{12}} & -\frac{6}{\sqrt{12}} & 0 \\
\frac{1}{\sqrt{20}} & \frac{1}{\sqrt{20}} & \frac{1}{\sqrt{20}} & \frac{1}{\sqrt{20}} & -\frac{4}{\sqrt{20}}
\end{bmatrix}
\begin{bmatrix} \beta_1 \\ \beta_2 \\ \beta_3 \\ \beta_4 \end{bmatrix} + \varepsilon
$$

Then

$$
X'X = U \Lambda U' =
\begin{bmatrix}
\frac{593}{60} & -\frac{367}{60} & -\frac{157}{60} & -\frac{19}{20} & -\frac{1}{5} \\
-\frac{367}{60} & \frac{593}{60} & -\frac{157}{760} & -\frac{19}{20} & -\frac{1}{5} \\
-\frac{157}{60} & -\frac{157}{60} & \frac{383}{60} & -\frac{19}{20} & -\frac{1}{5} \\
-\frac{19}{20} & -\frac{19}{20} & -\frac{19}{20} & \frac{61}{20} & -\frac{1}{5} \\
-\frac{1}{5} & -\frac{1}{5} & -\frac{1}{5} & -\frac{1}{5} & \frac{4}{5}
\end{bmatrix}
$$

where

$$
U' =
\begin{bmatrix}
\frac{1}{\sqrt{2}} & -\frac{1}{\sqrt{2}} & 0 & 0 & 0 \\
\frac{1}{\sqrt{6}} & \frac{1}{\sqrt{6}} & -\frac{2}{\sqrt{6}} & 0 & 0 \\
\frac{1}{\sqrt{12}} & \frac{1}{\sqrt{12}} & \frac{1}{\sqrt{12}} & -\frac{3}{\sqrt{12}} & 0 \\
\frac{1}{\sqrt{20}} & \frac{1}{\sqrt{20}} & \frac{1}{\sqrt{20}} & \frac{1}{\sqrt{20}} & -\frac{4}{\sqrt{20}}
\end{bmatrix}
$$

and

$$\Lambda = \begin{bmatrix} 16 & 0 & 0 & 0 \\ 0 & 9 & 0 & 0 \\ 0 & 0 & 4 & 0 \\ 0 & 0 & 0 & 1 \end{bmatrix}.$$

Thus, in the formulation of the JS the transformations of the LS estimators is given by

$$\begin{bmatrix} a_1 \\ a_2 \\ a_3 \\ a_4 \end{bmatrix} = \begin{bmatrix} 4 & 0 & 0 & 0 \\ 0 & 3 & 0 & 0 \\ 0 & 0 & 2 & 0 \\ 0 & 0 & 0 & 1 \end{bmatrix} \begin{bmatrix} \frac{1}{\sqrt{2}} & -\frac{1}{\sqrt{2}} & 0 & 0 & 0 \\ \frac{1}{\sqrt{6}} & \frac{1}{\sqrt{6}} & -\frac{2}{\sqrt{6}} & 0 & 0 \\ \frac{1}{\sqrt{12}} & \frac{1}{\sqrt{12}} & \frac{1}{\sqrt{12}} & -\frac{3}{\sqrt{12}} & 0 \\ \frac{1}{\sqrt{20}} & \frac{1}{\sqrt{20}} & \frac{1}{\sqrt{20}} & \frac{1}{\sqrt{20}} & -\frac{4}{\sqrt{20}} \end{bmatrix} \begin{bmatrix} b_1 \\ b_2 \\ b_3 \\ b_4 \\ b_5 \end{bmatrix}$$

Thus,

$$a_1 = \frac{4}{\sqrt{2}}(b_1 - b_2), \ a_2 = \frac{3}{\sqrt{6}}(b_1 + b_2 - 2b_3),$$

$$a_3 = \frac{2}{\sqrt{12}}(b_1 + b_2 + b_3 - 3b_4)$$

and

$$a_4 = \frac{1}{\sqrt{20}}(b_1 + b_2 + b_3 + b_4 - 4b_5).$$

Also notice in this case $a_1 = Y_1$, $a_2 = Y_2$, $a_3 = Y_3$ and $a_4 = Y_4$. Furthermore

$$a_. = \frac{1}{\sqrt{2}}(b_1 - b_2) + \frac{3}{4\sqrt{6}}(b_1 + b_2 - 2b_3) + \frac{1}{2\sqrt{2}}(b_1 + b_2 + b_3 - 3b_4)$$

$$+ \frac{(b_1 + b_2 + b_3 + b_4 - 4b_5)}{4\sqrt{20}}$$

and

$$
b_* = \begin{bmatrix}
\frac{1}{\sqrt{2}} & \frac{1}{\sqrt{6}} & \frac{1}{\sqrt{12}} & \frac{1}{\sqrt{20}} \\
-\frac{1}{\sqrt{2}} & \frac{1}{\sqrt{6}} & \frac{1}{\sqrt{12}} & \frac{1}{\sqrt{20}} \\
0 & -\frac{2}{\sqrt{6}} & \frac{1}{\sqrt{12}} & \frac{1}{\sqrt{20}} \\
0 & 0 & -\frac{3}{\sqrt{12}} & \frac{1}{\sqrt{20}} \\
0 & 0 & 0 & -\frac{4}{\sqrt{20}}
\end{bmatrix}
\begin{bmatrix}
\frac{1}{4} a_* \\
\frac{1}{3} a_* \\
\frac{1}{2} a_* \\
a_*
\end{bmatrix}
$$

Example 4.2.2. A standardized mathematics examination has a mean score of 75 with a standard deviation of 10. During six different years a sample of size six was taken of the grades.

1	2	3	4	5	6
62	77	67	75	76	65
95	67	60	76	70	65
73	76	82	91	51	80
89	53	74	66	80	73
68	96	77	82	66	78
53	80	88	87	60	74

To obtain the least square estimators, the between and within sum of squares so that the numerical values of the JS and JSL could be obtained an Analysis of Variance (ANOVA) was performed using Minitab. The ANOVA Table is given below.

Table 4.2.1
ANALYSIS OF VARIANCE

Source	DF	SS	MS	F	p
Factor	5	509	102	0.74	0.601
Error	30	4137	138		
Total	35	4646			

Treatment Means

Year	1	2	3	4	5	6
	73.33	74.83	74.67	79.50	67.17	76.67

For the linear model[1]

$$Y = (I_6 \otimes 1_6)\beta + \varepsilon$$

the treatment means are the LS estimates. Thus $\alpha_i = \sqrt{6}\beta_i$, $1 \le i \le 6$. Also

$$b'X'Xb = \text{FactorSS} + \frac{(\text{Grand Mean})^2}{36} = 509 + 199,068 = 199,577.$$

The JS is

$$\begin{bmatrix} \hat{\beta}_1 \\ \hat{\beta}_2 \\ \hat{\beta}_3 \\ \hat{\beta}_4 \\ \hat{\beta}_5 \\ \hat{\beta}_6 \end{bmatrix} = \begin{bmatrix} 73.14 \\ 74.64 \\ 74.47 \\ 79.29 \\ 67.00 \\ 76.47 \end{bmatrix}$$

The contraction factor for the JS is

$$1 - \frac{(30)(4)(138)}{(32)(199,577)} = .9974$$

The contraction factor for the JSL is

$$1 - \frac{(30)(3)(138)}{(32)(509)} = .2375.$$

Now $\overline{Y}.. = 74.35$. The JSL is

---

[1] The matrix $A \otimes B$ is the matrix with blocks $a_{ij}B$. Thus, $I_6 \otimes 1_6$ is 36x6 matrix consisting of blocks of 6x1 matrices. The blocks on the diagonal are 1_6. The off diagonal blocks are zero. Extensive use of this notation will be made in Chapter IX.

$$
\begin{bmatrix} \hat{\beta}_1 \\ \hat{\beta}_2 \\ \hat{\beta}_3 \\ \hat{\beta}_4 \\ \hat{\beta}_5 \\ \hat{\beta}_6 \end{bmatrix} = \begin{bmatrix} 74.12 \\ 74.47 \\ 74.43 \\ 75.58 \\ 72.65 \\ 74.91 \end{bmatrix}.
$$

Obtaining the ``distance'' of the estimator with the prior value of 75, i.e., calculating $(\hat{\beta} - 75)X'X(\hat{\beta} - 75)$ for the LS, JS and JSL, the values 87.27, 88.44 and 7.25 are obtained. Thus, of the three types of estimators the JSL is the closest to the true value.

Example 4.2.3. The JS and the JSL. The data used (see Table 4.2.2) is taken from Andrews (1985) with permission from Springer Verlag. It is concerned with insurance availability in Chicago. The dependent variable is market activity(MA), the independent variables include racial composition (RC), fire rates (FR), theft rates (TR), average age (AA) and income (I).

Table 4.4.2

RC	FR	TR	AA	I	MA
10.0	6.2	29	60.4	11744	5.3
22.2	9.5	44	76.5	9323	3.3
19.6	10.5	36	73.5	9948	6.0
17.3	7.7	37	66.9	10656	6.2
24.5	8.6	53	81.4	9730	6.6
54.0	34.1	68	52.6	8231	4.3
4.9	11.0	75	42.6	21480	7.9
7.1	6.9	18	78.5	11104	6.9
5.3	7.3	31	90.1	10694	8.0
21.5	15.1	25	89.8	9631	4.2

The JS and the LJS will be calculated using the data in Table 4.2.2. The LS estimators are

$$b = \begin{bmatrix} 9.8851 \\ -0.2536 \\ 0.1596 \\ 0.0865 \\ -.00023 \\ -0.0004 \end{bmatrix}$$

The ANOVA table is given in Table 4.2.3

Table 4.2.3

ANOVA

Source	DF	SS	MS	F	p
Model	5	13.435	2.678	1.178	.4496
Error	4	9.126	2.281		
Total	9	22.561			

For the SVD the diagonal elements of $\Lambda^{\frac{1}{2}}$ are 3304.773, 107.363, 69.340, 19.688, 7.108, and .146. Furthermore $U'=$

$$\begin{bmatrix} -0.000081 & -0.001318 & -0.000897 & -0.003565 & -0.005538 & -1.000000 \\ 0.008603 & 0.354500 & 0.149100 & 0.140200 & 0.912300 & -0.006154 \\ -0.000625 & -0.537900 & -0.310200 & -0.693100 & 0.366200 & 0.001431 \\ 0.004131 & 0.454500 & 0.548300 & -0.683300 & -0.161200 & 0.002240 \\ -0.999800 & 0.016030 & -0.009603 & -0.004370 & 0.005439 & 0.000054 \\ -0.017510 & -0.615000 & 0.762200 & 0.182000 & 0.086620 & -0.001000 \end{bmatrix}$$

Now $a = U'\Lambda^{\frac{1}{2}} b$. Thus,

$$a = \begin{bmatrix} -19.458 \\ 3.109 \\ 1.381 \\ -0.898 \\ 0.854 \\ 1.446 \end{bmatrix}$$

and $\sum_{i=1}^{6}(a_i - a_\bullet)^2 = 348.7$. The contraction factor for the JSL is

$$1 - \frac{4(3)(228)}{6(348.669)} = .9869$$

and the JSL for the α parameters is

$$\hat{\alpha} = \begin{bmatrix} -19.239 \\ 3.032 \\ 1.327 \\ -0.922 \\ 0.807 \\ -1.463 \end{bmatrix} \text{ with } a_\bullet = -2.743.$$

Furthermore for the β parameters the JSL is

$$\hat{\beta}_L = \begin{bmatrix} 0.000039 \\ -0.005141 \\ -0.002608 \\ 0.018173 \\ 0.472180 \\ 131.4290 \end{bmatrix}$$

For the JS the contraction factor is .9930 and the resulting JS for the β parameters is

$$\hat{\beta}_J = \begin{bmatrix} 9.7170 \\ -0.2492 \\ 0.1569 \\ 0.0849 \\ -0.0023 \\ 0.0004 \end{bmatrix}.$$

Exercise 4.2.5. Given the additional data on insurance availability in Chicago below:
A. Find the LS estimators.
B. Find the JS and the JSL.
C. Assume the LS estimators in Example 3.2.3 are prior values of the parameters. Compute for the LS, JS and JSL. Which appear to do better?

	RC	FR	TR	AA	I	MA
1	50.2	39.7	147	83.0	7459	6.1
2	74.2	18.5	22	78.3	8014	3.7
3	55.5	23.3	29	79.0	8177	3.6
4	62.3	12.2	46	48.0	8212	4.0
5	4.4	5.6	23	71.5	11230	8.3
6	46.2	21.8	4	73.1	8330	3.9
7	99.7	21.6	31	65.0	5583	1.4
8	73.5	9.0	39	75.4	8564	3.1
9	10.7	3.6	15	20.8	12102	9.1
10	1.5	5.0	32	61.8	11876	11.6
11	48.8	28.6	27	78.1	9742	5.4
12	98.9	17.4	32	68.6	7520	1.7

Exercise 4.2.6. The error in measurement of seven measuring instruments is being compared. Ten measurements are taken and compared with the known true value. The error in thousands of an inch are given below.

1	2	3	4	5	6	7
0.0298	0.1565	-1.4962	1.8497	-0.2670	-1.0174	10.2973
-0.5894	1.1051	0.3672	-1.1131	0.1053	-0.9088	10.3112
-2.4009	-1.4055	-0.9690	0.2644	-1.1245	1.6557	9.0634
-0.4714	-0.3030	1.4163	-1.1481	1.8296	0.8061	10.3311
-0.1671	0.2041	-0.6965	0.5178	-1.6531	0.8248	10.3928
-0.9725	0.5277	0.0544	-0.8095	-0.1533	-1.6045	10.2999
0.9408	-0.0085	0.5932	1.8524	-0.5950	-0.8681	9.8575
0.3750	0.9727	0.2202	0.8192	0.2461	-0.2122	6.2198
0.9366	1.0680	1.1418	1.1218	-0.6478	-2.8220	8.8527
0.2076	1.1350	0.2805	1.0510	-0.0861	0.8585	9.7122

A. Perform and ANOVA on

1. Instruments 1-6.
2. All seven instruments.
B. For each ANOVA find the LS, JS and JSL for $\mu + \tau_i$ i.e., the basic estimable parametric functions. Compare the results with a prior mean of zero for instruments 1-6 and a prior mean of ten thousands of an inch for instrument 7.
C. Find the JS and JSL for
1. The contrast

$$\tau_1 + \tau_2 + \tau_3 + \tau_4 + \tau_5 - 5\tau_6 \quad \text{in ANOVA 1}$$

2. The contrast

$$\tau_1 + \tau_2 + \tau_3 + \tau_4 + \tau_5 + \tau_6 - 6\tau_7 \quad \text{in ANOVA 2}$$

4.3   The Non-Bayesian Formulation of the JS and the JSL

4.3.1 The JS

The JS will now be obtained as an approximate MMSE. First the scalar multiple of the LS with the smallest MSE of prediction is found, i.e., the MMSE. The minimizing scalar is a fraction with numerator and denominator that are functions of population parameters. The approximate MMSE is the MMSE with unbiased sample estimates in place of the functions of the population parameters in the scalar multiple of the JS. The mathematical development follows.
Consider an estimator of the form

$$p'\beta = hp'b \tag{4.3.1}$$

h is such that

$$m = E_\beta (\hat{\beta} - \beta)'X'X(\hat{\beta} - \beta) \tag{4.3.2}$$

is minimized. The quadratic loss function (4.3.2) is called the predictive loss function because if $\hat{\beta}$ is an estimator of β then $X\hat{\beta}$ is the predicted value of Y. Equation (4.3.2) is the mean square error of prediction. Observe that the MSE in (4.3.2) is obtained conditional on β i.e., without averaging over the prior distribution.

The objective is to obtain the scalar h that minimizes (4.3.2). Notice that

$$m = h^2 E(b'X'Xb) - 2hE(b'X'X\beta) + \beta'X'X\beta. \qquad (4.3.3)$$

By differentiation

$$\frac{dm}{dh} = 2hE(b'X'Xb) - 2E(b'X'Xb) = 0. \qquad (4.3.4)$$

Thus,

$$h = \frac{E(b'X'X\beta)}{E(b'X'Xb)} = \frac{\beta'X'X\beta}{\beta'X'X\beta + \sigma^2 s} = (1 - \frac{s\sigma^2}{\beta'X'X\beta + \sigma^2 s}). \qquad (4.3.5)$$

By substitution of (4.3.5) into (4.3.1) the MMSE is

$$p'\hat{\beta} = (1 - \frac{s\sigma^2}{s\sigma^2 + \beta'X'X\beta})b \qquad (4.3.6)$$

Of course if β were known there would be no point in estimating it. Denote b'X'Xb by $\hat{f}$ Notice that

$$E(\hat{f}) = E(b'X'Xb) = \beta'X'X\beta + s\sigma^2 = f \qquad (4.3.7)$$

i.e., that $\hat{f}$ is an unbiased estimator of f. Furthermore

$$\hat{\sigma}^2 = \frac{1}{(n-s)}(Y'Y - Y'Xb) \qquad (4.3.8)$$

is an unbiased estimator of σ^2. Replace σ^2 and f by their unbiased estimators. Thus,

$$p'\hat{\beta} = (1 - \frac{cs\hat{\sigma}^2}{b'X'Xb})b, \qquad (4.3.9)$$

the JS is obtained as an approximate MMSE.

4.3.2 The JSL

A derivation along similar lines to that of Section 4.2.1 may be used to show that
the JSL is also an approximate MMSE. Consider an estimator of the form

$$p'\hat{\beta} = p'b_\bullet + (1-h)p'(b-b_\bullet) \tag{4.3.10}$$

where $b_\bullet = a_\bullet U\Lambda^{-\frac{1}{2}}$ and $a_\bullet = \sum_{i=1}^s \lambda_i^{\frac{1}{2}} U_i b/s$. Observe that $a_\bullet$ is a scalar. Again the
goal is to minimize m in (4.3.2) by finding an optimum h. Observe that the
optimization problem is equivalent to that of finding an estimator of the form

$$\hat{\alpha} = a - h(a - a_\bullet) \tag{4.3.11}$$

that minimizes

$$
\begin{aligned}
m &= E(\hat{\alpha}-\alpha)'(\hat{\alpha}-\alpha) = E_\beta(a-\alpha)'(a-\alpha) \\
&\quad -2hE_\beta(a-\alpha)'(a-a_\bullet 1_s) + h^2 E_\beta(a-a_\bullet 1_s)'(a-a_\bullet 1_s)'.
\end{aligned} \tag{4.3.12}
$$

Obtaining the expectations in (4.3.12) yields

$$m = s\sigma^2 - 2h(s-1)\sigma^2 + h^2[(s-1)\sigma^2 + (\alpha - \overline{\alpha}1_s)'(\alpha - \overline{\alpha}1_s)] \tag{4.3.13}$$

where

$$\overline{\alpha} = \frac{1}{s}\sum_{i=1}^s \alpha_i .$$

Differentiating m, setting the result equal to zero and solving the resulting equation
for h,

$$
\begin{aligned}
h &= \frac{(s-1)\sigma^2}{(s-1)\sigma^2 + (\alpha - \overline{\alpha}1_s)'(\alpha - \overline{\alpha}1_s)} \\
&= \frac{(s-1)\sigma^2}{(s-1)\sigma^2 + (\beta - \overline{\beta})'X'X(\beta - \overline{\beta})}
\end{aligned}
$$

$$= \frac{(s-1)\sigma^2}{T} \qquad (4.3.14)$$

where

$$\bar{\beta} = U\Lambda^{-\frac{1}{2}}\bar{\alpha}.$$

The MMSE is

$$p'\hat{\beta}_i = p'b_i - \frac{(s-1)\sigma^2}{T}p'(b_i - b). \qquad (4.3.15)$$

Now an unbiased estimator of T is

$$\hat{T} = (b - b_*)'X'X(b - b_*). \qquad (4.3.16)$$

Thus, the AMMSE is

$$p'\hat{\beta}_i = p'b_i - \frac{c(s-1)\sigma_*^2}{(b - b_*)'X'X(b - b_*)}p'(b - b_*). \qquad (4.3.17)$$

Exercise 4.3.1.
A. Find the estimator of the form

$$p'\hat{\beta} = p'(I - q\Sigma^+(X'X))b \qquad (4.3.18)$$

that minimizes

$$L = E(\hat{\beta} - \beta)'\Sigma(\hat{\beta} - \beta) \qquad (4.3.19)$$

assuming $\Sigma = UHU' + UU'$ where H is any positive definite matrix.
B. Use the methodology of this section to find an approximate MMSE. Compare the result to the solution of Exercise (4.2.3).
C. Give the form of the estimator obtained for $\Sigma = X'X$ and $\Sigma = I$.

4.4 The Average Mean Square Error (MSE)

The goal of this section is to obtain the MSE of the JS averaging over Dempster's prior. The main result is contained in Theorem 4.4.1 below.

Theorem 4.4.1. Assume that the distribution of β and ε conditional on β is multivariate normal. If $s>2$ then the estimable parametric functions of the JS (4.2.10) has average MSE for optimal $c = \dfrac{(n-s)(s-2)}{(n-s+2)s}$

$$m = p'E(\hat{\beta}-\beta)(\hat{\beta}-\beta)' = p'(X'X)^+p\sigma^2$$
$$-\frac{(n-s)(s-2)}{(n-s+2)s}\frac{\sigma^4}{(\sigma^2+\omega^2)}p'(X'X)^+p. \tag{4.4.1}$$

The proof of the theorem requires the following simple lemmas.

Lemma 4.4.1. Let v_1, v_2,...,v_s be independent identically distributed random variables. Then

$$E\left[\frac{v_i^2}{[v'v]^a}\right] = \frac{1}{s}E\left[\frac{1}{[v'v]^{a-1}}\right] \tag{4.4.2}$$

Proof. Observe that

$$E\left[\frac{1}{[v'v]^{a-1}}\right] = E\left[\frac{v'v}{[v'v]^a}\right] = \sum_{i=1}^{s}E\left[\frac{v_i^2}{[v'v]^a}\right] = sE\left[\frac{v_i^2}{[v'v]^a}\right]. \tag{4.4.3}$$

Divide Equation (4.3.3) through by s to obtain (4.4.2). From (4.4.3) when a=1

$$E\left[\frac{v_i^2}{v'v}\right] = \frac{1}{s}. \tag{4.4.4}$$

When $v_i \sim N(0,1)$ and a=2 it follows from (2.7.11) that

$$E\left[\frac{v_i^2}{[v'v]^2}\right] = \frac{1}{s(s-2)}. \tag{4.4.5}$$

Lemma 4.4.2. Let v_1, v_2, ...,v_s be independent identically distributed random variables. Then

$$E\left[\frac{v_i v_j}{[v'v]^m}\right] = 0 \text{ if } i \neq j \tag{4.4.6}$$

Proof. Observe that

$$I = E\left[\frac{v_i v_j}{[v'v]^m}\right] = \frac{1}{(2\pi)^{s/2}} \int_R \frac{v_i v_j}{[v'v]^2} \exp[-\frac{1}{2}v'v]dv \tag{4.4.7}$$

where I is an s dimensional integral over the real numbers. Rearrange the order of integration so that v_i and v_j are integrated out first. Now the partial integral

$$\int_{-\infty}^{\infty}\int_{-\infty}^{\infty} \frac{v_i v_j}{(v'v)^2} \exp[\frac{1}{2}v'v]dv_i dv_j = 0 \tag{4.4.8}$$

because $(v_i/(v'v)^2)\exp[-v'v/2]$ is an odd function of v_i.

Proof of Theorem 4.4.1. The strategy of the proof is:

1. reparametize the non-full rank model to one of full rank as was done in Section 4.1;
2. calculate the needed expectations for functions of the LS estimators for the full rank model;
3. obtain the MSE for the reparametized full rank model;
4. use the definition of estimable functions and the SVD of the design matrix and its MP to obtain the expression for the non-full rank model.

Since $p'\beta$ is estimable let $\hat{\gamma} = U'\hat{\beta}$. Then

$$m = d'E(\hat{\gamma} - \gamma)(\hat{\gamma} - \gamma)'d = d'E(g - \gamma)(g - \gamma)'d$$

$$-cs\sigma^2 d'E\left[\frac{g(g-\gamma)'}{g'\Lambda g}\right]d - cs\sigma^2 d'E\left[\frac{(g-\gamma)g'}{g'\Lambda g}\right]d \tag{4.4.9}$$

$$+\frac{c^2 s^2(n-s+2)\sigma^4}{(n-s)}d'E\left[\frac{gg'}{[g'\Lambda g]^2}\right]d.$$

Let E_γ denote expectation conditional on γ. Thus,

$$E(g-\gamma)(g-\gamma)' = EE_\gamma(g-\gamma)(g-\gamma)' = \sigma^2\Lambda^{-1} \tag{4.4.10}$$

since $D(g)= \sigma^2\Lambda^{-1}$. Now

$$E_g\left[\frac{g(g-\gamma)'}{g'\Lambda g}\right] = \frac{gg'}{g'\Lambda g} - \frac{gE_g(\gamma')}{g'\Lambda g}. \tag{4.4.11}$$

The LS is a sufficient statistic for the parameters of the linear model. As was explained in Chapter II a sufficient statistic contains all of the information about the parameters being estimated. Thus, it seems reasonable that the posterior distribution of a parameter conditional on a sufficient statistic would be the same as that conditional on the observations themselves. This is the content of the sufficiency principle that has already been mentioned in Chapter II (see Lindley (1965)) for more details). Thus, the expected value of a parameter would be the mean of the posterior distribution.

Since g is a sufficient statistic for γ,

$$E_g(\gamma) = g - \frac{\sigma^2}{\sigma^2 + \omega^2}g \tag{4.4.12}$$

and by substituting (4.4.12) into (4.4.11),

$$E_g\left[\frac{g(g-\gamma)'}{g'\Lambda g}\right] = \frac{\sigma^2}{\sigma^2+\omega^2}\frac{gg'}{g'\Lambda g}. \tag{4.4.13}$$

Thus,

$$N = E\left[\frac{g(g-\gamma)'}{g'\Lambda g}\right] = \frac{\sigma^2}{\sigma^2+\omega^2}E\left[\frac{gg'}{g'\Lambda g}\right]. \tag{4.4.14}$$

By taking the transpose of (4.4.14)

$$E\left[\frac{g-\gamma)'g}{g'\Lambda g}\right] = \frac{\sigma^2}{\sigma^2+\omega^2}E\left[\frac{gg'}{g'\Lambda g}\right]. \tag{4.4.15}$$

Since

$$g \sim N(0, (\sigma^2 + \omega^2)\Lambda^{-1}), \tag{4.4.16}$$

$$v = \frac{\Lambda^{\frac{1}{2}}g}{(\sigma^2 + \omega^2)^{\frac{1}{2}}} \sim N(0, I) \tag{4.4.17}$$

and

$$g = (\sigma^2 + \omega^2)^{\frac{1}{2}}\Lambda^{-\frac{1}{2}}v. \tag{4.4.18}$$

Thus, from (4.4.14) and (4.4.18), Lemmas 4.4.1 and 4.4.2

$$N = N' = \frac{\sigma^2}{\sigma^2 + \omega^2}\Lambda^{-\frac{1}{2}}E\left[\frac{vv'}{v'v}\right]\Lambda^{-\frac{1}{2}} = \frac{\sigma^2}{\sigma^2 + \omega^2}\Lambda^{-1}. \tag{4.4.19}$$

Likewise, from (4.4.17)

$$E\left[\frac{gg'}{[g'\Lambda g]^2}\right] = \frac{1}{\sigma^2 + \omega^2}\Lambda^{-\frac{1}{2}}E\left[\frac{vv'}{[v'v]^2}\right]\Lambda^{-\frac{1}{2}}$$

$$= \frac{1}{(s-2)s(\sigma^2 + \omega^2)}\Lambda^{-1}. \tag{4.4.20}$$

Substitution of (4.4.10), (4.4.19) and (4.4.20) into (4.4.9) yields

$$m = d'\Lambda^{-1}d - \frac{2c\sigma^4}{\sigma^2 + \omega^2}d'\Lambda^{-1}d$$

$$+ \frac{c^2s(n-s+2)\sigma^4}{(s-2)(n-s)(\sigma^2 + \omega^2)}d'\Lambda^{-1}d. \tag{4.4.21}$$

Since

$$d'\Lambda^{-1}d = d'U\Lambda^{-1}U'd = p'(X'X)^+p, \tag{4.4.22}$$

$$m = p'(X'X)^+p\sigma^2 - \frac{2c\sigma^4}{(\sigma^2 + \omega^2)}p'(X'X)^+p$$

$$+ \frac{c^2s(n-s+2)\sigma^4}{(s-2)(n-s)(\sigma^2 + \omega^2)}p'(X'X)^+p. \tag{4.4.23}$$

To find optimal c, let

$$h(c) = -2c + \frac{c^2 s(n-s+2)}{(n-s)(s-2)}$$ (4.4.24)

Differentiate (3.4.21), set the result equal to zero, i.e.,

$$h'(c) = -2 + \frac{2cs(n-s+2)}{(s-2)(n-s)}$$ (4.4.25)

to obtain optimal

$$c = \frac{(n-s)(s-2)}{(n-s+2)s}.$$ (4.4.26)

The result (4.4.1) follows from the substitution of (4.4.26) into (4.4.23).

Exercise 4.4.1. Find the average MSE of the generalized JS. In particular, consider the cases where $\Sigma = X'X$, $\Sigma = I$.

Exercise 4.4.2. Let $\hat{\beta}_b$ denote the BE and $\hat{\beta}_{eb}$ denote the EBE. For the generalized JS find

$$RSL = \frac{MSE(p'\hat{\beta}_{eb}) - MSE(p'\hat{\beta}_b)}{MSE(p'b) - MSE(p'\hat{\beta}_b)}$$

and show that it is independent of Σ.

Exercise 4.4.3. Let X and Y be independent random variables where $X \sim \chi^2(k)$ and $Y \sim \chi^2(l)$. Find an expression for

$$C = E\left[\frac{X^p}{(X+Y)^{p+q}}\right]$$

in terms of gamma functions. Hence prove Lemma 4.4.1 by letting $k = 1$, $l = s-1$ and assigning appropriate values to p and q.

Exercise 4.4.4. Find the MSE of the JSL. (This problem will be much easier with the results of Chapter VII. See Exercise 7.3.19.)

4.5 The Conditional MSE

The properties of the conditional MSE of the JS are quite different from those of the average MSE. Unlike the average MSE, the conditional MSE is uniformly smaller than the LS for some, not all, estimable linear parametric functions. In Gruber (1990) it was shown that the comparison of the MDE of two estimators is the same as the comparison of the MSE of two estimators averaging over any quadratic loss function with a non-negative definite matrix. Thus, for the full rank case comparison of the average MSE for parametric functions is the same as that averaging over any quadratic loss function; for the non-full rank case comparison of the average MSE of the estimable parametric functions is the same as comparison averaging over a loss function with matrix UHU′ with H a non-negative definite matrix. Since the conditional MSE of the JS is smaller than that of the LS only for some parametric functions the same holds true only for certain quadratic loss functions. Some specific conditions will be given in Section 4.6. An important particular case where the JS has a uniformly smaller MSE than the LS is when it is averaged over the predictive loss function, i.e, the MSE

$$m_{X'X} = E(\hat{\beta} - \beta)'X'X(\hat{\beta} - \beta) \qquad (4.5.1)$$

the MSE of prediction(MSEP).

The MSEP of the JS will now be obtained from the MSE of the parametric functions and shown to be uniformly smaller than that of the LS. The MSE of the JS of the parametric functions is not uniformly smaller than that of the LS. A necessary and sufficient condition will be derived for the JS of the parametric functions to be less than that of the LS will be derived.

4.5.1 *The MSE of the Estimable Parametric Functions*

The MSE is obtained in Theorem 4.5.1.

Theorem 4.5.1. The conditional MSE of the JS for the estimable parametric functions is

$$m = p'(X'X)^+p - 2csp'(X'X)^+p\sigma^4 E\left[\frac{1}{b'X'Xb}\right]$$

$$+\left[4cs + \frac{c^2s^2[(n-s)+2]}{(n-s)}\right]p'E\left[\frac{bb'}{[b'X'Xb]^2}\right]p\sigma^4 \qquad (4.5.2)$$

The proof will depend on the following lemma.

Lemma 4.5.1. Let v_i, $1 \le i \le s$ be $N(\mu_i, 1)$ independent random variables. Let D be a positive definite diagonal matrix. Then

$$E\left[\frac{v_i(v_i - \mu_i)}{v'Dv}\right] = E\left[\frac{1}{v'Dv}\right] - 2E\left[\frac{v_i^2 d_i}{[v'Dv]^2}\right] \qquad (4.5.3)$$

and

$$E\left[\frac{v_b(v_c - \mu_c)}{v'Dv}\right] = -2E\left[\frac{v_b v_c d_c}{[v'Dv]^2}\right]. \qquad (4.5.4)$$

The proof is by integration by parts.

Proof of the Theorem. Rewrite Equation (4.4.9) replacing the expectation with expectations conditional on β hence γ. Let $v = \Lambda^{1/2}g/\sigma$ hence $\delta = \Lambda^{1/2}\gamma/\sigma$. Then

$$E\left[\frac{g(g-\gamma)'}{g'\Lambda g}\right] = \Lambda^{-\frac{1}{2}}E\left[\frac{(v-\delta)v'}{v'v}\right]\Lambda^{-\frac{1}{2}}$$

$$= E\left[\frac{1}{v'v}\right]\Lambda^{-1} - 2\Lambda^{-\frac{1}{2}}E\left[\frac{vv'}{[v'v]^2}\right]\Lambda^{-\frac{1}{2}}$$

$$= E\left[\frac{1}{g'\Lambda g}\right]\Lambda^{-1}\sigma^2 - 2E\left[\frac{gg'}{[g'\Lambda g]^2}\right] \qquad (4.5.5)$$

The result follows after substitution into Equation (4.4.9) with conditional expectations replacing ordinary expectations and the fact that $d'\Lambda^{-1}d = p'(X'X)^+p$.

4.5.2 The MSE Averaging over a Predictive Loss Function

The MSEP of the JS may be obtained easily from the result of Theorem 4.5.1 because it can be shown to be the MSE of a particular parametric function of the regression parameters. The result is obtained in Corollary 4.5.1 by using this technique.

Corollary 4.5.1. For optimum $c = (n-s)(s-2)/(n-s+2)s$ the predictive MSE (4.5.1) is

$$m_{X'X} = \sigma^2 s - \frac{(n-s)}{(n-s+2)}(s-2)^2 \sigma^4 E\left[\frac{1}{b'X'Xb}\right]. \tag{4.5.6}$$

The first term of (4.5.6) is the MSEP of the LS. The second term is the improvement that comes from using the JS.

Proof. Let U_i be the columns of U where U is the column orthogonal matrix of eigenvectors of the non-zero eigenvalues of $X'X$. Then

$$E(\hat\beta - \beta)'X'X(\hat\beta - \beta) = \sum_{i=1}^{s} c_i' E(\hat\beta_i - \beta_i)(\hat\beta_i - \beta_i)'c_i$$

with $c_i = \lambda_i^{\frac{1}{2}} U_i$. Thus the MSEP of $\hat\beta$ is the same as that of sum of the parametric functions $c_i'\hat\beta_i$ so that Theorem 4.5.1 may be applied resulting in

$$m = \sigma^2 s - 2cs(s-2)\sigma^4 E\left[\frac{1}{b'X'Xb}\right] + \frac{c^2 s^2 [(n-s)+2]}{(n-s)}\sigma^4 E\left[\frac{1}{b'X'Xb}\right]. \tag{4.5.7}$$

Optimum c is obtained from minimization of

$$h(c) = -2cs(s-2) + \frac{c^2 [(n-s)+2]}{(n-s)}$$

as was done in the proof of Theorem 4.4.1. The result (4.5.6) follows when optimum c is substituted into Equation (4.5.7).

Exercise 4.5.1. Carry through the details of the proof of Lemma 4.5.1.

Exercise 4.5.2. Fill in the details of the proof of Corollary 4.5.1.

Exercise 4.5.3.
A. Show that $b'X'Xb$ is a complete sufficient statistic for $\sigma^2+\omega^2$.
B. Use the result of A to obtain (4.5.6) from (4.4.1).

Exercise 4.5.4. Obtain the MSE of the JSL. This will be an easier problem in the context of the results of Chapter IV.

4.5.3 The Comparison of the MSE of the Parametric Functions with the LS

The MSE of the parametric functions of the JS is not uniformly smaller than that of the LS. For example the individual MSE of the JS is not uniformly smaller than that of the LS. Theorem 4.5.2 gives conditions on the β parameters for the parametric functions of the JS to have a smaller MSE than the corresponding LS.

Theorem 4.5.2. If

1.
$$0 < c < \frac{2(n-s)(s-2)}{(n-s+2)s} \tag{4.5.8}$$

and

2.
$$\alpha = \frac{p'\beta\beta'p}{\beta'X'X\beta p'(X'X)^+p} \leq \frac{2}{4+cs(n-s+2)/(n-s)} \tag{4.5.9}$$

then the conditional MSE of the linear parametric functions of the JS is less than that of the LS. For optimum $c = (n-s)(s-2)/(n-s+2)s$ obtained in Theorem 4.4.1 and Corollary 4.5.1 (4.5.9) reduces to

$$\alpha < \frac{2}{s+2} \tag{4.5.10}$$

Proof. The proof consists of:

1. obtaining expressions for the expectations in (4.5.2);
2. rewriting (4.5.2) with these expressions substituted in;
3. solving the inequality that results from comparison of (4.5.2) with the LS.

The following notation is defined to make some of the expressions that follow less cumbersome. Define

$$S_{1,s}(\delta) = \sum_{k=0}^{\infty} \frac{e^{-\delta}\delta^k}{k!(2k+s-2)}, \tag{4.5.11}$$

$$S_{2,s}(\delta) = \sum_{k=0}^{\infty} \frac{e^{-\delta}\delta^k}{k!(2k+s)(2k+s-2)}, \tag{4.5.12}$$

and

$$S_{3,s}(\delta) = \sum_{k=0}^{\infty} \frac{e^{-\delta}\delta^k}{k!(2k+s)(2k+s+2)}, \tag{4.5.13}$$

with $\delta = \beta'X'X\beta/2\sigma^2$. The needed expectations will be obtained in Lemma 4.5.2.

Lemma 4.5.2. The expectations in Equation 4.5.2 are

1.

$$E\left[\frac{1}{b'X'Xb}\right] = \frac{1}{\sigma^2}S_{1,s}(\delta) \tag{4.5.14}$$

2.

$$E\left[\frac{p'bb'p}{[b'X'Xb]^2}\right] = \frac{1}{\sigma^4}\{\sigma^2 p'(X'X)^+ pS_{2,s}(\delta) + p'\beta\beta'p[S_{3,s}(\delta)]\} \tag{4.5.15}$$

Proof. Let $v = \Lambda^{1/2}U'b/\sigma$ Since $b \sim N(\beta, \sigma^2(X'X)^+)$, $v \sim N(\Lambda^{1/2}U'\beta/\sigma, I)$. From the computations in (2.6.7)-(2.6.12)

$$E\left[\frac{1}{b'X'Xb}\right] = \frac{1}{\sigma^2}E\left[\frac{1}{v'v}\right] = \frac{1}{\sigma^2}S_{1,s}(\delta)$$

establishing (4.5.14). To establish (4.5.15) observe that

$$E\left[\frac{p'bb'p}{[b'X'Xb]^2}\right] = d'\Lambda^{-\frac{1}{2}}E\left[\frac{vv'}{[v'v]^2}\right]\Lambda^{-\frac{1}{2}}d.\frac{1}{\sigma^2} \ . \tag{4.5.16}$$

Let $\theta = \Lambda^{\frac{1}{2}}U'\beta/\sigma$. The objective is to calculate the expectation in (4.5.16) by calculating the four expectations in

$$M = E\left[\frac{vv'}{[v'v]^2}\right] = E\left[\frac{(v-\theta)(v-\theta)'}{[v'v]^2}\right] + E\left[\frac{\theta v'}{[v'v]^2}\right]$$
$$+E\left[\frac{v\theta'}{[v'v]^2}\right] - E\left[\frac{\theta\theta'}{[v'v]^2}\right]. \tag{4.5.17}$$

Define

$$m(\theta) = E\left[\frac{1}{[v'v]^2}\right]$$
$$= \frac{1}{(2\pi)^{\frac{s}{2}}}\int_R \frac{1}{[v'v]^2}\exp\left[-\frac{1}{2}(v-\theta)'(v-\theta)\right]dv. \tag{4.5.18}$$

Then the vector

$$\frac{\partial m(\theta)}{\partial\theta} = E\left[\frac{(v-\theta)}{[v'v]^2}\right] \tag{4.5.19}$$

and the matrix

$$A(\theta) = \left(\frac{\partial^2 m(\theta)}{\partial\theta_i\theta_j}\right)_{\substack{1\le i\le s \\ 1\le j\le s}} = E\left[\frac{(v-\theta)(v-\theta)'}{[v'v]^2}\right] - E\left[\frac{1}{[v'v]^2}\right]I. \tag{4.5.20}$$

Repetition of the calculations in (2.6.4)-(2.6.12) with $1/u^2$ in place of $1/u$ yields

$$m(\theta) = S_{2,s-2}(\delta) \tag{4.5.21}$$

where $\delta = \theta'\theta/2$ and

$$\frac{\partial m}{\partial\theta} = \theta[S_{2,s}(\delta) - S_{2,s-2}(\delta)] \tag{4.5.22}$$

From (4.5.22) and some algebra

$$E\left[\frac{v\theta'}{[v'v]^2}\right] = \theta\theta'S_{2,s}(\delta).$$ (4.5.23)

Also

$$A(\theta) = I[S_{2,s}(\delta) - S_{2,s-2}(\delta)]$$
$$+\theta\theta'[S_{3,s}(\delta) - 2S_{2,s}(\delta) + S_{2,s-2}(\delta)].$$ (4.5.24)

From (4.5.20) and (4.5.21)

$$E\left[\frac{(v-\theta)(v-\theta)'}{[v'v]^2}\right] = I[S_{2,s}(\delta)]$$
$$+\theta\theta'[S_{3,s}(\delta) - 2S_{2,s}(\delta) + S_{2,s-2}(\delta)].$$ (4.5.25)

From (4.5.17), (4.5.21), (4.5.23) and (4.525)

$$M = S_{2,s}(\delta)I + \theta\theta'[S_{3,s}(\delta)].$$ (4.5.26)

Equation (4.5.15) follows from (4.5.16) after substitution of (4.5.26).

The proof of the theorem consists of substituting (4.5.14) into (4.5.2) and solving the inequality that compares it with the MSE of the LS. The MSE of the JS is

$$mjs = p'(X'X)^+p\sigma^2 - 2csp'(X'X)^+p\sigma^2S_{2,s}(\delta)$$
$$+\left[4cs + \frac{c^2s^2[(n-s)+2]}{(n-s)}\right][\sigma^2p'(X'X)^+pS_{2,s}(\delta) + p'\beta\beta'pS_{3,s}(\delta)].$$ (4.5.27)

The solution to the inequality

$$mjs \le p'(X'X)^+p\sigma^2 = mls$$ (4.5.28)

is

$$\alpha \le \frac{1}{2\delta S_{3,s}(\delta)}\left[\frac{2S_{1,s}(\delta)}{4+cs[(n-s)+2]/(n-s)}-S_{2,s}(\delta)\right]. \qquad (4.5.29)$$

The reader may verify that

$$2\delta S_{3,s}(\delta)+sS_{2,s}(\delta)=S_{1,s}(\delta). \qquad (4.5.30)$$

Substitution of (4.5.30) into (4.5.29) yields the equivalent inequality

$$\alpha \le \frac{2}{4+cs[(n-s)+2]/(n-s)} \\ +\frac{S_{2,s}(\delta)}{2\delta S_{3,s}(\delta)}\left[\frac{2s}{4+cs[(n-s)+2]/(n-s)}-1\right]. \qquad (4.5.31)$$

Equation (4.5.29) is a necessary and sufficient condition for the MSE of the parametric functions of the JS to be less than that of the LS. When condition (4.5.8) holds true the second term of the right hand side of Inequality (4.5.31) is positive. Thus, mjs ≤ mls when (4.5.9) holds true. For optimum c (4.5.9) reduces to (4.5.10).

Exercise 4.5.5. Show that $\delta S_{3,s}(\delta) = \displaystyle\sum_{k=0}^{\infty}\frac{e^{-\delta}k\delta^k}{k!(s+2k)(s+2k-2)}.$

Exercise 4.5.6. Verify (4.5.30) and then do the necessary algebra to obtain (4.5.31) from (4.5.29).

Exercise 4.5.7. A. Show that

$$S_{2,s}(\delta)=\frac{1}{4\delta^{s/2+k}e^{\delta}}\int_0^{\delta}\int_0^{x}e^y y^{s/2-2}dydx.$$

B. Obtain a similar expression for $S_{3,s}(\delta)$.

4.6 The MSE of the JS for Different Quadratic Loss Functions

As has already been pointed out the JS has a uniformly smaller conditional MSE than the LS for some but not all quadratic loss functions. Berger (1976a,b) obtained conditions on the quadratic loss function matrix for the JS to have a smaller MSE than LS. These conditions will be derived in Theorem 4.6.2.

Let $\hat{\beta}$ denote an estimator of β. The loss functions that have been considered include:

1. $\qquad M = E(\hat{\beta} - \beta)(\hat{\beta} - \beta)'$; (Matrix Loss MDE) $\qquad\qquad$ (4.6.1)

2. $\qquad p'Mp = p'E(\hat{\beta} - \beta)(\hat{\beta} - \beta)'p$; (MSE of p'$\hat{\beta}$) $\qquad\qquad$ (4.6.2)

3. $\qquad trM\Sigma = E(\hat{\beta} - \beta)'\Sigma(\hat{\beta} - \beta)$. (Quadratic Loss Function) $\qquad$ (4.6.3)

The important choices of Σ include $\Sigma = X'X$ the predictive loss function and $\Sigma = I$ the loss function used to obtain the total MSE.

The comparative behavior of pairs of estimators with respect to loss functions (4.6.2) and (4.6.3) may be compared using Theobald's (1974) result (Theorem 3.4.1) that was obtained in Chapter III. Recall that if $\hat{\beta}_1$ and $\hat{\beta}_2$ are two estimators with MDE M_1 and M_2 respectively then $M_1 \leq M_2$, i.e., is non-negative definite iff $p'M_1p \leq p'M_2p$. Corollary 3.2.3 will be used in what follows. Another important consequence of Theobald's result that will be used here is:

Suppose $\Sigma = UHU'$ where H is a non-negative definite matrix. The inequality $p'M_1p \leq p'M_2p$ holds for the estimable parametric functions iff $trM_1UHU' \leq trM_2UHU'$ holds for all non-negative definite H.

Theorem 4.6.1 below is a characterization of the quadratic loss functions and the parametric functions where the JS has a uniformly smaller MSE than the LS. This kind of result is important because the risk of the JS is not uniformly for every quadratic loss function or parametric function $p'\beta$.

Theorem 4.6.1. Let Σ be the quadratic loss function matrix in the form UHU' where H is non-negative definite. Let θ_1 be the largest relative eigenvalue of H and Λ, i.e., the largest root of

$$\det(H - \theta\Lambda) = 0 \qquad\qquad (4.6.4)$$

where Λ is the diagonal matrix of eigenvalues of X'X. Let θs be the smallest of the relative eigenvalues. If

$$\theta_L < \frac{1}{2}tr(X'X)^+\Sigma \qquad\qquad (4.6.5)$$

then for the JS (4.2.10) there exists a c where

$$m_\Sigma = E(\hat{\beta} - \beta)'\Sigma(\hat{\beta} - \beta) \le E(b - \beta)'\Sigma(b - \beta), \tag{4.6.6}$$

i.e., the JS averaging over the quadratic loss function (4.6.6) has a smaller MSE than LS. On the other hand if

$$\theta_s \ge \frac{1}{2}\mathrm{tr}(X'X)^+\Sigma \tag{4.6.7}$$

no such c exists.

Proof. Observe that

$$m_\Sigma = \mathrm{tr}(X'X)^+\Sigma - 2cs\mathrm{tr}(X'X)^+\Sigma E\left[\frac{1}{b'X'Xb}\right]$$
$$+ \left[4cs + c^2 s^2 \frac{[(n-s)+2]}{(n-s)}\right]E\left[\frac{b'\Sigma b}{[b'X'Xb]^2}\right]. \tag{4.6.8}$$

The first term of the right hand side of (4.6.8) is the MSE of the LS. Thus, in order that (4.6.8) be smaller than the LS

$$0 < c < \frac{(n-s)}{s[(n-s)+2]}\left[\frac{2\mathrm{tr}(X'X)^+\Sigma E[\frac{1}{b'X'Xb}]}{E[\frac{b'\Sigma b}{b'X'Xb}]} - 4\right] \tag{4.6.9}$$

The existence of a positive c where (4.6.9) holds true is possible only when the right hand side of Inequality (4.6.9) is positive. This is possible iff

$$1 = E\left[\frac{2\mathrm{tr}(X'X)^+\Sigma b'X'Xb - 4b'\Sigma b}{[b'X'Xb]^2}\right] > 0. \tag{4.6.10}$$

Inequality (4.6.10) holds true iff

$$\frac{b'\Sigma b}{b'X'Xb} < \frac{1}{2}\mathrm{tr}(X'X)^+\Sigma \text{ almost surely}. \tag{4.6.11}$$

On the other hand if $1 \le 0$ the less than sign in (4.6.11) is replaced by a greater than or equals sign. The result will follow after the Courant Fisher minimax theorem is applied. (Exercise 4.6.5) The Courant-Fisher minimax theorem states that for two quadratic forms X'AX and X'BX assuming that B is positive definite

$$\lambda_s \leq \frac{X'AX}{X'BX} \leq \lambda_L \qquad\qquad . \qquad\qquad (4.6.12)$$

where , λ_L and λ_s are the largest and smallest relative eigenvalues respectively of A and B. From the assumption about Σ , the singular value decomposition of X'X and the fact that the LS in orthogonal coordinates g = U'b (4.6.11) may be rewritten

$$\frac{b'\Sigma b}{b'X'Xb} = \frac{g'Hg}{g'\Lambda g} \leq \frac{1}{2}tr\Lambda^{-1}H. \qquad\qquad (4.6.13)$$

But

$$\theta_s \leq \frac{g'Hg}{g'\Lambda g} \leq \theta_L. \qquad\qquad (4.6.14)$$

A simple corollary of Theorem (4.6.2) for the MSE of estimable parametric functions is:

Corollary 3.6.1. If

$$\theta_L < \frac{1}{2}p'(X'X)^+p \qquad\qquad (4.6.15)$$

then for estimable parametric functions there exists a positive c where the MSE of the JS is smaller than that of the LS. On the other hand if

$$\theta_L \geq \frac{1}{2}p'(X'X)^+p \qquad . \qquad\qquad (4.6.16)$$

no such c exists.

Proof. The result follows from letting H=dd'.

Exercise 4.6.1. Formulate Theorem 4.6.1 for the case where X'X is of full rank.

Exercise 4.6.2. A. Assume that the LS in Equation (4.2.10) is derived using the MP. Extend Theorem 4.6.1 to the case $\Sigma = UHU' + UU'$. In particular interpret conditions (4.6.5) and (4.6.7) for the case $\Sigma = I$.

B. Can the proposed extension be used when any generalized inverse is used in the LS? Explain.

Exercise 4.6.3. What do conditions (4.6.5), (4.6.7) and (4.6.9) reduce to for the MSEP?

Exercise 4.6.4. Obtain a result similar to that of Theorem 4.6.1. when the generalized JS is averaged over a quadratic loss function with matrix N.

Exercise 4.6.5. Prove the Courant Fisher Minimax Theorem.

4.7 The Efficiency of the JS as Compared with Optimum Estimators

The principal goals of this section are:

1. to define and study the properties of the relative loss (RL) both analytically and numerically;
2. to use the properties of the RL to derive upper and lower bounds for the frequentist risk of the JS and hence obtain conditions for the EBE to have a smaller MSE than the JS;
3. to obtain conditions for the frequentist risk of the JS to be less than that of the BE using an inequality of C.R. Rao and Shinozaki (1978);
4. to compare the worthiness of the conditions described in 3 and 4;
5. to do some numerical computations to illustrate the analytical results that are obtained.

4.7.1 The RL and its Properties

The RSL is a measure of efficiency of an approximate BE as compared with the BE. For the MSE averaging over the prior the BE is the optimal estimator, i.e., the estimator with smallest average MSE. The RL is a measure of the relative efficiency of JS as compared with the MMSE. As has already been seen the JS is an approximate MMSE and the MMSE is the estimator with the smallest

MSE.without averaging over the prior, i.e., the conditional MSE. By analogy with the RSL the RL is defined by the equation

$$RL = \frac{MSE(JS) - MSE(MMSE)}{MSE(MLE) - MSE(MMSE)}. \tag{4.7.1}$$

The properties of the RL will be used to compare the conditional MSE of the JS and the BE.

By substitution of optimal h obtained in (4.3.5) into (4.3.3) the conditional MSE of the MMSE is

$$m = \sigma^2 s - \frac{\sigma^4 s^2}{\beta' X' X \beta + \sigma^2 s} = \sigma^2 s - \frac{\sigma^2 s^2}{2\delta + s} \tag{4.7.2}$$

where $\delta = \beta' X' X \beta / 2\sigma^2$.

Substituting (4.7.2) and (3.5.6) into (4.7.1) the RL as a function of δ is given by

$$RL(\delta) = 1 - \frac{(s-2)^2 (s+2\delta)(n-s)}{s^2 (n-s+2)} S_{1,s}(\delta). \tag{4.7.3}$$

Recall that

$$S_{1,s}(\delta) = \sum_{k=0}^{\infty} \frac{e^{-\delta} \delta^k}{k!(s+2k-2)} = \frac{1}{2\delta^{\frac{s}{2}-1} e^{\delta}} \int_0^{\delta} e^x x^{\frac{s}{2}-1} dx \tag{4.7.4}$$

The RL is an increasing function of δ bounded above and below. The lower bound is the RSL. These facts are established in Theorem 4.7.1 below.

Theorem 4.7.1. The RL is an increasing function with lower bound

$$RL(0) = RSL = 1 - \frac{(s-2)(n-s)}{s(n-s+2)} \tag{4.7.5}$$

and upper bound

$$RL(\infty) = \lim_{\delta \to \infty} RL(\delta) = 1 - \frac{(s-2)^2(n-s)}{s^2(n-s+2)}. \tag{4.7.6}$$

Proof. To establish that $R(\delta)$ is increasing it suffices to show that

$$h(\delta) = (s+2\delta)S_{1,s}(\delta) \tag{4.7.7}$$

is decreasing. This will be established by showing that $h'(\delta) < 0$. Now

$$h'(\delta) = -16\delta \sum_{k=0}^{\infty} \frac{e^{-\delta}\delta^k}{k!(s+2k+2)(s+2k)(s+2k-2)}. \tag{4.7.8}$$

To find the lower bound notice that

$$S_{1,s}(\delta) = \frac{1}{s-2} \tag{4.7.9}$$

Equation (4.7.5) then follows by substitution of (4.7.9) into (4.7.3).To establish (4.7.6) it is necessary to establish Lemma 4.7.1 below.

Lemma 4.7.1. If $s > 2$ then

$$\lim_{\delta \to \infty} (s+2\delta)S_{1,s}(\delta) = 1.$$

Proof. It will first be shown that

$$\lim_{\delta \to \infty} S_{1,s}(\delta) = 0.$$

The result in (4.7.11) follows by L'Hôpital's rule. After differentiating the numerator and the denominator of the last expression in (4.7.4)

$$\lim_{\delta \to \infty} S_{1,s}(\delta) = \lim_{\delta \to \infty} \frac{e^{\delta}\delta^{\frac{1}{2}-2}}{2e^{\delta}\delta^{\frac{1}{2}-1} + (s-2)\delta^{\frac{1}{2}-2}e^{\delta}}$$
$$= \lim_{\delta \to \infty} \frac{1}{2\delta + s - 2} = 0. \tag{4.7.12}$$

The lemma will first be established for s=3,4 and then for all s > 2 by mathematical induction. For s = 3, by L'Hospital's rule

$$L = \lim_{\delta \to \infty} \frac{(2\delta + 3)}{2\delta^{1/2} e^{\delta}} \int_0^{\delta} e^x x^{-1/2} dx$$

$$= \lim_{\delta \to \infty} \frac{2\delta + 3}{2\delta + 1} + \lim_{\delta \to \infty} \frac{2e^{-\delta} \delta^{1/2} \int_0^{\delta} e^x x^{-1/2} dx}{2\delta + 1} \tag{4.7.13}$$

$$= L_1 + L_2.$$

Clearly $L_1 = 1$. It will be shown that $L_2 = 0$. Rewrite

$$L_2 = \lim_{\delta \to \infty} \frac{2e^{-\delta} \delta^{1/2} \int_0^1 e^x x^{-1/2} dx}{2\delta + 1} + \lim_{\delta \to \infty} \frac{2e^{-\delta} \delta^{1/2} \int_1^{\delta} e^x x^{-1/2} dx}{2\delta + 1} \tag{4.7.14}$$

$$= L_3 + L_4.$$

Now it will be shown that L_3 and $L_4 = 0$. To show that $L_3 = 0$ observe that

$$\lim_{\delta \to \infty} \frac{\delta^{1/2}}{e^{\delta}(2\delta + 1)} = 0 \tag{4.7.15}$$

and

$$\int_0^1 \frac{e^x}{\sqrt{x}} dx \le e \int_0^1 \frac{dx}{\sqrt{x}} = 2e. \tag{4.7.16}$$

When $x \ge 1$, $1/\sqrt{x} \le 1$, hence

$$0 \le \int_1^{\delta} \frac{e^x}{\sqrt{x}} dx \le \int_1^{\delta} e^x dx = e^{\delta} - 1. \tag{4.7.17}$$

Consequently $L_3 = 0$ so

$$0 \le L_2 \le \lim_{\delta \to \infty} \frac{\delta^{1/2}(2 - e^{-\delta})}{2\delta + 1} = 0. \tag{4.7.18}$$

Consequently $L_2 = 0$ and the lemma is proved for $s = 3$. The result is established for $s = 4$ by observing that

$$\lim_{\delta\to\infty}\frac{(2\delta+4)e^{-\delta}}{2\delta}\int_0^\delta e^x dx = \lim_{\delta\to\infty}\left[1+\frac{2}{\delta}\right][1-e^{-\delta}]=1. \qquad (4.7.19)$$

When $s > 4$ assume the result holds for $S_{1,s}(\delta)$. To establish the Lemma for $s \geq 2$ by induction it must be shown that the result holds for $S_{1,s+2}(\delta)$. From Exercise 2.6.4 observe that

$$(s+2+2\delta)S_{1,s+2}(\delta)$$
$$=\frac{(s+2+2\delta)}{2\delta}-(s-2).\frac{(s+2+2\delta)}{2\delta}S_{1,s-2}(\delta) \qquad (4.7.20)$$
$$= t_1(\delta)-t_2(\delta).$$

Clearly

$$\lim_{\delta\to\infty}t_1(\delta)=1. \qquad (4.7.21)$$

Now $\lim_{\delta\to\infty}t_2(\delta)=0$ because

$$\lim_{\delta\to\infty}\frac{1}{2\delta}(s+2+2\delta)S_{1,s-2}(\delta)=0 \qquad (4.7.22)$$

Table 4.7.1

s | δ	0	0.5	1.5	3.5	5.5	∞
4	0.500	0.568	0.668	0.714	0.727	0.750
6	0.333	0.393	0.473	0.516	0.529	0.556
8	0.250	0.300	0.365	0.401	0.414	0.438
10	0.200	0.242	0.296	0.328	0.339	0.360

from (4.7.21) and (4.7.12). The result (4.7.6) then follows from Lemma 4.7.1 and (4.7.3).

An obvious but important consequence of Theorem 4.7.1 is that $RL(\delta) > 0$. Since the MSEP of the LS is greater than that of the MMSE, the MSEP of the approximate MMSE, i.e., the JS also has a larger MSEP than the MMSE.

The MSEP is the MSE averaging over the predictive loss function. The RL may be defined for other approximate MMSEs with MSE averaged over other loss functions provided that this MSE is uniformly smaller than LS.

Table 4.7.1 illustrates the properties of the RL by giving its numerical values at zero, infinity and some numbers in between. It is assumed that n is large so that $(n-s)/(n-s+2)$ is close to one. The RL is plotted in Figure 4.7.1 for a few values of s.

The objective is to compare the conditional MSEP of the JS with that of the BE. One way to do this is to first obtain upper and lower bounds on the MSE of the JS.

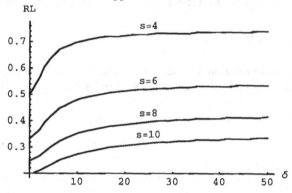

Figure 4.7.1 RL of JS for Different s

By solving the inequalities that result from comparing the MSE of the BE with these upper and lower bounds conditions can be obtained where the MSEP of the JS is less that that of the BE because the mathematical form of the bounds is less cumbersome than the exact MSEP. The upper and lower bounds on the MSEP are given by Corollary 4.7.1.

4.7.2 Comparing the BE and the JS using the RL

The objective is to compare the conditional MSEP of the JS with that of the BE. One way to do this is to first obtain upper and lower bounds on the MSE of the JS. By solving the inequalities that result from comparing the MSE of the BE with these upper and lower bounds conditions can be obtained where the MSEP of the JS is less that that of the BE because the mathematical form of the bounds is less cumbersome than the exact MSEP. The upper and lower bounds on the MSEP that are given by Corollary 4.7.1 are based on the properties of the RL and the RSL. The next section will derive more precise upper and lower bounds based on a result of C.R. Rao and Shinozaki (1978).

Corollary 4.7.1. Upper and lower bounds on the MSEP of the JS are given by the inequality

$$1 - \frac{(n-s)(s-2)}{(n-s+2)(2\delta+s)} \le \frac{m_j}{s\sigma^2} \le 1 - \frac{(n-s)(s-2)^2}{(n-s+2)(2\delta+s)s} \qquad (4.7.23)$$

where m_j denotes MSE (JS).

Proof. First some notation. In what follows m_m denotes the MMSE of the MMSE and m_l denotes the MSE of the LS. From Theorem 4.7.1,

$$RSL \le RL(\delta) \le RL(\infty) \qquad (4.7.24)$$

or written another way

$$1 + RSL - 1 \le \frac{m_j - m_m}{m_l - m_m} \le 1 + RL(\infty) - 1 \qquad (4.7.25)$$

From (4.7.25)

$$
\begin{aligned}
& m_l - m_m + (m_l - m_m)(RSL - 1) \\
& \le m_j - m_m \\
& \le m_l - m_m + (m_l - m_m)(RL(\infty) - 1)
\end{aligned}
\qquad (4.7.26)
$$

To obtain bounds on the risk of the JS averaged over the β parameters add m_m to all three members of Inequality (4.7.26) and divide the result by $s\sigma^2$. Thus, an equivalent form of Inequality (4.7.26) is

$$
\begin{aligned}
& \frac{1}{s\sigma^2}[m_l + (m_l - m_m)(RSL - 1)] \\
& \le \frac{m_j}{s\sigma^2} \le \frac{1}{s\sigma^2}[m_l + (m_l - m_m)(RL(\infty) - 1)]
\end{aligned}
\qquad (4.7.27)
$$

The result being established is obtained by substituting the expressions for the RSL and RL(∞) into Inequality (4.7.27).

Observe that the width of the interval between the lower and the upper bounds of the JS is (s-2)s/(2δ+s)s .This interval is narrower and the bounds are sharper for larger s and δ.

As a result of Corollary 4.7.1 the conditional MSE of the BE derived from Dempster's prior may be compared with that of the JS. Theorem 4.7.2 gives sufficient conditions for:

1. the JS to have a smaller conditional MSEP than the BE;
2. the BE to have a smaller conditional MSEP than the JS.

Furthermore Theorem 4.7.2 gives intervals that contain the crossing points of the MSEP of the JS and the BE.

Theorem 4.7.2. If

$$1. \max\left(0, \frac{s(c - \sqrt{RSL})}{2(1-c)}\right) < \delta < \frac{s(c + \sqrt{RSL})}{2(1-c)} \qquad (4.7.28)$$

with c = $\omega^2/(\sigma^2 + \omega^2)$ then the BE has a smaller MSE than the JS.
If either

$$a. \delta > \frac{s(c + \sqrt{RL(\infty)})}{2(1-c)} \qquad (4.7.29)$$

or

$$b. \sqrt{RL(\infty)} < c \text{ and } \delta < \frac{s(c - \sqrt{RL(\infty)})}{2(1-c)} \qquad (4.7.30)$$

then the JS has a smaller MSE than the BE.

Proof. From Corollary 4.7.1 the BE has a smaller MSE than the JS if

$$c^2 + (1-c)^2 \frac{2\delta}{s} < 1 - \frac{(n-s)(s-2)}{(n-s+2)s(2\delta/s+1)} \qquad (4.7.31a)$$

or equivalently

$$(1-c)^2 \left(\frac{2\delta}{s}\right)^2 - 2c(1-c)\left(\frac{2\delta}{s}\right) + c^2 < RSL. \qquad (4.7.31b)$$

Solving the quadratic inequality (4.7.31b) for $2\delta/s$ yields (4.7.28).
    Likewise the JS has a smaller MSE than the BE if

$$1 - \frac{(n-s)(s-2)^2}{(n-s+2)s^2(2(\delta/s)+1)} < (1-c)^2 \left(\frac{2\delta}{s}\right) + c^2 \qquad (4.7.32a)$$

or

$$(1-c)^2 \left(\frac{2\delta}{s}\right)^2 - [2c(1-c)]\left(\frac{2\delta}{s}\right) + c^2 > RL(\infty). \qquad (4.7.32b)$$

The results in (4.7.29) and (4.7.30) are obtained by solving (4.7.32b) for $2\delta/s$.
When the prior parameters are not too large as compared with the population
parameters the two estimators cross in the interval

$$I_c = \left(\frac{s(c+\sqrt{RSL})}{2(1-c)}, \frac{s(c+\sqrt{RL(\infty)})}{2(1-c)}\right)$$

    The MSEP of the JS is:

1. an increasing function;
2. a convex function, i.e., its second derivative is negative;
3. equal to the RSL when $\delta=0$.

    The reader is invited to establish these properties of the MSEP of the JS in Exercise
4.7.1.
    Summing up the consequences of Theorem 4.7.2 and Corollary 4.7.1:

1. If $c^2 < RSL$ the MSEP of the BE is less than that of the JS. However, there
exists a δ_0 such that if $\delta > \delta_0$ the JS has a smaller MSEP than the BE.
2. If $RSL < c^2$ there is an interval $[\delta_1, \delta_2]$ where the BE has a smaller MSEP than
the JS. Outside of this interval the JS has a smaller risk than the BE.

In other words when $c^2 < RSL$ the MSEPs of the BE and the JS cross once, but
when $RSL < c^2$ the MSEPs of the BE and the JS cross twice.

Some values of the MSEP of the JS for different values of δ and s are given in Table 4.7.2. Figure 4.7.2 illustrates pictorially the fact that the MSEP is a convex increasing function of δ.

Table 4.7.3 gives the crossing points of the MSEP of the JS and BE for different values of the prior parameter and s together with the upper and lower bounds of Theorem 4.7.2. Figure 4.7.3 illustrates situations where the MSEP of the JS and the BE cross once or twice

Exercise 4.7.1. Show that:
A. The MSEP of the JS is an increasing function of δ.
B. The MSEP of the JS is a convex function of δ, i.e., concave down.
C. The MSEP of the JS is the RSL and tends to 1 as $\delta \to \infty$.

Exercise 4.7.2. Generate Tables 4.7.1-4.7.3 for $s = 3, 5, 7, 9$ using Mathematica or some other software package.

Exercise 4.7.3. Show that if $s > 2/(1 - \sqrt{1 - c^2})$ or $s > 2/c^2$ then the MSEP of the JS is less than that of the BE outside of a finite interval $[\delta_0, \delta_1]$. When $s < 2/c^2$ there exists a δ_1 such that the MSEP of the BE is less than that of the JS for $\delta < \delta_1$.

Exercise 4.7.4. A. Establish that for a Poisson random variable r with mean δ

$$\frac{1}{p + 2\delta} \leq E\left[\frac{1}{p + 2r}\right] \leq \frac{1}{p - 2 + 2\delta}.$$

B. Let $p = s - 2$. Obtain upper and lower bounds for the MSEP of the JS.
C. If possible obtain conditions for the MSEP of the BE to be less than that of the JS and vice versa. Compare your results to that of Theorem 4.7.1.

4.7.3 Comparing the JS and the Ridge Estimator using the Result of C.R.Rao and Shinozaki

Bounds on the risk of the JS that are generally sharper than those obtained in the previous subsection may be obtained using the inequality of C.R. Rao and Shinozaki (1978). This leads in most cases to a narrower interval inside of which the JS and the EBE cross. Inferences can then be made on how many times the JS and EBE cross and when one estimator dominates the other. The inequality of C.R. Rao and

Shinozaki is established in Theorem 4.7.3 . The bounds on the JS are contained in Corollary 4.7.2.

Theorem 4.7.3. (Rao and Shinozaki). Let r be a Poisson random variable with $E(r) = b$. Assume that $q \geq 0$. Then

$$(q + 2b)^{-1} \leq E[(q + 2r)^{-1}] \leq (q - 2 + 2b)^{-1}. \qquad (4.7.33)$$

Proof. Jensen's inequality (see for example Feurguson (1969)) states that if X is a random variable and f is a convex function , i.e., has a positive second derivative, then

$$f(E[X]) \leq E[f(X)] \qquad . \qquad (4.7.34)$$

Now for non-negative q , $(q+2r)^{-1}$ is a convex function of r. Thus, the left-hand inequality follows from (4.7.34) because $E(q + 2r) = q + 2b$.

The right hand inequality is obtained by observing that

$$(q - 2 + 2b)E\left[\frac{1}{q + 2r}\right] = 1 - 2(q - 2)E\left[\frac{1}{(q + 2r)(q + 2r - 2)}\right] \leq 1.$$

$$(4.7.35)$$

Corollary 4.7.2. Bounds on the risk of the JS are given in the inequality (4.7.36) below.

$$1 - \frac{(n - s)(s - 2)^2}{s(n - s + 2)(s - 4 + 2\delta)} \leq \frac{m}{s\sigma^2} \leq 1 - \frac{(n - s)(s - 2)^2}{s(n - s + 2)(s - 2 + 2\delta)}.$$

$$(4.7.36)$$

Proof. Let $b = \delta$, $q = s - 2$ and $r = k$ in (4.7.33).

The upper bound in (4.7.36) is a good approximation to the risk of the JS. Comparison of the bounds in (4.7.36) with the MSE of the BE leads to another set of conditions for the JS to dominate the BE. These are given in Theorem 4.7.4. below.

Theorem 4.7.4. Assume that n is large compared to s so that $(n - s)/(n - s+2)$ is very close to 1.

1. If

$$1 + \frac{sc}{2(1-c)} - \frac{1}{(1-c)}\sqrt{c(s+c-2)}$$
$$< \delta < 1 + \frac{sc}{2(1-c)} + \frac{1}{(1-c)}\sqrt{c(s+c-2)} \tag{4.7.37}$$

Observe that when $0 < c < 1$ the left-hand side of (4.7.37) is positive.

2. If $\sqrt{RSL} < c$ and either

$$\delta < \frac{1}{2} + \frac{sc}{2(1-c)} - \frac{1}{2(1-c)}\sqrt{(c+1)(c+2s-3)} \tag{4.7.38}$$

or

$$\delta > \frac{1}{2} + \frac{sc}{2(1-c)} + \frac{1}{2(1-c)}\sqrt{(c+1)(c+2s-3)} \tag{4.7.39}$$

then the BE has a larger risk than the JS.

Proof. Compare the risk of the BE to the upper and lower bounds on the risk of the JS. Thus, the risk of the BE is smaller than that of the JS when

$$c^2 + (1-c)^2 \frac{2\delta}{s} < 1 - \frac{(s-2)^2}{s(s-4+2\delta)} \tag{4.7.40}$$

The range of values of δ given by Inequality (4.7.37) is the solution to Inequality (4.7.40). Likewise the range of values where the BE has a larger risk than the JS given by inequalities (4.7.39) and (4.7.40) result from the solution to

$$1 - \frac{(s-2)^2}{s(s-2+2\delta)} < c^2 + (1-c)^2 \frac{2\delta}{s}. \tag{4.7.41}$$

Again both bounds are positive quantities when $0 < c < 1$.

Exercise 4.7.5. Show that $f(b) = 1/(q + 2b)$ is a convex function of b when $b \geq 0$. Is this still true if $b < 0$?

Exercise 4.7.6. Establish (4.7.35).

Exercise 4.7.7. Show that the left hand side of (4.7.31a) is indeed the MSE/$s\sigma^2$ of the BE $\hat{\beta} = cb$.

4.7.4   Numerical Illustration of the Comparative Properties of Bayes and JS Estimators

The goal of this section is to numerically illustrate the properties of the risk of the JS and the BE and the RL of the JS.

It has already been noted that the JS is an increasing convex function of δ. Some numerical illustrations of this are provided by Table 4.7.2 and Figure 4.7.2 below.

Table 4.7.2
Risk of the JS

s/δ	0	1	2	3	4	5
4	0.500	0.684	0.784	0.842	0.877	0.901
6	0.333	0.509	0.622	0.696	0.748	0.786
8	0.250	0.405	0.514	0.592	0.650	0.694
10	0.200	0.337	0.438	0.514	0.574	0.621

Theorems 4.7.2 and 4.7.4 may be used together to obtain:

1. intervals where the BE has a smaller risk than the JS;
2. intervals where the JS has a smaller risk than the BE.

The idea is to use the endpoints of the intervals that give the widest possible interval where one estimator dominates another. The widest region where the BE has a smaller risk than the JS is obtained using the theorem whose results give the smaller left endpoint and the larger right endpoint. If $0 < c < (1/2)(1+\sqrt{2/s})$ then Theorem 4.7.2 yields the smaller left endpoint, otherwise use the result of Theorem 4.7.4. Likewise if $c > (1/2)(1-\sqrt{2/s})$ Theorem 4.7.4 yields the larger right endpoint, otherwise use Theorem 4.7.2. The widest intervals where the JS has a smaller risk than the BE are obtained using Theorem 4.7.4 for all c where $0 < c < 1$. Values obtained for different values of s are illustrated by Table 4.7.2 .

Exercise 4.7.8. Verify the statements of the above paragraph by solving the relevant inequalities.

The results of Theorems 4.7.2 and 4.7.4 lead to the following general conclusions:

1. If $c < \sqrt{RSL}$ the risk of the BE is less than that of the JS on an interval $[0, \delta_0]$ and greater than that of the JS elsewhere. There is one value where the two estimators cross.

2. If $\sqrt{RL(\infty)} < c$ there is an interval $[\delta_1, \delta_2]$ with both endpoints greater than zero where the BE has a smaller risk than the JS and a greater risk elsewhere. Thus, there are two values where the estimators cross. This behavior of the estimators is illustrated in Table and Figure 4.7.3.

The crossing values were found using the FindRoot command in Mathematica using the endpoints of the intervals given in Theorems 4.7.2 and 4.7.4 as initial values. Figures 4.7.3a, b and c illustrate for s = 6 the case of one crossing point, two crossing points with one at $\delta = 0$ and two crossing points. The crossing values also could have been found using the fsolve command in Maple.

The appendix contains some of the Mathematica commands that were used to generate the graphs. Corresponding commands in Maple are also included for Maple users. The reader who does not have access to Mathematica or Maple can also further investigate the comparison of the JS and the BE using a TI 82 or TI 85 graphing calculator. Software is available to store programs in a Macintosh or IBM compatible computer so that the results can be printed out. A sample program for obtaining the risks of the JS and BE is included in the appendix. The functions can them be plotted and the solver used to find the crossing values.

Exercise 4.7.9. A. For the cases s = 4 and s = 6 verify a few of the values of the risk of the JS doing the computations by hand and with a desk calculator roughly equivalent to a TI 30.
B. Show that of odd s the risks of the JS cannot be computed easily by hand. For s = 3 and s = 7 generate the graphs and tables above using either Mathematica, a graphing calculator or some other software package.

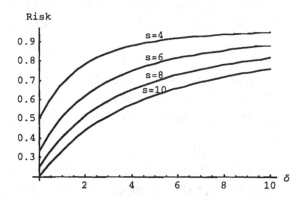

Figure 4.7.2 Risk of the JS

Table 4.7.3

4	0.2	0.3	0.4	$\sqrt{\frac{1}{5}}$	0.5
Crossing	2.42	3.08	3.98	4.51	5.24
BE < JS	(0,2.33)	(0,3.04)	(0,3.97)	(0,4.51)	(0,5.24)
JS < BE	(2.56,∞)	(3.23,∞)	(4.13,∞)	(4.66,∞)	(5.37,∞)

4	$\sqrt{\frac{1}{3}}$	0.6	0.7
Crossing	6.62	7.12	10.25
BE < JS	(0,6.62)	(0,7.12)	(0,10.25)
JS < BE	(6.74,∞)	(7.24,∞)	(10.35,∞)

4	$\sqrt{\frac{1}{2}}$	0.8
Crossing	0,10.55	0.801,16.48
BE < JS	(0,10.55)	(0.929,16.48)
JS < BE	(10.66,∞)	(0,0.422),(16.58,∞)

6	0.2	0.3	0.4	$\sqrt{\frac{1}{5}}$	0.5
Crossing	3.11	4.04	5.30	6.05	7.05
BE < JS	(0,2.92)	(0,3.91)	(0,5.21)	(0,5.98)	(0,7)
JS < BE	(3.33,∞)	(4.27,∞)	(5.52,∞)	(6.27,∞)	(7.28,∞)

6	$\sqrt{\frac{1}{2}}$	0.8
Crossing	1.18,14.49	2.68,22.81
BE < JS	(1.33,14.47)	(3.20,22.79)
JS < BE	(14.69,∞)	(0,2),(23,∞)

6	$\sqrt{\frac{1}{3}}$	0.6	0.7
Crossing	0,0.898	0.167,9.69	1.10,14.07
BE < JS	(0,8.94)	(0.170,9.65)	(1.23,14.05)
JS < BE	(9.20,∞)	(9.90,∞)	(14.27,∞)

Table 4.7.3(continued)

8	0.2	0.3	0.4	$\sqrt{\frac{1}{5}}$	0.5
Crossing	3.72	4.90	6.48	7.43	8.71
BE < JS	(0,3.5)	(0,4.68)	(0,6.33)	(0,7.31)	(0,8.61)
JS < BE	(3.98,∞)	(5.18,∞)	(6.78,∞)	(7.23,∞)	(9,∞)

8	$\sqrt{\frac{1}{3}}$	0.6	0.7
Crossing	0.698,11.55	0.939,12.04	2.34,17.59
BE < JS	(0.732,11.07)	(1,11.97)	(2.67,17.55)
JS < BE	(11.44,∞)	(12.33,∞)	(0,1.79),(17.88,∞)

8	$\sqrt{\frac{1}{2}}$	0.8
Crossing	2.46,18.13	4.86,28.68
BE < JS	(2.83,18.09)	(16.34,28.66)
JS < BE	(0,1.90),(18.41,∞)	(0,4.04),(128.96,∞)

10	0.2	0.3	0.4	$\sqrt{\frac{1}{5}}$
Crossing	4.28	5.70	7.60	0,8.73
BE < JS	(0,4.05)	(0,5.40)	(0,7.39)	(0,8.56)
JS < BE	(4.59,∞)	(6.03,∞)	(7.95,∞)	(9.09,∞)

10	0.5	$\sqrt{\frac{1}{3}}$	0.6
Crossing	0.514,10.27	1.44,13.20	1.77,14.28
BE < JS	(0.528,10.12)	(1.53,13.10)	(1.91,14.18)
JS < BE	(0,3.77),(10,∞)	(0,3.1),(13.56,∞)	(0,1.37),(14.63,∞)

10	0.7	$\sqrt{\frac{1}{2}}$
Crossing	3.18,20.95	3.68,21.60
BE < JS	(14.21,20.89)	(4.44,21.54)
JS < BE	(0,3.02),(12.31,∞)	(0,3.19),(21.96,∞)

Table 4.7.3 (continued)

	10	0.8
Crossing		7.26,34.30
BE < JS		(7.73,34.30)
JS < BE		(0,6.35),(34.65,∞)

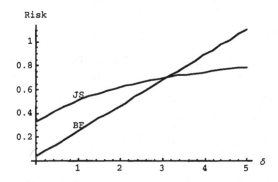

Figure 4.7.3A s = 6 c = 0.2

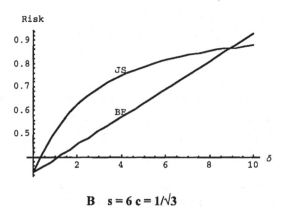

B   s = 6 c = 1/√3

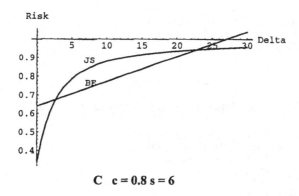

C   c = 0.8 s = 6

4.8 Summary

In this chapter:

1. Some basic ideas about linear models and least square estimation were reviewed.

2. The James-Stein estimator was formulated for a single linear model as:

a. an empirical Bayes estimator;

b. an approximate minimum mean square error estimator.

3. The mean square error of the James-Stein estimator was obtained with and without averaging over a prior distribution.

4. The average and conditional mean square error of the James-Stein estimator was compared with that of the least square estimator for estimable parametric functions and different quadratic loss functions.

5. The properties of two measures of efficiency the RL and the RSL were studied analytically and numerically.

6. Conditions were obtained for the Bayes estimator to have a smaller mean square error averaging over a predictive loss function than the James-Stein estimator and some crossing values of their mean square errors were obtained numerically.

4.9 Appendix

4.9.1 Mathematica Program for Obtaining the Risks of the JS and the BE

```
a[s_,x]=Integrate[Exp[y] y^((s/2)-2),{y,0,x}];
```

```
b[s_,x_]=(s-2)^2/s;
c[s_,x_]=Exp[-x]/(2 x^(s/2-1));
d[s_,x_]=a[s,x] b[s,x] c[s,x];
R[s_,x_]=1-d[s,x];
m[c_,s_,x_]=c^2+(1-c)^2 2 x/s;
```

4.9.2 Maple Program for Obtaining the Risks of the JS and the BE

```
> a:='a':b:='b':c:='c':d='d':r='r':m:='m':
a:=(s,x)->int(exp(y)*y^(s/2-2),y=0..x):
b:=(s,x)->1/((2*x^(s/2-1))*exp(x)):
c:=(s,x)->a(s,x)*b(s,x):
d:=s->((s-2)^2)/s:
r:=(s,x)->(1-d(s,x)*c(s,x)):
m:=(g,s,x)->g^2+((1-g)^2)*2*x/s:
```

4.9.3 TI 82 Program for Obtaining the Risks of the JS and the BE

FINDING WHERE THE JS AND THE BE CROSS
```
Prompt S,C,G
"fnInt(X^(S/2-2)*e^X,X,0,D)"→Y₁
"1/(2D^(S/2-1)*e^D)"→Y₂
(S-2)^2/S→F
"1-Y₁*Y₂*F"→Y₃
"C^2+((1-C)^2)*2*D/S"→Y₄
solve(Y₃-Y₄,D,G)→M
Disp M
```

GRAPHING THE JS AND THE BE
```
Disp "S?"
Disp "C?"
Input S
Input C
"fnInt(H^(S/2-2)*e^H,H,0,K)"→Y₁
"1/(2*H^(S/2-1)*e^H)"→Y₂
(S-2)^2/S→F
FnOff
"1-(Y₁*Y₂*F)"→Y₃
```

```
"C^2+((1-C)^2)*2*H/S"→Y₄
.01→Hmin
Disp "Hmax?"
Input Hmax
5→Hscl
0→Ymin
1→Ymax
.1→Yscl
DispGrap
```

4.9.4 TI 92 Script for finding The Risks of the JS and the BE

```
C:Define a(s,H)=∫(e^(y)*y^(s/2-2),y,0,H)
C:Define b(s,H)=(s-2)^2/s
C:Define c(s,H)=e^(-H)/(2*H^(s/2-1))
C:Define d(s,H)=a(s,H)*b(s,H)*c(s,H)
C:Define r(s,H)=1-d(s,H)
C:Define m(e,s,H)=e^2+(1-e)^2*(2*H/s)
```

The intersection points of the BE and the JS may now be obtained two ways. One way is to use the nSolve command and use the initial values suggested by Theorem 4.7.2 and 4.7.4. . For example to find the intersection for s =4 c = .3 do

$$nSolve(r(4,H)=m(.3,4,H),H)|H≥3.04 \text{ and } H≤3.2.3$$

The answer is x = 3.08259. The other way would be to graph the two functions for the values of s and c and use the Math Menu to find where the graphs intersect.

Chapter V

Ridge Estimators from Different Points of View

5.0 Introduction

The statistical literature contains a number of different mathematical formulations of ridge regression estimators. These include showing how the ridge estimator is:

1. a Bayes estimator (BE);
2. a the mixed estimator;
3. a minimax estimator;
4 . a point on an ellipsoid closest to a point in the parameter space.

In Chapter III the original formulation of the ordinary ridge estimator of Hoerl and Kennard was given; that is by finding the point on an ellipsoid that is closest to the origin. It was explained how for the full rank model a larger class of ridge regression estimators could be formulated for the full rank model by using a weighted sum of squares. The ordinary ridge estimator was also obtained as a special case of the BE. The mixed estimator was mentioned in Exercise 3.1.9.

What is interesting about these different approaches is that although they come from philosophically different points of view they yield estimators with similar mathematical forms. In fact for certain kinds of prior information and design matrices the estimators are the same. Except for the author's own work, Gruber

227

(1990), most articles and books describe the different estimators from only one of the aforementioned points of view. This chapter will provide a unified treatment by:

1. giving different mathematical formulations of ridge type estimators from both the frequentist and the Bayesian point of view;
2. exploring the relationship between the estimators obtained from different kinds of prior information and comparing their efficiencies.

For the sake of generality the non-full rank model will be used. To facilitate this Section 5.1 will extend and elaborate on some of the results on linear models that were presented in Section 4.1. The formulation of least square and restricted least square estimator will be reviewed.

The generalized ridge regression estimator was derived in Section 3.1 for the full rank model by finding the point on an ellipsoid whose distance from the origin is a minimum. In Section 5.2 the derivation of the generalized ridge estimator contained in Section 3.1 will be extended to the non-full rank case. A more general class of estimators will be obtained by minimizing the distance from any point in the parameter space to a point on an ellipsoid.

Section 5.3 deals with mixed estimators obtained from the combination of sample and stochastic prior information. What is interesting here is that the prior information can often be shown to be equivalent to taking additional observations.

The linear minimax estimator will be studied in Section 5.4.

Section 5.5 deals with the formulation of the linear Bayes estimator. Various equivalent forms of the linear Bayes estimator will be derived. These forms will be used to obtain conditions for the various ridge type estimators, the mixed estimator and the linear minimax estimator to be special cases of the Bayes estimator.

The conditions obtained in Chapter III for the ridge estimator to have a smaller MSE than the LS without averaging over the prior distribution will be considered in Section 5.6. Where possible comparisons of BEs derived from different priors will be presented.

The computation of the Bayes risk and the comparison of Bayes estimators with each other and the LS will be presented in Section 5.7.

In Section 5.8 the Jackknifed ridge estimator will be presented. It will be shown that although this estimator reduces bias it does poorly when compared with the ordinary ridge estimator for other measures of efficiency.

All of the results of this chapter are for non-stochastic k.

Section 5.9 is a summary.

Chapter III was intended to be an introduction to the main ideas about ridge type estimators that are commonly used for the solution of practical problems. The present chapter is more concerned with the theoretical aspects of ridge estimators. Much of the material in this chapter is also contained in Chapters III-VII of Gruber (1990) but some of the results are derived in a different way here. The interested reader might like to read this chapter in conjunction with Chapters III-VII of Gruber (1990) to obtain a broader viewpoint.

5.1 Some More Review of Linear Models

Two important facts about linear models are:

1. For the non-full rank case only estimable parametric functions of LS estimators are unique.
2. The LS is the minimum variance linear unbiased estimator (BLUE) (Gauss Markov Theorem).

The LS estimator was derived for the non-full rank case in Chapter IV for the non-full rank model and the concept of estimability was defined. Four different equivalent formulations of estimability will be presented. These will be used interchangeably in the sequel. The Gauss Markov Theorem will be proved. This geometric interpretation of the proof of this Theorem is that the LS is the point on a plane closest to the origin. Recall that the ridge estimator was the point on a sphere or ellipsoid closest to the origin. From a geometric standpoint a similar kind of optimization problem is solved to obtain both the ridge and the LS estimator. Most of the estimators to be studied in this chapter are solutions to one kind or other of constrained optimization problem. The least square estimator subject to an exact prior constraint will be studied in this section.

5.1.1 The LS and the Estimability Concept

The concept of estimability will be defined again this time by stating four equivalent conditions. The reader should recall the role of the matrices U and V in the singular value decomposition given by Definition 4.1.4.

Definition 5.1.1. A parametric function $p'\beta$ is estimable iff one of the following three equivalent conditions hold:

1. $p' V = 0$;
2. $p = Ud$ for some s dimensional vector d;
3. $p'\beta = E(c'Y)$ for some n dimensional vector c;
4. $p' = c'X$ for some n dimensional vector c.

The first of these says that the vector p is orthogonal to the eigenvectors of zero. The second and fourth result says that p is in the column space of X. The third result says that there exists a linear unbiased estimator of $p'\beta$ that is also a linear function of the observations.

Exercise 5.1.1. Prove the equivalence of conditions 1 to 4 in Definition 5.5.1.

Exercise 5.1.2. Given the linear model

$$Y = \begin{bmatrix} 1_4 & 1_4 & 0 & 0 \\ 1_4 & 0 & 1_4 & 0 \\ 1_4 & 0 & 0 & 1_4 \end{bmatrix} \begin{bmatrix} \mu \\ \alpha_1 \\ \alpha_2 \\ \alpha_3 \end{bmatrix} + \varepsilon$$

Show that $\mu + \alpha_i$, $i = 1,2,3$ are estimable functions but the α_i, $i = 1,2,3$ are not.

For estimable parametric functions the LS is independent of the choice of G inverse. This is established formally in Theorem 5.1.1 below.

Theorem 5.1.1. Let G_1 and G_2 be generalized inverses of X'X. For estimable parametric functions $p'\beta$ the LS obtained using these generalized inverses is the same, i.e.,

$$p'G_1X'Y = p'G_2X'Y. \tag{5.1.1}$$

The theorem is very easy to establish using the following simple lemma.

Lemma 5.5.5. Let U be the matrix of eigenvectors corresponding to the non-zero eigenvalues of X'X, that is the U matrix from the singular value decomposition. Then

$$UG_1U' = UG_2U' \tag{5.1.2}$$

Proof. Since G_1 and G_2 are generalized inverses of X'X.

$$X'X = X'XG_1X'X = X'XG_2X'X. \tag{5.1.3}$$

Using the singular value decomposition of X'X,

$$UAU'G_1UAU' = UAU'G_2UAU'. \qquad (5.1.4)$$

Multiplication of both sides of (5.1.4) by $\Lambda^{-1}U'$ yields (5.1.2).

Proof of the Theorem. Using the second of the estimability conditions in Definition 5.1.1 it follows that

$$p'G_1X'Y = d'U'G_1UU'X'Y = d'U'G_2UU'X'Y = p'G_2X'Y. \quad (5.1.5)$$

This uniqueness does not hold when the parametric functions are not estimable. An example is provided by Exercise 5.1.3.

Exercise 5.1.3. A. For the model in Exercise 5.1.1 show that

$$G_1 = \begin{bmatrix} 0 & 0 \\ 0 & \frac{1}{4}I_4 \end{bmatrix}$$

and

$$G_2 = \begin{bmatrix} \frac{1}{16} & 0 \\ -\frac{1}{16} & \frac{1}{4}I_4 \end{bmatrix}$$

are generalized inverses of the X'X matrix.

B. Find the LS for the α_i and μ for each of the G inverses in part A. Are they the same?

C. Find the parametric function of the LS corresponding to $\mu + \alpha_i$ for each of the LS obtained in B. Are they the same?

Exercise 5.1.4. Show that XGX' is independent of the choice of generalized inverse.

5.1.2 The Gauss Markov Theorem

The Gauss Markov Theorem (GM) states that the best linear unbiased estimator of the parametric functions of the parameters of a linear regression model is the least square estimator. An estimator $\hat{\theta}$ is unbiased for a parameter θ iff

$$E(\hat{\theta}) = \theta. \tag{5.1.6}$$

Linear estimators take the form $L'Y$. The formal statement of the GM will now be given. The proof follows that of Gruber (1990).

Theorem 5.1.2 (Gauss Markov). Consider the classical linear regression model

$$Y = X\beta + \varepsilon \tag{5.1.7a}$$

where the Rank $(X) \le m$ and

$$E(\varepsilon) = 0 \text{ and } D(\varepsilon) = \sigma^2 I \tag{5.1.7b}$$

Then:

1. The linear estimator $L'Y$ is unbiased for the parametric function $p'\beta$ iff $p'\beta$ is estimable or the matrix X is of full rank.
2. The minimum variance unbiased linear estimator or best linear unbiased estimator (BLUE) for the estimable parametric functions is the LS.

Proof. The result in 1 is obtained by observing that

$$E[L'Y] = L'X\beta = p'\beta \tag{5.1.8}$$

iff $p' = L'X$ [1] the fourth condition for estimability in Definition 5.1.1.

  To establish the second part of the GM notice that the variance of $L'Y$ is given by

$$V(L'Y) = L'L\sigma^2 \tag{5.1.9}$$

The proof consists of showing that the linear estimator that minimizes (5.1.9) subject to the constraint $p' = L'X$ is the LS. This may be done using Lagrange multipliers. Let

$$H = L'L\sigma^2 + (L'X - p')\lambda. \tag{5.1.10}$$

---

[1] This condition holds automatically when X is of full rank.

where λ is an m dimensional column vector of Lagrange multipliers. The expression in (5.1.10) is differentiated with respect to L and the result is set equal to zero. Thus,

$$L\sigma^2 + X\lambda = 0. \tag{5.1.11}$$

Differentiating (5.1.10) with respect to λ, setting the result equal to zero and transposing it gives the transpose of the estimability condition

$$X'L - p = 0. \tag{5.1.12}$$

The vectors λ and L will be obtained by solving (5.1.11) and (5.1.12) simultaneously. To accomplish this multiply (5.1.11) by X'. Obtain

$$X'L\sigma^2 + X'X\lambda = 0. \tag{5.1.13}$$

From (5.1.12) and (5.1.13)

$$\lambda = -GX'L\sigma^2 = -Gp\sigma^2 \tag{5.1.14}$$

where G is any generalized inverse of $X'X$. The optimum L is obtained by substituting the expression for λ in (5.1.14) into (5.1.11). Thus,

$$L = XGp. \tag{5.1.15}$$

Thus, the minimum variance unbiased linear estimator is

$$p'\beta = p'G'X'Y. \tag{5.1.16}$$

Equation (5.1.16) is the LS independent of G because $p'\beta$ is estimable. Note that if G is a generalized inverse of $X'X$ so is G'.

The proof of the GM amounts to finding the point in n dimensional space on the hyperplane $L'X = p'$ that lies closest to the origin.

Exercise 5.1.5. Show that if G is a generalized inverse of $X'X$ so is G'. When is G a symmetric matrix?

Exercise 5.1.6. A. For the model in Exercise 5.1.1 is it possible to find the BLUE of α_2 ?

B. Find the BLUE of $\alpha_1 - \alpha_2$.

Example 5.1.1. A Geometric Interpretation of the BLUE
Consider the linear model

$$X_i = \mu + \varepsilon_i, \ 1 \le i \le n.$$

Consider the geometric problem of finding the minimum distance of the a point in n dimensional space from the origin that lies on the hyperplane

$$Y_1 + Y_2 + Y_3 + \ldots + Y_n = 1$$

This is accomplished by differentiating

$$H = Y_1^2 + Y_2^2 + Y_3^2 + \ldots Y_n^2 - \lambda(Y_1 + Y_2 + Y_3 + \ldots Y_n - 1)$$

with respect to the Y_i s and setting the result equal to zero. Thus,

$$\frac{\partial H}{\partial Y_i} = 2Y_i - \lambda = 0$$

Then $Y_i = \lambda/2$. Substitution into the constraint equation yields $\lambda = 2/n$ and the point on the plane that is closest to the origin is $Y=(1/n, \ 1/n,\ldots,1/n)$. This equivalent to finding the unbiased linear estimator

$$\hat{\mu} = a_1 X_1 + a_2 X_2 + \ldots a_n X_n$$

with minimum variance because:

1. The variance can be shown to be

$$v = (a_1^2 + a_2^2 + a_3^2 + \ldots + a_n^2)\sigma^2.$$

2. The estimator $\hat{\mu}$ is unbiased for μ if

$$a_1 + a_2 + a_3 + \ldots a_n = 1.$$

The BLUE is the sample mean . This is the LS estimator for the parameters of the linear model of this example. When $n = 3$ this reduces to the problem of finding the point on the plane $X + Y + Z = 1$ that is closest to the origin. Likewise when $n = 2$ it reduces to finding the point on the line $X + Y = 1$ that is closest to the origin. See Exercise 5.1.7.

In dealing with regression data often there is correlation between the observations or some of the observations should receive more weight than others. One way to deal with these situations is to use the weighted least square estimator. The weighted least square error is obtained by minimizing the weighted sum of squares of the differences between the observed and the predicted values. Thus, the quantity

$$H = (Y - X\beta)'V^{-1}(Y - X\beta) \tag{5.1.17}$$

is minimized resulting in

$$p'b = p'G_V X'V^{-1}Y. \tag{5.1.18}$$

where G_V is a generalized inverse of $(X'V^{-1}X)$. One possible generalized inverse is the Moore-Penrose inverse denoted by $G_V = (X'V^{-1}X)^+$.

If the conditions (5.1.7b) are replaced by

$$E(\varepsilon) = 0 \text{ and } D(\varepsilon) = V \tag{5.1.19}$$

and the estimability condition by $p' = t'XV^{-\frac{1}{2}}$ the Gauss-Markov theorem holds true.

C.R. Rao (1973) considers the problem of finding the best linear unbiased estimator (BLUE) for the case where V is singular.

Exercise 5.1.7. Consider a linear estimator $\hat{\mu} = a_1 X_1 + a_2 X_2 + a_3 X_3$ of a parameter μ. Assume that:

1. The X_i are uncorrelated.
2. The mean of the X_i is μ.
3. The X_i have unequal variances σ_i^2.

Find the BLUE.

Exercise 5.1.8. Prove the GM assuming the conditions in (5.1.19) and the estimability condition $p' = t'XV^{-1/2}$ holds true.

5.1.3 The Constrained LS

The least square estimator may also be obtained together with prior constraints; i.e. exact or stochastic linear relationships between the β parameters. The case of exact linear relationships will be considered here. The case of stochastic prior constraints will be considered in Section 5.3 when the mixed estimator is derived. The optimization problem to be solved is that of minimizing

$$L = (Y - X\beta)'(Y - X\beta) \tag{5.1.20}$$

subject to the constraint $R\beta = r$ where R is an txm dimensional matrix and r is a t dimensional vector. Let λ be a t dimensional column vector of Lagrange multipliers . Differentiate

$$H = (Y - X\beta)'(Y - X\beta) + 2\lambda'(R\beta - r) \tag{5.1.21}$$

with respect to β and set the result equal to zero to obtain

$$X'X\beta - X'Y + R'\lambda = 0 \tag{5.1.22}$$

Let G be a generalized inverse of $X'X$. Then a solution to (5.1.22) is

$$\hat{\beta} = GX'Y - GR'\lambda = b - GR'\lambda. \tag{5.1.23}$$

The solution in (5.1.23) must satisfy the constraint equation $R\beta = r$. Thus, substitution of (5.1.23) into the constraint equation yields

$$RGR'\lambda = r - Rb. \tag{5.1.24}$$

Let M be a generalized inverse of RGR'. Solving (5.1.24) for λ yields

$$\lambda = M(r - Rb). \tag{5.1.25}$$

Thus,
$$\hat{\beta} = b - GR'M(r - Rb). \tag{5.1.26}$$

The estimator (5.1.26) is the constrained LS.

An important special case is when $R = AU'$ with A a txs matrix of full row rank, that is, the constraints are linearly independent and estimable. Then RGR' =

$AU'(X'X)^+UA' = A\Lambda^{-1}A'$, a nonsingular matrix and $M = (R(X'X)^+R')^{-1}$. For estimable parametric functions it is easy to show that (5.1.26) is independent of the choice of G inverse of $X'X$. Thus, if $p'\beta$ is estimable,

$$p'\hat{\beta} = p'b - p'(X'X)^+R'(R(X'X)^+R')^{-1}(r - Rb). \qquad (5.1.27)$$

Example 5.1.2 Consider a linear model with X matrix as given in Exercise 5.1.1 together with the constraint $R\beta = r$ where

$$R = \begin{bmatrix} 0 & 1 & -1 & 0 \\ 0 & 1 & 1 & -2 \end{bmatrix}$$

The matrix R satisfies the condition $R = AU'$ where

$$A = \begin{bmatrix} 0 & 1.225 & -0.707 \\ 0 & -1.225 & -2.121 \end{bmatrix}$$

$$U = \begin{bmatrix} -0.866 & 0 & 0 \\ -0.289 & 0.408 & -0.707 \\ -0.289 & -0.816 & 0 \\ -0.289 & 0.408 & 0.707 \end{bmatrix}.$$

The reader may show that

$$(X'X)^+ = \begin{bmatrix} \frac{3}{64} & \frac{1}{64} & \frac{1}{64} & \frac{1}{64} \\ \frac{1}{64} & \frac{11}{64} & -\frac{5}{64} & -\frac{5}{64} \\ \frac{1}{64} & -\frac{5}{64} & \frac{11}{64} & -\frac{5}{64} \\ \frac{1}{64} & -\frac{5}{64} & -\frac{5}{64} & \frac{11}{64} \end{bmatrix}.$$

Then

$$R(X'X)^+R' = \begin{bmatrix} \frac{1}{2} & 0 \\ 0 & \frac{3}{2} \end{bmatrix},$$

$$(R(X'X)^+R')^{-1} = \begin{bmatrix} 2 & 0 \\ 0 & \frac{2}{3} \end{bmatrix}$$

and

$$(X'X)^+R'(R(X'X)^+R')^{-1} = \begin{bmatrix} 0 & 0 \\ \frac{1}{2} & \frac{1}{6} \\ -\frac{1}{2} & \frac{1}{6} \\ 0 & -\frac{1}{3} \end{bmatrix}.$$

Thus, the basic estimable parametric functions would be estimated by

$$\begin{bmatrix} \mu \hat{+} \alpha_1 \\ \mu \hat{+} \alpha_2 \\ \mu \hat{+} \alpha_3 \end{bmatrix} = \begin{bmatrix} \overline{Y}_{.1} + \frac{1}{2}(r_1 - \frac{1}{4}(\overline{Y}_{.1} - \overline{Y}_{.2})) + \frac{1}{6}(r_2 - \frac{1}{4}(\overline{Y}_{.1} + \overline{Y}_{.2} - 2\overline{Y}_{.3})) \\ \overline{Y}_{.2} - \frac{1}{2}(r_1 - \frac{1}{4}(\overline{Y}_{.1} - \overline{Y}_{.2}) + \frac{1}{6}(r_2 - \frac{1}{4}(\overline{Y}_{.1} + \overline{Y}_{.2} - 2\overline{Y}_{.3})) \\ \overline{Y}_{.3} - \frac{1}{3}(r_2 - \frac{1}{4}(\overline{Y}_{.1} + \overline{Y}_{.2} - 2\overline{Y}_{.3}) \end{bmatrix}.$$

Exercise 5.1.9. Show that (5.1.26) is independent of the choice of G inverse.

Exercise 5.1.10. In Example 5.1.2 above
A. Verify the numerical value of the Moore Penrose inverse of $X'X$.
B. What are the constrained LS estimators of $\alpha_1 - \alpha_2$ and $\alpha_1 + \alpha_2 - \alpha_3$?
C. Show that if the constraint equations hold true the estimators of the basic estimable parametric functions are unbiased.

5.1.4 Some Useful Matrix Theory Results

This subsection will summarize a number of useful results from matrix theory that will be used at various points in the sequel.

The following consequence of Theorem 3.2.4 due to Baksalary and Kala(1983) is one such result.

Theorem 5.1.3. Let A be a symmetric n x n matrix, a an n x 1 vector and d a positive scalar. Then dA - aa' is positive semidefinite iff:

1. The matrix A is non-negative definite.
2. The vector a belongs to the range of A.
3. The inequality $a'A^+a \leq d$ holds true.

Proof. The matrix A has as its singular value decomposition

$$A = [P \quad Q]\begin{bmatrix} \Lambda & 0 \\ 0 & 0 \end{bmatrix}\begin{bmatrix} P' \\ Q' \end{bmatrix} = P\Lambda P'. \tag{5.1.28}$$

Now a necessary and sufficient condition for dA - aa' to be positive semidefinite is that the matrix

$$M = d\begin{bmatrix} \Lambda & 0 \\ 0 & 0 \end{bmatrix} - \begin{bmatrix} P'aa'P & P'aa'Q \\ Q'aa'P & Q'aa'Q \end{bmatrix} \tag{5.1.29}$$

is positive definite. This is possible iff both $d\Lambda$ - P'aa'P is positive semidefinite and Q'aa'Q = 0. However Q'aa'Q = 0 iff Q'a = 0 or equivalently a belongs to the range of A. Since M is positive semidefinite iff PMP' is positive semidefinite the result follows from 5.1.29.

The following result will also be used in the sequel.

Theorem 5.1.4. Let A and B be positive definite matrices. Assume that B − A is non-negative definite. Then $A^{-1} - B^{-1}$ is non-negative definite.
Proof. First observe that since B is positive definite from finding its singular value decomposition and the properties of orthogonal matrices

$$B = P_B \Lambda_B P_B' = P_B \Lambda^{1/2} P_B' P_B \Lambda^{1/2} P_B' = B^{1/2} B^{1/2} . \tag{5.1.30}$$

Now $B^{1/2}$ is also a positive definite matrix. Thus, $B^{-1/2}AB^{-1/2}$ is a positive definite matrix with singular value decomposition

$$B^{-1/2}AB^{-1/2} = Q\Delta Q'. \tag{5.1.31}$$

and

$$A = B^{1/2}Q\Delta Q'B^{1/2} = W'\Delta W \tag{5.1.32}$$

where $W = Q'B^{1/2}$. Then

$$B = W'W. \tag{5.1.33}$$

Now $B - A = W'(I - \Delta)W \geq 0$. Thus $I - \Delta \geq 0$ and $\Delta^{-1} - I \geq 0$. But then

$$A^{-1} - B^{-1} = W^{-1}\Delta^{-1}W'^{-1} - W^{-1}W'^{-1} = W^{-1}(\Delta^{-1} - I)W'^{-1} \geq 0. \tag{5.1.34}$$

Exercise 5.1.10. Let $A = \begin{bmatrix} 3 & 1 \\ 1 & 3 \end{bmatrix}$ and $B = \begin{bmatrix} 2 & 1 \\ 1 & 2 \end{bmatrix}$. Show that

A. The matrix $A - B$ is non-negative definite.
B. The matrix $B^{-1} - A^{-1}$ is non-negative definite.

Exercise 5.1.11. Let $A = \begin{bmatrix} \frac{3}{2} & \frac{1}{2} \\ \frac{1}{2} & \frac{3}{2} \end{bmatrix}$, $a = \begin{bmatrix} 1 \\ 1 \end{bmatrix}$ and $d = 2$. Show that

A. $dA - aa'$ is positive semidefinite.
B. The three conditions of Theorem 5.1.3 hold true.

5.2 The Generalized Ridge Regression Estimator

In Chapter III the ridge regression estimator was obtained by minimizing the distance from the vector B to the origin subject to the side condition that it lay on a certain ellipsoid. Different weighted distances yielded the different kinds of shrinkage estimators , e.g., the ordinary ridge estimator of Hoerl and Kennard and the contraction estimator of Mayer and Willke. These ridge estimators will now be generalized to include the cases where:

1. the linear model need not be of full rank;
2. the distance to minimized will be from B to some point in the parameter space denoted by the vector θ.

The generalized ridge regression estimator is obtained here by minimizing the weighted distance

$$D = (B - \theta)'H(B - \theta) \qquad (5.2.1)$$

subject to the side condition that B lies on the ellipsoid

$$(B - b_1)'X'X(B - b_1) = \phi_0 \qquad (5.2.2)$$

with H a positive semidefinite matrix.

Let λ be a scalar Lagrange multiplier. Differentiating

$$L = (B - \theta)'H(B - \theta) + \lambda[(B - b_1)'X'X(B - b_1) - \phi_0] \qquad (5.2.3)$$

yields

$$H(B - \theta) + \lambda X'X(B - b_1) = 0. \qquad (5.2.4)$$

Solving Equation (5.2.4) for B, the optimum estimator is

$$\hat{\beta} = [H + \lambda X'X]^{\dagger}[H\theta + \lambda X'Xb_1], \qquad (5.2.5)$$

the generalized ridge estimator.

Exercise 5.2.1. In deriving the estimator in (5.2.5) the Moore Penrose inverse of the matrix $H + \lambda X'X$ was used. Show that:

A. The estimator (5.2.5) is independent of the generalized inverse of $H + \lambda X'X$ for estimable parametric functions.

B. Let $H = P\Delta P'$ be the singular value decomposition of H. Show that if there is a d and e such that $p' = d'U' + e'P'$ then the parametric functions are independent of the choice of G inverse of $X'X + \lambda H$.

C. Are all parametric functions of the type described in B estimable? Are all estimable functions members of the class of functions that were presented in B?

5.3 The Mixed Estimators

In Section 5.1 least square estimators with exact prior constraints were considered. In this Section stochastic prior constraints will be considered. The stochastic prior assumptions correspond to additional observations based on either additional information or prior knowledge about the parameters and take the form of an additional linear model. The resulting least square estimators obtained using both the original and the additional linear model are called mixed estimators because they are derived using a "mixture" of sample and prior information.

Consider the augmented linear model

$$\begin{bmatrix} Y \\ r \end{bmatrix} = \begin{bmatrix} X \\ R \end{bmatrix} \beta + \begin{bmatrix} \varepsilon \\ \phi \end{bmatrix} \tag{5.3.1}$$

where

$$E\begin{bmatrix} \varepsilon \\ \phi \end{bmatrix} = 0 \text{ and } D\begin{bmatrix} \varepsilon \\ \phi \end{bmatrix} = \begin{bmatrix} \sigma^2 I & 0 \\ 0 & \tau^2 I \end{bmatrix}. \tag{5.3.2}$$

The augmented linear model is made up of two uncorrelated [2] classical regression models. The first is the one that has been considered all along. For the second model R is a txm matrix, r is a t dimensional vector of observations and ϕ is a t dimensional error vector.

The mixed estimator is the weighted LS that is obtained by minimizing

$$F(\beta) = (Y - X\beta)'(Y - X\beta)\frac{1}{\sigma^2} + (r - R\beta)'(r - R\beta)\frac{1}{\tau^2} \tag{5.3.3}$$

yielding the normal equations

$$(X'X\frac{1}{\sigma^2} + R'R\frac{1}{\tau^2})\beta = \frac{1}{\sigma^2}X'Y + \frac{1}{\tau^2}R'r. \tag{5.3.4}$$

The solution to the normal equation (5.3.4) is the mixed estimator

$$\begin{aligned} \hat{\beta} &= (X'X\tau^2 + R'R\sigma^2)^+ (\tau^2 X'Y + \sigma^2 R'r) \\ &= (X'X\tau^2 + R'R\sigma^2)^+ (\tau^2 X'Xb_1 + \sigma^2 R'Rb_2) \end{aligned} \tag{5.3.5}$$

---

[2] Independent when both the error terms are normally distributed.

with $b_1 = (X'X)^+ X'Y$ and $b_2 = (R'R)^+ R'r$. The mixed estimator is the LS with respect to augmented model weighted by the error variances of both models.

For certain parametric functions the choice of generalized inverse does not matter. To explain when this is the case the following generalization of the concept of estimability is needed.

Definition 5.3.1. Given the augmented model (5.3.1) the parametric function $p'\beta$:

1. is X estimable if it is estimable for the first of the two models that make up (5.3.1);
2. is R estimable if it is estimable for the second of the two models that make up (5.3.1);
3. is (X, R) estimable if it is estimable for the augmented model.

It is not hard to show that a parametric function that is either X estimable or R estimable, not necessarily both, is (X, R) estimable. Exercise 5.3.1 gives the reader the opportunity to do this and to see an example that shows an (X,R) estimable parametric function need not be either X estimable or R estimable.

Exercise 5.3.1. A. Show that an (X, R) estimable parametric function is X estimable and R estimable.

B. Consider the augmented linear model

$$Y = \begin{bmatrix} 1 & 1 & 0 \\ 1 & 1 & 0 \\ 1 & 0 & 1 \\ 1 & 0 & 1 \\ 0 & 1 & 0 \\ 0 & 0 & 1 \end{bmatrix} \begin{bmatrix} \mu \\ \alpha \\ \beta \end{bmatrix} + \varepsilon$$

with $X = \begin{bmatrix} 1 & 1 & 0 \\ 1 & 1 & 0 \\ 1 & 0 & 1 \\ 1 & 0 & 1 \end{bmatrix}$ and $R = \begin{bmatrix} 0 & 1 & 0 \\ 0 & 0 & 1 \end{bmatrix}$. Show that $\mu + \alpha + \beta$ is (X,R) estimable but that it is neither X estimable nor R estimable.

Exercise 5.3.2. What is the form of the mixed estimator when $D \begin{bmatrix} \varepsilon \\ \phi \end{bmatrix} = \begin{bmatrix} V_1 & 0 \\ 0 & V_2 \end{bmatrix}$ where V_1 and V_2 are nonsingular matrices? The other assumptions in (5.3.1) and (5.3.2) remain the same.

The mixed estimator is a generalized ridge estimator with $H = R'R$, $\lambda = \tau^2/\sigma^2$ and $\theta = (R'R)^+ R'Rb_2$. Likewise the generalized ridge estimator is a mixed estimator where $\tau^2 = \lambda\sigma^2$, $R = \Delta^{1/2}P'$ from the singular value decomposition of H (see Exercise (5.2.1) and $r = \Delta^{1/2}P'\theta$.

Let $k = \sigma^2/\tau^2$. Suppose that $r = 0$. When R is an orthogonal matrix the ridge estimator of Hoerl and Kennard results. If $R = X$ then the contraction estimator of Mayer and Willke is obtained. If $R = (\tau^2/\sigma^2) S'K^{1/2}U'$, where U and S is from the singular value decomposition of X the generalized ridge estimator of Hoerl and Kennard (1970) results. Thus, the ridge regression estimators are special cases of mixed estimators.

By making the substitutions suggested above the ridge estimators can be obtained for specific values of k or K using the routines in Statistical Software packages like SAS or Minitab for obtaining least square estimators.

For the (X, R) estimable parametric functions the mixed estimator is the BLUE as a result of the GM. Thus, for an appropriate augmented model the generalized ridge estimator derived in Section 5.2 is unbiased. The ordinary and generalized ridge estimator of Hoerl and Kennard and the contraction estimator of Mayer and Willke are biased for the assumptions of the augmented model.

Example 5.3.1. Consider an augmented linear model with X and R as in Example 5.2.1. Let $\sigma^2 = \tau^2 = 1$. Then

$$X'X + R'R = \begin{bmatrix} 12 & 4 & 4 & 4 \\ 4 & 4 & 0 & 0 \\ 4 & 0 & 4 & 0 \\ 4 & 0 & 0 & 4 \end{bmatrix} + \begin{bmatrix} 0 & 0 & 0 & 0 \\ 0 & 2 & 0 & -2 \\ 0 & 0 & 2 & -2 \\ 0 & -2 & -2 & 4 \end{bmatrix}$$

$$= \begin{bmatrix} 12 & 4 & 4 & 4 \\ 4 & 6 & 0 & -2 \\ 4 & 0 & 6 & -2 \\ 4 & -2 & -2 & 8 \end{bmatrix}$$

and

$$(X'X + R'R)^+ = \frac{1}{960} \begin{bmatrix} 45 & 15 & 15 & 15 \\ 15 & 101 & -59 & -27 \\ 15 & -59 & 101 & -27 \\ 15 & -27 & -27 & 69 \end{bmatrix}.$$

Then the mixed estimator is

$$\begin{bmatrix} \hat{\beta}_1 \\ \hat{\beta}_2 \\ \hat{\beta}_3 \\ \hat{\beta}_4 \end{bmatrix} = \frac{1}{960} \begin{bmatrix} 60Y_{..} \\ 15Y_{..} + 101Y_{1.} - 59Y_{2.} - 27Y_{3.} + 160r_1 + 96r_2 \\ 15Y_{..} - 59Y_{1.} + 101Y_{2.} - 27Y_{3.} - 160r_1 + 96r_2 \\ 15Y_{..} - 27Y_{1.} - 27Y_{2.} + 69Y_{3.} - 192r_2 \end{bmatrix}.$$

The least square estimators are

$$b_1 = (X'X)^+ X'Y = \frac{1}{64} \begin{bmatrix} 3 & 1 & 1 & 1 \\ 1 & 11 & -5 & -5 \\ 1 & -5 & 11 & -5 \\ 1 & -5 & -5 & 11 \end{bmatrix} \begin{bmatrix} 1_4' & 1_4' & 1_4' \\ 1_4' & 0 & 0 \\ 0 & 1_4' & 0 \\ 0 & 0 & 1_4' \end{bmatrix} Y$$

$$= \frac{1}{64} \begin{bmatrix} 4Y_{..} \\ Y_{..} + 11Y_{1.} - 5Y_{2.} - 5Y_{3.} \\ Y_{..} - 5Y_{1.} + 11Y_{2.} - 5Y_{3.} \\ Y_{..} - 5Y_{1.} - 5Y_{2.} + 11Y_{3.} \end{bmatrix}$$

and

$$b_2 = (R'R)^+ R'r = \frac{1}{18} \begin{bmatrix} 0 & 0 & 0 & 0 \\ 0 & 5 & -4 & -1 \\ 0 & -4 & 5 & -1 \\ 0 & -1 & -1 & 2 \end{bmatrix} \begin{bmatrix} 0 & 0 \\ 1 & 1 \\ -1 & 1 \\ 0 & -2 \end{bmatrix} \begin{bmatrix} r_1 \\ r_2 \end{bmatrix}$$

$$= \begin{bmatrix} 0 \\ \frac{1}{2}r_1 + \frac{1}{6}r_2 \\ -\frac{1}{2}r_1 + \frac{1}{6}r_2 \\ -\frac{1}{12}r_2 \end{bmatrix}$$

Example 5.3.3 The data below is taken from the Economic Report of the President (1988). It is concerned with the relationship between the dependent variable, Y (personal consumption expenditures) in billions of dollars and two other independent variables X_1, the Gross National Product and Personal Income in billions of dollars X_2.

Table 5.3.1

Obs	Year	Y	X_1	X_2
1	1975	1012.8	1598.4	1313.4
2	1976	1129.3	1782.8	1451.4
3	1977	1257.2	1990.5	1607.5
4	1978	1403.5	2249.7	1812.4
5	1979	1566.8	2508.2	2034.0
6	1980	1732.6	2732.0	2258.5
7	1981	1915.1	3052.6	2520.9
8	1982	2050.7	3166.0	2670.8
9	1983	2234.6	3405.7	2838.6
10	1984	2430.5	3772.2	3108.7

The relationship between the dependent and the independent variable could be studied using data from earlier years, if available, as prior information, e.g., 1965-1969. The prior information, also taken from the same source is given in Table 5.3.2.

Table 5.3.2

Obs	Year	Y	X_1	X_2
1	1965	440.7	705.1	552.0
2	1966	477.3	772.0	600.8
3	1967	503.6	816.4	644.5
4	1968	552.5	892.7	707.2
5	1969	579.9	963.9	772.9

For the data in Tables 5.3.1 and 5.3.2 the least square estimators are respectively

$$b_1 = \begin{bmatrix} -23.6082 \\ 0.143937 \\ 0.610175 \end{bmatrix} \text{ and } b_2 = \begin{bmatrix} 24.4237 \\ 0.968207 \\ -0.484003 \end{bmatrix}.$$

The second expression in (5.3.5) with sample estimators s_1^2 and s_2^2 in place of σ^2 and τ^2 is

$$\hat{\beta} = (X'Xs_2^2 + R'Rs_1^2)^+ (s_2^2 X'Xb_1 + s_1^2 R'Rb_2).$$

The estimates of the error variances from the within sums of squares are $s_1^2 = 545.933$ and $s_2^2 = 32.7883$. After substitution of these values and the matrices X and R obtained from Tables 5.3.1 and 5.3.2 the numerical value of the mixed estimator is

$$\hat{\beta} = \begin{bmatrix} -11.1703 \\ 0.341109 \\ 0.364529 \end{bmatrix}.$$

Exercise 5.3.3. A. Show that relative to the assumptions in (5.3.2) the mixed estimator is unbiased for the (X,R) estimable parametric functions and obtain its variance. Show that this variance is smaller than that of the LS.
B Show that if the expectation of the mixed estimator is obtained conditional on the random variable r then it is not unbiased. Compare the conditional variance and the conditional MSE with that of the LS. Obtain conditions for the conditional MSE to be smaller than that of the LS.

Exercise 5.3.4. For Example 5.3.2 above show that
A. The mixed estimator is the same when the second expression of Equation 5.3.5 is used.
B. The mixed estimator of the basic estimable functions $\mu + \alpha_i$, $i = 1,2,3$ is unbiased. Thus, it is unbiased for all estimable parametric functions.
C. Show that the parametric function $\mu + 3\alpha_1 - \alpha_2 - \alpha_3$ is both X estimable and R estimable.
D. If possible give an example of an (X,R) estimable parametric function that is not X estimable or R estimable.

Exercise 5.3.5. In the model in Example 5.3.2 above replace R by any orthogonal   4
x 4 matrix.
A. Find the mixed estimator.
B. Explain why every parametric function is R estimable and (X, R) estimable but
need not be X estimable.

Exercise 5.3.6. Redo the computation for the mixed estimator in Example 5.3.3
using the first expression in (5.3.5).

Exercise 5.3.7. Find the mixed estimator in Example 5.3.3 by using the prior
information below in place of Table 5.3.2. Why is your numerical value different
from that of the example?

Obs	Year	Y	X_1	X_2
1	1970	640.0	1015.5	831.8
2	1971	691.6	1102.7	894.0
3	1972	757.6	1212.8	981.6

5.4 The Linear Minimax Estimator

The linear minimax estimator is found by minimizing the maximum value of a class
of linear estimators on an ellipsoid. The class of linear estimators takes the form

$$p'\hat{\beta} = p'\theta + L'(Y - X\theta) \qquad (5.4.1)$$

where θ is an m dimensional vector of parameters. Let G be a positive definite mxm
matrix and let β be an m dimensional vector of parameters. The ellipsoid takes the
form

$$\Omega = \{\beta : (\beta - \theta)'G(\beta - \theta) \le 1\}. \qquad (5.4.2)$$

The following steps are needed to find the minimax estimator.

1. A expression for the risk of (5.4.1) will be obtained.
2. The maximum value of the risk on the ellipsoid (5.4.2) in terms of L will be
obtained.
3. The vector L that minimizes the expression obtained in 2 above will be obtained.

The expression for the risk or MSE of $p'\beta$ is

$$p'Rp = p'E(\hat{\beta} - \beta)(\hat{\beta} - \beta)'p$$
$$= E[p'(\theta - \beta) + L'(Y - X\theta)][(\theta - \beta)'p + (Y' - \theta'X')L]$$
$$= (L'X - p')(\beta - \theta)(\beta - \theta)'(XL - p) + \sigma^2 L'L. \qquad (5.4.3)$$

An equivalent form of the ellipsoid (5.4.2) is

$$(\beta - \theta)(\beta - \theta)' \le G^{-1} = F. \qquad (5.4.4)$$

Thus, the maximum of (5.4.3) on the ellipsoid (5.4.4) is given by

$$p'Rp = (L'X - p')F(X'L - p) + \sigma^2 L'L. \qquad (5.4.5)$$

To minimize (5.4.5) obtain the derivative of (5.4.5) with respect to L and set the result equal to zero. Thus,

$$2XFX'L - 2p'FX' + L\sigma^2 = 0. \qquad (5.4.6)$$

The derivative of the right hand side of (5.4.6) with respect to L is $2XFX' + I\sigma^2$, a positive definite matrix. Thus, (5.4.5) is minimized when

$$L' = p'FX'(XFX' + \sigma^2 I)^{-1}. \qquad (5.4.7)$$

Then the minimax estimator is

$$p'\hat{\beta} = p'\theta + p'FX'(XFX' + \sigma^2 I)^{-1}(Y - X\theta). \qquad (5.4.8)$$

The results can be extended to the non-full rank case by replacing the ellipsoid (5.4.2) by a region of the form

$$(\beta - \theta)(\beta - \theta)' \le F \qquad (5.4.9)$$

where the matrix F is non-negative definite not necessarily of full rank. By Theorem 5.1.3 the region (5.4.9) is an ellipsoid of the form (5.4.2) with $G = F^+$. The special case where $U'FU$ is positive definite is dealt with in Exercise 5.4.1.

Exercise 5.4.1. Assume that $U'GU$ is positive definite.
A. Show that

$$(\beta - \theta)'(UU'GUU')^+ (\beta - \theta) \le 1 \qquad (5.4.10)$$

iff

$$U'(\beta - \theta)(\beta - \theta)'U \le U'GU. \qquad (5.4.11)$$

B. Show that when $p'\beta$ is an estimable parametric function Equation (5.4.3) may be rewritten

$$p'Rp = (L'X - p)UU'(\beta - \theta)(\beta - \theta)'UU'(XL - p) + \sigma^2 LL'. \quad (5.4.12)$$

C. Show that (5.4.5) is the maximum of $p'Rp$ on the ellipsoid in (5.4.10) and consequently the minimax estimator is still (5.4.8).

Exercise 5.4.2. Assume that $X'X$ and G are nonsingular.

A. Find the estimator of the form

$$p'\hat{\beta} = p'\theta + c'(b - \theta)$$

whose maximum on the ellipse (5.4.2) is minimized.
B. Do the same problem for an estimator of the form

$$p'\hat{\beta} = p'\theta + L'X'(Y - X\theta).$$

C. Show that the results obtained in A and B are algebraically equivalent to (5.4.8).

5.5 The Linear Bayes Estimator

In this section:

1. The linear Bayes estimator (BE) will be derived following C.R. Rao (1973) (Section 5.5.1).
2. A number of algebraically equivalent forms of the BE will be obtained (Section 5.5.2).
3. These equivalent forms will be used to give conditions for mixed and ridge regression estimators to be special cases of the BE (Section 5.5.3).

5.5.1 The Derivation of the Linear Bayes Estimator

Assume that the vector of regression parameters to be estimated in the classical linear model is a random variable with known prior mean and dispersion given by the equations

$$E(\beta) = \theta \text{ and } D(\beta) = F \qquad (5.5.1)$$

where θ is an m dimensional vector and F is a positive semidefinite matrix. Also assume that

$$E(\varepsilon \mid \beta) = 0 \text{ and } D(\varepsilon \mid \beta) = \sigma^2 L \qquad (5.5.2)$$

Conditional expectations are used in (5.5.2) because the β parameters are random variables. The expectations computed below are obtained averaging over the prior distribution.

The linear Bayes estimator is obtained as the solution to the optimization problem of finding the linear estimator

$$\hat{\beta} = a + L'Y \qquad (5.5.3)$$

with minimum variance averaging over the prior distribution that satisfies the extended unbiasedness condition

$$E(p'\beta - a - L'Y) = 0. \qquad (5.5.4)$$

The variance to be minimized , subject to (5.5.4), is

$$\begin{aligned} v &= \mathrm{Var}(p'\beta - a - L'Y) \\ &= \mathrm{Var}(p'\beta) + \mathrm{Var}(L'Y) - 2\mathrm{cov}(p'\beta, L'Y) \qquad (5.5.5) \\ &= p'Fp + L'(XFX' + \sigma^2 I)L - 2p'FX'L. \end{aligned}$$

Calculating the expectation in (5.5.4) and solving the resultant equation for a gives

$$a = (p' - L'X)\theta. \qquad (5.5.6)$$

Differentiate Equation (5.5.5) with respect to L, set the result equal to zero, and isolate the expression containing L on one side of the equation.Thus,

$$(XFX' + \sigma^2 I)L = XFp \qquad (5.5.7)$$

and

$$L' = p'FX'(XFX' + \sigma^2 I)^{-1}. \qquad (5.5.8)$$

Observe that $XFX' + \sigma^2 I$ is positive definite. From Equations (5.5.6) and (5.5.8) the optimum estimator for both the full and the non-full rank case is

$$p'\hat{\beta} = p'\theta + p'FX'(XFX' + \sigma^2 I)^{-1}(Y - X\theta). \qquad (5.5.9)$$

The estimator (5.5.9) is the best linear unbiased estimator where unbiasedness here is in the extended sense relative to the prior assumptions (5.5.1). Notice that the mathematical form of the linear BE and the linear minimax estimator is the same.

When θ is unknown a cannot be expressed in terms of it. One possibility is to impose the additional condition that $p' = L'X$, i.e. $p'\beta$ is estimable . Then the variance in (5.5.5) reduces to

$$v = L'L\sigma^2. \qquad (5.5.10)$$

Minimization of v together with the condition that $p' = L'X$ yields the LS by the proof of the GM.

Notice that for the case when θ is known the minimization problem is solved for all parametric functions and is thus valid for both estimable and non-estimable parametric functions. In solving the minimization problem the additional constraint of estimability was not applied. However once the optimum estimator is obtained for this case estimable parametric functions can be considered. However these estimable parametric functions are not unbiased estimators of $p'\beta$. On the other hand when θ is unknown the optimization problem is solved assuming estimability, an additional restriction. Estimability means that there is a linear function of the observations that is an unbiased estimator of the parametric functions $p'\beta$. The result therefore is an unbiased estimator for $p'\beta$.

Instead of minimizing the linear functions of the observations subject to extended unbiasedness linear functions of the LS could be minimized. To do this minimize

$$v = Var(p'\beta - a - c'b) \qquad (5.5.11)$$

with a a constant and c a vector subject to

$$E(p'\beta - a - c'b) = 0. \tag{5.5.12}$$

From (5.5.12)

$$a = p'\theta - c'UU'\theta. \tag{5.5.13}$$

The variance to be minimized is

$$v = \text{Var}(p'\beta) + \text{Var}(c'b) - 2\text{cov}(p'\beta, c'b)$$
$$= p'Fp + c'[UU'FUU' + \sigma^2(X'X)^+]c - 2p'FUU'c \tag{5.5.14}$$

Differentiating (5.5.14), setting the result equal to zero and isolating c on one side of the equation yields

$$c'[UU'FUU' + \sigma^2(X'X)^+] = p'FUU'. \tag{5.5.15}$$

From (5.5.13) and (5.5.15)

$$p'\hat{\beta} = p'\theta + p'FUU'[UU'FUU' + \sigma^2(X'X)^+]^+(b - \theta). \tag{5.5.16}$$

Conditions for the equivalence of (5.5.16) and (5.5.9) will be given in Section (5.5.2).

If θ is unknown and $p' = c'UU'$ then $a = 0$ in (5.5.13). For this case the optimization problem is that of minimizing

$$v = c'(X'X)^+c\sigma^2 \tag{5.5.17}$$

subject to

$$p' - c'UU' = 0. \tag{5.5.18}$$

The method of Lagrange multipliers will be used again. Let λ be a vector of Lagrange multipliers. Let

$$T = c'(X'X)^+c - (p' - c'UU')\lambda. \tag{5.5.19}$$

After differentiating T with respect to c' set the result equal to zero to obtain

$$(X'X)^+c + UU'\lambda = 0. \tag{5.5.20}$$

Equation (5.5.20) must be solved for λ. To do this multiply (5.5.20) by $X'X$. Thus, since $X'X(X'X)^+ = UU'$ and $XUU' = X$,

$$UU'c + X'X\lambda = 0 \tag{5.5.21}$$

and

$$\lambda = -(X'X)^+UU'c = -(X'X)^+p. \tag{5.5.22}$$

Substitution of the second form of λ into (5.5.21) gives

$$UU'(p - c) = 0. \tag{5.5.23}$$

Then $c'UU'b = p'UU'b$ and consequently $c'b = p'b$. It has been shown that the optimal estimator of the form $c'b$ of $p'\beta$ is $p'b$ when θ is unknown and $p' = c'UU'$.

The linear BE obtained by finding the minimum variance unbiased in the extended sense estimator corresponds to the BE if and only if the prior distribution is multivariate normal (see Kagan, Linnik and Rao (1973)).

Exercise 5.5.1. For the classical linear model of full rank together with the prior assumptions (5.5.1) and assumptions (5.5.2) minimize

$$v = Var(p'\beta - a - L'X'Y)$$

where L is an m dimensional column vector subject to

$$E(p'\beta - a - L'X'Y) = 0.$$

Show that the resulting optimum estimator is equivalent to (5.5.9) and (5.5.16).

Exercise 5.5.2. Repeat Exercise 5.5.1 for the non-full rank model. Under what conditions is the optimum estimator equivalent to (5.5.9) and (5.5.16).

5.5.2 Alternative Forms of the BE

Different algebraically equivalent forms of the BE were obtained by C.R. Rao (1975) when X and F have full rank. These forms will be generalized to the non-full rank case in this section and conditions for their equivalence will be derived. The different alternative forms of the BE to be considered are important because they can be used to:

1. derive ridge type estimators as special cases of BEs;
2. study the relationship between the mixed estimator and the BE;
3. consider the relative weight of the sample and prior information;
4. derive JS estimators as special cases of EBEs.

The five alternative forms of the BE will be given in Theorem 5.5.1 below for the full rank case. The proof of Theorem 5.5.1 will be left to the reader. Theorem 5.5.2 contains the alternative forms of the BE for the non-full rank case together with the conditions for their equivalence.

Theorem 5.5.1. Let X be full rank and let F be a positive definite matrix. Then five equivalent alternative forms of the BE are

$$
\begin{aligned}
p'\hat{\beta} &= p'\theta + p'[X'X + \sigma^2 F^{-1}]^{-1} X'(Y - X\theta) \\
&= p'[X'X + \sigma^2 F^{-1}]^{-1} [X'Y + F^{-1}\theta\sigma^2] \\
&= p'\theta + p'F[\sigma^2 (X'X)^{-1} + F]^{-1} (b - \theta) \\
&= p'(X'X)^{-1} [F + \sigma^2 (X'X)^{-1}]^{-1} \theta\sigma^2 + p'F[F + \sigma^2 (X'X)^{-1}]^{-1} b \\
&= p'b - p'(X'X)^{-1} [F + \sigma^2 (X'X)^{-1}]^{-1} (b - \theta)\sigma^2.
\end{aligned} \tag{5.5.24}
$$

The matrix identity

$$
(A + BCD)^{-1} = A^{-1} - A^{-1}B(C^{-1} + DA^{-1}B)^{-1}DA^{-1} \tag{5.5.25}
$$

is helpful in establishing Theorem 5.5.1.

Exercise 5.5.3. A. Verify the matrix identity (5.5.25).
B. Prove Theorem 5.5.1.

Example 5.5.1 . The Forms of the BE for the Means Model

Consider the linear model

$$y_i = \mu + \varepsilon_i, \ 1 \le i \le n$$

together with the assumptions that conditional on μ the ε_i are uncorrelated with mean zero and variance σ^2. The parameter μ is assumed to have prior mean θ and dispersion τ^2. Using the forms derived above the Linear Bayes estimator may then be written in the following equivalent forms illustrating the roles of sample and prior information.

$$
\begin{aligned}
\hat{\mu} &= \theta + \tau^2 1_n'(\tau^2 J + \sigma^2 I)^{-1}(Y - 1_n \theta) \\
&= \theta + \frac{n\tau^2}{n\tau^2 + \sigma^2}(\overline{Y} - \theta) \\
&= \frac{n\tau^2 \overline{Y} + \sigma^2 \theta}{n\tau^2 + \sigma^2} \\
&= \overline{Y} - \frac{\sigma^2}{n\tau^2 + \sigma^2}(\overline{Y} - \theta).
\end{aligned}
$$

The first of these forms may also be shown to be algebraically equivalent to the others by observing that

$$(\tau^2 J + \sigma^2 I)^{-1} = \frac{1}{\sigma^2} I - \frac{\tau^2}{\sigma^2(\sigma^2 + n\tau^2)} J.$$

Exercise 5.5.4. Verify the matrix inverse in Example 5.5.1 above.

When $\theta = 0$ the first form of (5.5.24) is the generalized ridge regression estimator of C.R. Rao (1975). Let $G = \sigma^2 F^{-1}$. The form of the generalized ridge estimator then becomes

$$p'\hat{\beta} = p'[X'X + G]^{-1}X'Y. \tag{5.5.26}$$

In Chapter III it was shown how different kinds of ridge regression estimators could be obtained for different choices of G. From (5.5.24) and (5.5.26) it is readily apparent that these different kinds of ridge estimators are all special cases of the BE.

The second form of (5.5.24) may be used to study the relationship between Bayes and mixed estimators and to establish an extended GM Theorem.

The third form of (5.5.24) is an expression for the BE in terms of the LS and the linear function of the BE that is minimum variance unbiased in the extended sense.

The relative importance of sample and prior information is expressed by the fourth form of (5.5.24) because it is a weighted average of the prior mean θ and the LS b.

The final form of (5.5.24) has already been used in Chapter IV to formulate the JS for a single linear model and will be used to formulate EBEs in Chapter VII.

Theorem 5.5.2 below gives the alternative forms of the BE for the non- full rank case and conditions for them to be equivalent. The forms are similar to those of 5.5.24 with $(X'X)^+$ replacing $(X'X)^{-1}$ and $UU'FUU'$ replacing F. Basically the two kinds of expressions are linear functions of the observations and the prior mean and linear functions of the LS and the prior mean, i.e. $a + L'Y$ and $a + c'b$

Theorem 5.5.2. 1. If $U'FU$ is positive definite and either $p'\beta$ is an estimable parametric function or $V'FU = 0$ and $V'\theta = 0$ alternative forms of the BE of the kind $a + L'Y$ equivalent to (5.5.9) are given by

$$p'\hat{\beta} = p'\theta + p'[X'X + (UU'FUU')^+]^+ X'(Y - X\theta) \qquad (5.5.27)$$

and

$$p'\hat{\beta} = p'[X'X + \sigma^2(UU'FUU')^+]^+[X'Y + (UU'FUU')^+\theta\sigma^2]. \qquad (5.5.28)$$

2. If the conditions in 1 above are satisfied the following equivalent alternative forms in of the BE of the kind $a + c'$ b may be written

$$p'\hat{\beta} = p'\theta + p'F[(X'X)^+\sigma^2 + UU'FUU']^+(b-\theta), \qquad (5.5.29)$$

$$\begin{aligned} p'\hat{\beta} = &\; p'(X'X)^+[UU'FUU' + \sigma^2(X'X)^+]^+\theta\sigma^2 \\ &+ p'UU'FUU'[UU'FUU' + \sigma^2(X'X)^+]^+(b-\theta)\sigma^2 \end{aligned} \qquad (5.5.30)$$

and

$$p'\hat{\beta} = p'b - p'(X'X)^+[UU'FUU' + \sigma^2(X'X)^+]^+(b-\theta)\sigma^2. \quad (5.5.31)$$

Proof. Recall that

$$X = S'\Lambda^{\frac{1}{2}}U' = S'\Lambda^{\frac{1}{2}}U'UU' = XUU'. \quad (5.5.32)$$

Since U'FU is positive definite it follows from (5.5.25) and (5.5.32) that

$$(\sigma^2I + XFX')^{-1} = (\sigma^2I + XUU'FUU')^{-1}$$

$$= \frac{1}{\sigma^2}I - \frac{1}{\sigma^2}XU\left[\frac{U'X'XU}{\sigma^2} + (U'FU)^{-1}\right]^{-1}\frac{U'X'}{\sigma^2}. \quad (5.5.33)$$

Substitution of (5.5.33) into (5.5.9) yields

$$p'\hat{\beta} = p'\theta + p'FU(U'FU)^{-1}\left[\frac{U'X'XU}{\sigma^2} + (U'FU)^{-1}\right]^{-1}\frac{U'X'}{\sigma^2}(Y - X\theta).$$

$$(5.5.34)$$

If $p'\hat{\beta}$ is estimable or V'FU = 0 since UU' + VV' = I (5.5.34) may be written in the form (5.5.27).

To obtain (5.5.28) from (5.5.27) observe that

$$I - [X'X + \sigma^2(UU'FUU')^+]^+X'X$$
$$= UU' + VV' - U[\Lambda + \sigma^2(U'FU)^{-1}]^+\Lambda U'$$
$$= U[I - [\Lambda + \sigma^2(U'FU)^{-1}]^+\Lambda]U' + VV' \qquad (5.5.35)$$
$$= U[\Lambda + \sigma^2(U'FU)^{-1}]^+(U'FU)^{-1}U'\sigma^2 + VV'$$
$$= [X'X + \sigma^2(UU'FUU')^+]^+(UU'FUU')^+\sigma^2 + VV'.$$

Since V'X = 0 substitution of (5.5.35) into (5.5.27) yields (5.5.28).

To establish (5.5.29) after letting $C = \sigma^2\Lambda^{-1} + U'FU$ observe that from (5.5.27)

$$p'\hat{\beta} = p'\theta + p'FU(U'FU)^{-1}[\Lambda C(U'FU)]^{-1}U'X'(Y - X\theta)$$
$$= p'\theta + p'FUC^{-1}\Lambda^{-1}U'X'(Y - X\theta). \qquad (5.5.36)$$

However

$$UC^{-1}\Lambda^{-1} = UC^{-1}U'U\Lambda^{-1}U'U$$
$$= [(X'X)^+\sigma^2 + (UU'FUU')]^+(X'X)^+U$$

(5.5.37)

Substitution of (5.5.37) into (5.5.36) gives (5.5.29).

The result (5.5.30) and (5.5.31) may be obtained easily after noticing that

$$UU'FUU'[UU'FUU' + \sigma^2(X'X)^+]^+$$
$$= UU'FUU'U(U'FU + \sigma^2\Lambda^{-1})^{-1}U'$$
$$= UU'FU(U'FU + \sigma^2\Lambda^{-1})^{-1}U'$$
$$= UU' - U\Lambda^{-1}(U'FU + \sigma^2\Lambda^{-1})^{-1}U'\sigma^2$$
$$= UU' - (X'X)^+[UU'FUU' + \sigma^2(X'X)^+]^+\sigma^2.$$

(5.5.38)

5.5.3 The Equivalence of the Generalized Ridge Estimator and the BE

In Section 5.2 and earlier in Chapter III the generalized ridge regression estimator was obtained by generalizing the frequentist argument of Hoerl and Kennard. It will be shown by exhibiting the form of the prior dispersion how the generalized ridge regression estimator is a special case of the BE. The alternative form (5.5.28) will be used. Equation (5.2.5) may be written

$$p'\hat{\beta} = p'[H + \lambda X'X]^+[H\theta + \lambda X'Y].$$

(5.5.39)

Assume that $U'FU$ is positive definite. Then since

$$F = (UU' + VV')F(UU' + VV')$$
$$= [U \quad V]\begin{bmatrix} U'FU & U'FV \\ V'FU & V'FV \end{bmatrix}\begin{bmatrix} U' \\ V' \end{bmatrix},$$

(5.5.40)

the matrix F may be written in the form

$$F = [U \quad V]\begin{bmatrix} A^{-1} & C_1 \\ C_1' & C_2 \end{bmatrix}\begin{bmatrix} U' \\ V' \end{bmatrix}\lambda\sigma^2,$$

(5.5.41)

with A a positive definite matrix. Thus,

$$U'FU = U'[U \quad V]\begin{bmatrix} A^{-1} & C_1 \\ C_1' & C_2 \end{bmatrix}\begin{bmatrix} U' \\ V' \end{bmatrix}U\lambda\sigma^2$$

$$= [I \quad 0]\begin{bmatrix} A^{-1} & C_1 \\ C_1' & C_2 \end{bmatrix}\begin{bmatrix} I \\ 0 \end{bmatrix}\lambda\sigma^2 = A^{-1}\lambda\sigma^2. \tag{5.5.42}$$

Then

$$A = (U'FU)^{-1}\lambda\sigma^2. \tag{5.5.43}$$

If

$$H = UAU' = (UU'FUU')^+\lambda\sigma^2, \tag{5.5.44}$$

then (5.5.39) is the alternative form of the BE (5.5.28). For the above choice of H the generalized ridge estimator is Bayes with respect to the prior with mean θ and dispersion F. Also, the BE (5.5.28) can be derived as the generalized ridge estimator by the method of Section 5.2 where H would take the form (5.5.44).

Exercise 5.5.5 Is the choice of H above unique? What other choices if any are available?

4.5.4 The Equivalence of the Mixed Estimator and the BE

Conditions for the equivalence of the mixed estimator described in Section 4.3 and the mixed estimator will be obtained. Specifically it will be shown that if $U'FU$ is positive definite and $R = NU'$ with N a nonsingular sxs matrix:

1. The BE and the mixed estimator are equivalent.
2. Sample and prior information may be exchanged.
3. Alternative forms of the mixed estimator may be derived and interpreted.

The model (4.3.1) may be rewritten as two separate models. The first model is

$$Y = X\beta + \varepsilon \qquad\qquad \text{Model I} \qquad\qquad (5.5.45a)$$

with

$$E(\varepsilon) = 0 \text{ and } D(\varepsilon) = \sigma^2 I. \tag{5.5.45b}$$

The second model is

$$r = R\beta + \phi \qquad\qquad \text{Model II} \qquad\qquad (5.5.46a)$$

with

$$E(\phi) = 0 \text{ and } D(\phi) = \tau^2 I. \tag{5.5.46b}$$

Three kinds of estimable parametric functions will be considered here. To facilitate their definition observe that the singular value decomposition of

$$R = [A' \quad B'] \begin{bmatrix} \Delta^{\frac{1}{2}} & 0 \\ 0 & 0 \end{bmatrix} \begin{bmatrix} P' \\ Q' \end{bmatrix} = A' \Delta^{\frac{1}{2}} P'. \tag{5.5.47}$$

Now:

1. $p'\beta$ is estimable with respect to Model I (X estimable) if there is a vector d where $p' = d'U'$;

2. $p'\beta$ is estimable with respect to Model II (R estimable) if there is a vector c where $p' = c'P'$;

3. $p' \beta$ is estimable with respect to the augmented model if there is a vector e where

$$p' = e' \begin{bmatrix} U' \\ P' \end{bmatrix}.$$

If $p'\beta$ is X estimable and $R = NU'$ then Equation (5.3.5) is algebraically equivalent to

$$p'\hat{\beta} = p'b_2 + p'(R'R)^+ X'[X(R'R)^+ X'\tau^2 + \sigma^2 I]^{-1}(Y - Xb_2)\tau^2. \tag{5.5.48}$$

Estimator (5.5.48) is algebraically the same as the BE with b_2 as the prior mean and $(R'R)^+$ as the prior dispersion.

Similarly if $p'\beta$ is R estimable and $P'X'XP$ is positive definite it can be shown the (5.3.5) is algebraically equivalent to (5.5.48) with the roles of X and R interchanged. Thus,

$$p'\hat{\beta} = p'b_1 + p'(X'X)^+ R'[R(X'X)^+ R'\sigma^2 + \tau^2 I]^{-1}(Y - Rb_1)\sigma^2.$$

$$(5.5.49)$$

Estimator (5.5.49) is Bayes with respect to a prior with mean b_1 and dispersion $(X'X)^+$.

From (5.5.48) and (5.5.49) when $U'(R'R)^+U$ and $P'(X'X)^+P$ are both positive definite the sample and prior information can be exchanged and equivalent estimators will result.

The following alternative forms of the mixed estimator may be obtained for Model I. Assume that $p'\beta$ is an estimable parametric function. Then

$$p'\hat{\beta} = p'b_2 + p'\left[\frac{X'X}{\sigma^2} + \frac{R'R}{\tau^2}\right]^+ \frac{X'}{\sigma^2}(Y - Xb_2)$$

$$p'\left[\frac{X'X}{\sigma^2} + \frac{R'R}{\tau^2}\right]^+ \left[\frac{X'X}{\sigma^2}b_1 + \frac{R'R}{\tau^2}b_2\right].$$

$$(5.5.50)$$

Furthermore

$$p'\hat{\beta} = p'b_2 + p'(R'R)^+[(X'X)^+\sigma^2 + (R'R)^+\tau^2]^+(b_1 - b_2)\tau^2$$
$$= p'(X'X)^+[(X'X)^+\sigma^2 + (R'R)^+\tau^2]^+ b_2\sigma^2 \qquad (5.5.51)$$
$$+ p'(R'R)^+[(X'X)^+\sigma^2 + (R'R)^+\tau^2]^+ b_1\tau^2.$$

Another alternative form is given by

$$p'\hat{\beta} = p'b_1 - p'(X'X)^+[\tau^2(R'R)^+ + \sigma^2(X'X)^+]^+(b_1 - b_2)\sigma^2. \quad (5.5.52)$$

Corresponding alternative forms may be obtained for Model II using the exchangeability of the sample and the prior information. Exchange X and R, b_1 and b_2, σ^2 and τ^2 in equations (5.5.0) and (5.5.2).

For model II for R estimable parametric functions

$$p'\hat{\beta} = p'b_1 + p'\left[\frac{X'X}{\sigma^2} + \frac{R'R}{\tau^2}\right]^+ \frac{R'}{\tau^2}(r - Rb_1)$$

$$= p'\left[\frac{X'X}{\sigma^2} + \frac{R'R}{\tau^2}\right]^+ \left[\frac{X'X}{\sigma^2}b_1 + \frac{R'R}{\tau^2}b_2\right]$$

$$= p'b_2 - p'(R'R)^+[\tau^2(R'R)^+ + \sigma^2(X'X)^+]^+(b_2 - b_1)\sigma^2. \qquad (5.5.53)$$

For both models the second form in (5.5.50) and (5.5.53) are the same since they are the weighted average of the LS estimators for Model I and Model II.

For the augmented linear model (5.3.1) the mixed estimator is the solution to the optimization problem for unknown prior mean provided that the parametric function $p'\beta$ is (X, R) estimable.

Five ways of considering the mixed estimator are:

1. as a weighted LS for an augmented linear model;
2. as equivalent to a BE with known prior mean;
3. as an optimum estimator with unknown prior mean if $p'\beta$ is (X,R) estimable;
4. as a minimax estimator;
5. as a generalized ridge estimator.

Exercise 5.5.6 Explain how the mixed estimator may be obtained
A. as a special case of the minimax estimator;
B. as a special case of the generalized ridge estimator.

Exercise 5.5.7 For the data in Example 5.3.3 verify that the same estimator results when the calculations are performed using any of the alternative forms in (5.5.49)-(5.5.53).

5.5.5 Well Known Ridge Estimators as Special Cases of BE, Minimax or Mixed Estimators

An example of how the fact that the Hoerl Kennard ridge estimator was a special case of the BE facilitated the choice of the ridge parameter k was given in Chapter III. The present goal is to explain how each of the well known ridge estimators are special cases of Bayes, minimax or mixed estimators. It was already explained that the generalized ridge estimator of C.R. Rao is a special case of the BE for the full rank case. By using the different choices of G given in Chapter III each of the ridge estimators described there can be shown to be special cases of the BE. The non-full rank case is considered here.

In (5.5.27) let $\theta = 0$ and $H = \sigma^2(UU'FUU')^+$. Then (5.5.27) can be rewritten as

$$p'\hat{\beta} = p'[X'X + H]^+X'Y, \qquad (5.5.54)$$

the generalized ridge estimator.

Now observe that for (5.527) with $\theta = 0$ it follows that

$$\begin{aligned} p'\hat{\beta} &= p'[X'X + \sigma^2(UU'FUU')^+]^+ X'Y \\ &= pU[\Lambda + \sigma^2(U'FU)^{-1}]^{-1}U'X'Y. \end{aligned} \tag{5.5.55}$$

Let T be any $(n-s) \times (n-s)$ non-negative definite matrix. Since $V'X = 0$

$$\begin{aligned} p'\hat{\beta} &= p'[U(\Lambda + \sigma^2(U'FU)^{-1}U' + VTV']X'Y \\ &= p'[U \quad V]\begin{bmatrix} (\Lambda + (U'FU)^{-1}\sigma^2)^{-1} & 0 \\ 0 & T \end{bmatrix}\begin{bmatrix} U' \\ V' \end{bmatrix}X'Y \\ &= p'[U \quad V]\begin{bmatrix} \Lambda + (U'FU)^{-1}\sigma^2 & 0 \\ 0 & T^+ \end{bmatrix}^+\begin{bmatrix} U' \\ V' \end{bmatrix}X'Y \\ &= p'[X'X + \sigma^2(UU'FUU')^+ + VT^+V']^+ X'Y. \end{aligned} \tag{5.5.56}$$

Now let $G = H + VT^+V'$. Then $s \leq \text{Rank}(G) \leq m$ and

$$p'\hat{\beta} = p'[X'X + G]^+ X'Y, \tag{5.5.57}$$

the generalized ridge estimator of C.R. Rao is obtained.

Let $G = kI$. The ridge estimator of Hoerl and Kennard

$$p'\hat{\beta} = p'(X'X + kI)^{-1}X'Y \tag{5.5.58}$$

results.

The generalized ridge estimator of Hoerl and Kennard is obtained by letting $G = UKU'$ with K a positive definite diagonal matrix. Thus,

$$p'\hat{\beta} = p'(X'X + UKU')^+ X'Y. \tag{5.5.59}$$

When $G = c\,X'X$ with c a positive constant the contraction estimator of Mayer and Willke is obtained. It takes the form

$$p'\hat{\beta} = \frac{p'b}{1+c}. \tag{5.5.60}$$

Exercise 5.5.8. Show that if in the contraction estimator of Mayer and Willke $c/(1+c)$ is estimated by $\hat{\sigma}^2 d / b'X'Xb$ where $\hat{\sigma}^2$ is the within sum of squares the JS results.

Exercise 5.5.9. Let $G = UC\Lambda U'$ with C a positive definite diagonal matrix. Show that

$$p'\hat{\beta} = p'[I + UCU']^{-1}b. \qquad (5.5.61)$$

This estimator will be called the generalized contraction estimator.

The above ridge estimators are also special cases of minimax estimators since that kind of estimator has the same form as the BE. It will now be shown how the ridge type estimators are special cases of mixed estimators.

Let $r = 0$ and choose R so that for the positive definite diagonal matrix Δ the matrix $\Delta^{-1/2}U'R$ is orthogonal. Hence $R'R = U\Delta U'$. Then if $\Delta = kI$, $k\Lambda$, K and KΛ respectively the ridge estimators (5.5.58)-(5.5.61) are obtained as special cases of the mixed estimators.

5.5.6 An Extension of the Gauss Markov Theorem

The Gauss Markov theorem of Section 5.5.1 noted that the LS is a best linear unbiased estimator (BLUE) because of all the unbiased estimators that were linear functions of the observations it was the one with the smallest variance. As has already been pointed out the linear BE was obtained as a minimum variance unbiased in an extended sense linear estimator. Conditions have been given for the BE and the mixed estimator to be equivalent using the alternative forms . These results suggest the following extended Gauss Markov theorem.

Theorem 5.5.3.
1. The linear BE is the minimum variance unbiased in the extended sense linear estimator.
2. When UFU' is positive definite there is an augmented linear model of the form (5.3.1) such that relative to this model the linear BE is the LS.

5.6 Comparing the Efficiency of the Ridge and the Least Square Estimator from the Frequentist Point of View

This Section will:

1. obtain different forms of the MSE of the BE without averaging over the prior distribution (Section 5.6.1);

2. compare these MSE to that of the LS (Section 5.6.2);

3. insofar as is possible, compare the MSE of different kinds of BE estimators with each other (Section 5.6.3);

4. investigate analogous results for mixed, generalized ridge and minimax estimators (Section 5.6.4).

5.6.1 The Frequentist MSE of the LS and the BE

The MSE of the parametric functions to be studied in this section is given by

$$m = p'E_\beta (\hat{\beta} - \beta)(\hat{\beta} - \beta)'p = p'Mp \qquad (5.6.1)$$

where M is a non-negative definite dispersion matrix of $\hat{\beta}$. The subscript β means that when a BE is being used its MSE is without averaging over the prior assumptions, i.e., since the β parameters are random variables a conditional MSE is being computed. The subscript β is not needed if the estimator is not derived assuming that the parameters are random variables. One estimator $p'\hat{\beta}_1$ will have a smaller MSE than another $p'\hat{\beta}_2$ if the difference between their dispersion matrices is non-negative definite. It is well known and easy to show that the MSE is the sum of the conditional variance

$$v = p'E_\beta [\hat{\beta} - E_\beta(\hat{\beta}))(\hat{\beta} - E_\beta(\hat{\beta}))] = p'Vp \qquad (5.6.2)$$

and the squared bias

$$bi = p'[\beta - E_b(\hat{\beta})][\beta - E_b(\hat{\beta})]'p = p'Bp. \qquad (5.6.3)$$

The variances and biases of estimators will be compared using the same criterion as the MSE. For an estimator of the form $a + L'Y$ of $p'\beta$

$$v = L'D(Y)L = L'L\sigma^2 \qquad (5.6.4)$$

where $D(Y)$ is the variance covariance matrix of Y. The squared bias in the context of estimating the parameters of the classical linear model is given by

$$bi = (p'\beta - a - L'E(Y))^2 = ((p' - L'X)\beta - a)^2. \qquad (5.6.5)$$

For the LS from (5.6.4) and (5.6.5) since $L' = (X'X)^+ X'$ and $a = 0$

$$v = p'(X'X)^+ (X'X)(X'X)^+ p\sigma^2 = p'(X'X)^+ p\sigma^2. \qquad (5.6.6)$$

and

$$bi = p'(I - (X'X)^+ X'X)\beta\beta'(I - (X'X)^+ X'X)'p. \qquad (5.6.7)$$

When $p'\beta$ is estimable $bi = 0$ and the MSE is given by (5.6.6).

Using the methodology suggested above two forms of the conditional MSE are given by

$$m = p'\Gamma_1^+[\sigma^2 X'X + (UU'FUU')^+ (\beta - \theta)(\beta - \theta)'(UU'FUU')^+ \sigma^4 \, \Gamma_1^+ p. \qquad (5.6.8)$$

and

$$m = p'(X'X)^+ p\sigma^2 - 2p'(X'X)^+ C_1^+ (X'X)^+ p\sigma^4$$
$$+ p'(X'X)^+ C_1^+[(X'X)^+ \sigma^2 + (\beta - \theta)(\beta - \theta)']C_1^+ (X'X)^+ p\sigma^4 \qquad (5.6.9)$$

where

$$\Gamma_1 = X'X + \sigma^2(UU'FUU')^+$$

and

$$C_1 = UU'FUU' + \sigma^2(X'X)^+.$$

5.6.2 Comparison of the Frequentist Risk of the BE with that of the LS

Theorem 3.2.6 gave necessary and sufficient conditions for the linear parametric functions of the ridge estimator of C.R. Rao to have a smaller MSE than that of the LS for the full rank case. Theorem 5.6.1 below gives a similar comparison of a BE with an LS that is valid for both the full and the non-full rank case.

Theorem 5.6.1. Let $p'\hat{\beta}$ and $p'b$ be the BE and the LS respectively for the estimable parametric function $p'\beta$. The parametric functions of the BE have a smaller MSE than that of the LS ,i.e.,

$$p'E(\hat{\beta}-\beta)(\hat{\beta}-\beta)'p \le p'E(b-\beta)(b-\beta)'p \qquad (5.6.10)$$

is that

$$(\beta-\theta)'(2UU'FUU'+\sigma^2(X'X)^+)^+(\beta-\theta) \le 1. \qquad (5.6.11)$$

Proof. Using Equations (5.6.6) and (5.6.9) Inequality (5.6.10) holds true iff

$$p'(X'X)^+p\sigma^2 - 2p'(X'X)^+C_1^+(X'X)^+p\sigma^4$$
$$+p'(X'X)^+C_1^+[(X'X)^+\sigma^2 + (\beta-\theta)(\beta-\theta)']C_1^+(X'X)^+p\,\sigma^4 \qquad (5.6.12)$$
$$\le \sigma^2p'(X'X)^+p.$$

Since $p'\beta$ is estimable, $p' = d'U'$. Recall that for any nonsingular matrix A, $(UAU')^+$ $= UA^{-1}U'$ and that $(X'X)^+ = U\Lambda^{-1}U'$. Using these facts and some algebra Inequality (5.6.12) is equivalent to

$$d'\sigma^2\Lambda^{-1}(U'FU+\sigma^2\Lambda^{-1})^{-1}(\sigma^2\Lambda^{-1}+(\gamma-U'\theta)(\gamma-U'\theta)')$$
$$(U'FU+\sigma^2\Lambda^{-1})^{-1}\Lambda^{-1}\sigma^2d \le 2\sigma^4d'\Lambda^{-1}(U'FU+\sigma^2\Lambda^{-1})^{-1}\Lambda^{-1}d. \qquad (5.6.13)$$

for all d. Inequality (5.6.13) holds true iff

$$\sigma^2\Lambda^{-1}(U'FU+\sigma^2\Lambda^{-1})^{-1}(\sigma^2\Lambda^{-1}+(\gamma-U'\theta)(\gamma-U'\theta)')$$
$$(U'FU+\sigma^2\Lambda^{-1})^{-1}\Lambda^{-1}\sigma^2 \le 2\sigma^4\Lambda^{-1}(U'FU+\sigma^2\Lambda^{-1})^{-1}\Lambda^{-1} \qquad (5.6.14)$$

where the $\leq$ sign means that the difference between the larger and the smaller matrix is positive semidefinite. Positive semidefiniteness is invariant under linear transformations so pre and post multiplication by $(U'FU + \sigma^2\Lambda^{-1})\Lambda$ gives

$$(\gamma - U'\theta)(\gamma - U'\theta)' \leq 2U'FU + \sigma^2\Lambda^{-1} \tag{5.6.15}$$

From Theorem 3.2.4 Inequality (5.6.15) is equivalent to

$$(\gamma - U'\theta)'(2U'FU + \sigma^2\Lambda^{-1})^{-1}(\gamma - U'\theta) \leq 1 \tag{5.6.16}$$

Since $\gamma = U'\beta$ Inequality (5.6.16) is equivalent to

$$(\beta - \theta)'U(2U'FU + \sigma^2\Lambda^{-1})^{-1}U'(\beta - \theta) \leq 1 \tag{5.6.17}$$

Since $U(2U'FU + \sigma^2\Lambda^{-1})^{-1}U' = (2UU'FUU' + \sigma^2(X'X)^+)^+$ inequalities (5.6.17) and (5.6.11) are equivalent and the result follows.

Exercise 5.6.1. Show that the result of Theorem 5.6.1 could have been obtained using the form of the MSE in (5.6.8) instead of the form in (5.6.9).

A result of Farebrother (1976) is obtained from Theorem 5.6.1 in Corollary 5.6.1.

Corollary 5.6.1. For the non-full rank model the parametric functions of the ridge estimator of Hoerl and Kennard obtained using the prior with mean 0 and dispersion $F = \dfrac{\sigma^2}{k}I$ has smaller MSE than the LS iff

$$\beta'\left[\frac{2}{k}I + (X'X)^+\right]^{-1}\beta \leq \sigma^2 + \frac{k}{2}\delta'\delta \tag{5.6.18}$$

where $\delta = V'\beta$.

Proof. Specializing Inequality (5.6.16) the ridge estimator has a smaller MSE than the LS iff

$$\gamma'\left(\frac{2}{k}I + \Lambda^{-1}\right)^{-1}\gamma \leq \sigma^2 \tag{5.6.19}$$

Using (5.6.19) the result (5.6.18) follows because

$$\beta'\left[\frac{2}{k}I+(X'X)^+\right]^{-1}\beta$$

$$= \beta'[U \quad V]\left[\begin{bmatrix}U'\\V'\end{bmatrix}\left[\frac{2}{k}I+(X'X)^+\right][U \quad V]\right]^{-1}\begin{bmatrix}U'\\V'\end{bmatrix}\beta$$

$$= \gamma'\left[\frac{2}{k}I+\Lambda^{-1}\right]^{-1}\gamma+\frac{k}{2}\delta'\delta \le \sigma^2+\frac{k}{2}\delta'\delta. \qquad (5.6.20)$$

Let $G = \sigma^2(UU'FUU')^+$ with $U'FU$ positive definite. When $\theta = 0$ the generalized ridge estimator of C. R. Rao is obtained, i.e.,

$$p'\hat{\beta} = p'(X'X+G)^+X'Y. \qquad (5.6.21a)$$

Since $G = \sigma^2(UU'FUU')^+ = \sigma^2U(U'FU)^{-1}U'$, $U'GU = \sigma^2(U'FU)^{-1}$ so that $UU'FUU' = \sigma^2G^+$. Thus the MSE of (5.6.20) is smaller than that of the LS iff

$$\beta'[2G^+ + (X'X)^+]^+\beta \le \sigma^2. \qquad (5.6.21b)$$

The result (5.6.21b) may be specialized to the ridge and contraction estimators. To obtain the contraction estimator of Mayer and Willke let $G = k\ X'X$ and to obtain the ridge estimator of Hoerl and Kennard let $G = kUU'$ with k a positive constant. The generalized ridge estimator of Hoerl and Kennard is obtained by letting $G = UKU'$ where K is a positive definite diagonal matrix and the generalized contraction estimator is obtained by letting $G = U\Lambda KU'$. Thus, the contraction estimator takes the form

$$p'\hat{\beta} = \frac{1}{1+k}b \qquad (5.6.22)$$

and dominates the LS iff

$$\beta'(X'X)\beta \le \sigma^2\left(\frac{2}{k}+1\right). \qquad (5.6.23)$$

The ridge estimator takes the form

$$p'\ddot{\beta} = p'(X'X + kUU')^+ X'Y. \tag{5.6.24}$$

The MSE of its linear parametric functions is less than that of the LS iff

$$\beta'\left(\frac{2UU'}{k} + (X'X)^+\right)^+ \beta \le \sigma^2 \tag{5.6.25}$$

The form of the generalized ridge estimator of Hoerl and Kennard is

$$p'\ddot{\beta} = p'(X'X + UKU')^+ X'Y. \tag{5.6.26}$$

It dominates the LS iff

$$\beta'\left(2UK^{-1}U' + (X'X)^+\right)^+ \beta \le \sigma^2. \tag{5.6.27}$$

Finally the generalized contraction estimator takes the form

$$p'\ddot{\beta} = p'(X'X + U\Lambda KU')^+ X'Y. \tag{5.6.28}$$

The region of LS domination is

$$\beta'\left(2U\Lambda^{-1}K^{-1}U' + (X'X)^+\right)^+ \beta \le \sigma^2. \tag{5.6.29}$$

Exercise 5.6.2. A. Verify that the ordinary ridge estimators obtained using $G = kI$ and $G = kUU'$ are equivalent.

B. Verify the form of the generalized contraction estimator given in (5.6.22) and the form of the region of LS domination in (5.6.23).

C. What are the forms of the Hoerl Kennard generalized ridge and the generalized contraction estimators in orthogonal coordinates?

Exercise 5.6.3. A. Show that the variance of the generalized ridge estimator of C. R. Rao is smaller than that of the LS for any parametric function.

B. Is it possible to compare the bias of a ridge type estimator with that of a LS for non-estimable parametric functions? Explain.

Exercise 5.6.4. A. Let N be any non-negative definite matrix where U'NU is positive definite. Let

$$T = \frac{1}{\sigma^4} UU'F\Gamma_1 N\Gamma_1 FUU' - \frac{1}{\sigma^2} UU'FX'XFUU'.$$

Assume that U'TU is positive definite. Let p'β be an estimable parametric function. Prove that

$$MSE(p'\hat{\beta}) \le p'Np$$

iff

$$(\beta - \theta)'T^+ (\beta - \theta) \le 1.$$

B. Obtain the result of Theorem 5.6.1 as a special case of the result in A.

Exercise 5.6.5. A. Let

$$F = [U \quad V] \begin{bmatrix} A & 0 \\ 0 & B \end{bmatrix} \begin{bmatrix} U' \\ V' \end{bmatrix},$$

where A and B are positive definite matrices. Prove that a necessary and sufficient condition for the parametric functions of the BE to have a smaller MSE than the corresponding LS is that

$$(\beta - \theta)'(2F + \sigma^2 (X'X)^+)^+ (\beta - \theta) \le \sigma^2 + (\beta - \theta)'(VBV)^+ (\beta - \theta).$$

B. Show that when A = (σ^2/k)I and B = I the result in A specializes to Farebrother's(1976) result, Inequality (5.6.19).

Exercise 5.6.6 Establish the following result of Farebrother (1978). Let $\gamma = U'\beta$. If the matrix

$$W = A'\Lambda^{-1}A + A'\Lambda^{-1}(I - A) + (I - A)\Lambda^{-1}A$$

is positive definite then the MSE of

$$\ddot{\gamma} = g + A'(U'\theta - g)$$

is less than that of the LS $g = \Lambda^{-1}U'X'Y$ iff

$$(\gamma - U'\theta)'[\Lambda^{-1} + \Lambda^{-1}(A^{-1} - I) + (A^{-1} - I)'\Lambda^{-1}]^{-1}(\gamma - U'\theta) \le \sigma^2.$$

5.6.3 Comparison of BE with Each Other

The comparison of the parametric functions of two Bayes estimators is a more difficult problem than that of comparing a BE and a LS. The form of the region of comparison is much more complicated than an ellipsoid. Theorem 5.6.2 due to Trenkler (1985) gives necessary and sufficient conditions for BE $p'\hat{\beta}_1$ to have a smaller MSE than $p'\hat{\beta}_2$.

Theorem 5.6.2. The

$$MSE(p'\hat{\beta}_1) \le MSE(p'\hat{\beta}_2) \tag{5.6.30}$$

iff

1. The matrix $T = \text{Dispersion}(\hat{\beta}_2) - \text{Dispersion}(\hat{\beta}_1) + \text{Squared Bias}(\hat{\beta}_2)$ is non-negative definite.
2. The bias of $\hat{\beta}_1$ is in the range of T.
3. The inequality

$$(\text{Bias}\hat{\beta}_1)'T^+(\text{Bias}\hat{\beta}_1) \le 1. \tag{5.6.31}$$

When $p'\hat{\beta}_2 = p'b$ inequality (5.6.31) reduces to Farebrother's result. Observe that T depends on the β parameters as well as the prior dispersions, unlike comparing a BE to the LS where the matrix T depends only on the F_i. That is why the shape of the region is more complicated than an ellipsoid.

Proof. The result follows directly from Theorem 5.1.3.

The result in Theorem 5.6.2 is somewhat cumbersome because the matrix T depends on the β parameters. A necessary but not sufficient condition for one BE to dominate another in the form of the β parameters lying in an ellipsoid will be given in Theorem 5.6.3 below.

Theorem 5.6.3. Let $H = \text{Dispersion}(\hat{\beta}_2) - \text{Dispersion}(\hat{\beta}_1)$. Assume that H is non-negative definite. Let $p'\hat{\beta}_i$, $i = 1, 2$ be the BEs associated with the prior dispersions F_i, $i = 1, 2$ where $F_1 \leq F_2$. Suppose that the β parameters lie in the ellipsoid.

$$(\text{Bias}\hat{\beta}_1)'H^+(\text{Bias}\hat{\beta}_1) \leq 1 \qquad (5.6.32)$$

then

$$\text{MSE}(p'\hat{\beta}_1) \leq \text{MSE}(p'\hat{\beta}_2). \qquad (5.6.33)$$

Proof. Since $H \leq T$ the ellipsoid in (5.6.32) lies in that of (5.6.31).

Example 5.6.1. Consider two contraction estimators $\hat{\beta}_1$ and $\hat{\beta}_2$ with parameters k_1 and k_2 respectively where $k_1 \geq k_2$. Then from Theorem 5.6.3 $p'\hat{\beta}_1$ has a smaller MSE than $p'\hat{\beta}_2$ for those values of β that lie in the ellipsoid

$$\beta'(X'X)\beta \leq \frac{\sigma^2(k_1 - k_2)(k_1 + k_2 + 2)}{k_1^2(1 + k_2)^2}.$$

Simpler conditions for comparison of the parametric functions of estimators may be obtained when $F = UDU'$ where D is a diagonal matrix. This would include the ordinary and generalized ridge regression estimator of Hoerl and Kennard and the contraction estimators. Also it is sometimes easier to compare other measures of efficiency. These include the variance, the squared bias, the signal to noise ratio and the MSE of the estimator averaging over specific quadratic loss functions, e.g., the total MSE, the individual MSE and the predictive MSE. The matrices of these loss functions also take the form UDU' with D a diagonal matrix.

Theorem 5.6.4 gives conditions for the parametric functions of one ridge estimator to have a smaller MSE than the parametric functions of another ridge estimator.

Theorem 5.6.4. Let $p'\hat{\beta}_1$ and $p'\hat{\beta}_2$ be two Bayes estimators derived from priors with dispersions $F_1 = UD_1U'$ and $F_2 = UD_2U'$. Assume the $F_2 - F_1$ is a positive semidefinite matrix. Then $DV = \text{Variance}(p'\hat{\beta}_2) - \text{Variance}(p'\hat{\beta}_1) \geq 0$ for all p.

Proof. The variance term in (5.6.8) is given by

$$p'Vp = p'T_1X'XT_1p\sigma^2. \tag{5.6.34}$$

Thus, for j =1, 2

$$\text{Var}(p'\hat{\beta}_j) = d'R_jd \tag{5.6.35}$$

where

$$R_j = [\Lambda + \sigma^2 D_j^{-1}]^{-1}\Lambda[\Lambda + \sigma^2 D_j^{-1}]^{-1}. \tag{5.6.36}$$

The theorem will be proved once it is shown that if $D_2 - D_1 \geq 0$ implies $R_2 - R_1 \geq 0$. The matrix D_j is diagonal. Denote the non-zero elements by d_{ij}. Then the matrix R_j is diagonal with elements

$$f(d_{ij}) = \frac{\lambda_i d_{ij}^2}{(\lambda_i d_{ij} + \sigma^2)^2}. \tag{5.6.37}$$

Since d_{ij} and λ_i are positive numbers the derivative

$$f'(d_{ij}) = \frac{2d_{ij}\lambda_i\sigma^2}{(\lambda_i d_{ij} + \sigma^2)^3} > 0 \tag{5.6.38}$$

and $f(d_{ij})$ is an increasing function. It follows that if $D_2 - D_1 \geq 0$ then $R_2 - R_1 \geq 0$ and thus $DV = p'(R_2 - R_1)p \geq 0$.

Exercise 5.6.7. Show that the conclusion of Theorem 5.6.4 holds true when the prior dispersions are of the form

$$F_i = [U \quad V] \begin{bmatrix} D_i & C_i \\ C_i' & E_i \end{bmatrix} \begin{bmatrix} U' \\ V' \end{bmatrix}, i = 1, 2$$

where the D_i are diagonal matrices, the C_i and the E_i are arbitrary matrices and $F_2 - F_1 \geq 0$.

Exercise 5.6.8. A. Let $A = \begin{bmatrix} 2 & 1 \\ 1 & 1 \end{bmatrix}$ and $B = \begin{bmatrix} 2 & 1 \\ 1 & 1+\epsilon \end{bmatrix}$ with $\epsilon > 0$. Show that $B - A$ is positive semidefinite but that $B^2 - A^2$ is not positive semidefinite.

B. Let $X'X = I$, $\sigma^2 = 1$, $F_1 = A$ and $F_2 = B$. Show that the conclusion of Theorem 5.6.4 does not hold.

Exercise 5.6.9. Do the necessary algebra to establish that the inequality in Example 5.6.1 follows from Theorem 5.6.3.

Exercise 5.6.10. For the means model $Y = \mu + \epsilon$ what is the necessary condition of Theorem 5.6.3 for one ridge estimator to have a smaller MSE than another?

The difference between the bias of the parametric functions of two estimators is never non-negative definite (Exercise 3.2.6). Thus, for this criterion it is impossible to compare the biases of two estimators. However, comparisons are possible for a class of quadratic loss functions that includes the total MSE (TMSE), the predictive MSE(PMSE) and the individual MSE(IMSE) in orthogonal coordinates. With this objective in mind let A be a non-negative definite matrix, V and B denote the dispersion and bias of the parametric functions of a ridge type estimator. The weighted variance is given by

$$\begin{aligned} trAV &= trAE(\hat{\beta} - E(\hat{\beta}))(\hat{\beta} - E(\hat{\beta}))' \\ &= E(\hat{\beta} - E(\hat{\beta}))' A(\hat{\beta} - E(\hat{\beta})) \end{aligned} \tag{5.6.39}$$

and the weighted bias is given by

$$\begin{aligned} trAB &= trA(E(\hat{\beta}) - \beta)(E(\hat{\beta}) - \beta)' \\ &= [E(\hat{\beta}) - \beta]'A[E(\hat{\beta}) - \beta]. \end{aligned} \tag{5.6.40}$$

The weighted signal to noise ratio is given by

$$SN = \frac{trAB}{trAV}. \qquad (5.6.41)$$

From an engineering standpoint it would be desirable that the signal to noise ratio be as large as possible. Thus, one estimator will be considered more efficient than another if its signal to noise ratio is larger. For the ridge and contraction estimators most frequently considered in the literature if one estimator has a smaller prior dispersion than another the variance of the parametric functions also has this property. By Theobald's result Theorem 3.2.1 and Corollary 3.2.3 it follows that the weighted variances also have this property. Theorem 5.6.6 shows that for the TMSE, PMSE and IMSE a more precise prior means a larger weighted bias . It has already been shown that the weighted variance is smaller. Thus a more precise prior means a larger SN.

Theorem 5.6.6. Suppose F_i , $i = 1,2$ are in the form of Theorem 5.6.5. Let $A = U\Delta U' + VV'$ where Δ is a diagonal matrix with non-negative elements. Let B_i , $i =1,2$ be the squared matrix bias of $\hat{\beta}_i$, $i = 1,2$. Then if $F_1 \leq F_2$ it follows that $trB_1A \geq trB_2A$.

Proof. The matrix

$$H = \Gamma_1^+ (UU'FUU')^+ \sigma^2 + VV' = U\Theta U' \sigma^2 + VV' \qquad (5.6.42)$$

where $\Theta = (\Lambda + \sigma^2 D)^{-1} D^{-1}$ and D is a diagonal matrix. The individual diagonal elements of Θ are

$$\theta_j = \left(\lambda_j + \frac{\sigma^2}{d_j}\right)^{-1} \frac{1}{d_j} = \frac{1}{\lambda_j d_j + \sigma^2}. \qquad (5.6.43)$$

The matrices H_i , $i = 1,2$ have the same form as (5.6.42) with D_i and F_i replacing Θ and F. Let d_{ij} be the diagonal elements of D_i . Then $F_1 \leq F_2$ iff $d_{1j} \leq d_{2j}$ for all j. Then $\Theta_1 \geq \Theta_2$ and consequently $H_1 \geq H_2$. Thus,

$$trB_1A = (\beta - \theta)'H_1'AH_1(\beta - \theta) \geq (\beta - \theta)'H_2'AH_2(\beta - \theta) = trB_2A \qquad (5.6.44)$$

by virtue of

$$H_1AH_1 = U\Theta_1\Delta\Theta_1U' + VV' \leq U\Theta_2\Delta\Theta_2U' + VV' = H_2AH_2. \qquad (5.6.45)$$

Exercise 5.6.11. Why is the comparison of

$$SN = \frac{p'Bp}{p'Vp}$$

for two different estimators impossible?

Let A be a non-negative definite mxm matrix. The weighted MSE is the sum of the weighted variance and the weighted bias. Thus,

$$M_A = E(\hat{\beta} - \beta)'A(\hat{\beta} - \beta) = E(\hat{\beta} - E(\hat{\beta}))'A(\hat{\beta} - E(\hat{\beta}))$$
$$+(\beta - E(\hat{\beta}))'A(\beta - E(\hat{\beta})) \qquad (5.6.46)$$
$$= \text{Weighted Variance} + \text{Weighted Squared Bias.}$$

The important cases of the MSE are those with $A = X'X$ the PMSE , $A = I$, the IMSE and $A = U_j U_j'$ the individual MSE. Theorem 5.6.7 gives a necessary and sufficient condition for one ridge or contraction type estimator to have a smaller IMSE, PMSE or TMSE than another.

It was shown in Chapter III (Theorem 3.2.1, Theobald's result) that one estimator having a smaller mean dispersion error than another is equivalent to it having a smaller weighted MSE for all non-negative definite weight matrices. In Exercise 3.2.8 the reader was asked to prove that an estimator having a smaller IMSE was equivalent to an estimator having a smaller weighted MSE with a diagonal weight matrix. The result of Exercise 3.2.8 will be established in Theorem 5.6.7 below . It will be used for the proof of Theorem 5.6.6.

Theorem 5.6.7. Let D be any non-negative definite diagonal matrix. Let θ_i , $i = 1,2$ be an n dimensional vector with coordinates θ_{ij} , $1 \le j \le m$. Let $\hat{\theta}_{ij}$ be an estimator of θ_{ij}. Then the individual MSE s of $\hat{\theta}_{1j}$ is less than that of $\hat{\theta}_{2j}$, i.e.,

$$m_{1j} = E(\hat{\theta}_{1j} - \theta_{1j})^2 \le E(\hat{\theta}_{2j} - \theta_{2j})^2 = m_{2j}, \qquad (5.6.47)$$

iff the weighted MSE with weight matrix D of $\hat{\theta}_{1j}$ is less than that of $\hat{\theta}_{2j}$, i.e.,

$$m_1 = E(\hat{\theta}_1 - \theta_1)'D(\hat{\theta}_1 - \theta_1) \le E(\hat{\theta}_2 - \theta_2)'D(\hat{\theta}_2 - \theta_2) = m_2. \qquad (5.6.48)$$

Proof. The result in (5.6.48) follows from that of (5.6.47) by virtue of the equality

$$E(\hat{\theta}_i - \theta_i)'D(\hat{\theta}_i - \theta_i) = \sum_{j=1}^{m} d_j E(\hat{\theta}_{ij} - \theta_{ij})^2 = \sum_{j=1}^{m} d_j m_{ij}. \qquad (5.6.49)$$

The inequality (5.6.47) is the special case of (5.6.48) when $d_j = 1$ and $d_k = 0$ when $j \neq k$.

The following simple corollary allows the comparison of regression estimators using Theorem 5.6.7.

Corollary 5.6.2. Let $A = UDU'$. Then the weighted MSE of $\hat{\beta}_1$ is less than that of $\hat{\beta}_2$ iff for each j where $1 \leq j \leq s$ the IMSE of $\hat{\gamma}_{1j} = U_j'\hat{\beta}_1$ is less than that of $\hat{\gamma}_{j2} = U_j'\hat{\beta}_2$.

Proof. The result follows from Theorem 5.6.7 because

$$\begin{aligned}
E(\hat{\beta}_i - \beta)'A(\hat{\beta}_i - \beta) &= E(\hat{\beta}_i - \beta)'UDU'(\hat{\beta}_i - \beta) \\
&= E(\hat{\gamma}_i - \gamma_i)'D(\hat{\gamma}_i - \gamma_i).
\end{aligned} \qquad (5.6.50)$$

Theorem 5.6.8. Assume that $F_1 \leq F_2$ where $F_i = UD_iU' + VEV'$ where the D_i are positive definite diagonal matrices, $D_2 - D_1$ is positive definite and E is either the identity or the zero matrix. Then if $A = U\Delta U' + VV'$ where Δ is a non-negative definite diagonal matrix the weighted MSE of $\hat{\beta}_1$ is less than that of $\hat{\beta}_2$ iff γ_j lies in the s dimensional rectangle

$$\left| \gamma_j - U_j'\theta \right| \leq \left[\frac{(d_{1j} + d_{2j})\sigma^2 + 2\lambda_j d_{1j} d_{2j}}{2\sigma^2 + \lambda_j (d_{1j} + d_{2j})} \right]^{1/2}, \quad 1 \leq j \leq s. \qquad (5.6.51)$$

Proof. Let $\hat{\gamma} = U'\hat{\beta}$. Then the weighted MSE may be written as

$$M_A = E_\beta(\hat{\gamma} - \gamma)'D(\hat{\gamma} - \gamma) + E_\beta(\hat{\beta} - \beta)'VV'(\hat{\beta} - \beta). \qquad (5.6.52)$$

From (5.5.9), the form of F and $X'V = 0$ $V'\hat{\beta} = V'\theta$. Consequently

$$E_\beta(\hat{\beta} - \beta)'VV'(\hat{\beta} - \beta) = (\theta - \beta)'VV'(\theta - \beta). \qquad (5.6.53)$$

By Theorem 5.6.5 and its corollary it suffices to show that the IMSE of the components of $\hat{\gamma}_1$ is smaller than those of $\hat{\gamma}_2$. Let U_j be the columns of U. If $i = 1,2$,

$$E(\hat{\gamma}_{ij} - \gamma_{ij})^2 = U_j'E(\hat{\beta} - \beta)(\hat{\beta} - \beta)'U_j. \tag{5.6.54}$$

Thus, from (5.6.8)

$$E(\hat{\gamma}_{ij} - \gamma_{ij})^2 = \left(\frac{1}{\lambda_j + \sigma^2/d_{ij}}\right)^2 \left(\sigma^2\lambda_j + \frac{\sigma^4}{d_{ij}^2}(\gamma_j - U_j'\theta)^2\right)$$
$$= \left(\frac{1}{\lambda_j d_{ij} + \sigma^2}\right)^2 (\sigma^2\lambda_j d_{ij}^2 + \sigma^4(\gamma_j - U_j'\theta)^2). \tag{5.6.55}$$

Notice that $F_1 \leq F_2$ implies that $d_{1j} \leq d_{2j}$. Some algebra is necessary to simplify the expressions for the differences of the variances and the biases of the estimators. First observe that

$$\left(\frac{d_{2j}}{\lambda_j d_{2j} + \sigma^2}\right)^2 - \left(\frac{d_{1j}}{\lambda_j d_{1j} + \sigma^2}\right)^2$$
$$= \left(\frac{d_{2j}}{\lambda_j d_{2j} + \sigma^2} + \frac{d_{1j}}{\lambda_j d_{1j} + \sigma^2}\right)\left(\frac{d_{2j}}{\lambda_j d_{2j} + \sigma^2} - \frac{d_{1j}}{\lambda_j d_{1j} + \sigma^2}\right) \tag{5.6.56}$$
$$= \frac{[2\lambda_j d_{1j} d_{2j} + \sigma^2(d_{1j} + d_{2j})](d_{1j} - d_{2j})\sigma^2}{(\lambda_j d_{2j} + \sigma^2)^2(\lambda_j d_{1j} + \sigma^2)^2}.$$

Moreover

$$\left(\frac{1}{\lambda_j d_{1j} + \sigma^2}\right)^2 - \left(\frac{1}{\lambda_j d_{2j} + \sigma^2}\right)^2$$
$$= \left(\frac{1}{\lambda_j d_{1j} + \sigma^2} + \frac{1}{\lambda_j d_{2j} + \sigma^2}\right)\left(\frac{1}{\lambda_j d_{1j} + \sigma^2} - \frac{1}{\lambda_j d_{2j} + \sigma^2}\right) \tag{5.6.57}$$
$$= \frac{[2\sigma^2 + \lambda_j(d_{1j} + d_{2j})](d_{2j} - d_{1j})\lambda_j}{(\lambda_j d_{2j} + \sigma^2)^2(\lambda_j d_{1j} + \sigma^2)^2}.$$

Let

$$d = E(\hat{\gamma}_{2j} - \gamma_{2j})^2 - E(\hat{\gamma}_{1j} - \gamma_{1j})^2. \tag{5.6.58}$$

Use (5.6.57) and (5.6.58) to simplify the algebraic expression resulting from substitution of (5.6.52) into (5.6.58). Thus, $d \geq 0$ iff

$$\sigma^4 \lambda_j (d_{2j} - d_{1j})[\lambda_j(d_{1j} + d_{2j}) + 2\sigma^2](\gamma_j - U_j'\theta)^2$$
$$\leq \sigma^2 \lambda_j [2\lambda_j d_{1j} d_{2j} + \sigma^2(d_{1j} + d_{2j})](d_{2j} - d_{1j}) \tag{5.6.59}$$

When $d_{1j} < d_{2j}$ the inequality (5.6.51) is equivalent to (5.6.59).

Exercise 5.6.12. Most of the results above required that $F_2 - F_1$ be non-negative definite. What does this imply about the relationships between the k's and k_i s for the dispersion of the priors associated with the ridge, generalized ridge, contraction and generalized contraction estimator?

Exercise 5.6.13. What does the non-negative definiteness $F_2 - F_1$ of imply about the size of an ellipsoid where the BE has a smaller MSE than the LS?

Exercise 5.6.14. Consider all sixteen possible comparisons of ridge and contraction estimators with one another for $F = UDU'$. Write the condition implied by Inequality (5.6.55) in Theorem 5.6.8 for each case.

Exercise 5.6.15. A. Write a necessary and sufficient condition in the form of an ellipsoid to compare the TMSE of two ordinary ridges, two generalized ridges and an ordinary and generalized ridge estimator.
B. Do likewise for comparing the PMSE of contraction and generalized contraction estimators.

5.6.4 Comparison of the Mixed Estimators

When the BE was formulated it was assumed that β was a random variable with mean θ and dispersion F. The mixed estimator was formulated assuming that β was a constant and that additional observations were taken using the model $r = R\beta + \phi$ where ϕ had mean zero and variance $\tau^2 I$. These additional observations could be shown to correspond to assumptions about a prior variance in certain cases. The mathematical forms of the BE and the mixed estimators were similar and could be shown to be equivalent in certain cases.

The MSE of the BE without averaging over the prior distribution was obtained by finding the expectation conditional on the β parameters. The analogous computation for the mixed estimators would be to find the MSE conditioning on the additional observations r. Recall that the mathematical forms of the BE and the mixed estimator were the same with $(R'R)^+$ playing the analogous role in the mixed estimator to $UU'FUU'$ in the mixed estimator. The same is true for the mathematical forms of the conditional and average MSE. Two forms are given for the MSE below ((5.6.60)-(5.6.61)).Each of these forms is the sum of a variance and a bias term. When the prior assumptions are used the mixed estimator is unbiased. However , if the prior assumptions are neglected the mixed estimator is biased. This question will be discussed further later.

To obtain the conditional MSE of $p'\hat\beta$ define

$$\Gamma_2 = \tau^2 X'X + \sigma^2 R'R$$

and

$$C_2 = (X'X)^+ \sigma^2 + (R'R)^+ \tau^2.$$

The two forms of the MSE corresponding to (5.6.8) and (5.6.9) respectively are

$$\begin{aligned} M = {}& p'T_2^+ (X'X)\Gamma_2^+ p\sigma^2\tau^4 \\ & + p'T_2^+ R'(R\beta - r)(R\beta - r)'R\Gamma_2^+ p\sigma^4 \end{aligned} \tag{5.6.60}$$

and

$$\begin{aligned} M = {}& p'(X'X)^+ p\sigma^2 - 2p'(X'X)^+ C_2^+ (X'X)^+ p\sigma^4 \\ & + p'(X'X)^+ C_2^+ (X'X)^+ C_2^+ (X'X)^+ p\sigma^6 \\ & + p'(X'X)^+ C_2^+ (\beta - b_2)(\beta - b_2)'C_2^+ (X'X)^+ p\sigma^4. \end{aligned} \tag{5.6.61}$$

Results that are similar to that of Farebrother (1976) and Gruber (1979,1990) may be derived to compare the MSE of mixed estimators   neglecting the prior information in the stochastic prior constraints. One such result is given by Theorem 5.6.9.

Theorem 5.6.9. Consider an augmented linear model where $R = AU'$ where A is an sxs nonsingular matrix. Then the MSE of the parametric functions of the mixed estimator is less than that of the LS iff $R\beta$ lies in the ellipsoid

$$(R\beta - r)'(2\tau^2 I + \sigma^2 R(X'X)^+ R')^{-1}(R\beta - r) \leq 1. \tag{5.6.62}$$

Proof. The MSE M in (5.6.60) is less than that of the LS iff for all parametric functions $p'\beta$ iff

$$\begin{aligned} p'\Gamma_2^+ R'(R\beta - r)(R\beta - r)'R\Gamma_2^+ p\sigma^4 \\ \leq p'(X'X)^+ p\sigma^2 - p'\Gamma_2^+ (X'X)\Gamma_2^+ p\sigma^2\tau^4. \end{aligned} \tag{5.6.63}$$

Since $R = AU'$ and $\Gamma_2^+ = U(\Lambda\tau^2 + A'A\sigma^2)^{-1}U'$, (5.6.63) is equivalent to

$$\begin{aligned} (\Lambda\tau^2 + A'A\sigma^2)^{-1}A'(R\beta - r)(R\beta - r)'A(\Lambda\tau^2 + A'A\sigma^2)^{-1}\sigma^4 \\ \leq \Lambda^{-1}\sigma^2 - (\Lambda\tau^2 + A'A\sigma^2)^{-1}\Lambda(\Lambda\tau^2 + A'A\sigma^2)^{-1}\sigma^2\tau^4. \end{aligned} \tag{5.6.64}$$

Premultiply (5.6.68) by $C = A'^{-1}(\Lambda\tau^2 + A'A\sigma^2)$ and postmultiply by C'. After some matrix algebra it is seen that (5.6.64) is equivalent to

$$(R\beta - r)(R\beta - r)' \leq 2I\tau^2 + A\Lambda^{-1}A'\sigma^2. \tag{5.6.65}$$

The result follows from Theorem 3.2.4 and the fact that $A\Lambda^{-1}A' = AU'UA^{-1}U'U = R(X'X)^+R'$.

Exercise 5.6.16. Show that if an sxs matrix A is non-singular then A'A is positive definite.

Comparison of (5.6.61) with the LS estimator leads to a necessary and sufficient condition for the mixed estimator to have a smaller MSE than the LS that is similar to that of (5.6.11). The reader is asked to prove this result in Exercise 5.6.17.

Exercise 5.6.17. Given the hypothesis of Theorem 5.6.9 show that a necessary and sufficient condition for a mixed estimator to have a smaller MSE than the LS is

$$(\beta - b_2)'(2(R'R)^+\tau^2 + (X'X)^+\sigma^2)^+(\beta - b_2) \le 1. \qquad (5.6.69)$$

5.7 Comparing the MSE of Ridge-Type Estimators from the Bayesian Point of View

Section 5.6 made comparisons of the MSE of the different Bayes estimators without averaging over the prior distribution and the mixed estimator neglecting the prior assumptions. This section will make the same kinds of comparisons for the MSE averaged over the prior distribution. In general the results for this case are simpler. It will be shown that more precise prior information leads to an estimator with smaller average MSE. The MSE of the BE averaging over the prior will always be smaller than that of the LS. This was not the case for the conditional MSE; the conditional MSE of the BE was only less than the LS for a range of values of the β parameters. The average MSE of the BE and the maximum value of the minimax estimator on an ellipsoid of the form $\beta'F^+\beta \le 1$ will turn out to be the same. When the prior stochastic constraints are not neglected the MSE of the mixed estimator shares many of the mathematical properties of the BE in spite of the fact that the mixed estimator is derived from the frequentist point of view.

5.7.1 The MSE of the BE

Some expressions for the average MSE of the BE will be obtained by computing the expectation of the conditional MSE in (5.6.8) and (5.6.9). Recall that the basic assumptions about the prior distribution are

$$E(\beta) = \theta \text{ and } D(\beta) = E(\beta - \theta)(\beta - \theta)' = F. \qquad (5.7.1)$$

Finding the expectation of (5.6.8) and noticing that

$$\begin{aligned}
&(UU'FUU')^+ F(UU'FUU')^+ \\
&= U(U'FU)^{-1}U'FU(U'FU)^{-1}U' \\
&= U(U'FU)^{-1}U' = (UU'FUU')^+,
\end{aligned} \qquad (5.7.2)$$

$$m_a$$
$$= p'\Gamma_1^+ [\sigma^2 X'X + (UU'FUU')^+ E(\beta - \theta)(\beta - \theta)'(UU'FUU')^+ \sigma^4] \Gamma_1^+ p.$$
$$= p'\Gamma_1^+ \Gamma_1 \Gamma_1^+ p\sigma^2 = p'\Gamma_1^+ p\sigma^2 = p'(X'X + \sigma^2 (UU'FUU')^+)^+ p\sigma^2.$$

$$(5.7.3)$$

To obtain another form of the MSE which shows the improvement of the BE over the LS first let $A = \Lambda^{-1} + U'FU$. Then $C_1 = UAU'$. Consequently

$$C_1^+ [(X'X)\sigma^2 + F]C_1^+ = UA^{-1}U'[(X'X)\sigma^2 + F]UA^{-1}U'$$
$$= UA^{-1}[\Lambda^{-1} + \sigma^2 U'FU]A^{-1}U' = U'A^{-1}U = C_1^+.$$

$$(5.7.4)$$

Finding the expectation of (5.6.9) and then substituting (5.7.4) into the resulting expression and combining terms yields the form of the average MSE

$$m_a = p'(X'X)^+ p\sigma^2 - p'(X'X)^+ C_1^+ (X'X)^+ p\sigma^4$$
$$= p'(X'X)^+ p\sigma^2 - p'(X'X)^+ (UU'FUU' + (X'X)^+ \sigma^2)^+ (X'X)^+ p\sigma^4.$$

$$(5.7.5)$$

The first term of (5.7.5) is the average MSE of the LS; the second term represents the improvement from use of the BE. Thus the BE regardless of the choice of prior always has smaller MSE than the LS.

Another form of the MSE of the BE will be derived from (5.7.3) for estimable parametric functions of the β parameters. From this form it will be seen that :

1. The MSE of the BE after sampling is always less than that of the prior parameter.
2. The MSE of the BE is the difference between the dispersion of the prior distribution and the dispersion of the BE.

From (5.7.3)

$$m_a = p'(X'X + \sigma^2 (UU'FUU')^+)^+ p\sigma^2$$
$$= p'U(\Lambda + \sigma^2 (U'FU)^{-1})^{-1} U'p\sigma^2.$$

$$(5.7.6)$$

But

$$(\Lambda + \sigma^2 (U'FU)^{-1})^{-1} \sigma^2 = \sigma^2 \Lambda^{-1} (\sigma^2 \Lambda^{-1} + U'FU)^{-1} U'FU$$
$$= U'FU - U'FU(\sigma^2 \Lambda^{-1} + U'FU)^{-1} U'FU.$$

$$(5.7.7)$$

Substitute (5.7.7) into (5.7.6). Since for estimable parametric functions $p' = p'UU'$, an equivalent form of the MSE is

$$m_a = p'Fp - p'F[UU'FUU' + \sigma^2(X'X)^+]^+Fp. \qquad (5.7.8)$$

The first term is the dispersion of the prior distribution; the second term is the improvement due to sampling. Notice, that since for the observation vector in a classical linear regression model with the parameters assumed to be random variables with a prior distribution with mean θ and dispersion F,

$$D(Y) = XFX' + \sigma^2 I, \qquad (5.7.9)$$

the dispersion of the LS is

$$D(b) = UU'FUU' + \sigma^2(X'X)^+. \qquad (5.7.10)$$

For the form of the BE obtained in (5.5.29) , i.e.,

$$p'\hat{\beta} = p'\theta + p'F[(X'X)^+\sigma^2 + UU'FUU']^+(b - \theta), \qquad (5.7.11)$$

using (5.7.10) the variance is

$$Var(p'\hat{\beta}) = p'F[(X'X)\sigma^2 + UU'FUU']^+Fp. \qquad (5.7.12)$$

Thus,

$$m_a = Var(p'\beta) - Var(p'\hat{\beta}). \qquad (5.7.13)$$

Thus, the average MSE is the difference between the variance of the parametric functions of the parameters of the regression model and the variance of the parametric functions of the linear BE. When the data and the prior distribution are normal the linear BE is the BE. When the Bayes risk is calculated using a squared error loss function the BE is the mean of the posterior distribution. The last term of (5.7.13) is thus the variance of the mean of a posterior distribution. Since the BE is the expectation of the posterior mean

$$\begin{aligned} m_a &= p'E(\hat{\beta} - \beta)(\hat{\beta} - \beta)'p \\ &= p'E(\beta - \hat{\beta})(\beta - \hat{\beta})'p \\ &= E(PVar(p'\hat{\beta})). \end{aligned} \qquad (5.7.14)$$

The symbol Pvar means posterior variance. Let $\underline{x}$ represent a random sample from a population or a sufficient statistic for β. Then

$$\text{Var}(p'\beta) = E((\text{Var}_{\underline{x}}(p'\beta)) + \text{VarE}_{\underline{x}}(p'\beta) \tag{5.7.15}$$

The result in (5.7.15) may also be looked upon as a special case of the well known fact that for random variables X and Y

$$\text{Var}(Y) = E[\text{Var}(Y \mid X)] + \text{VarE}(Y \mid X). \tag{5.7.16}$$

Example 5.7.1. Consider the means model

$$Y = \mu + \varepsilon$$

where

$$E(\varepsilon|\mu) = 0 \text{ and } D(\varepsilon|\mu) = \sigma^2$$

and the prior assumptions about μ are

$$E(\mu) = \theta \text{ and } D(\mu) = \tau^2.$$

Some forms of the BE are

$$\hat{\mu} = \bar{x} - \frac{\sigma^2}{\sigma^2 + n\tau^2}(\bar{x} - \theta) = \frac{n\tau^2}{\sigma^2 + n\tau^2}\bar{x} + \frac{\sigma^2}{\sigma^2 + n\tau^2}\theta.$$

The three forms of the average MSE corresponding to (5.7.5), (5.7.6) and (5.7.8) are given by

$$m_a = \frac{\sigma^2}{n} - \frac{\sigma^4}{n(n\tau^2 + \sigma^2)} = \frac{\sigma^2\tau^2}{n\tau^2 + \sigma^2} = \tau^2 - \frac{n\tau^4}{n\tau^2 + \sigma^2}.$$

Clearly for all choices of prior parameter the BE has a smaller MSE than the sample mean.

An expression for the MSE without averaging over the prior distribution is

$$m_c = \frac{n\tau^4\sigma^2 + \sigma^4(\mu - \theta)^2}{(\sigma^2 + n\tau^2)^2}.$$

By solving the inequality $m_c \leq \sigma^2/n$ for $(\mu - \theta)$ the conditional MSE of the BE is less than that of the LS iff

$$|\mu - \theta| \leq \sqrt{\frac{\sigma^2}{n} + 2\tau^2}.$$

This is a specialization of Farebrother's result.

Example 5.7.2. Consider the problem of estimating the basic estimable functions $\tau_i = \mu + \alpha_i$ in the one way ANOVA model

$$Y = (I_a \otimes 1_n)\tau + \varepsilon$$

with prior assumptions

$$E(\tau) = 1_a \theta \text{ and } D(\tau) = \sigma_\alpha^2 I_a$$

where θ is a scalar and

$$E(\varepsilon \mid \alpha) = 0 \text{ and } D(\varepsilon \mid \alpha) = \sigma^2 I_n.$$

A form of the BE is

$$\hat{\tau} = h - \frac{\sigma^2}{\sigma^2 + n\sigma_\alpha^2}(h - 1_a\theta)$$

where h represents the LS of τ . The three forms of the average MSE of each of the uncorrelated coordinates of τ is given by

$$m = \frac{\sigma^2}{n} - \frac{\sigma^4}{n(n\sigma_\alpha^2 + \sigma^2)} = \frac{\sigma^2\sigma_\alpha^2}{n\sigma_\alpha^2 + \sigma^2} = \sigma_\alpha^2 - \frac{n\sigma_\alpha^2}{n\sigma_\alpha^2 + \sigma^2}.$$

The linear model may also be written in the form

$$Y_{ij} = \tau_i + \varepsilon_{ij}, \ 1 \leq i \leq a, \ 1 \leq j \leq n.$$

The coordinates of the LS would then be

$$\hat{\tau}_i = \overline{Y}_{i\cdot} = \frac{1}{n}\sum_{j=1}^{n} Y_{ij}.$$

The BE may thus be written in the form

$$\hat{\tau}_i = \overline{Y}_{i\cdot} - \frac{\sigma^2}{\sigma^2 + n\sigma_\alpha^2}(\overline{Y}_{i\cdot} - \theta), \ 1 \le i \le a.$$

When $\theta = 0$ the BE may be written in the form

$$\hat{\tau}_i = \frac{n\overline{Y}_{i\cdot}}{\sigma^2/\sigma_\alpha^2 + n},$$

the ridge regression estimator with $k = \sigma^2/\sigma_\alpha{}^2$.

In the random effects model where

$$\tau_i \sim N(\mu, \sigma_\alpha^2) \text{ and } \varepsilon_{ij} \sim N(0, \sigma^2),$$

the expectation of the Treatment Sum of Squares T divided by $a - 1$, the degrees of freedom, is $\sigma^2 + n\sigma_\alpha{}^2$ and the expectation of the error sum of squares W is the Within Sum of Squares divided by $(a-1)n$, the number of degrees of freedom used to estimate error. If θ is unknown it can be estimated by

$$\overline{Y}_{\cdot\cdot} = \frac{1}{an}\sum_{i=1}^{a}\sum_{j=1}^{n} Y_{ij}.$$

Thus, an empirical Bayes estimator for τ_i is given by

$$\hat{\tau}_i = \overline{Y}_{i\cdot} - \frac{c(a-1)W}{(n-1)aT}(\overline{Y}_{i\cdot} - \overline{Y}_{\cdot\cdot}).$$

The reader may obtain the average MSE if he or she so wishes.

Exercise 5.7.1. Obtain (5.7.5) from (5.7.6).

Exercise 5.7.2. Obtain the three forms of the average MSE given above by
A. Finding the MSE of the different forms of the BE using the assumptions of the prior mean and variance without finding a conditional probability first.
 B. By finding expressions for the variance of the different forms of the BE and employing Equation (5.7.13).

Exercise 5.7.3. What are the three forms of the average MSE for
A. The generalized ridge estimator of C.R. Rao?
B. The ordinary and the generalized ridge estimator of Hoerl and Kennard?
C. The contraction and the generalized contraction estimator ?

5..7.2 Comparing the Average MSE of Two BE

The main result for the comparison of the average MSE of two BE is that for a given value of the prior mean the MSE of a BE is smaller when the prior knowledge about the β parameter is more precise. This is the content of Theorem 5.7.1.

Theorem 5.7.1. Suppose that $F_2 - F_1 \geq 0$, i.e., $F_2 - F_1$ is non-negative definite. Let $p'\hat{\beta}_i$, i=1,2 be the BE associated with the prior mean θ_i and dispersion F_i. Then averaging over the respective prior distributions

$$\text{MSE}(p'\hat{\beta}_1) \leq \text{MSE}(p'\hat{\beta}_2).$$ (5.7.17)

Proof. From the non-negative definiteness of $F_2 - F_1$ it follows that

$$U'F_1U + \sigma^2\Lambda^{-1} \leq U'F_2U + \sigma^2\Lambda^{-1}.$$ (5.7.18)

Applying Theorem 5.1.4

$$(U'F_2U + \sigma^2\Lambda^{-1})^{-1} \leq (U'F_1U + \sigma^2\Lambda^{-1})^{-1}$$ (5.7.19)

and

$$\Lambda^{-1}(U'F_2U + \sigma^2\Lambda^{-1})^{-1}\Lambda^{-1} \leq \Lambda^{-1}(U'F_1U + \sigma^2\Lambda^{-1})^{-1}\Lambda^{-1} .$$ (5.7.20)

Then it follows that

$$p'(X'X)^+(UU'F_2UU' + \sigma^2(X'X)^+)^+(X'X)^+p$$

$$\leq p'(X'X)^+ (UU'F_iUU' + \sigma^2(X'X)^+)^+ (X'X)^+ p \tag{5.7.21}$$

and

$$MSE(p'\hat{\beta}_1)$$
$$= \sigma^2(X'X)^+ - p'(X'X)^+ (UU'F_1UU' + \sigma^2(X'X)^+)^+ (X'X)^+ p$$
$$\leq \sigma^2(X'X)^+ - p'(X'X)^+ (UU'F_2UU' + \sigma^2(X'X)^+)^+ (X'X)^+ p$$
$$= MSE(p'\hat{\beta}_2). \tag{5.7.22}$$

A simple consequence of Theorem 5.7.1 is that the average MSE for parametric functions of the ordinary ridge estimator of Hoerl and Kennard and the contraction estimator of Mayer and Willke is a decreasing function of k or c (k denotes the ridge parameter; c denotes the parameter of the contraction estimator). Let $p'\hat{\beta}_i$, i = 1, 2 be two generalized ridge estimators of Hoerl and Kennard with ridge parameter matrices K_i, i = 1,2 where $K_1 \geq K_2$ implying that the individual diagonal elements $k_{i1} \geq k_{i2}$, $1 \leq i \leq s$. Then $MSE(p'\hat{\beta}_1) \leq MSE(p'\hat{\beta}_2)$. A similar result may be established for the generalized contraction estimator.

Another application of Theorem 5.7.1 is the comparison of the average MSE of different ridge and contraction estimators. Denote the i'th latent root (eigenvalue) of X'X by λ_i, the i'th element of K by k_i and the i'th element of C by c_i. The following conditions for comparison of the estimators are easy consequences of Theorem 7.5.1:

1. If $k/c < \lambda_i$ then the contraction estimator has smaller average MSE than the ridge estimator.

2. The average MSE of the generalized contraction estimator is smaller than that of the generalized ridge estimator if $k_i/c_i < \lambda_i$.

3. The generalized ridge estimator has a smaller average MSE than the ridge estimator provided that $k_i > k$.

4. If $c_i > c$ then the generalized contraction estimator has a smaller average MSE than the contraction estimator.

5. The average MSE of the contraction estimator is smaller than that of the generalized ridge estimator provided that $k_i/c < \lambda_i$.

6. The ridge estimator has a smaller average MSE than the generalized contraction estimator provided that $k_i/c > \lambda_i$.

Exercise 5.7.4. State and prove a condition for one generalized contraction estimator to have a smaller average MSE than another.

It is possible to compare the conditional MSE of a BE to its average MSE. The range of β parameters where the conditional MSE of $p'\hat{\beta}$ is less than the average MSE is given by Theorem 5.7.2.

Theorem 5.7.2. The conditional MSE of the BE is less than or equal to its average MSE with respect to its associated prior for parametric functions iff

$$(\beta - \theta)'(UU'FUU')^+(\beta - \theta) \le 1. \tag{5.7.23}$$

Proof. Let m_a denote the average MSE and m_c denote the conditional MSE. From (5.6.9) and (5.7.5)

$$
\begin{aligned}
&m_a - m_c \\
&= p'(X'X)^+ C_1^+ (X'X)^+ p\sigma^4 \\
&\quad -p'(X'X)^+ C_1^+[(X'X)^+\sigma^2 + (\beta - \theta)(\beta - \theta)']C_1^+ (X'X)^+ p\sigma^4 \ge 0
\end{aligned}
\tag{5.7.24}
$$

iff

$$
\begin{aligned}
&p'(X'X)^+ C_1^+ (\beta - \theta)(\beta - \theta)' C_1^+ (X'X)^+ p \\
&\le p'(X'X)^+ C_1^+ (X'X)^+ p - p'(X'X)^+ C_1^+ (X'X)^+ C_1^+ (X'X)^+ p\sigma^2.
\end{aligned}
\tag{5.7.25}
$$

Inequality (5.7.25) holds true iff

$$
\begin{aligned}
&\Lambda^{-1}(U'C_1 U)^{-1} U'(\beta - \theta)(\beta - \theta)' U (U'C_1 U)^{-1}\Lambda^{-1} \\
&\le \Lambda^{-1}(U'C_1 U)^{-1}\Lambda^{-1} - \Lambda^{-1}(U'C_1 U)^{-1}\Lambda^{-1}(U'C_1 U)^{-1}\Lambda^{-1}\sigma^2.
\end{aligned}
\tag{5.7.26}
$$

Inequality (5.7.26) is equivalent to

$$U'(\beta - \theta)(\beta - \theta)'U \le U'C_1 U - \Lambda^{-1}\sigma^2 = U'FU \tag{5.7.27}$$

by pre and post multiplication by $U'C_1 U\Lambda$ and because $U'C_1 U = U'FU + \sigma^2\Lambda^{-1}$. The result then follows by application of Theorem 3.2.4.

Exercise 5.7.5. Prove Theorem 5.7.1 by comparing (5.6.8) with (5.7.6).

Using Theorem 5.7.2 the conditional MSEs of the ridge type estimators may be compared with their average MSEs. This is done in Corollary 5.7.1

Corollary 5.7.1.

1. The conditional MSE of the ordinary ridge regression estimator is less than or equal to the average MSE iff

$$\beta'UU'\beta \le \frac{\sigma^2}{k}. \qquad (5.7.28)$$

2. The conditional MSE of the generalized ridge regression estimator is less than or equal to the average MSE iff

$$\beta'UKU'\beta \le \sigma^2. \qquad (5.7.29)$$

3. The conditional MSE of the contraction estimator is less than or equal to the average MSE iff

$$\beta'X'X\beta \le \sigma^2. \qquad (5.7.30)$$

4. The conditional MSE of the generalized contraction estimator is less than or equal to the average MSE iff

$$\beta'UC\Lambda U'\beta \le \sigma^2. \qquad (5.7.31)$$

Proof. Substitute $\theta = 0$ and the appropriate form of F into (5.7.23).

Recall that the linear minimax estimator was derived by finding the linear estimator where the maximum value of the conditional MSE on an ellipsoid of the form obtained in Theorem 5.7.1 is minimized. The mathematical form of the resulting estimator is the same as that of the linear BE. Thus, one measure of efficiency of the minimax estimator is the supremum (least upper bound) of the conditional MSE. Thus, if $A = \{\beta : (\beta - \theta)'(UU'FUU')^+ (\beta - \theta) \le \sigma^2\}$, then

$$eff = \sup_A p'E_\beta(\hat{\beta} - \beta)(\hat{\beta} - \beta)'p. \qquad (5.7.32)$$

The mathematical form of the efficiency found in (5.7.32) is the same as that of the MSE averaging over the prior. This is easily seen by substitution of Inequality (5.7.23) into (5.6.8) or (5.6.9).

Theorem 5.7.3. Assume that $F_1 \le F_2$ and that $U'F_iU$, $i = 1, 2$ is positive definite. Let $p'\hat{\beta}_i$, $i = 1,2$ be the minimax estimator associated with F_i. Then

$$\text{eff}(p'\hat{\beta}_1) \le \text{eff}(p'\hat{\beta}_2) \ . \tag{5.7.33}$$

Proof. Since $F_1 \le F_2$, $U'F_1U \le U'F_2U$ and $(U'F_2U)^{-1} \le (U'F_1U)^{-1}$, it follows that $(UU'F_2UU')^+ \le (UU'F_1UU')^+$. Then the ellipsoid associated with F_1 is contained within that associated with F_2 because for F_2 there are more values of the β parameters where the associated quadratic form is less than σ^2 than for F_1. Thus the supremum of the minimum conditional MSE of the linear estimator of the form (5.4.1) for the ellipsoid associated with F_2 must be greater than or equal to that for the ellipsoid associated with F_1.

Equation (5.7.33) also follows from the fact that the efficiency of the minimax estimator associated with the ellipsoid with matrix F is equal to the average MSE of the BE associated with the prior dispersion F. Thus, Theorem 5.7.3 could have obtained from Theorem 5.7.1. Likewise Theorem 5.7.1 could have been obtained form Theorem 5.7.3. Furthermore from the argument given in the proof of Theorem 5.7.3 when $F_1 \le F_2$ the ellipsoid where the BE associated with F_1 has smaller than average conditional MSE is contained in that where the BE associated with F_2 has this property.

5.7.3 The MSE of the Mixed Estimator

Recall that the mixed estimator is the weighted least square estimator for the augmented linear model given in (5.3.1) subject to the assumptions in (5.3.2). By the Gauss Markov theorem it is the best linear unbiased estimator (BLUE) with respect to the linear model (5.3.1), best in the sense that among all of the unbiased estimators it has minimum variance. When $R = AU'$ three algebraically equivalent forms of the variance or MSE[3] are

$$\begin{aligned} v &= p'(X'X + R'R\sigma^2)^+ p\sigma^2\tau^2 \\ &= p'(X'X)^+ p\sigma^2 - p'(X'X)^+[(X'X)^+\sigma^2 + (R'R)^+\tau^2]^+(X'X)^+ p\sigma^4 \\ &= p'(R'R)^+ p\tau^2 - p'(R'R)^+[(X'X)^+\sigma^2 + (R'R)^+\tau^2]^+(R'R)^+ p\tau^4 \end{aligned}$$

$$\tag{5.7.34}$$

Notice that the forms of the estimator in (5.7.34) correspond to those of the BE with $(R'R)^+\tau^2$ replacing $(UU'FUU')^+$. The first term of the second form of the

---

[3] The variance and the MSE are the same in this case because the estimator is unbiased.

variance in (5.7.33) is the variance of $p'b_1$; the second term of the third form is the variance of $p'b_2$. Clearly the variance in (5.7.34) is smaller than that of both LS estimators. In the last two forms in (5.7.34) the roles of the sample and prior information are exchanged. The first term of the second form is the MSE of $p'b_2$; the second term is the improvement using the mixed estimator.

The MSE of the BE has been calculated with and without averaging over the prior distribution. The linear BE was obtained using the prior information about the mean and the dispersion of the β parameters. The mixed estimator was obtained using additional observations taken using an additional linear model. Just like the MSE of the BE was obtained neglecting the prior assumptions about the mean and the dispersion the MSE of the mixed estimator may be obtained neglecting the additional observations. This MSE is computed conditional on r, the additional observations. For these conditional expectations the mixed estimators are not unbiased. Two forms of this MSE will now be given as the sum of the variance and the bias. Let

$$\Gamma_2 = X'X\tau^2 + R'R\sigma^2$$

and

$$C_2 = (X'X)^+\sigma^2 + (R'R)^+\tau^2.$$

Two forms of the MSE neglecting the stochastic prior assumptions are

$$m = p'\Gamma_2^+(X'X)^+\Gamma_2^+p\sigma^2\tau^4 + p'\Gamma_2^+R'(R\beta - r)(R\beta - r)'R\Gamma_2^+p\sigma^4 \tag{5.7.35}$$

and

$$\begin{aligned} m = {} & p'(X'X)^+p\sigma^2 - 2p'(X'X)^+C_2^+(X'X)^+p\sigma^4 \\ & + p'(X'X)^+C_2^+(X'X)^+C_2^+(X'X)^+p\sigma^6 \\ & + p'(X'X)^+C_2^+(\beta - b_2)(\beta - b_2)'C_2^+(X'X)^+p\sigma^4. \end{aligned} \tag{5.7.36}$$

Exercise 5.7.6. Show that by obtaining the expectation of (5.7.35) and (5.7.36) the MSE of the mixed estimator in Equation (5.7.34) is obtained.

The MSE neglecting the prior assumptions may be compared with that of the LS and with the MSE using the stochastic prior assumptions. The reader may show how to do this in Exercises 5.7.7 and 5.7.8.

Exercise 5.7.7. Assume that $R = AU'$ where A is nonsingular. Prove that for the MSE neglecting the prior assumptions

$$MSE(p'\hat{\beta}) \le MSE(p'b_1)$$

iff either

$$(R\beta - r)'(\sigma^2 R(X'X)^+ R' + 2\tau^2 I)^{-1}(R\beta - r) \le 1 \tag{5.7.37a}$$

or

$$(\beta - b_2)'(2\tau^2(R'R)^+ + \sigma^2(X'X)^+)^+(\beta - b_2) \le 1. \tag{5.7.37b}$$

Exercise 5.7.8. For Example 5.1.2

A. Show that

$$R(X'X)^+ R' = \begin{bmatrix} \frac{1}{2} & 0 \\ 0 & \frac{3}{2} \end{bmatrix}$$

and consequently

$$(\sigma^2 R(X'X)^+ R' + 2\tau^2 I)^{-1}$$
$$= \begin{bmatrix} \dfrac{6\sigma^2 + 8\tau^2}{3\sigma^3 + 16\sigma^2\tau^2 + 16\tau^4} & 0 \\ 0 & \dfrac{2\sigma^2 + 4\tau^2}{3\sigma^3 + 16\sigma^2\tau^2 + 16\tau^4} \end{bmatrix}$$

B. Show that the ellipsoid in (5.7.37a) takes the form

$$\frac{\eta_1^2}{a_1^2} + \frac{\eta_2^2}{a_2^2} \le 1$$

where

$$a_1 = \sqrt{\frac{3\sigma^4 + 16\sigma^2\tau^2 + 16\tau^4}{3(2\sigma^2 + 4\tau^2)}},$$

$$a_2 = \sqrt{\frac{3\sigma^4 + 16\sigma^2\tau^2 + 16\tau^4}{2(\sigma^2 + 2\tau^2)}} \,,$$

$$\eta_i = \alpha_1 - \alpha_2 - r_1$$

and

$$\eta_2 = \alpha_1 + \alpha_2 - 2\alpha_3.$$

Exercise 5.7.9. A. Show that the MSE of the mixed estimator neglecting the stochastic prior assumptions is less than that using the stochastic prior assumptions iff

$$(\beta - b_2)'(R'R)^+(\beta - b_2) \le \tau^2. \tag{5.7.38a}$$

by

(1) Recalling that the mixed estimator can also be viewed as a BE with prior mean b_2 and prior dispersion $(R'R)^+\tau^2$ and using (5.7.23).

(2) Direct comparison of (5.7.36) to (5.7.34).

B. Show that when $R = AU'$ an equivalent form of (5.7.38a) is

$$(R\beta - r)'(R\beta - r) \le \tau^2. \tag{5.7.39b}$$

Exercise 5.7.10. Suppose that the sample and the prior information were interchanged , i.e. , the linear model $r = R\beta + \phi$ is given with the stochastic prior constraint $Y = X\beta + \epsilon$. The assumptions about the error terms are the same. What are the results analogous to those presented in Exercise 5.7.8 and 5.7.9. Are there any conditions that must be imposed on X and R?

The results of this section says something very interesting about the unbiasedness of the LS. Suppose that the classical linear model is partitioned into two linear submodels

$$Y_1 = X_1\beta + \epsilon_1 \tag{5.7.40a}$$

and

$$Y_2 = X_2\beta + \epsilon_2. \tag{5.7.40b}$$

The LS estimator then is the same as the mixed estimator where the first model is the linear model and the second model are the stochastic prior assumptions or vice versa. The LS is unbiased if both submodels are used in the computation of the expectations. However if the mean, variance or MSE is computed using only one of the submodels neglecting the other one then the LS with respect to the full linear model is no longer unbiased. Its MSE would be less than that of the LS obtained for a submodel for the β parameters inside an ellipsoid similar to that of Exercise 5.7.7. Also the variance of the LS for the full model would always be smaller than that of the LS for the submodel. The reason that the LS is minimum variance unbiased is that all of the information from the data is used to compute its efficiency; the picture can be quite different if this is not the case.

Exercise 5.7.11. Find the ellipsoid where the MSE of LS for the full linear model is less than that of the LS for the submodel (5.7.40a) assuming the submodel (5.7.40b) is neglected in the MSE computations. Do any conditions have to be imposed on X_1 and X_2 for the non-full rank case?

5.8 The Jack-knifed Ridge Estimator

The ridge regression estimators are biased. A natural question to ask is whether it is possible to find an estimator with a smaller bias that still retains the some of the worthwhile features of the ridge regression estimator. The jack-knifed ridge estimator will be studied in this section and compared to the ridge estimators of Hoerl and Kennard with respect to different measures of efficiency.

As was pointed out in the historical survey in Chapter I Quenouillle (1956) proposed the application of the jack-knife procedure to a biased estimator to reduce the bias. Singh, Chaubey and Dwividi (1986) explained how to jack-knife a ridge regression estimator. They showed that the jack-knifed ridge estimator had a substantially smaller MSE than the Hoerl-Kennard ridge estimator.

Some of the measures of goodness for comparing two ridge type estimators include:

1. the bias;
2. the variance;
3. the MSE;
4. the smallest possible MSE for optimum values of the parameters;
5. the average MSE when the estimators are viewed from a Bayesian point of view;
6. the signal to noise ratio;
7. the region where the MSE is less than the LS.

An estimator is better with respect to any of the measures of efficiency in 1-5 that are smaller. Likewise an estimator is better with respect to whatever measures in 6 and 7 that are larger.

Gruber (1991) compared the Hoerl-Kennard ridge estimator (UR) to the jack-knifed ridge regression estimator (JR) with respect to the measures of efficiency in 1-7 above. This section is based on the results of that article.

5.8.1 The Jack-knifed and the Usual Ridge Estimator

For convenience recall that in orthogonal coordinates the LS is given by

$$g = \Lambda^{-1}U'X'Y. \tag{5.8.1}$$

The usual generalized ridge estimator (UR) is

$$\hat{\gamma}^{R} = (\Lambda + K)^{-1}U'X'Y = (\Lambda + K)^{-1}\Lambda g \tag{5.8.2}$$

where K is a positive definite diagonal matrix with non-zero elements k_i. The jack-knifed version of the ridge estimator that was presented by Singh et al. (1986) is

$$\hat{\gamma}^{J} = (I + (\Lambda + K)^{-1}K)\hat{\gamma}^{R} = (I - ((\Lambda + K)^{-1}K)^{2})g \tag{5.8.3}$$

The comparison of the estimators will be facilitated by writing them in terms of their individual coordinates. The coordinates of the estimators in (5.8.1)-(5.8.3) are

$$\hat{\gamma}_{i} = [f(k_{i})]g_{i}, \ 1 \leq i \leq s, \tag{5.8.4}$$

where $0 < f(k_i) \leq 1$. For (5.8.1)

$$f(k_{i}) = 1, \tag{5.8.5}$$

for (5.8.2)

$$f(k_{i}) = \frac{\lambda_{i}}{\lambda_{i} + k_{i}} \tag{5.8.6}$$

and for (5.8.3)

$$f(k_i) = \frac{\lambda_i}{\lambda_i + k_i} + \frac{\lambda_i k_i}{(\lambda_i + k_i)^2}$$

$$= 1 - \frac{k_i^2}{(\lambda_i + k_i)^2} \qquad (5.8.7)$$

$$= \frac{\lambda_i^2 + 2\lambda_i k_i}{(\lambda_i + k_i)^2}.$$

In the literature that the author has read the form of (5.8.3) is given without any justification. A rationale for why it works and is an appropriate jack-knife estimator is given in the discussion that follows

The estimator in (5.8.3) is a simplification of what would be obtained if the method of Quenouille and Tukey were applied. This method is as follows. Let $\hat{\theta}$ be an estimator of θ of the form

$$\hat{\theta} = \hat{\theta}(x_1, x_2, \cdots, x_n). \qquad (5.8.8)$$

The technique is based on sequentially deleting points x_i and computing $\hat{\theta}$ each time with the remaining $n - 1$ observations. The value of the statistic then takes the form

$$\hat{\theta}_{(i)} = \hat{\theta}(x_1, x_2, \cdots, x_{i-1}, x_i, \cdots, x_n). \qquad (5.8.9)$$

Letting

$$\hat{\theta}_{(\cdot)} = \frac{1}{n} \sum_{i=1}^{n} \hat{\theta}_{(i)}, \qquad (5.8.10)$$

the estimate of bias proposed by Quenouille is

$$\hat{B} = (n - 1)(\hat{\theta}_{(\cdot)} - \hat{\theta}). \qquad (5.8.11)$$

Thus, the bias corrected jack-knifed estimate of θ is

$$\breve{\theta} = \hat{\theta} - \hat{B} = n\hat{\theta} - (n - 1)\hat{\theta}_{(\cdot)}. \qquad (5.8.12)$$

How does the estimator in (5.8.12) reduce the bias? To see this let E_n denote the expectation for a sample of size n, i.e., $E_n = E\hat{\theta}(X_1, X_2, \cdots, X_n)$. For many common statistics, most maximum likelihood estimators included,

$$E_n = \theta + \frac{a_1}{n} + \frac{a_2}{n^2} + \cdots, \tag{5.8.13}$$

where the a_i do not depend on n. Furthermore

$$E\hat{\theta}_{(\cdot)} = E_{n-1} = \theta + \frac{a_1}{n-1} + \frac{a_2}{(n-1)^2} + \cdots. \tag{5.8.14}$$

Thus,

$$\begin{aligned} E\breve{\theta} &= nE_n - (n-1)E_{n-1} \\ &= \theta - \frac{a_2}{n(n-1)} + a_3\left(\frac{1}{n^2} + \frac{1}{(n-1)^2}\right) + \cdots \end{aligned} \tag{5.8.15}$$

The original estimator was biased $O(1/n)$. The jack-knifed estimator is biased $O(1/n^2)$.

Example 5.8.1 illustrates the application of this procedure in the context of shrinkage estimation.

Example 5.8.1. Recall that for the means model

$$Y = \mu + \varepsilon$$

the ridge regression estimator takes the form

$$\hat{\mu} = \frac{n}{n+k}\bar{y}.$$

Deleting observations one at a time and computing the ridge estimator with $n - 1$ observations

$$\hat{\mu}_{(i)} = \frac{1}{n-1+k}(y_1 + y_2 + \cdots + y_{i-1} + y_{i+1} + \cdots + y_n).$$

Then

$$\hat{\mu}_{(\cdot)} = \frac{(n-1)\bar{y}}{n-1+k}$$

and the jack-knifed estimator is

$$\tilde{\mu} = \frac{n^2}{n+k}\bar{y} - \frac{(n-1)^2}{(n-1+k)}\bar{y}$$

$$= \frac{n}{n+k}\bar{y} + \frac{(n-1)k}{(n+k)(n-1+k)}\bar{y}$$

$$= \left(1 + \frac{(n-1)k}{n(n-1+k)}\right)\hat{\mu}.$$

In the context of the means model the jack-knifed estimator in (5.8.3) takes the form

$$\tilde{\mu} = \left(1 + \frac{k}{n+k}\right)\frac{n}{n+k}\bar{y}.$$

Thus, the form of the Jack-knifed estimator obtained by Quenouille's method might suggest the form suggested in (5.8.3).
Observe that

$$Bias(\tilde{\mu}) = \left(1 - \frac{n(n-1)}{n(n-1)+k}\right)\frac{k\mu}{n+k} \le \frac{k\mu}{n+k} = Bias(\hat{\mu}).$$

5.8.2 The Measures of Efficiency

The estimator (5.8.4) has:

1. bias

$$B_i = (1 - f(k_i))\gamma_i; \tag{5.8.16}$$

2. variance

$$V_i = [f(k_i)]^2 \frac{\sigma^2}{\lambda_i}; \tag{5.8.17}$$

3. individual MSE (IMSE) equal to the sum of the variance (5.8.16) and the square of the bias (5.8.17) that is

$$E(\hat{\gamma}_i - \gamma_i)^2 = [f(k_i)]^2 \frac{\sigma^2}{\lambda_i} + [1 - f(k_i)]^2 \gamma_i^2; \qquad (5.8.18)$$

4. signal to noise ratio

$$\frac{Signal}{Noise} = \frac{(Bias)^2}{Variance} = \frac{[1 - f(k_i)]^2 \gamma_i^2 \lambda_i}{[f(k_i)]^2 \sigma^2}; \qquad (5.8.19)$$

5. optimum MSE when

$$f(k_i) = \frac{\gamma_i^2}{\gamma_i^2 + \sigma^2 / \lambda_i}. \qquad (5.8.20)$$

Assume that the prior distributions of the γ_i , $1 \le i \le s$ and the distribution of the error terms in the linear model are independently normally distributed . The estimator (5.8.4) is Bayes with respect to a normal prior distribution where

$$E(\gamma_i) = 0 \text{ and } D(\gamma_i) = \frac{f(k_i)\sigma^2}{(1 - f(k_i))\lambda_i}. \qquad (5.8.21)$$

The MSE averaging over the prior distribution is obtained form the expectation of (5.8.18) using (5.8.21). Remember that if it is assumed that the γ_i have a prior distribution the MSE in (5.8.18) is a random variable. The average MSE is given by

$$MSE = f(k_i)\frac{\sigma^2}{\lambda_i} = \frac{\sigma^2}{\lambda_i} - \frac{\sigma^2}{\lambda_i}(1 - f(k_i)). \qquad (5.8.22)$$

5.8.3 Comparing the Measures of Efficiency of the UR, the JR and the LS

Consider two estimators of the form

$$\hat{\gamma}_{ij} = [f_j(k_i)]g_i, \, j = 1,2, \, 1 \le i \le s, \qquad (5.8.23)$$

where $0 \leq f_1(k_i) \leq f_2(k_2)$. From (5.8.16) - (5.8.19) it follows that for the estimator $\hat{\gamma}_{i1}$ when compared to $\hat{\gamma}_{i2}$

1. the bias is larger;
2. the variance is smaller;
3. the MSE is smaller provided that the condition

$$|\gamma_i| \leq \sqrt{\left(\frac{f_1(k_i)+f_2(k_i)}{2-f_1(k_1)-f_2(k_1)}\right)\frac{\sigma^2}{\lambda_i}} \qquad (5.8.24)$$

holds true.

4. the signal to noise ratio is larger;
5. the Bayes risk is smaller.

To compare the UR and the JR let $f_1(k_i)$ be the function in (5.8.6) and let $f_2(k_i)$ be the function in (5.8.7). This means that $\hat{\gamma}_{i1}$ is the UR and $\hat{\gamma}_{i2}$ is the JR. Clearly $f_1(k_i) \leq f_2(k_2)$. Thus, the UR is a better estimator than the JR in the following ways:

1. The UR has a smaller variance than the JR.
2. The signal to noise ratio of the UR is larger than that of the JR.
3. The average MSE is smaller (see Equation 5.8.22) hence on the average with respect to the prior distribution improvement over the LS is larger.

On the other hand

1. The JR does have a smaller bias than the UR.
2. The MSE of the JR is smaller than that of the LS for a larger rectangle of the form in (5.8.24).

To show that that the rectangle where the JR dominates the LS is larger than that of the UR first observe that

$$h(t) = \sqrt{\frac{1+t}{1-t}} \qquad (5.8.25)$$

is an increasing function. Then compare the size of the expression on the right hand side where $f_2(k_i) = 1$ and $f_1(k_i)$ are each of the expressions in (5.8.6) and (5.8.7).

If theoretically optimum k_i is obtained after substituting (5.8.6) and (5.8.7) into (5.8.10) the minimum MSE is the same for the JR and the UR. Thus, both estimators have the same potential for improving the conditional MSE.

Although the JR has a smaller bias than the UR its variance, average MSE and for some of the parameter values its conditional MSE is larger than that of the UR.

Thus, the bias can be reduced by jack-knifing without paying an appropriate price in terms of the other measures of efficiency.

Exercise 5.8.1. Verify in general that $\hat{\gamma}_{i1}$ has a smaller MSE than $\hat{\gamma}_{i2}$ when the condition (5.8.24) holds true.

Exercise 5.8.2. A. Show that h(t) is an increasing function.
B. Explicitly show that the JR has a smaller MSE than the LS for a larger rectangle than the UR.

Exercise 5.8.3. Find optimum k_i that minimizes the MSE (5.8.10) for both the UR and the JR. Verify that for this optimum k_i the MSE is the same for both estimators.

5.9 Summary

In this chapter it was explained how ridge regression type estimators could be derived using:

1. a generalization of the method of Hoerl and Kennard;
2. as a mixed estimator;
3. as a minimax estimator;
4. as a linear Bayes estimator.

Alternative forms of the linear Bayes estimator were derived and used to study the mathematical relationships between mixed, minimax, ridge type and Bayes estimators.

The efficiencies of the different Bayes, ridge, mixed, minimax and least square estimators were compared from both a Bayesian and a frequentist viewpoint.

The reduction of bias using a jack-knifed ridge regression estimator was studied. It was observed that although bias is reduced substantially the jack-knifed estimators do worse with respect to other measures of goodness.

Chapter VI

Improving the James-Stein Estimator: The Positive Parts

6.0 Introduction

The inadmissibility of the MLE was established by showing that the JS had a uniformly smaller risk. The JS is also inadmissible because there are a number of different estimators obtained by truncating the JS that have uniformly smaller Bayes and frequentist risk. The methods of truncation consist of altering the form of the contraction factors when they are negative. The different ways of doing this produce the different positive part estimators. These positive part estimators are the subject of this chapter. Two cases will be considered:

1. estimating the mean of a multivariate normal distribution;
2. the single linear model.

The possibilities for extending the results of this chapter to the case of r linear models and the multivariate model will be indicated in Chapters VII and IX respectively.

Judge and Bock (1978) propose five positive part estimators that will be denoted by PP_0 - PP_5. The estimator PP_0, the classical positive part, was introduced in Section 2.8. It is formulated by setting the contraction factor of the JS equal to zero when it is negative. The estimator PP_1 is formulated by replacing the contraction factor of the JS with its absolute value. The estimator PP_2 is obtained from PP_1 by replacing its contraction factor by 1 when its value is less than -1. The estimator PP_3 is formulated by using half of the contraction factor of the JS when the contraction factor has values between 0 and -1 and one for other negative values. The estimator PP_4 is the same as the JS if the contraction factor is greater than -1 and the same as the MLE or the LS elsewhere. These positive parts together with a general form of a positive part estimator that includes these as special cases are given in Section 6.1.

Section 6.2 obtains and evaluates the Bayes risk of the five positive part estimators in comparison with each other and with the JS. First it will be shown that all of the positive part estimators have a smaller Bayes risk than the MLE or the LS. Then an explicit formula will be derived for the Bayes risk and the relative savings loss of the general positive part estimator considered. For the estimation of the mean of a multivariate normal distribution this formula will be in terms of incomplete Gamma functions; for the parameters of the single linear model it will be in terms of incomplete Beta functions.

Because of the complicated nature of the formulae to be derived analytical comparisons among the different positive parts will be very difficult. Some numerical studies of the Bayes risk and the RSL are done. The Software package MATHEMATICA proved to be very useful for this investigation. The programs used to generate numerical values of the different positive parts are given as part of the exercises.

The formulae for the frequentist risk and the relative loss of the different positive parts of the JS is obtained in Section 6.3. These formulae are more complicated than those of Section 6.2 because they involve weighted sums of incomplete gamma and beta functions. Some numerical studies of these risks are given. The MATHEMATICA programs are included in the exercises. Some equivalent Maple programs are included for users of that software.

In general for small values of the parameters or the prior dispersion the classical positive part estimator PP_0 has the smallest risk. For larger values PP_2 appears to be the best. It has been known for a long time that the classical positive part estimator is inadmissible but an explicit estimator with smaller risk was obtained only recently by Shao and Strawderman (1994). The results of Shao and Strawderman will be summarized in Section 6.4 .

Section 6.5 is a short summary.

6.1 The Positive Parts of the James-Stein Estimator (PP$_0$ -PP$_4$)

The five positive parts of the JS will be formulated in this section first for the problem of estimating the mean of a multivariate normal distribution, then for estimating the parameters of a linear regression model. A general positive part estimator will then be given that includes these five positive parts as special cases.

6.1.1 The Multivariate Normal Model

In Chapter II a p variate multivariate normal distribution $N(\theta, I)$ was considered with θ a p dimensional parameter with components θ_j and I was the identity matrix. Recall that the MLE was given by

$$\hat{\theta}_i = x_i , \ 1 \le i \le p. \tag{6.1.1}$$

The JS is the MLE multiplied by a contraction factor estimated from the data. Thus the JS is

$$\hat{\theta}_i = \left(1 - \frac{(p-2)}{\sum\limits_{i=1}^{p} x_i^2} \right) x_i , \ 1 \le i \le p. \tag{6.1.2a}$$

The formulation of the positive parts will be easier if the following notation is used. Let

$$w_1 = \frac{(p-2)}{\sum\limits_{i=1}^{p} x_i^2}. \tag{6.1.2b}$$

Then the JS can be written

$$\hat{\theta}_i = \left(1 - w_1 \right) x_i , \ 1 \le i \le p. \tag{6.1.2c}$$

To facilitate the definition of the positive parts let A be an interval on the real line. Define

$$I_A(w_1) = 1 \text{ if } w_1 \in A$$
$$0 \text{ otherwise.}$$
(6.1.3)

The classical positive part is defined by setting the estimator equal to zero when the contraction factor is negative. This happens when $w_1 > 1$. Thus, the classical positive part PP_0 takes the form

$$\hat{\theta}_i = (1 - w_1)I_{[0,1)}(w_1)x_i , \ 1 \le i \le p.$$
(6.1.4)

It was shown in (2.8) that the frequentist risk of (6.1.4) was smaller than that of the JS. It seems reasonable that (6.1.4) will have a smaller MSE because in the computation of the MSE both the positive and the negative terms will be smaller.

Instead of setting the contraction factor equal to zero when it is negative maybe it could be forced to remain positive. One way to accomplish this is to use its absolute value. Then the resulting estimator coincides with the JS when the contraction factor is positive and is the negative of the JS elsewhere. Thus, the contraction factor always remains positive. This estimator PP_1 takes the form

$$\hat{\theta}_i = |1 - w_1|x_i$$
$$= (1 - w_1)I_{[0,1)}(w_1)x_i - (1 - w)I_{[1,\infty)}(w_1)x_i , \ 1 \le i \le p.$$
(6.1.5)

When $1 - w_1$ is very far away form zero in the negative direction it could add a lot to the MSE. In general PP_0 has a smaller MSE than PP_1. This motivates not using a the factor $1-w_1$ at all if its absolute value is greater than one because it is no longer a contraction factor. This suggests the form of PP_2 given by

$$\hat{\theta}_i = (1 - w_1)(I_{[0,1)}(w_1) - I_{[1,2)}(w_1))x_i + I_{[2,\infty)}(w_1)x_i , \ 1 \le i \le p. \quad (6.1.6)$$

Another idea is to make the contraction factor smaller, e.g. , cut it in half for part of the region where it is negative and then use the MLE. This is what is done in formulating PP_3. Thus, its form is given by

$$\hat{\theta}_i = (1 - w_1)(I_{[0,1)}(w_1) + \tfrac{1}{2}I_{[1,2)}(w_1))x_i + I_{[2,\infty)}(w_1)x_i , \ 1 \le i \le p.$$
(6.1.7)

Finally a positive part estimator is formulated that is the JS as long as estimator is a contraction factor, that is, its absolute value is less than one and the MLE elsewhere. This leads to the form of PP_4

$$\hat{\theta}_i = (1 - w_1) I_{[0,2)}(w_1) x_i + I_{[2,\infty)}(w_1) x_i, \quad 1 \leq i \leq p. \tag{6.1.8}$$

All of the five positive parts are special cases of the general form of the positive part given by

$$\hat{\theta}_i = (1 - w_1) I_{[0,1)}(w_1) x_i + a_1 (1 - w_1) I_{[1,\alpha)}(w_1) x_i$$
$$+ a_2 I_{[\alpha,\infty)}(w_1) x_i \tag{6.1.9}$$

for $1 \leq i \leq p$. The various positive parts PP_0 - PP_4 are obtained as special cases of (6.1.9) by specifying α, a_1 and a_2 as given in Table 6.1.1.

Table 6.1.1

Estimator	a_1	a_2	α
JS	1	0	∞
PP_0	0	0	0
PP_1	−1	0	∞
PP_2	−1	1	2
PP_3	1/2	1	2
PP_4	1	1	2

The positive parts of the JS of the parameters in a single linear regression model will now be taken up.

Exercise 6.1.1. Verify that substitution of the values in Table 6.1.1 into equation 6.19 yields the positive parts PP_0 -PP_4.

6.1.2 The Single Linear Model

Recall that the form of the JS for the estimable parametric functions of the regression parameters in the general linear model was given by

$$p'\hat{\beta} = \left(1 - \frac{(n-s)(s-2)\hat{\sigma}^2}{(n-s+2)b'X'Xb}\right)p'b. \qquad (6.1.10a)$$

Here b is the LS. Let

$$w_2 = \frac{(n-s)(s-2)\hat{\sigma}^2}{(n-s+2)b'X'Xb}. \qquad (6.1.10b)$$

Then the JS may be written more simply in the form

$$p'\hat{\beta} = (1 - w_2)p'b. \qquad (6.1.10c)$$

The general form of the positive part analogous to that of (6.1.9) is given by

$$\hat{\beta} = [(1-w_2)I_{[0,1)}(w_2) + a_1(1-w_2)I_{[1,\alpha)}(w_2) + a_2 I_{[\alpha,\infty)}(w_2)]b. \qquad (6.1.11)$$

The estimators PP_0 - PP_4 may be formulated using the values of a_1, a_2 and α given in Table 6.1.1.

Exercise 6.1.2. Write down the forms of the different positive parts by substituting the values in Table 6.1.1 into 6.1.11.

6.2 The Average Mean Square Error

This section will:

1. show that the Bayes risk (average MSE) of the positive part estimators is smaller than that of the JS;
2. obtain analytical formulae for the average MSE of the different positive parts;
3. make some numerical and analytical comparisons of the Bayes risk of the different positive parts with each other and with the JS.

The problem of estimating the mean of a multivariate normal distribution will be considered first; then the single linear model will be considered. The expressions obtained for the risk of the estimators of the mean of a multivariate normal distribution will be in terms of incomplete Gamma functions. The expressions obtained for the predictive MSE for the estimators of the parameters of a general linear model will be in terms of incomplete Beta functions. Also in the means problem the variance is known to be one; in the regression problem the variance is unknown. For these reasons the two cases deserve separate consideration.

6.2.1 The Multivariate Normal Distribution

First it will be shown in Theorem 6.2.1 that the average MSE of all of the positive parts is smaller than that of the JS. The numerical studies will show that no one positive part dominates the other.

Theorem 6.2.1. The positive parts PP_0 to PP_4 all have smaller Bayes risk than the JS.

Proof. Let $h = 1 - w_1$. Also let h_g be the contraction factor for the general positive part, that is

$$h_g = (1 - w_1)I_{[0,1)}(w_1) + a_1(1 - w_1)I_{[1,\alpha)}(w_1) + a_2I_{(\alpha,\infty)}(w_1) \qquad (6.2.1)$$

The average MSE of the positive parts is

$$m = \frac{1}{p}\sum_{i=1}^{p} E(\hat{\theta}_i - \theta_i)^2 = \frac{1}{p}\sum_{i=1}^{p}[E(h_g^2 x_i^2) - 2E(h_g x_i \theta_i) + E(\theta_i^2)]. \; (6.2.2)$$

Since

$$E(\theta_i \mid x_i) = (1 - \frac{1}{A+1})x_i = \frac{A}{A+1}x_i \, ,$$

$$m = \frac{1}{p}\sum_{i=1}^{p} E[(h_g^2 - \frac{2A}{A+1}h_g)x_i^2] + E(\theta_i^2) \qquad (6.2.3)$$

$$\leq \frac{1}{p}\sum_{i=1}^{p} E[(h^2 - \frac{2A}{A+1}h)x_i^2] + E(\theta_i^2)$$

because

$$h_g^2 - \frac{2A}{A+1}h_g \leq h^2 - \frac{2A}{A+1}h. \qquad (6.2.4)$$

The formula for the Bayes risk is given by Theorem 6.2.2.

Theorem 6.2.2. Let $h(v)$ be the probability density function of a chi-square distribution with p degrees of freedom. Define

$$A_r(\alpha,\beta) = \int_{\frac{(p-2)}{\beta(A+1)}}^{\frac{(p-2)}{\alpha(A+1)}} v^r h(v) dv = E\left[v^r I_{\left[\frac{(p-2)}{\beta(A+1)},\frac{(p-2)}{\alpha(A+1)}\right]}(v)\right]. \tag{6.2.5}$$

Then the average MSE of the generalized positive part of the JS is

$$\begin{aligned} \text{mpp} = \text{mjs} &- M(\infty) + a_1^2 M(\alpha) + 2AN(\infty) - a_1 N(\alpha) \\ &+ (\frac{a_2^2(1+A) - 2a_2 A}{p}) A_1(\alpha,\infty) \end{aligned} \tag{6.2.6}$$

where

$$\begin{aligned} M(\alpha) = \frac{1}{p}((A+1)A_1(1,\alpha) &- 2(p-2)A_0(1,,\alpha) \\ &+ \frac{(p-2)^2}{A+1} A_{-1}(1,\infty)) \end{aligned} \tag{6.2.7}$$

and

$$N(\alpha) = \frac{1}{p}(A_1(1,\alpha) - \frac{(p-2)}{(A+1)} A_0(1,\alpha)). \tag{6.2.8}$$

Proof. Recall that if $v \sim \chi_p^2$ with probability density function denoted by $h(v)$ then

$$\int_0^\infty h(v) dv = 1, \tag{6.2.9}$$

$$E(v) = \int_0^\infty v h(v) dv = p \tag{6.2.10}$$

and

$$E\left[\frac{1}{v}\right] = \int_0^\infty \frac{h(v)}{v} dv = \frac{1}{p-2} . \tag{6.2.11}$$

Since

$$\sum_1^p x_i^2 \sim (1+A)\chi_p^2, \tag{6.2.12}$$

it follows that

$$E\left[\sum_{i=1}^p x_i^2 I_{(\alpha,\beta)}(w_1)\right] = (1+A)A_1(\alpha,\beta), \tag{6.2.13}$$

$$E[I_{(\alpha,\beta)}(w_1)] = A_0(\alpha,\beta) \tag{6.2.14}$$

and

$$E[w_1 I_{(\alpha,\beta)}(w_1)] = \frac{(p-2)A_1(\alpha,\beta)}{1+A}. \tag{6.2.15}$$

To obtain the MSE of the Positive Parts the first term of the left-hand expression in (6.2.3) must be computed. This may be done in two parts. Thus, one must compute

$$
\begin{aligned}
E_1 &= \sum_{i=1}^p E(h_g^2 x_i^2) = E\left[\sum_{i=1}^p x_i^2(1-w_1)^2 I_{[0,1]}(w_1)\right] \\
&\quad + a_1^2 E\left[\sum_{i=1}^p x_i^2(1-w_1)^2 I_{(1,\alpha]}(w_1)\right] + a_2^2 E\left[\sum_{i=1}^p x_i^2 I_{(l\alpha,\infty]}(w_1)\right]
\end{aligned}
\tag{6.2.16}
$$

and

$$
\begin{aligned}
E_2 &= \sum_{i=1}^p E(h_g x_i^2) = E\left[\sum_{i=1}^p x_i^2(1-w_1)I_{[0,1]}(w_1)\right] \\
&\quad + a_1 E\left[\sum_{i=1}^p x_i^2(1-w_1)^2 I_{(1,\alpha]}(w_1)\right] + a_2 E\left[\sum_{i=1}^p x_i^2 I_{(l\alpha,\infty]}(w_1)\right].
\end{aligned}
\tag{6.2.17}
$$

Using (6.2.1-6.2.3) it follows that

$$E_1 = (1+A)A_1(0,1) - 2A_0(0,1)(p-2)$$

$$+(p-2)^2 \frac{A_{-1}(0,1)}{(1+A)}$$

$$+a_1^2 \left[(1+A)A_1(1,\alpha) - 2A_0(1,\alpha)(p-2) + (p-2)^2 \frac{A_{-1}(1,\alpha)}{(1+A)} \right]$$

$$+a_2^2[(1+A)A_1(\alpha,\infty)$$

$$= (1+A)(p - A_1(1,\infty)) - 2(p-2)(1 - A_0(1,\infty)) \qquad (6.2.18)$$

$$+\frac{(p-2)^2}{(1+A)} \left[\frac{1}{p-2} - A_{-1}(\alpha,\infty) \right]$$

$$+a_1^2 \left[(1+A)A_1(1,\alpha) - 2A_0(1,\alpha)(p-2) + (p-2)^2 \frac{A_{-1}(1,\alpha)}{(1+A)} \right]$$

$$+a_2^2[(1+A)A_1(\alpha,\infty)]$$

and

$$E_2 = (1+A)A_1(0,1) - (p-2)A_0(0,1)$$

$$+a_1[(1+A)A_1(1,\alpha) - (p-2)A_0(1,\alpha)]$$

$$+a_2A_1(\alpha,\infty)$$

$$= (1+A)(p - A_1(1,\infty)) - (p-2)(1 - A_1(1,\infty)) \qquad (6.2.19)$$

$$+a_1[(1+A)A_1(1,\alpha) - (p-2)A_0(1,\alpha)]$$

$$+a_2A_1(\alpha,\infty).$$

The result follows after substitution of E_1 and E_2 into (6.2.3).

Exercise 6.2.1. A. For each of the five positive parts show the explicit forms of the risk in Theorem 6.2. above are as follows

1. The JS[1]

$$m = 1 - \frac{(p-2)}{p(A+1)}; \qquad (6.2.20)$$

2. The classical positive part PP_0

---

[1] This formula was derived in Chapter II. It is mentioned here for completeness.

$$m_0 = m + \frac{1}{p}(A-1)A_1(1,\infty) + \frac{2(p-2)}{p(A+1)}A_0(1,\infty)$$
$$-\frac{(p-2)^2}{p(A+1)}A_{-1}(1,\infty); \tag{6.2.21}$$

3. The positive part PP$_1$

$$m_1 = m - \frac{4A(p-2)}{p(A+1)}A_0(1,\infty) + \frac{4A}{p}A_1(1,\infty); \tag{6.2.22}$$

4. The positive part PP$_2$

$$m_2 = m - \frac{(p-2)^2}{p(A+1)}A_{-1}(2,\infty) + \frac{2(p-2)}{p(A+1)}A_0(2,\infty)$$
$$-\frac{4A(p-2)}{p(1+A)}A_0(1,2) + \frac{4A}{p}A_1(1,2); \tag{6.2.23}$$

5. The positive part PP$_3$

$$m_3 = m + \left(\frac{1-3A}{4p}\right)A_1(1,2) + \frac{(A-1)(p-2)}{2(A+1)p}A_0(1,2)$$
$$+\frac{(p-2)^2}{4p(A+1)}A_{-1}(1,2) + \frac{(1-A)}{p}A_1(1,\infty); \tag{6.2.24}$$

6. The positive part PP$_4$

$$m_4 = m - \frac{2(p-2)^2}{p(A+1)}A_{-1}(2,\infty) + \frac{2(p-2)}{p}A_0(2,\infty)$$
$$-\frac{2(p-2)A}{p(A+1)}A_0(2,\infty). \tag{6.2.25}$$

B. Verify from the explicit formulae that these MSEs are all less than that of the JS.

Let BR denote the Bayes risk of an estimator. The RSL of the positive parts is given by

$$RSL = \frac{BR(PP) - BR(BE)}{BR(MLE) - BR(BE)}$$

$$= \frac{BR(PP) - A/(A+1)}{1 - (1 - 1/(A+1))} \qquad (6.2.26)$$

$$= (A+1)BR(PP) - A.$$

Exercise 6.2.2. Obtain explicit formulae for the RSL of the different positive parts using formulae (6.2.20)-(6.2.26).

The RSL of the JS is a constant function of the prior parameters. However the RSL of the different positive parts is not and as will be shown in the sequel has some interesting properties. Two other ratios worth considering are

$$reff(PP) = \frac{BR(PP)}{BR(JS)} \qquad (6.2.27)$$

and

$$RSLM = \frac{RSL(PP)}{RSL(JS)} = \frac{P}{2}[(A+1)BR(PP) - A]. \qquad (6.2.28)$$

These ratios help in comparing the relative improvement of the different positive parts. They also give an indication of the worthiness of the positive parts for different choices of the prior parameters. Indeed if these ratios are close to one then for practical purposes the JS will do just as well as its positive parts.

The complex nature of the formulae for the MSE make analytical comparisons of the efficiency difficult. That is why studies of the numerical values of the estimators is being resorted to. However the following analytical properties of the positive parts are not difficult to establish and are left as exercises for the reader.

1. If one positive part or JS estimator has a smaller RSL than another the same is true of its average MSE when averaged over the same prior distribution.

2. When $A = 0$ the following ordering of the Bayes risk of the positive parts holds true

$$m_0 \leq m_3 \leq m_2 = m_4 \leq m_1 = mjs. \qquad (6.2.29)$$

3. When the prior dispersion $A \to \infty$ the Bayes risk and the RSL of the positive parts all approach that of the JS.

Exercise 6.2.3. Verify the properties of the positive parts of the JS given in 1 to 3 above.

Some numerical values of the MSE of the JS and its positive parts are presented in Table 6.2.1.

It appears that the BR of all of the positive parts increase as the prior dispersion becomes less precise. All of the positive parts dominate the JS. It appears that that PP_2 dominates PP_4. Otherwise no one positive part dominates the other. It can be shown very easily for any prior variance that the ordering of the risk and the RSL of the JS and the positive parts is the same. The RSLs are more spread out than the MSEs making it easier to distinguish between estimators on a graph. The ratio of the RSL of the positive parts to that of the JS(RSLM) gives an indication of the improvement of the Positive parts as compared with the JS. The RSL of the JS is constant. When the RSLM is greater than 0.95 the positive part does not improve significantly on the JS. Since the RSL of the JS is a constant the shape of the RSLM of the positive parts is no different from that of the RSL because it is just the RSL multiplied by the scale factor 1/RSL. For these reasons the author decided to plot the RLSM in Figure 6.2.1. Table 6.2.2 gives the RSL of the JS and its positive parts for the same values as those in Table 6.2.1.

Table 6.2.1

A	0	0.2	0.5	0.8	1	3	5
JS	0.500	0.583	0.667	0.722	0.750	0.875	0.917
PP_0	0.368	0.475	0.586	0.660	0.697	0.858	0.909
PP_1	0.500	0.557	0.631	0.687	0.717	0.861	0.910
PP_2	0.393	0.485	0.586	0.657	0.693	0.856	0.907
PP_3	0.385	0.487	0.593	0.665	0.701	0.860	0.909
PP_4	0.393	0.496	0.602	0.673	0.708	0.862	0.911

$$p = 4$$

A	0	0.2	0.5	0.8	1	3	5
JS	0.250	0.375	0.500	0.583	0.625	0.813	0.875
PP_0	0.162	0.306	0.455	0.554	0.603	0.810	0.874
PP_1	0.250	0.341	0.462	0.554	0.601	0.809	0.874
PP_2	0.205	0.318	0.453	0.540	0.598	0.809	0.874
PP_3	0.187	0.323	0.465	0.561	0.608	0.810	0.875
PP_4	0.205	0.341	0.479	0.570	0.615	0.811	0.875

$$p = 8$$

A	0	0.2	0.5	0.8	1	3	5
JS	0.125	0.271	0.417	0.514	0.562	0.781	0.854
PP_0	0.075	0.235	0.399	0.505	0.557	0.781	0.854
PP_1	0.125	0.244	0.396	0.503	0.555	0.781	0.854
PP_2	0.115	0.240	0.395	0.502	0.555	0.781	0.854
PP_3	0.092	0.247	0.405	0.509	0.559	0.781	0.854
PP_4	0.115	0.264	0.414	0.513	0.562	0.781	0.854

$$p = 16$$

A	0	0.2	0.5	0.8	1	3	5
JS	0.063	0.219	0.375	0.479	0.531	0.766	0.844
PP_0	0.036	0.202	0.370	0.478	0.531	0.766	0.844
PP_1	0.063	0.202	0.368	0.477	0.530	0.766	0.844
PP_2	0.062	0.202	0.368	0.477	0.530	0.766	0.844
PP_3	0.044	0.209	0.372	0.478	0.531	0.766	0.844
PP_4	0.063	0.218	0.375	0.479	0.531	0.766	0.844

$$p = 32$$

From the graphs in Figure 6.2.1 it is apparent that:

1. For small values of A the classical positive part estimator PP_0 gives the most significant improvement relative to the JS; for larger values of A the alternative positive part estimator PP_2 gives the most significant improvement with PP_1 running a close second when A is larger.

2. When A = 0 the positive parts are more efficient as p increases. However the range of values where the positive parts are a significant improvement on the JS becomes smaller.

3. The RSLM and hence the RSL of the estimators PP_0 and PP_4 are increasing while that of PP_1, PP_2 and PP_3 decrease, pass through a minimum value and then increase.

4. The RSLM , RSL and hence the BR of all the Positive Parts approach that of the JS as A→∞.

Table 6.2.2 The RSL

A	0	0.2	0.5	0.8	1	3	5
JS	0.500	0.500	0.500	0.500	0.500	0.500	0.500
PP_0	0.368	0.370	0.378	0.388	0.393	0.433	0.452
PP_1	0.500	0.469	0.446	0.437	0.435	0.445	0.457
PP_2	0.393	0.383	0.379	0.383	0.386	0.422	0.443
PP_3	0.385	0.385	0.390	0.398	0.403	0.439	0.456
PP_4	0.393	0.396	0.403	0.411	0.416	0.448	0.463

p = 4

A	0	0.2	0.5	0.8	1	3	5
JS	0.250	0.250	0.250	0.250	0.250	0.250	0.250
PP_0	0.162	0.167	0.182	0.197	0.205	0.239	0.246
PP_1	0.250	0.209	0.194	0.197	0.202	0.235	0.244
PP_2	0.205	0.182	0.179	0.188	0.196	0.235	0.244
PP_3	0.187	0.187	0.198	0.210	0.216	0.242	0.247
PP_4	0.205	0.209	0.218	0.226	0.231	0.246	0.249

p = 8

A	0	0.2	0.5	0.8	1	3	5
JS	0.125	0.125	0.125	0.125	0.125	0.125	0.125
PP_0	0.075	0.082	0.098	0.110	0.115	0.125	0.125
PP_1	0.125	0.092	0.098	0.105	0.110	0.124	0.125
PP_2	0.115	0.088	0.093	0.104	0.110	0.124	0.125
PP_3	0.092	0.097	0.108	0.116	0.119	0.125	0.125
PP_4	0.115	0.117	0.121	0.123	0.124	0.125	0.125

p = 16

A	0	0.2	0.5	0.8	1	3	5
JS	0.0625	0.0625	0.0625	0.0625	0.0625	0.0625	0.0625
PP_0	0.0355	0.0429	0.0554	0.0603	0.0615	0.0625	0.0625
PP_1	0.0625	0.0425	0.0522	0.0590	0.0609	0.0625	0.0625
PP_2	0.0615	0.0423	0.0520	0.0590	0.0609	0.0625	0.0625
PP_3	0.0435	0.0504	0.0584	0.0613	0.0620	0.0625	0.0625
PP_4	0.0615	0.0620	0.0624	0.0625	0.0625	0.0625	0.0625

$$p = 32$$

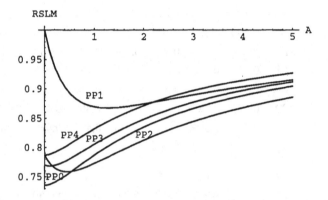

Figure 6.2.1 a   p = 4

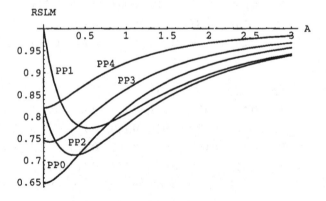

b   p = 8

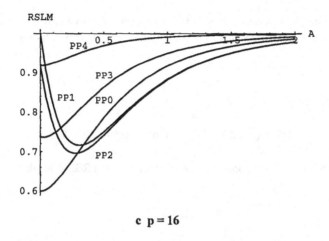

c p = 16

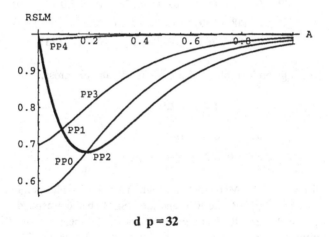

d p = 32

Important values of the RSL and RSLM are:

1. that of PP_0 at $A = 0$;
2. the minimum values for PP_1, PP_2 and PP_3 when they exist;
3. the crossing values of PP_0 and PP_2.

The value of the RSLM at 0 (see Table 6.2.3) is important for PP_0 because it represents the maximum possible improvement for these five estimators. Likewise the minima of $PP_1 - PP_3$ are the best improvement possible for these estimators (see Table 6.2.4). The crossing values of PP_0 and PP_2 represent where the two positive parts change places as optimal estimators (see Table 6.2.5). When A is very large

PP_1 has a smaller risk and RSL than PP_2. However the risks and RSL of the positive parts is so close to the JS that it does not matter for practical purposes.

Table 6.2.3

p	4	8	16	32
$RSLM(PP_0)(0)$	0.736	0.647	0.599	0.568

In Table 6.2.4 the first coordinate is A the second is the RSLM for that value of A.

Table 6.2.4

p	4	8	16	32
PP_1	(1.267,0.868)	(0.545,0.774)	(0.307,0.717)	(0.192,0.679)
PP_2	(0.448,0.759)	(0.370,0.712)	(0.276,0.696)	(0.189,0.677)
PP_3	(0.108,0.768)	(0.076,0.742)	(0.022,0.737)	none

The values of A given in Table 6.2.5 are where PP_0 and PP_2 cross.

Table 6.2.5

p	4	8	16	32
A	0.541	0.422	0.294	0.188

Exercise 6.2.4. Use the Mathematica program or the Maple program below to compute values of the BR, the RSL and the RSLM that correspond to the given tables for p = 6, 12, 24 ,40. Does the behavior of the measures of efficiency for these quantities appear to be similar to those for the cases illustrated?

Mathematica Program
```
mjs[p_,A_]=1-(p-2)/(p (A+1));
f[p_,al_,A_]=(p-2)/(2 al (A+1));
F[p_,g_,al_,A_]=Gamma[g,0,f[p,al,A]]/Gamma[p/2];
G1[p_,al_,A_]=
1/p (2 F[p,p/2+1,al,A] (A+1) -2 (p-2) F[p,p/2,al,A]
+(p-2)^2/(2 (A+1)) F[p,p/2-1,al,A]);
G2[p_,al_,A_]=G1[p,1,A]-G1[p,al,A];
H1[p_,al_,A_]=
1/p (2 F[p,p/2+1,al,A]-(p-2)/(A+1) F[p,p/2,al,A]);
```

```
H2[p_,al_,A_]=H1[p,1,A]-H1[p,al,A];
mpp[a1_,a2_,A_,p_,al_]=mjs[p,A]
+(a1^2 G2[p,al,A]-G1[p,1,A]
+(a2^2 (1+A)-2 a2 A)/p 2 F[p,p/2+1,al,A]
+2 A (H1[p,1,A]-a1 H2[p,al,A]));
rsl[a1_,a2_,A_,p_,al_]=(A+1) mpp[a1,a2,A,p,al]-A;
rslmjs[p_,A_]=(A+1) mjs[p,A]-A;
rsn[a1_,a2_,A_,p_,al_]=rsl[a1,a2,A,p,al]/rslmjs[p,A];
```

Maple Program
(a and b correspond to a1 and a2 in the Mathematica program)
```
> mjs:=(p,A)->1-(p-2)/(p*(A+1)):
> f:=(p,al,A)->(p-2)/(2*al*(A+1)):
>F:=(p,g,al,A)->(GAMMA(g)-GAMMA(g,f(p,al,A)))/GAMMA(p/2):
>G1:=(p,al,A)->(1/p)*(2*F(p,p/2+1,al,A)*(A+1)
-2*(p-2)*F(p,p/2,al,A)
+(((p-2)^2)/(2*(A+1)))*F(p,p/2-1,al,A)):
>G2:=(p,al,A)->G1(p,1,A)-G1(p,al,A):
>H1:=(p,al,A)->(1/p)*(2*F(p,p/2+1,al,A)-((p-
2)/(A+1))*F(p,p/2,al,A)):
>H2:=(p,al,A)->H1(p,1,A)-H1(p,al,A):
>mpp:=(a,b,A,p,al)->mjs(p,A)+(((a^2)*G2(p,al,A)-
G1(p,1,A)+(((b^2)*(1+A)-
2*b*A)/p)*(2*F(p,p/2+1,al,A))+2*A*(H1(p,1,A)-
a*H2(p,al,A)))):
> rsl:=(a,b,A,p,al)->(A+1)*mpp(a,b,A,p,al)-A:
> rslmjs:=(p,A)->(A+1)*mjs(p,A)-A:
>rsn:=(a,b,A,p,al)->rsl(a,b,A,p,al)/rslmjs(p,A);
```

Exercise 6.2.5. For the values of p illustrated in the text and suggested in Exercise 6.2.4:
A. Where does the RSLM of PP_1 cross that of PP_4, PP_1 and PP_2?
B. Where does the RSLM of PP_2 cross that of PP_3?

Hint: the following type of Mathematica command would be helpful.

```
FindRoot[rsn[-1,1,A,8,Infinity]-
rsn[1,1,A,8,2]==0,{A,.15}].
```

The corresponding Maple command is
```
> fsolve(rsn(-1,1,A,8,100000)-rsn(1,1,A,8,2)=0,A,.1...2);
```

6.2.2 The Single Linear Model

The mean square error of the positive parts of the JS (6.1.11) averaging over Dempster's prior and averaging over a predictive loss function. (the MSEP) will be studied in this section. The form of this MSEP can be shown to be

$$
\begin{aligned}
m &= E(\hat{\beta} - \beta)'X'X(\hat{\beta} - \beta) \\
&= E[((1 - w_2)^2 (I_{[0,1)}(w_2) + a_1^2 I_{[1,\alpha)}(w_2)) + a_2^2 I_{[\alpha,\infty)}(w_2))b'X'Xb] \\
&\quad - \frac{2\omega^2}{\sigma^2 + \omega^2} E[((1 - w_2)(I_{[0,1)}(w_2) - 2a_1 I_{[1,\alpha)}(w_2))b'X'Xb] \\
&\quad + \frac{2\omega^2}{\sigma^2 + \omega^2} a_2 E[I_{[\alpha,\infty)}(w_2))b'X'Xb] + \omega^2 E(\beta'X'X\beta).
\end{aligned}
$$

$$(6.2.30)$$

The random variable w_2 is expressible in terms of central chi square random variables. This is accomplished by writing

$$w_2 = \theta w_3 / w_4$$

where

$$w_3 = (n - s)\hat{\sigma}^2 / \sigma^2,$$

$$w_4 = b'X'Xb/(\sigma^2 + \omega^2)$$

and

$$\theta = (s - 2)\sigma^2 / (((n - s) + 2)(\sigma^2 + \omega^2)). \qquad (6.2.31)$$

For $0 \leq \phi \leq 1$ let the incomplete beta function be represented by

$$H_\phi(\gamma_1, \gamma_2) = \int_0^\phi x^{\gamma_1 - 1}(1 - x)^{\gamma_2 - 1} dx. \qquad (6.2.32)$$

The calculation of the MSE in (6.2.30) depends on the following Lemma.

Lemma 6.2.1. Let $X \sim \chi^2(\alpha), Y \sim \chi^2(\beta)$ and $A = \{(x,y): d < \frac{\theta x}{y} < c\}$. Let

$$I_A(x,y) = 1 \text{ if } (x,y) \in A$$
$$0 \text{ otherwise.}$$

Then

$$E\left[\frac{x^a}{y^b}I_A(x,y)\right]$$

$$= \frac{2^{a-b}\Gamma\left(\frac{\alpha+\beta}{2}+a-b\right)}{\Gamma\left(\frac{\alpha}{2}\right)\Gamma\left(\frac{\beta}{2}\right)}\left[H_{\frac{d}{\theta+d}}\left(\frac{\alpha}{2}+a,\frac{\beta}{2}-b\right)-H_{\frac{c}{\theta+c}}\left(\frac{\alpha}{2}+a,\frac{\beta}{2}-b\right)\right].$$

(6.2.33)

Proof. Observe that

$$E\left[\frac{x^a}{y^b}I_A(x,y)\right]=\int_0^\infty\int_{\frac{cx}{c}}^{\frac{dx}{d}}\frac{x^{\alpha/2+a-2}y^{\beta/2-b-1}e^{-(x+y)/2}}{\Gamma\left(\frac{\alpha}{2}\right)\Gamma\left(\frac{\beta}{2}\right)}\,dy\,dx.\qquad(6.2.34)$$

After some algebra the transformation $x = 2uv$, $y = 2u(1-v)$ with Jacobian

$$J = \begin{vmatrix} 2u & 2(1-v) \\ 2v & -2u \end{vmatrix} = -4u \qquad (6.2.35)$$

$$E\left[\frac{x^a}{y^b}I_A(x,y)\right]$$

$$= K\int_0^\infty u^{\alpha/2+\beta/2+a-b-1}e^{-u}du\int_{\frac{c}{\theta+c}}^{\frac{d}{\theta+d}}v^{\alpha/2+a-1}(1-v)^{\beta/2-b-1}dv$$

(6.2.36)

where $K = 2^{a-b}/\Gamma(\alpha/2)\Gamma(\beta/2)$. The result (6.2.33) follows from (6.2.36).

Some notation useful for simplifying the writing the expectations in (6.2.30) will now be defined. First let

$$F_\phi(\gamma_1,\gamma_2) = \frac{H(\gamma_1,\gamma_2)}{Be(\gamma_1,\gamma_2)} \qquad (6.2.37a)$$

where

$$Be(\gamma_1,\gamma_2) = \int_0^1 x^{\gamma_1-1}(1-x)^{\gamma_2-1}dx, \qquad (6.2.37b)$$

i.e., the Beta function. Also let $g = (n - s)/2$ and $h = s/2$. Define

$$G_\alpha(i,j) = F_{\frac{\alpha}{\alpha+\theta}}[i+g, j+h].$$

(6.2.38)

Lemma 6.2.2 gives the expectations in (6.2.30).

Lemma 6.2.2. The expectations in (6.2.30) are

$$E[(1 - w_2)^2 I_{\{0,1)}(w_2)b'X'Xb] = (\sigma^2 + \omega^2)[sG_1[0,1] - 2(n - s)\theta G_1[1,0]$$

$$+ \frac{\sigma^2}{\sigma^2 + \omega^2}(n - s)\theta G_1[2,-1]],$$

(6.2.39)

$$E[(1 - w_2)^2 I_{(1,\alpha)}(w_2)b'X'Xb]$$

$$= (\sigma^2 + \omega^{2)})[s(G_\alpha[0,1] - G_1[0,1]) - 2(n - s)\theta(G_\alpha[1,0] - G_1[1,0])$$

$$+ \frac{\sigma^2}{\sigma^2 + \omega^2}(n - s)\theta(G_\alpha[2,-1] - G_1[2,-1])],$$

(6.2.40)

$$E[I_{(\alpha,\infty)}(w_2)b'X'Xb] = s(\sigma^2 + \omega^2)(1 - G_\alpha[0,1]),$$

(6.2.41)

$$E[(1 - w_2)I_{\{0,1)}(w_2)b'X'Xb]$$

$$= (\sigma^2 + \omega^2)[sG_1[0,1] - (n - s)\theta G_1[1,0]]$$

(6.2.42)

and

$$E[(1 - w_2)I_{(1,\alpha)}(w_2)] = (\sigma^2 + \omega^2)[s(G_\alpha[0,1] - G_1[0,1])$$

$$-(n - s)\theta(G_\alpha[1,0] - G_1[1,0])].$$

(6.2.43)

Proof. Apply Lemma 6.2.1.

The formula for the MSE of the general positive part (6.1.11) is given in Theorem 6.2.3. It is obtained by substituting the expectations in Lemma 6.2.2 into (6.2.30).

Theorem 6.2.3. Let $f = (n - s)$. The MSE of (6.1.11) is given by

$$m = \sigma^2 s - \sigma^2 f\theta + (\sigma^2 - \omega^2)[s(G_1[0,1]-1)]$$
$$-2\sigma^2 f\theta[G_1[1,0]-1] + \sigma^2 f\theta[G_1[2,-1]-1]$$
$$+a_1^2[(\sigma^2 + \omega^2)s(G_\alpha[0,1]-G_1[0,1])$$
$$-2f\theta(\sigma^2 + \omega^2)(G_\alpha[1,0]-G_1[1,0])$$
$$+\sigma^2 f\theta(G_\alpha[2,-1]-G_1[2,-1])]$$
$$-2a_1[\omega^2 s(G_\alpha[0,1]-G_1[0,1])$$
$$-f\theta\omega^2(G_\alpha[1,0]-G_1[1,0])]$$
$$+(a_2^2(\sigma^2 + \omega^2) - 2a_2\omega^2)s[1-G_\alpha[0,1]].$$

(6.2.44)

The first two terms in (6.2.44) are the MSE of the JS denoted by m_{js}

Exercise 6.2.7. Show that the MSEs of PP_0-PP_4 are given by the formulae in (6.2.45)-(6.249) below.

A. For PP_0 the MSE is

$$m_0 = m_{js} + (\sigma^2 - \omega^2)[s(G_1[0,1]-1)]$$
$$-2\sigma^2 f\theta[G_1[1,0]-1] + \sigma^2 f\theta[G[2,-1]-1].$$

(6.2.45)

B. The MSE of PP_1 is given by

$$m_1 = m_{js} + 4\omega^2 s(1 - G_1[0,1]) - 4\omega^2 f\theta(1 - G_1[1,0]).$$

(6.2.46)

C. The MSE of PP_2 is given by

$$m_2 = m_{js} + 4\omega^2 s(G_2[0,1]-G_1[0,1])$$
$$-2\sigma^2 f\theta(G_2[1,0]-1) - 4f\theta\omega^2(G_2[1,0]-G_1[1,0])$$
$$+\sigma^2 f\theta(G_2[2,-1]-1).$$

(6.2.47)

D. The MSE of PP_3 is given by

$$m_3 = m_{js} + \sigma^2 f\theta + .25(\omega^2 - 3\sigma^2)s(G_2[0,1]-G_1[0,1])$$
$$-2f\theta(G_1[1,0]-1) + .5f\theta(\omega^2 - \sigma^2)(G_2[1,0]-G_1[1,0])$$
$$+.25f\theta\sigma^2(G_2[2,-1]+3G_1[2,-1]-4).$$

(6.2.48)

E. The MSE of PP_4 is given by

$$m_4 = m_{js} + \sigma^2 f\theta(1 - 2G_2[1,0] + G_2[2,-1]).$$ (6.2.49)

Exercise 6.2.7. Show that as the number of observations tends to infinity the MSE/sσ^2 the positive parts in (6.2.45)-(6.2.49) tends to the risks in (6.2.21)-(6.2.25). This is to be expected because the estimation of the parameters in a regression problem may be reduced to that of estimating the mean of a multivariate normal distribution with known variance.

The RSL is the ratio of the difference between the BR averaged over the predictive loss function of the positive part and the that of the BE to the difference between the BR of the LS and the BE.

Exercise 6.2.8. Let $k = \omega^2 / \sigma^2$.

A. Show that the RSL of the positive parts may be written

$$RSL(PP) = (1+k)BR(PP) - 1.$$ (6.2.51)

B. Show that the RSLM may be written

$$RSLM = \frac{RSL(PP)}{RSL(JS)} = \frac{s}{2}\left(1 - \frac{s-2}{n}\right)[(k+1)BR(PP) - k].$$ (6.2.52)

The behavior of the MSE, RSL and RSLM of the Positive Parts of the JS for the linear model with unknown error variance is similar to that for the mean of a multivariate normal distribution. However as would be expected since there is an additional unknown quantity the error variance the Risk and RSL are slightly larger. As the number of observations increase the Risk and the RSL of the positive parts for the linear model with unknown variance tends toward that for estimators of the mean of a multivariate normal distribution with known variance. The risk of the JS and its positive parts are studied in Table 6.2.6. The RSL is studied in Table 6.2.7 and the RSLM is plotted in Figure 6.2.2. For purposes of comparison with the known variance case the values n =20, 100, s = 4 and n = 80, 200, s = 16 are considered.

Notice that the behavior of the RSLM of the five positive parts for the linear model is very similar to that of the corresponding measure of efficiency for the case of a multivariate normal distribution with known variance. However, as was noted for the RSL, the RSLM is slightly larger; this may be attributed to estimating the variance. There does not appear to be much difference in the shape of the curves for different values of n; there is considerable difference for different values of s. Table 6.2.8 gives the RSLM of PP$_0$ when k = 0 because this represents the best possible improvement of the five positive parts over the JS. As was the case for the mean of the multivariate normal distribution the RSLM of PP_1-PP_3 decreases then increases. Table 6.2.9 gives the minimum values of the RSLM of PP_1 - PP_3. For small values of k PP_0 is optimal; for larger values PP_2 is the optimal estimator. The crossing values of the two positive parts is given in Table 6.2.10.

Table 6.2.6 MSE/sσ^2

k	0	0.2	0.5	0.8	1	3	5
JS	0.556	0.630	0.704	0.753	0.778	0.889	0.926
PP_0	0.430	0.527	0.627	0.693	0.727	0.873	0.918
PP_1	0.556	0.606	0.670	0.720	0.747	0.876	0.919
PP_2	0.454	0.537	0.627	0.691	0.724	0.870	0.917
PP_3	0.446	0.538	0.634	0.699	0.731	0.874	0.919
PP_4	0.454	0.546	0.641	0.705	0.737	0.876	0.920

$$n = 20 \quad s = 4$$

k	0	0.2	0.5	0.8	1	3	5
JS	0.510	0.592	0.673	0.728	0.755	0.878	0.918
PP_0	0.379	0.485	0.593	0.666	0.702	0.861	0.910
PP_1	0.510	0.566	0.638	0.693	0.723	0.864	0.911
PP_2	0.404	0.495	0.594	0.663	0.699	0.858	0.909
PP_3	0.396	0.496	0.601	0.671	0.707	0.862	0.911
PP_4	0.404	0.505	0.609	0.679	0.713	0.865	0.912

$$n = 100 \quad s = 4$$

k	0	0.2	0.5	0.8	1	3	5
JS	0.152	0.293	0.434	0.529	0.576	0.788	0.859
PP_0	0.095	0.252	0.413	0.518	0.569	0.788	0.859
PP_1	0.152	0.263	0.410	0.514	0.566	0.788	0.859
PP_2	0.138	0.259	0.409	0.514	0.566	0.788	0.859
PP_3	0.115	0.265	0.420	0.522	0.572	0.788	0.859
PP_4	0.138	0.284	0.431	0.527	0.575	0.788	0.859

$$n = 80 \quad s = 16$$

k	0	0.2	0.5	0.8	1	3	5
JS	0.134	0.279	0.423	0.519	0.567	0.784	0.856
PP_0	0.082	0.241	0.404	0.510	0.561	0.783	0.856
PP_1	0.134	0.250	0.401	0.507	0.559	0.783	0.856
PP_2	0.123	0.247	0.400	0.507	0.559	0.783	0.856
PP_3	0.100	0.254	0.411	0.513	0.564	0.784	0.856
PP_4	0.123	0.271	0.420	0.518	0.567	0.784	0.856

n = 200 s =16

Table 6.2.7 RSL

k	0	0.2	0.5	0.8	1	3	5
JS	0.556	0.556	0.556	0.556	0.556	0.556	0.556
PP_0	0.430	0.433	0.440	0.448	0.454	0.491	0.509
PP_1	0.556	0.527	0.506	0.497	0.494	0.502	0.514
PP_2	0.454	0.444	0.441	0.444	0.447	0.480	0.500
PP_3	0.446	0.446	0.451	0.457	0.462	0.496	0.512
PP_4	0.454	0.456	0.462	0.470	0.474	0.505	0.520

n = 20 s = 4

k	0	0.2	0.5	0.8	1	3	5
JS	0.510	0.510	0.510	0.510	0.510	0.510	0.510
PP_0	0.379	0.382	0.389	0.399	0.404	0.444	0.462
PP_1	0.510	0.480	0.457	0.448	0.446	0.455	0.468
PP_2	0.404	0.394	0.391	0.394	0.397	0.433	0.453
PP_3	0.396	0.396	0.401	0.409	0.414	0.449	0.466
PP_4	0.404	0.407	0.414	0.422	0.426	0.459	0.474

n = 100 s = 4

k	0	0.2	0.5	0.8	1	3	5
JS	0.152	0.152	0.152	0.152	0.152	0.152	0.152
PP_0	0.095	0.102	0.119	0.132	0.138	0.151	0.151
PP_1	0.152	0.116	0.115	0.126	0.133	0.150	0.151
PP_2	0.138	0.110	0.113	0.126	0.132	0.150	0.151
PP_3	0.115	0.118	0.130	0.139	0.143	0.151	0.151
PP_4	0.138	0.141	0.146	0.149	0.150	0.151	0.152

$$n = 80 \quad s = 16$$

k	0	0.2	0.5	0.8	1	3	5
JS	0.134	0.134	0.134	0.134	0.134	0.134	0.134
PP_0	0.082	0.089	0.106	0.118	0.123	0.134	0.134
PP_1	0.134	0.100	0.101	0.112	0.118	0.134	0.134
PP_2	0.123	0.096	0.100	0.112	0.118	0.134	0.134
PP_3	0.100	0.104	0.116	0.124	0.127	0.134	0.134
PP_4	0.123	0.125	0.130	0.132	0.133	0.134	0.134

$$n = 100 \quad s = 16$$

Figure 6.2.2 The RSLM

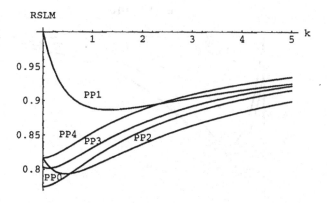

$$s = 4, n = 20$$

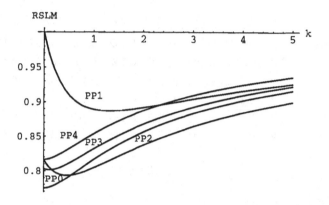

n = 100   s = 4

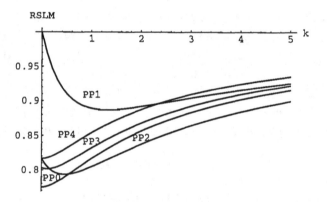

n = 80 s =16

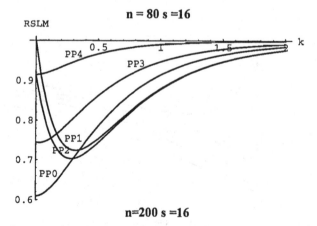

n=200 s =16

Table 6.2.8

n,s	20,4	100,4	80,16	200,16
RSLM(PP_0)(0)	0.775	0.743	0.628	0.610

The minimum points for the graph of RSLM for PP_1- PP_3 are given in Table 6.2.9. The first coordinate is the value of k; the second coordinate is the corresponding minimum value.

Table 6.2.9

n,s	20,4	100,4	80,16	200,16
PP_1	(1.321,0.887)	(1.277,0.871)	(0.335,0.738)	(0.317,0.725)
PP_2	(0.455,0.794)	(0.449,0.765)	(0.295,0.716)	(0.283,0.704)
PP_3	(0.111,0.802)	(0.109,0.802)	(0.035,0.726)	(0.027,0.744)

Table 6.2.10

n,s	20,4	100,4	80,16	200,16
k	0.548	0.479	0.313	0.301

Exercise 6.2.9. Use the Mathematica or the Maple program below to compute the average MSE and the RSL of the five positive parts of the JS for (n, s) = (40,8), (100,8), (160,32) and (400,32) for the same values of k given in Table 6.2.6.

Mathematica Program

```
th[s_,k_,n_]=(s-2)/((n-s+2) (1+k));
f[al_,s_,k_,n_]=1/(1+th[s,k,n]/al);
G[a_,b_,al_,s_,k_,n_]=Beta[f[al,s,k,n],a,b]/Beta[a,b];
G1[al_,s_,k_,n_]=G[(n-s)/2,s/2+1,al,s,k,n];
G2[al_,s_,k_,n_]=G[(n-s)/2+1,s/2,al,s,k,n];
:G3[al_,s_,k_,n_]=G[(n-s)/2+2,s/2-1,al,s,k,n];
A[s_,k_,n_]=((1-k) (G1[1,s,k,n]-1)
-2(n-s)/s th[s,k,n] (G2[1,s,k,n]-1)
+(n-s)/s th[s,k,n] (G3[1,s,k,n]-1));
H1[al_,s_,k_,n_]=G1[al,s,k,n]-G1[1,s,k,n];
H2[al_,s_,k_,n_]=G2[al,s,k,n]-G2[1,s,k,n];
H3[al_,s_,k_,n_]=G3[al,s,k,n]-G3[1,s,k,n];
B[al_,s_,k_,n_]=((1+k) H1[al,s,k,n]-
2 (n-s) (s-2)/((n-s+2) s) H2[al,s,k,n]+
(n-s)/s th[s,k,n] H3[al,s,k,n]);
```

```
K[al_,s_,k_,n_]=k H1[al,s,k,n]-
((n-s) (s-2))/((n-s+2) s) k/(1+k) H2[al,s,k,n];
mjs[s_,k_,n_]=1-(n-s)/s th[s,k,n];
m[al_,a2_,al_,s_,k_,n_]=(mjs[s,k,n]+(A[s,k,n]+
(al^2 B[al,s,k,n]-2 al K[al,s,k,n])+
(a2^2 (1+k)-2 a2 k) (1-G1[al,s,k,n])));
r[al_,a2_,al_,s_,k_,n_]=(1+k) m[al,a2,al,s,k,n]-k;
rjs[s_,k_,n_]=(1+k) mjs[s,k,n]-k;
rslm[al_,a2_,al_,s_,k_,n_]=r[al,a2,al,s,k,n]/rjs[s,k,n];
```

Maple Program

```
> th:=(s,k,n)->(s-2)/((n-s+2)*(1+k)):
f:=(al,s,k,n)->1/(1+th(s,k,n)/al):
G:=(a,b,al,s,k,n)->
int((x^(a-1))*((1-x)^(b-1)),x=0..f(al,s,k,n))/Beta(a,b):
G1:=(al,s,k,n)->G((n-s)/2,s/2+1,al,s,k,n):
> G2:=(al,s,k,n)->G(((n-s)/2)+1,s/2,al,s,k,n):
>G3:=(al,s,k,n)->G(((n-s)/2)+2,s/2-1,al,s,k,n):
>A:=(s,k,n)->((1-k)*(G1(1,s,k,n)-1)
-(2*(n-s)/s)*th(s,k,n)*(G2(1,s,k,n)-1)
+((n-s)/s)*th(s,k,n)*(G3(1,s,k,n)-1)):
> H1:=(al,s,k,n)->G1(al,s,k,n)-G1(1,s,k,n):
H2:=(al,s,k,n)->G2(al,s,k,n)-G2(1,s,k,n):
H3:=(al,s,k,n)->G3(al,s,k,n)-G3(1,s,k,n):
B:=(al,s,k,n)->((1+k)*H1(al,s,k,n)
-((2*(n-s)*(s-2))/((n-s+2)*s))*H2(al,s,k,n)
+((n-s)/s)*th(s,k,n)*H3(al,s,k,n)):
K:=(al,s,k,n)->k*H1(al,s,k,n)
-(((n-s)*(s-2))/((n-s+2)*s))*(k/(1+k))*H2(al,s,k,n):
>mjs:=(s,k,n)->1-((n-s)/s)*th(s,k,n):
m:=(l,w,al,s,k,n)>(mjs(s,k,n)+(A(s,k,n)
+((l^2)*B(al,s,k,n)-2*l*K(al,s,k,n))+((w^2)*(1+k)-
2*w*k)*(1-G1(al,s,k,n))))):
r:=(l,w,al,s,k,n)->(1+k)*m(l,w,al,s,k,n)-k:
> rjs:=(s,k,n)->(1+k)*mjs(s,k,n)-k:
```

Exercise 6.2.10. For the values of n and s illustrated in the text and suggested in Exercise 6.2.9

A. Where does the RSLM of PP_1 cross that of PP_4, PP_1 and PP_2?

B. Where does the RSLM of PP_2 cross that of PP_3?

6.3 The Conditional MSE

The properties of the average (Bayes) risk and the RSL of the positive part estimators were considered in Section 6.2. The objective of this section will be to study the analogous properties of the conditional (frequentist) risk and the relative loss (RL). With this objective in mind:

1. Analytical formulae will be obtained for the conditional MSE and the RL of PP_0 $-PP_4$.

2. Some numerical and analytical comparisons of the frequentist risk and the RL of the different positive parts will be made for the same values of p, n and s that were considered in Section 6.2.

As was done in Section 6.2 the problem of estimating the mean of a multivariate normal distribution will be considered first; then the single linear model will be considered. The expressions for the predictive MSE will be in terms of non-central chi square distributions for the problem of estimating the mean of a multivariate normal distribution; for the linear model the expressions will be in terms of non-central F distributions.

6.3.1 The Multivariate Normal Distribution

As was done in Section 6.2.1 the analytical results will be established, some tables of numerical values will be given, some graphs will be plotted and some comments will be made. The formula for the risk of the positive parts is given in Theorem 6.3.1.

Theorem 6.3.1. Let $u_2 = \chi^2_{p+2}$, $u_4 = \chi^2_{p+4}$ and $g = p - 2$. The MSE of the general positive part estimator 6.1.9 is given by

$$
m = m_{js} - E\left[\left(1 - \frac{g}{u_2}\right)^2 I_{[0,g)}(u_2)\right] - \frac{2\delta}{p}E\left[\left(1 - \frac{g}{u_4}\right)^2 I_{[0,g)}(u_4)\right]
$$
$$
+ \frac{4\delta}{p}E\left[\left(1 - \frac{g}{u_2}\right)I_{[0,g)}(u_2)\right]
$$

$$+a^2 E\left[\left(1-\frac{g}{u_2}\right)^2 I_{[\frac{s}{\alpha},g)}(u_2)\right]+\frac{2a^2\delta}{p}E\left[\left(1-\frac{g}{u_4}\right)^2 I_{[\frac{s}{\alpha},g)}(u_4)\right]$$

$$-\frac{4a\delta}{p}E\left[\left(1-\frac{g}{u_2}\right)I_{[\frac{s}{\alpha},g)}(u_2)\right] \tag{6.3.1}$$

$$+\left(b^2-\frac{4b\delta}{p}\right)E[I_{[0,\frac{s}{\alpha})}(u_2)]+\frac{2b^2\delta}{p}E[I_{[0,\frac{s}{\alpha})}(u_4)].$$

Before proving this theorem some important special cases will be taken up. Letting $a = 0$ and $\alpha = 1$ the MSE of PP$_0$ is obtained. This result was obtained by Judge and Bock (1978). Equation (6.3.1) may be used to find the MSE of the other positive parts (See Exercise 6.3.2.).

The MSE of PP$_0$ is given by

$$m_0 = m_{js} - E\left[\left(1-\frac{g}{u_2}\right)^2 I_{[0,g)}(u_2)\right]-\frac{2\delta}{p}E\left[\left(1-\frac{g}{u_4}\right)I_{[0,g)}(u_4)\right]$$

$$+\frac{4\delta}{p}E\left[\left(1-\frac{g}{u_2}\right)I_{[0,g)}(u_2)\right]. \tag{6.3.2}$$

Exercise 6.3.1. Show that the general positive part estimator (6.1.9) has a smaller frequentist risk than the JS by methods similar to those used in Section 6.1.

Exercise 6.3.2 A. Show that the positive parts PP$_1$ – PP$_4$ have frequentist risks given by

$$m_1 = m_{js} + \frac{8\delta}{p}E\left[\left(1-\frac{g}{u_2}\right)I_{[0,g)}(u_2)\right], \tag{6.3.3}$$

$$m_2 = m_{js} - E\left[\left(1-\frac{g}{u_2}\right)^2 I_{[0,\frac{s}{2})}(u_2)\right]-\frac{2\delta}{p}E\left[\left(1-\frac{g}{u_4}\right)^2 I_{[0,\frac{s}{2})}(u_4)\right]$$

$$+\frac{4\delta}{p}E\left[\left(1-\frac{g}{u_2}\right)I_{[0,\frac{s}{2})}(u_2)\right]+\frac{8\delta}{p}E\left[\left(1-\frac{g}{u_2}\right)I_{[\frac{s}{2},g)}(u_2)\right] \tag{6.3.4}$$

$$+\left(1-\frac{4\delta}{p}\right)E[I_{[0,\frac{s}{2})}(u_2)]+\frac{2\delta}{p}E[I_{[0,\frac{s}{2})}(u_4)],$$

$$m_3 = m_{js} - 0.75E\left[\left(1-\frac{g}{u_2}\right)^2 I_{[\frac{8}{2},8)}(u_2)\right]$$

$$-\frac{1.5\delta}{p}E\left[\left(1-\frac{g}{u_4}\right)^2 I_{[\frac{8}{2},8)}(u_4)\right]$$

$$-E\left[\left(1-\frac{g}{u_2}\right)^2 I_{[0,\frac{8}{2})}(u_2)\right]-\frac{2\delta}{p}E\left[\left(1-\frac{g}{u_4}\right)^2 I_{[0,\frac{8}{2})}(u_4)\right] \quad (6.3.5)$$

$$+\frac{2\delta}{p}E\left[\left(1-\frac{g}{u_2}\right)I_{[\frac{8}{2},8)}(u_2)\right]+\frac{4\delta}{p}E\left[\left(1-\frac{g}{u_2}\right)I_{[\frac{8}{2},8)}(u_2)\right]$$

$$+\left(1-\frac{2\delta}{p}\right)E[I_{[0,\frac{8}{2})}(u_2)+\frac{0.5\delta}{p}E[I_{[0,\frac{8}{2})}(u_4)$$

and

$$m_4 = m_{js} - E\left[\left(1-\frac{g}{u_2}\right)^2 I_{[0,\frac{8}{2})}(u_2)\right]-\frac{2\delta}{p}E\left[\left(1-\frac{g}{u_4}\right)^2 I_{[0,\frac{8}{2})}(u_4)\right]$$

$$+\frac{4\delta}{p}E\left[\left(1-\frac{g}{u_2}\right)I_{[0,\frac{8}{2})}(u_2)\right] \quad (6.3.6)$$

$$+\left(1-\frac{4\delta}{p}\right)E[I_{[0,\frac{8}{2})}(u_2)+\frac{2\delta}{p}E[I_{[0,\frac{8}{2})}(u_4).$$

B. From the above expressions show that the risk of PP_0 and PP_1 is smaller than that of the MLE.

C. Show that the risk of PP_2 is never greater than that of PP_4.

E. Show that when $\delta = 0$ the ordering of the frequentist risks of the JS is the same as that of the Bayes risk when $A = 0$.

Proof of Theorem 6.3.1. The proof of Theorem 6.3.1 depends on Lemmas 6.3.1 and 6.3.2 that follow. These are the special cases of results obtained in Bock and Judge (1978) needed to prove Theorem 6.3.1.

Lemma 6.3.1. Suppose v has a non-central chi square distribution with noncentrality parameter δ, i.e., $v\sim\chi^2(p, \delta)$. Then for any integer r and a, b with $0\leq a \leq b$

$$E[\,v^{r+1}I_{(a,b)}(v)] = E[(u_2)^r I_{(a,b)}(u_2)]p + E[(u_4)^r I_{(a,b)}(u_4)].2\delta. \qquad (6.3.7)$$

Proof. The form of the non-central chi square distribution with p degrees of freedom is

$$f(v) = e^{-\delta}\sum_{j=0}^{\infty}\frac{\delta^k v^{p/2+j-1}e^{-v/2}}{j!2^{n/2+j}\Gamma(p/2+j)}. \qquad (6.3.8)$$

Thus, taking into account the uniform convergence properties of the infinite series it can be shown that

$$E[\,v^{r+1}I_{(a,b)}(v)] = \sum_{j=0}^{\infty}\frac{e^{-j}\delta^j}{j!2^{j+p/2}\Gamma\left(\frac{p+2j}{2}\right)}A_j \qquad (6.3.9)$$

where

$$A_j = \int_a^b v^{p/2+j+r}e^{-v/2}dv. \qquad (6.3.10)$$

Recall the functional equation of the Gamma function is $\Gamma(\alpha+1) = \Gamma(\alpha)$. Applying it with $\alpha=(p + 2j)/2$ it follows that

$$\frac{A_j}{2^{j+p/2}\Gamma\left(\frac{p+2j}{2}\right)} = \frac{(2j+p)A_j}{2^{j+1+p/2}\Gamma\left(\frac{p+2j+2}{2}\right)}. \qquad (6.3.11)$$

Also

$$\sum_{j=0}^{\infty}\frac{e^{-\delta}j\delta^j A_j}{j!2^{j+p/2+1}\Gamma\left(\frac{p+2j+2}{2}\right)} = \delta\sum_{j=0}^{\infty}\frac{e^{-\delta}\delta^j A_j}{j!2^{j+p/2+2}\Gamma\left(\frac{p+2j+4}{2}\right)}. \qquad (6.3.12)$$

Substitution of (6.3.11) and (6.3.12) into (6.39) and some algebra yields (6.3.7).

Lemma 6.3.2. Let $w\sim N_p(\theta,I)$, i.e., the random variable u is p-variate normal with mean vector θ and identity dispersion matrix. Then the random variable $v = w'w$ has a non-central chi square distribution with p degrees of freedom and noncentrality parameter $\delta = \theta'\theta/2$, i.e., $v \sim \chi^2(p, \delta)$. Then

$$E[(w'w)^r w I_{[a,b)}(w'w)] = \theta E[(u_2)' I_{[a,b)}(u_2)].$$ (6.3.13)

Proof. It suffices to prove the result when a = 0 because

$$I_{[a,b)}(v) = I_{[0,b)}(v) - I_{[0,a)}(v).$$ (6.3.14)

The column vector

$$\begin{aligned}
H &= E[(w'w)^r w I_{[0,b)}(w'w)] \\
&= \frac{1}{(2\pi)^{p/2}} \int_{-\sqrt{b}}^{\sqrt{b}} (w'w)^r w \exp\left[-\tfrac{1}{2}(w-\theta)'(w-\theta)\right] dw \\
&= \frac{e^{-\delta}}{(2\pi)^{p/2}} \int_{-\sqrt{b}}^{\sqrt{b}} (w'w)^r w \exp\left[-\tfrac{1}{2}(w'w - \theta'w - w'\theta)\right] dw \\
&= \frac{e^{-\delta}}{(2\pi)^{p/2}} \frac{\partial}{\partial\theta}\left[\int_{-\sqrt{b}}^{\sqrt{b}} (w'w)^r \exp\left[-\tfrac{1}{2}(w'w - \theta'w - w'\theta)\right] dw\right] \\
&= \frac{e^{-\delta}}{(2\pi)^{p/2}} \frac{\partial}{\partial\theta}\left[e^{\delta} \int_{-\sqrt{b}}^{\sqrt{b}} (w'w)^r \exp\left[-\tfrac{1}{2}(w-\theta)'(w-\theta)\right] dw\right] \\
&= e^{-\delta} \frac{\partial}{\partial\theta}\left[e^{\delta} E[v^r I_{[0,b)}(v)]\right].
\end{aligned}$$ (6.3.15)

But

$$E[v^r I_{[0,b)}(v)] = \sum_{j=0}^{\infty} \frac{\delta^j e^{-\delta}}{j!} B_j$$ (6.3.16)

where

$$B_j = \int_0^b \frac{v^{n/2+j+r-1} e^{-v/2}}{\Gamma\left(\frac{n+2j}{2}\right) 2^{(n+2j)/2}} dv$$ (6.3.17)

with B_j a constant function of θ. Thus,

$$H = \theta \sum_{j=0}^{\infty} \frac{\delta^j e^{-\delta}}{j!} B_j + \theta \sum_{j=1}^{\infty} \frac{j\delta^{j-1} e^{-\delta}}{j!} B_j - \theta \sum_{j=0}^{\infty} \frac{\delta^j e^{-\delta}}{j!} B_j$$

$$= \theta \sum_{j=0}^{\infty} \frac{\delta^j e^{-\delta}}{j!} B_{j+1} = \theta E[(u_2)^r I_{[0,b)}(u_2)].$$

(6.3.18)

The result then follows by virtue of (6.3.14).

Proof of Theorem 6.3.1. The form of the generalized positive part (6.1.9) is

$$\hat{\theta}_i = hx_i, \ 1 \le i \le p$$

(6.3.19a)

with

$$h = (1 - w_1) I_{[0,1)}(w_1) + a_1 (1 - w_1) I_{[1,\alpha)}(w_1) + a_2 I_{[\alpha,\infty)}(w_1).$$

(6.3.19b)

Then the risk takes the form

$$m = \frac{1}{p} E\left[\sum_{i=1}^{p} (hx_i - \theta_i)^2 \right]$$

$$= \frac{1}{p} \left[E\left(h^2 \sum_{i=1}^{p} x_i^2 \right) - \sum_{i=1}^{p} 2\theta_i E(hx_i) + \sum_{i=1}^{p} \theta_i^2 \right]$$

(6.3.20)

Let $v = \sum_{i=1}^{p} x_i^2$. Now $v \sim \chi^2(p, \delta)$ where $\delta = \theta'\theta/2$. The first term in the second expression of (6.3.20) may be expanded as

$$E(h^2 v) = E\left[\left(1 - \frac{g}{v}\right)^2 v I_{[0,1)}(v) \right] + a_1^2 E\left[\left(1 - \frac{g}{v}\right)^2 v I_{[1,\alpha)}(v) \right]$$

$$+ a_2^2 E[v I_{[\alpha,\infty)}(v)].$$

(6.3.21)

Letting $x' = [x_1, x_2,..., x_p]$ and $\theta = [\theta_1, \theta_2,..., \theta_p]$ so that $x'\theta = \sum_{i=1}^{p} x_i \theta_i$ expanding the second term of the second expression of (6.3.20) yields

$$E(hx'\theta) = E\left[\left(1-\frac{g}{v}\right)x'\theta I_{(0,1)}(v)\right] + a_1 E\left[\left(1-\frac{g}{v}\right)x'\theta I_{[1,\alpha)}(v)\right] \quad (6.3.22)$$
$$+ a_2 E[x'\theta I_{[\alpha,\infty)}(v)].$$

The result then follows by application of Lemma 6.3.1. and 6.3.2.

Exercise 6.3.3. Expand (6.3.21) and (6.3.22) and show how to apply Lemmas 6.3.1 and 6.3.2 to obtain the result (6.3.1).

Examples of the numerical values of the risk of the JS and its PP are given in Table 6.3.1. The risk of the MMSE is also given because it represents the smallest attainable risk.

Table 6.3.1

δ	0	0.2	0.5	0.8	1	2	3
JS	0.500	0.547	0.607	0.656	0.684	0.784	0.842
PP_0	0.368	0.428	0.505	0.570	0.608	0.743	0.820
PP_1	0.500	0.529	0.573	0.614	0.640	0.746	0.818
PP_2	0.393	0.443	0.511	0.569	0.604	0.734	0.814
PP_3	0.385	0.442	0.515	0.578	0.614	0.746	0.822
PP_4	0.393	0.451	0.525	0.587	0.623	0.752	0.825
mmse	5×10^{-5}	0.091	0.200	0.286	0.333	0.500	0.600

$$p = 4$$

δ	0	0.2	0.5	0.8	1	2	3
JS	0.250	0.286	0.335	0.379	0.405	0.514	0.592
PP_0	0.162	0.202	0.259	0.310	0.341	0.471	0.565
PP_1	0.250	0.272	0.306	0.341	0.364	0.472	0.560
PP_2	0.205	0.235	0.278	0.320	0.347	0.465	0.557
PP_3	0.187	0.225	0.278	0.327	0.357	0.481	0.571
PP_4	0.205	0.244	0.297	0.345	0.374	0.494	0.581
MMSE	2.5×10^{-5}	0.048	0.111	0.167	0.200	0.333	0.429

$$p = 8$$

δ	0	0.2	0.5	0.8	1	2	3
JS	0.125	0.146	0.177	0.205	0.223	0.303	0.369
PP_0	0.075	0.098	0.130	0.162	0.182	0.271	0.346
PP_1	0.125	0.139	0.161	0.184	0.199	0.274	0.343
PP_2	0.115	0.130	0.154	0.178	0.194	0.272	0.342
PP_3	0.092	0.114	0.146	0.176	0.196	0.282	0.355
PP_4	0.115	0.136	0.167	0.197	0.215	0.298	0.366
MMSE	1.25×10^{-5}	0.024	0.059	0.091	0.111	0.200	0.273

$p = 16$

δ	0	0.2	0.5	0.8	1	2	3
JS	0.063	0.074	0.091	0.107	0.118	0.167	0.212
PP_0	0.036	0.048	0.065	0.082	0.934	0.147	0.195
PP_1	0.063	0.071	0.084	0.097	0.106	0.151	0.195
PP_2	0.062	0.070	0.083	0.096	0.105	0.150	0.195
PP_3	0.044	0.056	0.073	0.090	0.102	0.154	0.202
PP_4	0.062	0.073	0.090	0.106	0.117	0.167	0.211
MMSE	6.24×10^{-6}	0.012	0.030	0.048	0.059	0.111	0.158

$p = 32$

Table 6.3.2 The RL

δ	0	0.2	0.5	0.8	1	2	3
JS	0.5	0.502	0.508	0.518	0.526	0.568	0.604
PP_0	0.368	0.370	0.381	0.398	0.411	0.485	0.551
PP_1	0.500	0.482	0.466	0.459	0.460	0.492	0.544
PP_2	0.393	0.388	0.388	0.397	0.406	0.469	0.535
PP_3	0.385	0.386	0.394	0.409	0.421	0.491	0.555
PP_4	0.393	0.396	0.406	0.422	0.435	0.504	0.564

$p = 4$

δ	0	0.2	0.5	0.8	1	2	3
JS	0.250	0.250	0.252	0.255	0.257	0.270	0.285
PP_0	0.162	0.163	0.166	0.172	0.177	0.207	0.239
PP_1	0.250	0.236	0.220	0.210	0.206	0.208	0.230
PP_2	0.205	0.197	0.188	0.184	0.184	0.198	0.226
PP_3	0.187	0.186	0.188	0.192	0.196	0.221	0.250
PP_4	0.205	0.206	0.209	0.214	0.217	0.241	0.266

$$p = 8$$

δ	0	0.2	0.5	0.8	1	2	3
JS	0.125	0.125	0.125	0.126	0.126	0.129	0.133
PP_0	0.075	0.075	0.076	0.078	0.079	0.089	0.101
PP_1	0.125	0.118	0.109	0.102	0.099	0.092	0.097
PP_2	0.115	0.109	0.101	0.096	0.093	0.090	0.096
PP_3	0.092	0.092	0.093	0.094	0.095	0.103	0.113
PP_4	0.115	0.0115	0.115	0.116	0.117	0.122	0.128

$$p = 16$$

δ	0	0.2	0.5	0.8	1	2	3
JS	0.063	0.063	0.063	0.063	0.063	0.063	0.063
PP_0	0.036	0.036	0.036	0.036	0.037	0.040	0.044
PP_1	0.063	0.059	0.055	0.052	0.050	0.044	0.044
PP_2	0.062	0.058	0.054	0.051	0.049	0.044	0.044
PP_3	0.044	0.044	0.044	0.045	0.045	0.048	0.052
PP_4	0.062	0.062	0.062	0.062	0.062	0.063	0.063

$$p = 32$$

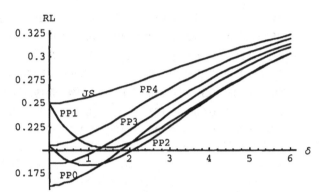

Figure 6.3.1 a RL p=8

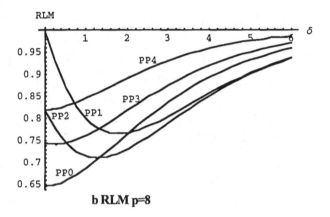

b RLM p=8

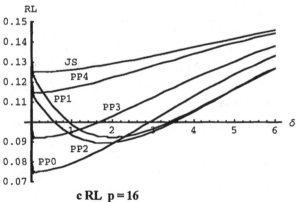

c RL p = 16

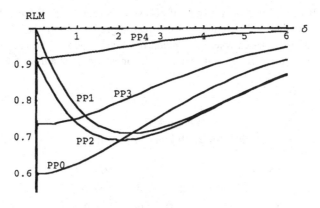

d RLM p=16

There are a number of similarities between the behavior of the conditional risk and the RL as a function of δ and the corresponding Bayes risk and the RSL as a function of A. The similarities include:

1. The conditional risk function for the JS and its positive parts is an increasing function of δ.

2. When $\delta = 0$ the ordering of the conditional risks and the RL of the JS and its positive parts is the same as that of the RSL and the average risk when A =0. That is

$$m_0 \leq m_3 \leq m_2 = m_4 \leq m_1 = m_{js} \qquad (6.3.23)$$

3. When $\delta \to \infty$ the risk and the RL of the positive parts approach that of the JS although this appears to happen more slowly.

4. For small values of δ PP_0 is the dominant estimator for larger values PP_2 is the dominant estimator. Some crossing values are contained in Table 6.3.4.

5. All of the positive part estimators dominate the JS.

6. The estimator PP_2 dominates PP_4; otherwise no one estimator dominates the other. The crossing points of estimators for the special cases plotted will be given below (see Table 6.3.5).

7. The RL of PP_0 and PP_4 appear to be increasing functions of δ.

8. The RL of PP_1, PP_2 and PP_3 decreases, pass through a minimum value and then increase. The minimum values will also be given for the cases where graphs were plotted(See Table 6.3.3.)

An important difference between the RL and the RSL of the JS, established in Chapter III is that the RL is an increasing function with values between the RSL and

$1 - (1- RSL)^2$. Thus the measure of efficiency analogous to the RSLM the RLM may behave slightly differently. For this reason graphs of both the RL and the RLM are plotted for the cases where $p = 8$ and $p=16$ are plotted. The reader may do the same computations for $p = 4$ and $p = 32$ or any other values of p if he or she wishes.

Table 6.3.3
Minimum Points of PP_1–PP_3

p	8		16	
	δ	RL	δ	RL
PP_1	1.396	0.203	1.969	0.092
PP_2	0.974	0.184	1.785	0.089
PP_3	0.179	0.186	0.178	0.092

Table 6.3.4
Crossing Points of PP_0 and PP_2

p	8		16	
	δ	RL	δ	RL
Crossing Point	1.324	0.186	2.049	0.090

The estimator PP_1 has RL equal to that of the JS at $\delta = 0$ decreases and then increases. It crosses the other positive parts in the order PP_4, PP_3, PP_0 and finally PP_2. However in the regions where it is better than PP_2 the improvement appears to be negligible. Table 6.2.5 lists the crossing points in the order mentioned above.

Table 6.3.5
Crossing Points of PP_1 and the Other Positive Parts

p	8		16	
	δ	RL	δ	RL
JS	0	0.250	0	0.125
PP_4	0.714	0.212	0.282	0.115
PP_3	1.312	0.203	1.189	0.096
PP_0	2.054	0.209	2.316	0.093
PP_2	8.482	0.339	12.84	0.168

Exercise 6.3.4. Use the Mathematica or the Maple program below to compute the MSE and the RL of the JS and its positive parts for s = 4 and s = 32 when δ = 0, 0.2, 0.5, 0.8, 1, 2, 3. Compare your results with those in the Tables and graphs of this section.

Mathematica Program

```
f[u_,v_,c_,j_,n_]=2^c
Gamma[n/2+j+c,u/2,v/2]/Gamma[n/2+j];
g[d_,j_]=Exp[-d] d^j/j!;
h[u_,v_,c_,d_,j_,n_]=f[u,v,c,j,n] g[d,j];
s[u_,v_,c_,d_,n_,m_]=Sum[h[u,v,c,d,j,n],{j,0,m}];
f1[u_,v_,d_,n_,m_]=(s[u,v,0,d,n,m]-2 (p-2) s[u,v,-
1,d,n,m]
+(p-2)^2 s[u,v,-2,d,n,m]);
f2[u_,v_,d_,n_,m_]=s[u,v,0,d,n,m]-(p-2) s[u,v,-1,d,n,m];
a[d_,p_]=Exp[-d]/(2 d^(p/2-1));
b[d_,p_]=Integrate[Exp[x] x^(p/2-2),{x,0,d}];
c[d_,p_]=a[d,p] b[d,p];
mjs[d_,p_]=1-(p-2)^2/p c[d,p];
mpp[a_,b_,d_,p_,al_,m_]=(mjs[d,p]-f1[0,p-2,d,p+2,m]
-2 d/p f1[0,p-2,d,p+4,m]
+4 d/p f2[0,p-2,d,p+2,m]
+a^2 (f1[(p-2)/al,p-2,d,p+2,m]
+2 d/p f1[(p-2)/al,p-2,d,p+4,m])
-4 a d/p f2[(p-2)/al,p-2,d,p+2,m]
+(b^2-4 b d/p) s[0,(p-2)/al,0,d,p+2,m]
+b^2 s[0,(p-2)/al,0,d,p+4,m] 2 d/p);
mms[d_,p_]=1-1/(2 d/p+1);
rl[a_,b_,d_,p_,al_,m_]=(mpp[a,b,d,p,al,m]-mms[d,p])/(1-
mms[d,p]);
rljs[d_,p_]=(mjs[d,p]-mms[d,p])/(1-mms[d,p]);
```

Maple Program

```
> f:=(u,v,c,j,n)->(2^c)*(GAMMA(n/2+j+c,u/2)-
GAMMA(n/2+j+c,v/2))/GAMMA(n/2+j):
g:=(d,j)->exp(-d)*(d^j)/j!:
h:=(u,v,c,d,j,n)->f(u,v,c,j,n)*g(d,j):
s:=(u,v,c,d,n,m)->sum(h(u,v,c,d,j,n),j=0..m):
f1:=(u,v,d,p,n,m)->(s(u,v,0,d,n,m)
-2*(p-2)*s(u,v,-1,d,n,m)+((p-2)^2)*s(u,v,-2,d,n,m)):
f2:=(u,v,d,p,n,m)->s(u,v,0,d,n,m)-((p-2))*s(u,v,1,d,n,m):
a:=(d,p)->exp(-d)/(2*(d^(p/2-1))):
> b:=(d,p)->int(exp(x)*(x^(p/2-2)),x=0..d):
c:=(d,p)->a(d,p)*b(d,p):
> mjs:=(d,p)->1-(((p-2)^2)/p)*c(d,p):
```

```
> mpp:=(a,b,d,p,al,m)->(mjs(d,p)-f1(0,p-2,d,p,p+2,m)
-2*(d/p)*f1(0,p-2,d,p,p+4,m)
+(4*d/p)*f2(0,p-2,d,p,p+2,m)
+(a^2)*(f1((p-2)/al,p-2,d,p,p+2,m)
+(2*d/p)*f1((p-2)/al,p-2,d,p,p+4,m))
-(4*a*d/p)*f2((p-2)/al,p-2,d,p,p+2,m)
+(b^2-(4*b*d/p))*s(0,(p-2)/al,0,d,p+2,m)
+(b^2)*(2*d/p)*s(0,(p-2)/al,0,d,p+4,m)):
> mms:=(d,p)->1-1/(((2*d)/p)+1):
> rl:=(a,b,d,p,al,m)->(mpp(a,b,d,p,al,m)
-mms(d,p))/(1-mms(d,p)):
```

The MSE of the positive parts of the JS averaging over the predictive loss functions will now be obtained. Numerical studies similar to those of the Section 6.3.1 will then be undertaken.

The computation of the risk of the JS for the single linear model may be reduced to the problem of computing the risk of a JS for the mean of the multivariate normal distribution with an unknown error variance. The general form of this risk is given in Theorem 6.3.2.

The JS and its positive parts are expressible as a function of $b'X'Xb \sim \chi^2_{p,\delta}$ and $\hat{\sigma}^2 \sim \chi^2_{n-s}$. The form of the non-central chi-square distribution in Equation (6.3.8) is a weighted infinite series of central chi square distributions with $p + 2j$ degrees of freedom. The weights are the probabilities associated with the integer values of a Poisson random variable. Thus, the notation $\chi^2_{p,\delta}$ may be replaced by χ^2_{p+2K} where K is a Poisson random variable.

Theorem 6.3.2. Let $\hat{\beta} = h\left(\dfrac{\chi^2_{p+2K}}{\chi^2_N}\right)b$ where b is the LS estimator for the parameters of a linear model and the numerator and denominator in the fraction inside of h are independent random variables. Then

$$R(\hat{\beta},\theta,\sigma^2) = E[h^2\left(\frac{\chi^2_{p+2K}}{\chi^2_N}\right)\chi^2_{p+2K}] - 4E[h\left(\frac{\chi^2_{p+2K}}{\chi^2_N}\right)K] + \beta'X'X\beta$$

$$(6.3.24)$$

where K is Poisson distributed with parameter $\delta = \beta'X'X\beta/2\sigma^2$.

Proof. The result will follow once it is established that

$$E[h\left(\frac{\chi^2_{p+2K}}{\chi^2_N}\right)b'\beta] = 2E[h\left(\frac{\chi^2_{p+2K}}{\chi^2_N}\right)K]\qquad(6.3.25)$$

The rest of the proof may then be left as an exercise. From previous considerations it suffices to consider the case where $X'X$ is the identity matrix. The expectations will be obtained conditional on $Q = \chi^2_N$. Let $Y \sim N(\theta,1)$. Then

$$E_Q[h\left(\frac{\chi^2_{p+2K}}{Q}\right)Y'\theta]$$

$$= E_Q[h\left(\frac{\chi^2_{p+2K}}{Q}\right)(Y-\theta)'\theta] + E_Q[h\left(\frac{\chi^2_{p+2K}}{Q}\right)\theta'\theta]$$

$$= \frac{\partial}{\partial\theta}E_Q[h\left(\frac{\chi^2_{p+2K}}{Q}\right)]'\theta + E_Q[h\left(\frac{\chi^2_{p+2K}}{Q}\right)\theta'\theta]\qquad(6.3.26)$$

$$= \frac{\partial}{\partial\theta}\sum_{k=0}^{\infty}\frac{e^{-\delta}\delta^k}{k!}E_Q[h\left(\frac{\chi^2_{p+2K}}{Q}\right)]'\theta + 2\delta\sum_{k=0}^{\infty}\frac{e^{-\delta}\delta^k}{k!}E_Q[h\left(\frac{\chi^2_{p+2K}}{Q}\right)]$$

$$= \sum_{k=0}^{\infty}\frac{e^{-\delta}\delta^k 2k}{k!}E_Q[h\left(\frac{\chi^2_{p+2K}}{Q}\right)]'.$$

Exercise 6.3.5. A. Fill in the details in 6.3.26.
B. Complete the proof of Theorem 6.3.2.

Let

$$c_1(p,k,n) = E\left[h^2\left(\frac{\chi^2_{p+2K}}{\chi^2_N}\right)\chi^2_{p+2K}\right]$$

and

$$c_2(p,k,n) = E\left[h\left(\frac{\chi^2_{p+2K}}{\chi^2_N}\right)\right].$$

Conditioning on K,

$$e_1 = E\left[h^2\left(\frac{\chi^2_{p+2K}}{\chi^2_N}\right)\chi^2_{p+2K}\right] = EE\left[\left[h^2\left(\frac{\chi^2_{p+2K}}{\chi^2_N}\right)\chi^2_{p+2K}\right]K\right]$$

$$= \sum_{k=0}^{\infty}\frac{e^{-\delta}\delta^k}{k!}c_1(p,k,n)$$

(6.3.27)

and

$$e_2 = E\left[h\left(\frac{\chi^2_{p+2K}}{\chi^2_N}\right)K\right] = \sum_{k=0}^{\infty}\frac{ke^{-\delta}\delta^k}{k!}c_2(p,k,N) = \sum_{k=0}^{\infty}\frac{e^{-\delta}\delta^k}{k!}c_2(p,k+1,N).$$

(6.3.28)

A single linear model of rank s is being considered here. Thus, $N = n - s$ and $p = s$. For $g = (s-2)/(n-s+2)$ and $w_2 = g\chi^2_{n-s}/\chi^2_{s+2K}$, recall that for the positive parts

$$h(w_2) = (1-w_2)I_{[0,1)}(w_2) + a_1(1-w_2)I_{[1,\alpha)}(w_2) + a_2 I_{[\alpha,\infty)}(w_2).$$

(6.3.29)

Then

$$c_1(s,k,n-s) = E[(\chi^2_{s+2K} - 2g\chi^2_{n-s} + g^2\frac{(\chi^2_{n-s})^2}{\chi^2_{s+2K}})I_{[0,1)}(w_2)]$$

$$+a_1^2 E[(\chi^2_{s+2K} - 2g\chi^2_{n-s} + g^2\frac{(\chi^2_{n-s})^2}{\chi^2_{s+2K}})I_{[1,\alpha)}(w_2)] \quad (6.3.30)$$

$$a_2^2 E[I_{[\alpha,\infty)}(w_2)]$$

and

$$c_2(s,k,n-s) = E\left[\left(1 - g\frac{\chi^2_{n-s}}{\chi^2_{s+2K}}\right)I_{[0,1)}(w_2)\right]$$

$$+a_1 E\left[\left(1 - g\frac{\chi^2_{n-s}}{\chi^2_{s+2K}}\right)I_{[1,\alpha)}(w_2)\right] \quad (6.3.31)$$

$$+a_2 E[I_{[\alpha,\infty)}(w_2)].$$

The expectations in (6.3.30) and (6.3.31) are computed using the following formulae:

$$E[I_{[0,1)}(w_2)] = \frac{1}{Be\left(\frac{n-s}{2}, \frac{s+2k}{2}\right)} \cdot H_{\frac{1}{1+\theta}}\left[\frac{n-s}{2}, \frac{s+2k}{2}\right], \quad (6.3.32)$$

$$E[I_{[1,\alpha)}(w_2)]$$
$$= \frac{1}{Be\left(\frac{n-s}{2}, \frac{s+2k}{2}\right)}\left[.H_{\frac{\alpha}{\alpha+\theta}}\left[\frac{n-s}{2}, \frac{s+2k}{2}\right] - H_{\frac{1}{1+\theta}}\left[\frac{n-s}{2}, \frac{s+2k}{2}\right]\right],$$

$$(6.3.33)$$

$$E[I_{[\alpha,\infty)}(w_2)] = 1 - \frac{1}{Be\left(\frac{n-s}{2}, \frac{s+2k}{2}\right)} \cdot H_{\frac{\alpha}{\alpha+\theta}}\left[\frac{n-s}{2}, \frac{s+2k}{2}\right], \quad (6.3.34)$$

$$E\left[\frac{\chi^2_{n-s}}{\chi^2_{s+2k}} I_{[0,1)}(w_2)\right] = \frac{1}{Be\left(\frac{n-s}{2}, \frac{s+2k}{2}\right)} \cdot H_{\frac{1}{1+\theta}}\left[\frac{n-s}{2}-1, \frac{s+2k}{2}+1\right],$$

$$(6.3.35)$$

$$E\left[\frac{\chi^2_{n-s}}{\chi^2_{s+2k}} I_{[1,\alpha)}(w_2)\right]$$
$$= \frac{1}{Be\left(\frac{n-s}{2}, \frac{s+2k}{2}\right)} \cdot \quad (6.3.36)$$

$$\left[.H_{\frac{\alpha}{\alpha+\theta}}\left[\frac{n-s}{2}-1, \frac{s+2k}{2}+1\right] - H_{\frac{1}{1+\theta}}\left[\frac{n-s}{2}-1, \frac{s+2k}{2}+1\right]\right],$$

$$E[\chi_{s+2k}^2 I_{[0,1)}(w_2)] = \frac{(n+2k)}{Be\left(\frac{s+2k}{2}, \frac{n-s}{2}\right)} H_{\frac{1}{1+\theta}}\left[\frac{s+2k}{2}+1, \frac{n-s}{2}\right],$$

$$(6.3.37)$$

$$E[\chi_{n-s}^2 I_{[0,1)}(w_2)] = \frac{(n+2k)}{Be\left(\frac{s+2k}{2}, \frac{n-s}{2}\right)} H_{\frac{1}{1+\theta}}\left[\frac{n-s}{2}+1, \frac{s+2k}{2}\right],$$

$$(6.3.38)$$

$$E\left[\frac{(\chi_{n-s}^2)^2}{\chi_{s+2k}^2} I_{[0,1)}(w_2)\right]$$
$$= \frac{(n+2k)}{Be\left(\frac{s+2k}{2}, \frac{n-s}{2}\right)} H_{\frac{1}{1+\theta}}\left[\frac{n-s}{2}+2, \frac{s+2k}{2}-1\right], \qquad (6.3.39)$$

$$E\left[\frac{(\chi_{n-s}^2)^2}{\chi_{s+2k}^2} I_{[0,\alpha)}(w_2)\right] = \frac{(n+2k)}{Be\left(\frac{s+2k}{2}, \frac{n-s}{2}\right)} \cdot \left[H_{\frac{\alpha}{\alpha+\theta}}\left[, \frac{n-s}{2}+2, \frac{s+2k}{2}-1\right]\right.$$
$$\left. - H_{\frac{1}{1+\theta}}\left[\frac{n-s}{2}+2, \frac{s+2k}{2}-1\right]\right],$$

$$(6.3.40)$$

$$E\left[\chi_{s+2k}^2 I_{(0,\alpha)}(W_2)\right] =$$
$$\frac{(n+2k)}{Be\left(\frac{s+2k}{2}, \frac{n-s}{2}\right)} \left[H_{\frac{\alpha}{\alpha+\theta}}\left[\frac{s+2k}{2}+1, \frac{n-s}{2}\right] - H_{\frac{1}{1+\theta}}\left[\frac{s+2k}{2}+1, \frac{n-s}{2}\right]\right],$$

$$(6.3.41)$$

$$E\left[\chi^2_{n-s} I_{[0,\alpha)}(w_2)\right] = \frac{(n+2k)}{Be\left(\dfrac{s+2k}{2}, \dfrac{n-s}{2}\right)}$$

$$\left[H_{\frac{\alpha}{\alpha+\theta}}\left[\frac{n-s}{2}+1, \frac{s+2k}{2}\right] - H_{\frac{1}{1+\theta}}\left[\frac{n-s}{2}+1, \frac{s+2k}{2}\right]\right]$$

$$(6.3.42)$$

and

$$E[I_{(\alpha,\infty)}(w_2)] = 1 - \frac{(n+2k)}{Be\left(\dfrac{s+2k}{2}, \dfrac{n-s}{2}\right)} H_{\frac{\alpha}{\alpha+\theta}}\left[\frac{n-s}{2}, \frac{s+2k}{2}+1\right], (6.3.43)$$

The above results are summarized in Theorem 6.3.3 below.

Theorem 6.3.3. The risk of the general positive part is given by

$$R = \sum_{k=0}^{\infty} \frac{e^{-\delta}\delta^k}{k!} c_1(s,k,n-s) - 4\sum_{k=0}^{\infty} \frac{e^{-\delta}\delta^k}{k!} c_2(s,k+1,n-s) + 2\sigma^2\delta.$$

$$(6.3.44)$$

where the functions c_1 and c_2 are given by Equations (6.3.30) and (6.3.31) with expectations given by Equations (6.3.31)-(6.3.43).

Exercise 6.3.6. Find the formula for the risk of each of PP_0 - PP_4 using Equation (6.3.44).

The numerical values of the risk for the same values of n and s that were used for the average risk are given in Table 6.36. Table 6.3.7 contains the RL.

Table 6.3.6 The Risk of the Positive Parts

δ	0	0.2	0.5	0.8	1	2	3
JS	0.556	0.597	0.650	0.694	0.719	0.808	0.859
PP_0	0.430	0.484	0.554	0.613	0.646	0.768	0.838
PP_1	0.556	0.581	0.619	0.655	0.678	0.772	0.836
PP_2	0.454	0.499	0.559	0.612	0.643	0.760	0.832
PP_3	0.446	0.497	0.564	0.620	0.652	0.771	0.840
PP_4	0.454	0.505	0.572	0.628	0.661	0.777	0.844
MMSE	5×10^{-5}	0.091	0.200	0.286	0.333	0.500	0.600

$$n = 20 \, , s = 4$$

δ	0	0.2	0.5	0.8	1	2	3
JS	0.510	0.556	0.615	0.663	0.690	0.788	0.845
PP_0	0.379	0.438	0.514	0.578	0.615	0.747	0.824
PP_1	0.510	0.539	0.581	0.621	0.647	0.751	0.821
PP_2	0.404	0.453	0.519	0.577	0.611	0.739	0.817
PP_3	0.396	0.452	0.524	0.586	0.621	0.750	0.825
PP_4	0.404	0.461	0.533	0.595	0.630	0.756	0.829
MMSE	5×10^{-5}	0.091	0.200	0.286	0.333	0.500	0.600

$$n = 100, \, s = 4$$

δ	0	0.2	0.5	0.8	1	2	3
JS	0.152	0.172	0.202	0.229	0.247	0.324	0.389
PP_0	0.095	0.117	0.149	0.180	0.199	0.287	0.361
PP_1	0.152	0.165	0.185	0.206	0.221	0.292	0.358
PP_2	0.138	0.153	0.176	0.199	0.214	0.289	0.357
PP_3	0.115	0.136	0.167	0.196	0.215	0.300	0.370
PP_4	0.138	0.159	0.189	0.218	0.236	0.317	0.383
MMSE	1.25×10^{-5}	0.024	0.059	0.091	0.111	0.200	0.273

$$n = 80, \, s = 16$$

δ	0	0.2	0.5	0.8	1	2	3
JS	0.134	0.156	0.186	0.214	0.232	0.311	0.376
PP_0	0.082	0.105	0.137	0.168	0.188	0.277	0.351
PP_1	0.134	0.148	0.170	0.192	0.207	0.280	0.348
PP_2	0.123	0.138	0.162	0.185	0.201	0.278	0.347
PP_3	0.100	0.122	0.153	0.183	0.202	0.288	0.360
PP_4	0.123	0.144	0.175	0.204	0.223	0.305	0.372
MMSE	1.25×10^{-5}	0.024	0.059	0.091	0.111	0.200	0.273

$$n = 200, s = 16$$

Table 6.3.6 The RL

δ	0	0.2	0.5	0.8	1	2	3
JS	0.556	0.557	0.563	0.572	0.579	0.616	0.648
PP_0	0.430	0.433	0.442	0.458	0.470	0.536	0.596
PP_1	0.556	0.539	0.524	0.517	0.517	0.544	0.589
PP_2	0.454	0.449	0.449	0.457	0.465	0.521	0.580
PP_3	0.446	0.447	0.454	0.468	0.479	0.542	0.599
PP_4	0.454	0.456	0.465	0.480	0.491	0.554	0.609

$$n = 20, s = 4$$

δ	0	0.2	0.5	0.8	1	2	3
JS	0.510	0.512	0.518	0.528	0.536	0.576	0.612
PP_0	0.379	0.382	0.392	0.409	0.422	0.495	0.559
PP_1	0.510	0.493	0.476	0.470	0.470	0.502	0.552
PP_2	0.404	0.399	0.399	0.408	0.416	0.478	0.543
PP_3	0.396	0.397	0.405	0.420	0.432	0.500	0.563
PP_4	0.404	0.407	0.417	0.433	0.445	0.513	0.572

$$n = 100, s = 4$$

δ	0	0.2	0.5	0.8	1	2	3
JS	0.152	0.152	0.152	0.152	0.153	0.156	0.159
PP_0	0.095	0.095	0.096	0.098	0.099	0.109	0.121
PP_1	0.152	0.144	0.134	0.127	0.123	0.115	0.118
PP_2	0.138	0.132	0.124	0.119	0.116	0.111	0.116
PP_3	0.115	0.115	0.115	0.116	0.117	0.125	0.134
PP_4	0.138	0.138	0.139	0.140	0.140	0.146	0.152

n = 80, s = 16

δ	0	0.2	0.5	0.8	1	2	3
JS	0.134	0.134	0.135	0.135	0.136	0.139	0.142
PP_0	0.082	0.082	0.083	0.085	0.086	0.096	0.108
PP_1	0.134	0.127	0.118	0.111	0.107	0.100	0.104
PP_2	0.123	0.117	0.109	0.104	0.101	0.097	0.102
PP_3	0.100	0.100	0.101	0.102	0.103	0.111	0.120
PP_4	0.123	0.123	0.124	0.125	0.125	0.131	0.137

n = 200, s = 16

Figure 6.3.1 contains some examples of plots of the RL of the JS and its positive parts and the ratio of the RL of the positive parts to that of the JS. The behavior in terms of the ordering of the risk functions and the RL, whether the RL increases or decreases as the parameter δ increases is very similar to that of the estimators for the mean of a multivariate normal distribution with known variance. Table 6.3.7 gives the crossing points of the different positive parts and the minimum points of the RL of $PP_1 - PP_3$.

As was the case for the mean of a multivariate normal distribution with known variance and for the average MSE with the value of the prior value replacing δ in the discussion:

1. All of the positive parts dominate the JS; PP_0 and PP_2 dominate PP_4.
2. For smaller values of δ PP_0 has the smallest risk, RL and RLM.
3. For some larger values of δ PP_2 dominates the other positive parts; however for very large values of δ PP_1 is the dominant estimator.

Figure 6.3.2 The RL

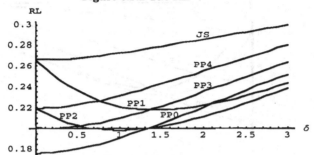

The RL for n = 100 s= 8

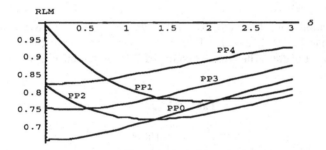

The RLM for n = 100 s = 8

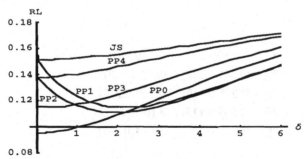

The RL for n = 80 s = 16

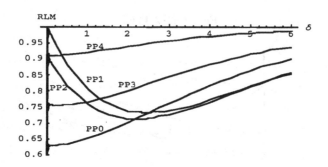

The RLM for n = 80 s = 16

4. The RL and the RLM estimators PP_0 and PP_4 appear to be increasing functions of δ while the RL of PP_1 ,PP_2 and PP_3 first decrease, pass through a minimum point and then increase.

5. When δ = 0 the ordering of the risk, RL and RLM is as follows $PP_0 \le PP_3 \le PP_2 = PP_4 \le PP_1 = JS$.

The minimum values of the RL of PP_1–PP_3 for the special cases in Figure 6.3.1 are given in Table 6.3.7a below. Table 6.3.7b shows the crossing points of PP_1 and the other estimators. The crossing points of PP_2 and the other estimators are given in Table 6.3.7c.

Table 6.3.7a

n,s	80,16	100,8
PP_1	(2.167,0.303)	(1.435,0.424)
PP_2	(1.922,0.283)	(0.987,0.357)
PP_3	(0.214,0.138)	(0.186,0.236)

b

n,s	80,16	100,8
PP_4	(0.367,0.138)	(0.752,0.227)
PP_3	(1.315,0.119)	(1.358,0.218)
PP_0	(2.519,0.115)	(2.1049,0.224)
PP_2	(14.10,0.195)	(17.54,0.397)

c

n,s	80,16	100,8
PP_4	(0,0.138)	(0,0.219)
PP_3	(0.948,0.117)	(0.522,0.202)
PP_0	(2.185,0.111)	(1.347,0.200)
PP_1	(14.69,0.197)	(17.53,0.397)

The Mathematica program given below was used to evaluate the risk, RL and the RLM of the JS and its positive parts. The evaluation was carried out using m = 15, that is fifteen terms of the infinite series were summed. It was found after comparing the results for some special cases to evaluations using up to fifty terms that comparison of the result to three decimal places yielded the same result as for fifteen terms. The reader using the program may use more or fewer values of the series depending on the level of accuracy desired.

Mathematica Program to Evaluate the Risk and the Relative Loss of Estimators in a Single Linear Model

```
a[d_,s_]=Exp[-d]/(2 d^(s/2-1));
b[d_,s_]=Integrate[Exp[x] x^(s/2-2),{x,0,d}];
c[d_,s_]=a[d,s] b[d,s];
mjs[d_,s_,n_]=1-((n-s)/(n-s+2)) (s-2)^2/s c[d,s];
e[s_,n_]=(s-2)/(n-s+2);
h[al_,s_,n_]=al/(al+e[s,n]);
ask[s_,k_]=(s+2 k)/2;
asn[s_,n_]=(n-s)/2;
f[x_,y_,z_,s_,k_,n_]=Beta[z,x,y]/Beta[ask[s,k],asn[s,n]];
a11[s_,k_,n_,al_]=
((n+2 k) f[asn[s,n],ask[s,k]+1,h[al,s,n],s,k,n]
-2 e[s,n] (n+2 k) f[asn[s,n]+1,ask[s,k],h[al,s,n],s,k,n]+
 e[s,n]^2 (n+2 k) f[asn[s,n]+2,ask[s,k]-
1,h[al,s,n],s,k,n]);
b11[s_,k_,n_,al_]=a11[s,k,n,al]-a11[s,k,n,1];
c11[s_,k_,n_,al_]=
(s+2 k-(n+2 k) f[asn[s,n],ask[s,k]+1,h[al,s,n],s,k,n]);
c12[s_,k_,n_,al_]=1-f[asn[s,n],ask[s,k],h[al,s,n],s,k,n];
h1[al_,a2_,s_,k_,n_,al_]=(a11[s,k,n,1]+a1^2 b11[s,k,n,al]
+a2^2 c11[s,k,n,al]);
a21[s_,k_,n_,al_]=(f[asn[s,n],ask[s,k],h[al,s,n],s,k,n]
-e[s,n] f[asn[s,n]+1,ask[s,k]-1,h[al,s,n],s,k,n]);
b21[s_,k_,n_,al_]=(a21[s,k,n,al]-a21[s,k,n,1]);
h2[al_,a2_,s_,k_,n_,al_]=(a21[s,k,n,1]+a1 b21[s,k,n,al]+
a2 c12[s,k,n,al]);
e1[al_,a2_,s_,n_,al_,d_,m_]=Sum[(Exp[-d] (d^k)/k!)*
```

```
h1[a1,a2,s,k,n,al], {k,0,m}];
e2[a1_,a2_,s_,n_,al_,d_,m_]= Sum[(Exp[-d] (d^(k+1))/k! )*
h2[a1,a2,s,k+1,n,al], {k,0,m}];
r[a1_,a2_,s_,n_,al_,d_,m_]=(e1[a1,a2,s,n,al,d,m]
-4 e2[a1,a2,s,n,al,d,m] +2 d)/s;
mms[d_,s_]=1-1/(2 d/s+1);
rl[a1_,a2_,s_,n_,al_,d_,m_]=(r[a1,a2,s,n,al,d,m]
-mms[d,s])/(1-mms[d,s]);
rljs[d_,s_,n_]=(mjs[d,s,n]-mms[d,s])/(1-mms[d,s]);
rlm[a1_,a2_,s_,n_,al_,d_,m_]
=rl[a1,a2,s,n,al,d,m]/rljs[d,s,n];
```

In the above program functions were defined as integrals and the summation of the series is done afterward. This program was used to do the computation and plot the graphs. A program with corresponding statements in the Maple syntax produced the same answer for those cases tested when n and s were small but would not run due to memory problems for intermediate and large values of n. Another Mathematica program that did not use the special functions packages for the incomplete Beta function where the summation of the series was done before the integration was written and for the tested cases gave the same results. The corresponding Maple program works fine on both the Macintosh Performa 6115 and the Macintosh 7200. However it tends to be a bit slow for larger values of n and s but it works both for evaluation and plotting graphs and is thus usable. Most of the computers currently on the market are faster than the ones the author has access to. The reader may find the program very slow if he/she does not have access to a power Macintosh or Pentium PC. The program is given below.

Maple Program to Evaluate the Risk and the Relative Loss of Estimators in a Single Linear Model

```
> a:=(d,s)->exp(-d)/(2*d^(s/2-1)):
> b:=(d,s)->int(exp(x)*x^(s/2-2),x=0..d):
> c:=(d,s)->a(d,s)*b(d,s):
>mjs:=(d,s,n)->1-((n-s)/(n-s+2))*((s-2)^2/s)*c(d,s):
> g:=(s,n)->(s-2)/(n-s+2):
> h:=(al,s,n)->al/(al+g(s,n)):
> ask:=(s,k)->(s+2*k)/2:
> asn:=(s,n)->(n-s)/2:
>w:=(a,b,k,v,s,n)->(v^(a-1))
*(1-v)^(b-1)/Beta(ask(s,k),asn(s,n)):
>f01:=(al,d,m,n,s)
->evalf(int(sum((n+2*k)*w(asn(s,n),ask(s,k)+1,k,v,s,n)
*exp(-d)*(d^k)/k!,k=0..m),v=0..h(al,s,n))):
```

```
>f10:=(al,d,m,n,s)->
evalf(int(sum((n+2*k)*w(asn(s,n)+1,ask(s,k),k,v,s,n)
*exp(-d)*(d^k)/k!,k=0..m),v=0..h(al,s,n))):
>f2n1:=(al,d,m,n,s)->
evalf(int(sum((n+2*k)*w(asn(s,n)+2,ask(s,k)-1,k,v,s,n)
*exp(-d)*(d^k)/k!,k=0..m),v=0..h(al,s,n))):
>f00:=(al,d,m,n,s)
->evalf(int(sum(w(asn(s,n),ask(s,k),k,v,s,n)
*exp(-d)*(d^k)/(k-1)!,k=1..m+1),v=0..h(al,s,n))):
>f1n1:=(al,d,m,n,s)->evalf(int(sum(w(asn(s,n)+1,ask(s,k)-
1,k,v,s,n)
*exp(-d)*(d^k)/(k-1)!,k=1..m+1),v=0..h(al,s,n))):
>a11:=(al,d,m,n,s)->(f01(al,d,m,n,s)-
2*g(s,n)*f10(al,d,m,n,s)
+(g(s,n)^2)*f2n1(al,d,m,n,s)):
>b11:=(al,d,m,n,s)->a11(al,d,m,n,s)- a11(1,d,m,n,s):
> c11:=(al,d,m,n,s)->(s+2*d-f01(al,d,m,n,s)):
> c12:=(al,d,m,n,s)->d-f00(al,d,m,n,s):
>h1:=(a1,a2,al,d,m,n,s)
->(a11(1,d,m,n,s)+(a1^2)*b11(al,d,m,n,s)
 +(a2^2)*c11(al,d,m,n,s)):
>a21:=(al,d,m,n,s)->(f00(al,d,m,n,s)-
g(s,n)*f1n1(al,d,m,n,s)):
>b21:=(al,d,m,n,s)->a21(al,d,m,n,s)
-a21(1,d,m,n,s):
>h2:=(a1,a2,al,d,m,n,s)
->(a21(1,d,m,n,s)+a1*b21(al,d,m,n,s)
+a2*c12(al,d,m,n,s)):
>r:=(a1,a2,al,d,m,n,s)->(h1(a1,a2,al,d,m,n,s)
-4*h2(a1,a2,al,d,m,n,s)+2*d)/s:
> mms:=(d,s)->1-1/(2*d/s+1):
>rl:=(a1,a2,al,d,m,n,s)->(r(a1,a2,al,d,m,n,s)
-mms(d,s))/(1-mms(d,s)):
>rljs:=(d,s,n)->(mjs(d,s,n)-mms(d,s))/
(1-mms(d,s)):
> rsl:=(s,n)->1-((n-s)*(s-2))/((n-s+2)*s):
>rlm:=(a1,a2,al,d,m,n,s)
->rl(a1,a2,al,d,m,n,s)/rljs(d,s,n):
```

Exercise 6.3.7. For n = 100, s = 8 and n = 160, s = 32 compute the values of the risk and the RL for the JS and its positive parts using the above Mathematica program for $\delta = 0$, 0.2, 0.5, 0.8, 1, 2 , 3.

Exercise 6.3.8. Write a Mathematica program along similar lines to the above Maple program.

6.4 The Estimators of Shao and Strawderman

None of the positive part estimators studied in Sections 6.1-6.3 had the smallest Bayes or frequentist risk for all priors or points of the parameter space. Although it was known for a long time that PP_0 was inadmissible, an explicit estimator that dominated it was not available until the work of Shao and Strawderman (1994). The objective of this section is to summarize the work of Shao and Strawderman and to suggest some directions for future research in light of the results already described in this chapter.

Two basic kinds of estimators are studied by Shao and Strawderman. Let $v = \sum_{i=1}^{p} x_i^2$. The first kind of estimator takes the form

$$\hat{\eta}_i = \hat{\theta}_i - \frac{ag(v)}{v} I_{[p-2,p]}(v)x_i, \ 1 \le i \le p. \tag{6.4.1}$$

where $\hat{\theta}_i$ is the estimator PP_0. The function $g(.)$ is an even symmetric piecewise linear function about $v = p-1$ with

$$g(p\text{-}2) = g(p) = 0, \ g'(p-2) < 0 \text{ and } |g'(t)| \ \equiv 1 \text{ a.e on } [p-2,p].$$

Functions with these properties are called "W" shaped because they   are shaped like a W. The full specification of the estimator requires the specification of the constant a and a value p^* in (p-1,p) such that $g(p^*)$ attains its minimum value. There are values of a and p^* where (6.4.1) dominates PP_0. The second class of estimators takes the form

$$\hat{\eta}_i = \hat{\theta}_i - a\left[\frac{g(v)}{v}I_{[q,\infty)}(v) + kh(v)I_{[0,q)}(v)\right]x_i, \ 1 \le i \le p. \tag{6.4.2}$$

Conditions on a, k, q and g will be given where (6.4.2) dominates PP_0.

For each of the two kinds of estimators the conditions for dominance over PP_0 will be given. The results will be stated without proof. The interested reader may consult Shao and Strawderman (1994). A W shaped function is illustrated by Figure 6.4.1 below.

6..4.1 The First Class of Estimators

Before the conditions for the estimator (6.4.1) to dominate PP_0 can be stated some notation and mathematical functions need to be defined. First for $p \geq 3$, let

$$s(t) = 2(p-1) - t \qquad (6.4.3a)$$

and

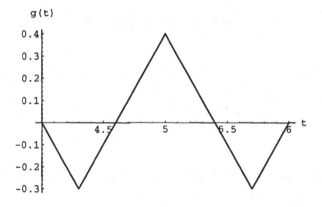

Figure 6.4.1 A W Shaped Function p = 6

$$F_n(t) = \exp\left(-\frac{t}{2}\right) t^{p/2+n-1} - \exp\left(-\frac{s(t)}{2}\right) s(t)^{p/2+n-1} \qquad (6.4.3b)$$

with domain $[p-1, p]$. Shao and Strawderman prove that if $p>2$ and $n \geq j(p)$ where $j(p)$ is the smallest integer that is greater than $(5 + \sqrt{1+6p})/2$ and $t \in [p-1,p]$ then

$F_a''(t) \geq 0$. They also establish the existence of a point $c_0 \in (p-1, p-1+\sqrt{2}/2)$ such that

$$\int_{p-1}^{c_0} F_0(t)dt = \int_{c_0}^{p} F_0(t)dt. \tag{6.4.4}$$

They define a point $c_j \in (p-1, p)$ where

$$\int_{p-1}^{c_j} F_j(t)dt = \int_{c_j}^{p} F_j(t)dt. \tag{6.4.5}$$

Their conditions for (6.4.1) to dominate PP_0 are given by Theorem 6.4.1 below. The theorem is stated without proof.

Theorem 6.4.1. Let $p \geq 3$ and let p^* be any value in the interval $(c_0, \min(c_1, c_2, \ldots, c_{j(p)-1}, p-1+\sqrt{2}/2))$ where

$$b = 1 - \sqrt{2}(p^* - (p-1)) < \min[\tfrac{1}{2}, p - p^*]. \tag{6.4.6}$$

Define

$$g(t) = \begin{cases} t-p, & \text{if } p^* \leq t \leq p, \\ 2p^* - p - t, & \text{if } p-1 \leq t \leq p^*. \end{cases} \tag{6.4.7a}$$

Extend the definition of g(t) to [p-2, p-1) so that g(t) is symmetric about $t = p - 1$. Then

$$g(t) = \begin{cases} t - 3p + 2p^* + 2, & \text{if } 2(p-1) - p^* \leq t \leq (p-1) \\ p - 2 - t, & \text{if } (p-2) \leq 2(p-1) - p^*. \end{cases} \tag{6.4.7b}$$

Figure 6.4.1 is an illustration of the graph of (6.4.7) when $p = 6$ and $p^* = 5.7$. Let

$$A = 1 - \exp(1-b)\left(\frac{p-2+b}{p-b}\right)^{p/2+j(p)-1} \tag{6.4.8}$$

and

$$
B = \min\left\{ \frac{4\int_{p-2}^{p} g'(t)\exp(-t/2)t^{p/2+j-1}dt}{\int_{p-2}^{p} g^2(t)\exp(-t/2)t^{p/2+j-2}dt} \text{ for } j = 0,1,\ldots j(p)-1 \right\}.
$$

(6.4.9)

If

$$
0 < a < \min\{B, 2(p-2)bA\}
$$

(6.4.10)

then the estimator (6.4.1) dominates PP_0.

6.4.2 The Second Class of Estimators

The estimators presented in Section 6.4.1 only change the values of PP_0 on the compact set $\{p - 2 \le \sum_{i=1}^{p} x_i^2 \le p\}$. They are not admissible. Shao and Strawderman also present a class of estimators of the form (6.4.2) that allow for changes in the sum of the squares of the x_i coordinates on the positive real line. They believe but have not proved that this class of estimators contain admissible improvements. The conditions for the existence of these estimators is presented in Theorem 6.4.2 also stated without proof. The interested reader may consult Shao and Strawderman.

Theorem 6.4.2. Let $g(t)$ be a continuous and piecewise differentiable W shaped function defined on $[q,\infty)$ with $g(q) = g(\infty) = 0$ where $p - 2 < q < p - 1$. That means that there exists numbers $t_1, t_2, \ldots, t_6$ where $p-2 < q < t_1 < t_2 < t_3 < t_4 < t_5 < t_6 < \infty$, where

$$
g(t_2) = g(t_4) = 0
$$

(6.4.11a)

and

$$
g'(t) = \begin{cases} \ge 0, & \text{if } t_1 \le t \le t_3 \text{ or } t_5 \le t < \infty, \\ \le 0, & \text{if } q \le t < t_1 \text{ or } t_3 \le t \le t_5. \end{cases}
$$

(6.4.11b)

Let h(t) be a continuous non-positive function on [0,q] with bounded absolute value with h(q) = 0. Let

$$b_j = 2^{p/2+j} j! \Gamma(p/2+j).$$ (6.4.12)

Assume that there exists a positive integer J such that

$$\int_q^{t_6} g'(t) \exp(-t/2) t^{p/2+J+1} dt > 0$$ (6.4.13)

and

$$\int_{t_3}^{t_6} g'(t) \exp(-t/2) t^{p/2+J-1} dt \geq 0.$$ (6.4.14)

Also assume that for all $t \geq t_6$,

$$g^2(t) \leq 4t g'(t).$$ (6.4.15)

Let $B_1 < B_2$ be two constants where if $w \geq 2B_2$ or $0 \leq w \leq 2B_1$ for some integer $N > J$,

$$\int_q^\infty \frac{g'(t)\exp(-t/2)t^{p/2-1}dt}{b_0} + 2\sum_{j=1}^n \frac{\int_q^{t_6} g'(t)\exp(-t/2)t^{p/2+j-1}dt}{b_j}\left(\frac{w}{2}\right)^j \geq 0.$$ (6.4.16)

Assume that

$$k \geq \frac{4 b_1 \sum_{j=1}^{J-1}\int_q^{t_6} g'(t)\exp(-t/2)t^{p/2+j-1}dt[(B_2)^j]/b_j}{B_1 \int_0^{p-2} h(t)\exp(-t/2)t^{p/2}dt},$$ (6.4.17)

$$2\int_q^\infty g'(t)\exp(-t/2)t^{p/2-1}dt$$
$$+k(q-p+2)\int_{p-2}^q h(t)t^{p/2-1}\exp(-t/2)dt > 0,$$ (6.4.18)

and $0 < a < \min\{0.5, A_1, A_2, 1/(qkM), B_1/(kM)\}$. The quantity A_1 is given by

$$A_1 = \frac{2G_1 + k(q-p+2)H_1}{M_1 + k^2N_1} \qquad (6.4.19a)$$

with

$$G_1 = \int_q^\infty g'(t)\exp(-t/2)t^{p/2-1}dt, \qquad (6.4.19b)$$

$$H_1 = \int_{p-2}^q h(t)\exp(-t/2)t^{p/2-1}dt, \qquad (6.4.19c)$$

$$M_1 = \int_q^\infty g(t)^2 \exp(-t/2)t^{p/2-1}dt \qquad (6.4.19d)$$

and

$$N_1 = \int_0^q h(t)^2 \exp(-t/2)t^{p/2}dt. \qquad (6.4.19e)$$

The quantity A_2 is given by

$$A_2 = \frac{4G_2}{4M_2 + N_2} \qquad (6.4.20a)$$

with

$$G_2 = \int_{t_6}^\infty g'(t)\exp(-t/2)t^{p/2-1}dt, \qquad (6.4.20b)$$

$$M_2 = \int_{t_6}^\infty g'(t)\exp(-t/2)t^{p/2-1}dt \qquad (6.4.20c)$$

and

$$N_2 = \int_q^{t_6} g(t)^2 \exp(-t/2)t^{p/2-1}dt. \qquad (6.4.20d)$$

Under the above assumptions and conditions the estimator of the form (6.4.2) dominates PP_0.

Exercise 6.4.1. Show that the estimators PP_1 to PP_4 are of the form (6.4.2) but do not satisfy all of the conditions of Theorem 6.4.2.

Since PP_2 dominates the classical positive part PP_0. for a large portion of the parameter space it would be worthwhile to search for an estimator that uniformly dominates PP_2 and PP_0.

6.5 Summary

In this chapter for the mean of a multivariate normal distribution and for the parameters of a single linear regression model:

1. Five different positive part estimators were defined PP_0 -PP_4.
2. The Bayes risks, Relative Saving Loss, frequentist risk and Relative Loss were evaluated.
3. Some numerical, graphical and, where possible, analytical studies of the estimators were made.
4. In general it was found that although none of the positive part estimators were uniformly better than the others PP_0 and PP_2 were usually the best performers with respect to the measures of efficiency evaluated.

In addition the estimators of Shao and Strawderman that have a uniformly smaller frequentist risk than the classical positive part PP_0 were presented and their properties were summarized.

Chapter VII

The Simultaneous Estimation Problem

7.0 Introduction

The JS has the curious property that all of the variables are used to estimate a single coordinate of a multidimensional parameter. It is indeed remarkable that the resulting estimator has a smaller MSE than the conventional MLE.

Often industrial experiments that are analyzed using linear models are replicated, i.e., done more than once. Simultaneous estimation of the parameters for a setup with r linear models may actually produce a better estimator, i.e., one with a smaller average and conditional MSE. In a manner similar to estimation of the mean of a multivariate normal distribution using all of the variables to estimate a single coordinate estimation of the parameters of one linear model uses the information supplied by all r linear models.

In Chapter IV the EBEs were obtained for prior distributions whose dispersions were known up to a constant. Thus, a single parameter was estimated for the prior dispersion. The setup with r linear models will facilitate estimation of the prior parameters given different degrees of knowledge about them because all of the models will be used in the estimation procedure. For example Wind (1973) formulates an EBE for the case where the relative eigenvalues of X'X and F are unknown. The estimator of Wind may be formulated when F = UDU' with D a diagonal matrix with unknown diagonal elements. C.R. Rao (1975) formulates EBEs when the prior dispersion is completely unknown. The simultaneous estimation problem is indeed a rich one because many different EBEs may be formulated, evaluated and compared.

The simultaneous estimation procedures in this chapter will enable the estimation of an optimal ridge parameter matrix that produces generalized ridge regression estimators of the type considered by Hoerl and Kennard (1971) and C.R.Rao (1975) with uniformly smaller MSE than that of the LS.

Section 7.1 summarizes a portion of C.R.Rao's 1975 Biometrics paper and extends the results therein to the non-full rank case. It includes:

1. the setup with r linear models and the assumptions made about them;
2. the linear BE as the solution to an optimization problem;
3. the formulation of the EBE when the prior mean and dispersion are completely unknown;
4. the computation of the conditional and the average MSE;
5. the formulation of a non-Bayesian approximate MMSE whose mathematical form turns out to be the same as the EBE;
6. the generalized ridge estimator of C.R. Rao;
7. a technique for estimating the ridge parameters of the optimum generalized ridge estimator of C.R.Rao that leads to an estimator with uniformly smaller MSE than LS.

The formulation of the linear BE, EBE, MMSE and approximate MMSE makes no assumption about the form of the population distribution; i.e. whether it is normal, exponential, logistic, etc. However the MSE computations assume a normal population distribution and for the EBE a normal prior distribution.

The estimator of Wind (1973) is formulated from both the Bayesian and the frequentist point of view in Section 7.2. The average and the conditional MSE are evaluated. The estimation of the optimal K matrix for the generalized ridge estimator of Hoerl and Kennard (1970) using all of the linear models is taken up, the resulting estimators are formulated and their MSEs are shown to be uniformly smaller than that of the LS.

The estimators of Chapter IV, i.e., those derived from Dempster's (1973) prior distribution, and their efficiencies are considered in Section 7.3. Unlike Chapter IV the prior mean of the estimators in this chapter need not be zero, the same for all coordinates of the estimated parameters or completely known; they may be formulated even when the prior mean is completely unknown because all r linear models can be used for the estimation process.

Section 7.4 develops a general form of the EBE and a general formula for its MSE averaging over the prior distribution. The estimators considered in Sections 7.1 to 7.3 are special cases of this general form of EBE. The exercises consider the formulation of estimators where the level of prior knowledge is intermediate between that of Dempster and Wind and between that of Wind and Rao. These results and those in Sections 7.1 to 7.3 led the author to the formulation of the general type of EBE.

The properties of the RL and the RSL are taken up in Section 7.5 extending and generalizing the results of Section 4.7.

An outline of how the results about the positive part estimators that were studied in Chapter VI can be generalized to r linear models is taken up in Section 7.6.

Section 7.7 is a summary.

7.1 The EBE of C. R. Rao

7.1.1 The Estimator

The setup to be considered consists of r independent linear models

$$Y_i = X\beta_i + \varepsilon_i \quad 1 \le i \le r. \tag{7.1.1}$$

In each case Y_i is an n dimensional vector of observations. Also X is a fixed nxm matrix of rank $s \le m$ and ε is an m dimensional error vector. It is assumed that has a prior distribution with

$$E(\beta_i) = \theta \text{ and } D(\beta_i) = F. \tag{7.1.2}$$

The mean θ is an m dimensional vector and F is an m x m non-negative definite matrix. Also it will be assumed that

$$E(\varepsilon_i \mid \beta_i) = 0 \text{ and } D(\varepsilon_i \mid \beta_i) = \sigma^2 L \tag{7.1.3}$$

The EBE of C.R. Rao will be obtained for the full rank case. The reader is invited to extend it to the non-full rank case in Exercise 7.1.1. Recall that the linear BE for the full rank case is

$$\begin{aligned}
p'\hat{\beta}_i &= p'\theta + p'FX'(XFX + \sigma^2 I)^{-1}(Y_i - X\theta) \\
&= p'b_i - p'(X'X)^{-1}(F + \sigma^2(X'X)^{-1})^{-1}(b_i - \theta)\sigma^2.
\end{aligned} \tag{7.1.4}$$

Recall that the estimators in (7.1.4) were obtained assuming that σ^2, θ and F were completely known. The EBE methodology, i.e., replacing functions of the prior parameters with sample estimates in the BE, will now be used to formulate an estimator of β_i for the case where σ^2, θ and F are unknown. To estimate the parameters for each of the individual models all r of the linear models will be used. Unbiased estimators of θ and F will be substituted into the last expression in (7.1.4). These unbiased estimators are for σ^2

$$r(n-m)\sigma_*^2 = W = \sum_{i=1}^{r}(Y_i - Xb_i)(Y - Xb_i)', \qquad (7.1.5)$$

for θ

$$b_* = \frac{1}{r}\sum_{i=1}^{r} b_i \qquad (7.1.6)$$

and for F

$$(r-1)(F_* + \sigma_*^2(X'X)^{-1}) = B = \sum_{i=1}^{r}(b_i - b_*)(b_i - b_*)'. \qquad (7.1.7)$$

Thus, the EBE is

$$p'\hat{\beta}_i = p'b_i - cW(X'X)^{-1}B^{-1}(b_i - b_*),\ 1 \le i \le r. \qquad (7.1.8)$$

The constant c will be chosen to minimize the MSE.

Example 7.1.1. A Replicated Industrial Experiment. In Gruber (1990) Example 10.1.1 some data was given for an industrial experiment where the experimental design was that of a one way ANOVA model . In order to compare how closely least square and empirical Bayes estimators approximate the true value of the parameters in a linear regression model and to illustrate a possible application of a setup involving r linear models some data was simulated using Minitab for the one way ANOVA model. The experiment that the data was simulated for was performed as follows. Three different units were chosen from a large number of units of a certain type of camera. For each unit the time from the first flash until the camera's ready light went back on was measured. Six different readings were taken for each camera. The experiment was replicated eight times. The data is given below in Table 7.1.1.

Table 7.1.1

Replicate 1				Replicate 5			
Camera	1	2	3	Camera	1	2	3
	4.98	6.28	5.05		5.01	6.39	4.82
	5.49	6.48	6.52		4.88	5.46	5.17
	4.18	6.30	6.87		5.02	5.81	4.58
	5.49	6.50	6.17		4.52	5.43	5.72
	5.35	5.51	4.69		5.42	5.40	6.41
	4.79	6.84	6.28		4.54	6.30	6.40

Replicate 2				Replicate 6			
Camera	1	2	3	Camera	1	2	3
	5.25	6.11	4.52		4.66	6.15	5.66
	4.70	6.84	3.61		5.05	6.91	5.44
	5.85	5.60	5.43		4.22	5.48	5.43
	4.23	5.98	6.06		5.38	5.54	5.02
	5.41	6.04	5.67		4.59	6.71	6.62
	5.17	6.35	5.52		5.50	6.39	5.35

Replicate 3				Replicate 7			
Camera	1	2	3	Camera	1	2	3
	5.20	4.86	4.70		4.48	7.42	5.85
	5.32	6.73	5.85		4.69	5.61	5.03
	4.54	6.16	6.47		5.37	6.56	7.14
	4.66	5.92	4.82		4.95	5.30	5.27
	4.39	6.59	6.64		5.41	5.47	6.41
	4.29	6.50	5.74		4.78	6.01	5.66

	Replicate 4				Replicate 8		
Camera	1	2	3	Camera	1	2	3
	5.59	5.68	5.44		5.83	6.41	4.94
	4.62	5.78	4.78		5.40	6.65	6.33
	5.02	6.58	5.11		5.14	5.52	5.55
	4.92	5.60	6.15		4.59	6.63	5.70
	4.66	7.06	5.50		4.70	5.53	4.75
	4.14	5.90	5.23		4.48	5.92	5.83

Consider the eight linear models

$$Y_i = (I_3 \otimes 1_6)\beta_i + \varepsilon_i, \ 1 \le i \le 8$$

The LS for each of the replicates is

$$b_1 = \begin{bmatrix} 5.047 \\ 6.138 \\ 5.930 \end{bmatrix} \quad b_2 = \begin{bmatrix} 5.102 \\ 6.153 \\ 5.135 \end{bmatrix} \quad b_3 = \begin{bmatrix} 4.733 \\ 6.127 \\ 5.703 \end{bmatrix} \quad b_4 = \begin{bmatrix} 4.825 \\ 6.100 \\ 5.368 \end{bmatrix}$$

$$b_5 = \begin{bmatrix} 4.898 \\ 5.798 \\ 5.517 \end{bmatrix} \quad b_6 = \begin{bmatrix} 4.900 \\ 6.197 \\ 5.587 \end{bmatrix} \quad b_7 = \begin{bmatrix} 4.947 \\ 6.062 \\ 5.893 \end{bmatrix} \quad b_8 = \begin{bmatrix} 5.023 \\ 6.110 \\ 5.517 \end{bmatrix}$$

Thus,

$$b_. = \begin{bmatrix} 4.934 \\ 6.108 \\ 5.581 \end{bmatrix},$$

$$B = \begin{bmatrix} 0.10355 & 0.03605 & -0.03650 \\ 0.03605 & 0.15262 & 0.06301 \\ -0.03650 & 0.06301 & 0.48675 \end{bmatrix}$$

and

$$B^{-1} = \begin{bmatrix} 11.1973 & -3.1602 & 1.2487 \\ -3.1602 & 7.8140 & -1.2485 \\ 1.2487 & -1.2485 & 2.3097 \end{bmatrix}.$$

Also W = 43.673 is the sum of the within sums of squares for ANOVAs performed on each replicate, c = 3/122 and the EBE of C.R. Rao is:

$$\hat{\beta}_1 = \begin{bmatrix} 4.863 \\ 6.166 \\ 5.808 \end{bmatrix} \quad \hat{\beta}_2 = \begin{bmatrix} 4.892 \\ 6.085 \\ 5.292 \end{bmatrix} \quad \hat{\beta}_3 = \begin{bmatrix} 5.119 \\ 6.014 \\ 5.702 \end{bmatrix} \quad \hat{\beta}_4 = \begin{bmatrix} 5.087 \\ 6.002 \\ 5.479 \end{bmatrix}$$

$$\hat{\beta}_5 = \begin{bmatrix} 4.810 \\ 6.197 \\ 5.482 \end{bmatrix} \quad \hat{\beta}_6 = \begin{bmatrix} 5.018 \\ 6.055 \\ 5.612 \end{bmatrix} \quad \hat{\beta}_7 = \begin{bmatrix} 4.826 \\ 6.203 \\ 5.751 \end{bmatrix} \quad \hat{\beta}_8 = \begin{bmatrix} 4.861 \\ 6.143 \\ 5.524 \end{bmatrix}.$$

7.1.2 The Average MSE

The computation of the average MSE of the EBE of C.R. Rao (7.1.9) assuming a normal prior and population distribution is contained in the proof of Theorem 7.1.1 below. The full rank case will be dealt with; the generalization to the non-full rank case is left to the exercises.

Theorem 7.1.1. If for the setup (7.1.1) with X of full rank:

1. the prior distribution of and the distribution of the are multivariate normal;
2. the number of linear models r > m+2 where m is the number of regression parameters being estimated;
then optimum

$$c = \frac{r - m - 2}{r(n - m) + 2} \tag{7.1.9}$$

and for optimum c the parametric functions have MSE

$$p'Mp = rp'(X'X)^{-1}p\sigma^2$$
$$-\frac{(r-m-2)}{[r(n-m)+2]}p'(X'X)^{-1}(F+\sigma^2(X'X)^{-1})^{-1}(X'X)^{-1}p\sigma^2.$$

$$(7.1.10)$$

Proof. Observe that

$$p'Mp = \sum_{i=1}^{r}p'E(\hat{\beta}_i-\beta_i)(\hat{\beta}_i-\beta_i)'p = \sum_{i=1}^{r}p'E(b_i-\beta_i)(b_i-\beta_i)'p$$
$$-cp'(X'X)^{-1}E\left[B^{-1}\sum_{i=1}^{r}(b_i-b_{\bullet})(b_i-\beta_i)\right]pr(n-m)\sigma^2$$
$$-cp'E\left[\sum_{i=1}^{r}(b_i-\beta_i)(b_i-b_{\bullet})'B^{-1}\right](X'X)^{-1}pr(n-m)\sigma^2$$
$$+c^2p'(X'X)^{-1}E[B^{-1}](X'X)^{-1}p[r(n-m)+2]r(n-m)\sigma^4$$

$$(7.1.11)$$

The first term of (7.1.12) is the MSE of the LS $p'(X'X)^+p\sigma^2$. The second term may be written

$$E\left[\sum_{i=1}^{r}(b_i-\beta_i)(b_i-b_{\bullet})'B^{-1}\right] = I-\sum_{i=1}^{r}E[(\beta_i-b_{\bullet})(b_i-b_{\bullet})'B^{-1}].$$

$$(7.1.12)$$

The LS b_i is a sufficient statistic for β_i (see for Example Graybill (1976). The sufficiency principle (see Lindley (1965) and the discussion in Chapter II) states that the posterior distribution conditioning on a sufficient statistic is the same as that conditioning on the data. Let E_b be the expectation conditional on b where $b = [b_1, b_2, ..., b_r]$. Then

$$\sum_{i=1}^{r} E[(\beta_i - b_.)(b_i - b_.)B^{-1}]$$

$$= \sum_{i=1}^{r} EE_b[(\beta_i - b_.)(b_i - b_.)'B^{-1}]$$

$$= \sum_{i=1}^{r} E[(b_i - \sigma^2(X'X)^{-1}(F + \sigma^2(X'X)^{-1})^{-1}(b_i - \theta))(b_i - b_.)'B^{-1}]$$

$$= I - \sigma^2(X'X)^{-1}(F + \sigma^2(X'X)^{-1})^{-1}.$$

$$(7.1.13)$$

Substituting (7.1.13) into (7.1.12)

$$\sum_{i=1}^{r} E[(\beta_i - b_.)(b_i - b_.)B^{-1}] = \sigma^2(X'X)^{-1}(F + \sigma^2(X'X)^{-1})^{-1}. \quad (7.1.14)$$

Observe that B has a Wishart distribution with r-1 degrees of freedom, that is

$$B \sim W_m((r - 1), F + \sigma^2(X'X)^{-1}) \qquad (7.1.15)$$

Let ℓ be a vector. Thus, from C.R. Rao (1973) p.538

$$\frac{\ell'(F + \sigma^2(X'X)^{-1})^{-1}\ell}{\ell'B^{-1}\ell} \sim \chi^2(r - m). \qquad (7.1.16)$$

From (7.1.14) and (7.1.16)

$$p'E[(X'X)^{-1}B^{-1}(X'X)^{-1}]p$$
$$= (r - m - 2)^{-1}p'(X'X)^{-1}(F + \sigma^2(X'X)^{-1})^{-1}(X'X)^{-1}p\sigma^2. \qquad (7.1.17)$$

The MSE is now obtained by substitution of (7.1.14) and (71.17) and the MSE of p'b into (7.1.11). Thus

$$p'Mp = p'(X'X)^{-1}pr\sigma^2$$
$$-2cp'(X'X)^{-1}(F + \sigma^2(X'X)^{-1})^{-1}(X'X)^{-1}pr(n - m)\sigma^4$$
$$+c^2p'(X'X)^{-1}(F + \sigma^2(X'X)^{-1})^{-1}(X'X)^{-1}p. \qquad (7.1.18)$$
$$\frac{r(n - m)[r(n - m) + 2]}{r(n - m)}\sigma^4$$

To obtain optimum c let

$$h(c) = -2cr(n-m) + \frac{c^2[r(n-m)+2]r(n-m)}{r-m-2} \qquad (7.1.19)$$

Differentiate h(c) in (7.1.19) ,set the derivative equal to zero to obtain optimum c (7.1.10). Substitution of optimum c into (7.1.18) yields (7.1.11).

Exercise 7.1.1. Suppose X is of rank s < m, that is of non-full rank. Let $\gamma=U'\beta$. Reparametize (7.1.1) to the full rank model.
A. Obtain the EBE of $d'\gamma$.
B. Show that the EBE for the estimable parametric functions is

$$p'\hat{\beta}_i = p'b_i - cWp'(X'X)^+(UU'BUU')^+(b_i - b_.). \qquad (7.1.20)$$

C. Write down an expression for the MSE of $d'\hat{\gamma}$.
D. Show that for optimal c the average MSE of the estimable parametric functions is given by

$$p'Mp = rp'(X'X)^+p\sigma^2 - \frac{r(n-s)(r-s-2)}{[r(n-s)+2]}\sigma^4 p'(X'X)^+p \qquad (7.1.21)$$

with $D = UU'FUU' + \sigma^2(X'X)^+$.

Exercise 7.1.2. Let $\theta = 0$, F be an arbitrary non-negative definite matrix.
A. Obtain the BE and the EBE for the full and the non-full rank cases.
B. Obtain the average MSE of the BE for the full and the non-full rank cases.

Exercise 7.1.3. Consider r linear models of the form

$$Y_i = \mu + \varepsilon_i, \ 1 \le i \le r.$$

Assume that $E(\varepsilon_i|\mu) = 0$, $D(\varepsilon_i|\mu) = \sigma^2$, $E(\mu) = \theta$, $D(\mu) = \tau^2$. What is the EBE when θ, σ^2 and τ^2 are all unknown? What is the average MSE of the EBE?

Exercises 7.1.4 and 7.1.5 are based on results of Efron and Morris (1972).

Exercise 7.1.4. Consider r independent problems of estimating a vector with p coordinates. Assume that for all r problems the independent data vectors are

$$x_i \sim N(\theta_i, I_p) \qquad (7.1.22)$$

Assume that the prior assumptions are

$$\theta_i \sim N(0,A). \tag{7.1.23}$$

A. Find the BE $\hat{\delta}^b$.
B. Find the EBE $\hat{\delta}_i^{eb}$.
C. Find the average risk of the BE and the EBE

$$R(A,\hat{\delta}) = E\left[\frac{1}{pr} tr(\hat{\delta} - \theta)(\hat{\delta} - \theta)'\right]. \tag{7.1.24}$$

D. Find the RSL.

Exercise 7.1.5. Replace (7.1.22) with

$$x_i \sim N(\theta_i,D).$$

Using the loss function

$$L(\theta,\hat{\delta}) = \frac{1}{pr}(\hat{\delta} - \theta)'D^{-1}(\hat{\delta} - \theta),$$

repeat Exercise 7.1.4.

Exercise 7.1.6. A. Show that for the EBE of C.R.Rao (7.1.8)

$$RSL = 1 - \frac{r(n-m)(r-m-2)}{r[r(n-m)+2]}. \tag{7.1.25}$$

B. What is the RSL when X is of rank s<m?

Exercise 7.1.7. Consider six linear models of the form

$$Y_i = \begin{bmatrix} 1 & -1 \\ 1 & 0 \\ 1 & 1 \end{bmatrix} \begin{bmatrix} \alpha_i \\ \beta_i \end{bmatrix} + \varepsilon_i, \ 1 \le i \le 6.$$

Suppose the observation vectors are

$$Y_1 = \begin{bmatrix} 1.3 \\ 1.4 \\ 1.1 \end{bmatrix}, \ Y_2 = \begin{bmatrix} 1.6 \\ 1.7 \\ 1.9 \end{bmatrix}, \ Y_3 = \begin{bmatrix} 2.1 \\ 1.9 \\ 1.7 \end{bmatrix},$$

$$Y_4 = \begin{bmatrix} 3.5 \\ 3.7 \\ 3.2 \end{bmatrix}, \ Y_5 = \begin{bmatrix} 2.1 \\ 2.5 \\ 2.0 \end{bmatrix}, \ Y_6 = \begin{bmatrix} 2.0 \\ 2.2 \\ 2.7 \end{bmatrix}.$$

A. Find the numerical value of the LS for each model.
B. Find the numerical value of the EBE of C.R. Rao.

7.1.2 The Approximate MMSE

In Sections 4.2 and 5.5 the linear estimator was found that minimized the variance of the difference between the parametric function and the estimator subject to the bias averaging over the prior being zero. In this section a similar form of optimization problem will be solved from the frequentist point of view. The main difference between the method of this section and that of Section 5.5 is:

1. The MSE will be computed conditional on the β_i parameters.
2. The MSE and the bias is averaged over the β_i parameters instead of the prior distribution.

Once the optimization problem is solved and the MMSE obtained unbiased estimates will be substituted for functions of the unknown parameters. The resulting approximate MMSE will be the same as the EBE (7.1.9). As was done in Section 7.1.1 the full rank case will be considered; the non-full rank case will be left to the exercises.

With this in mind it seems reasonable to solve the optimization problem of minimizing

$$V = \sum_{i=1}^{r} E_\beta (p'\beta_i - a - L'Y_i)^2 \tag{7.1.26}$$

subject to

$$\sum_{i=1}^{r} E_\beta (p'\beta_i - a - L'Y_i) = 0. \tag{7.1.27}$$

From (7.1.27)

$$a = p'\bar{\beta} - L'X\bar{\beta} . \tag{7.1.28}$$

Substituting (7.1.28) into (7.1.26) the optimization problem becomes that of minimizing

$$
\begin{aligned}
V &= \sum_{i=1}^{r} E_\beta [\, p'(\beta_i - \bar{\beta}) - L'(Y - X\bar{\beta})]^2 \\
&= \sum_{i=1}^{r} p'(\beta_i - \bar{\beta})(\beta_i - \bar{\beta})'p - \sum_{i=1}^{r} p'(\beta_i - \bar{\beta})(\beta_i - \bar{\beta})'X'L \\
&\quad - \sum_{i=1}^{r} L'X(\beta_i - \bar{\beta})(\beta_i - \bar{\beta})'p \\
&\quad + \sum_{i=1}^{r} L'[(r-1)\sigma^2 I + X\sum_{i=1}^{r} L'[(r-1)\sigma^2 I + X\sum_{i=1}^{r} (\beta_i - \bar{\beta})(\beta_i - \bar{\beta})X']L
\end{aligned}
$$

$$\tag{7.1.29}$$

To find the optimum estimator, i.e., the MMSE obtain the vector of derivatives and set them equal to zero. Thus, the optimum estimator is

$$
\begin{aligned}
p'\hat{\beta}_i &= p'\bar{\beta} + p'HX'(\sigma^2 I + XHX')^{-1}(Y_i - \bar{Y}) \\
&= p'b_i - (r-1)p'(X'X)^{-1}T^{-1}(b_i - \bar{\beta})\sigma^2
\end{aligned}
\tag{7.1.30}
$$

where

$$H = \frac{1}{r-1} \sum_{i=1}^{r} (\beta_i - \bar{\beta})(\beta_i - \bar{\beta})'$$

and

$$T = (r-1)[(X'X)^{-1}\sigma^2 + H].$$

The unbiased estimators of σ^2, $\bar{\beta}$ and T are given by (7.1.6), (7.1.7) and (7.1.8) respectively. Substitution of (7.1.6), (7.1.7) and (7.1.8) into (7.1.30) yields (7.1.9).

Exercise 7.1.7. Establish the algebraic equivalence of the two expressions in (7.1.30).

Exercise 7.1.8. When X is of non-full rank and $p'\beta_i$ is an estimable parametric function show that

$$\begin{aligned}
p'\hat{\beta}_i &= p'\bar{\beta} + p'HX'(\sigma^2 I + XHX')^{-1}(Y_i - \overline{Y}) \\
&= p'b_i - (r-1)p'(X'X)^+ T^+ (b_i - \bar{\beta})\sigma^2
\end{aligned} \tag{7.1.31}$$

with

$$H = \frac{1}{r-1} \sum_{i=1}^{r} (\beta_i - \bar{\beta})(\beta_i - \bar{\beta})'$$

and

$$T = (r-1)[(X'X)^+\sigma^2 + UU'HUU'].$$

Formulate the approximate MMSE and observe that it takes the same form as (7.1.20).

Exercise 7.1.9. Show that for (7.1.31) with b_* in place of $\bar{\beta}$

$$\text{MSE}(p'\hat{\beta}_i) = rp'(X'X)^+\sigma^2 - (r-1)^2 p'(X'X)^+ T^+ (X'X)^+ p\sigma^4 \tag{7.1.32}$$

7.1.3 The Conditional MSE

To obtain the conditional MSE of the parametric functions $p'\beta_i$ first observe that B is a complete sufficient statistic for $F + \sigma^2(X'X)^{-1}$. Furthermore

$$E(B^{-1}) = \frac{1}{(r-m-2)}(F + \sigma^2(X'X)^{-1})^{-1}. \tag{7.1.33}$$

The conditional MSE of is from (7.1.18)

$$p'Mp = rp'(X'X)^{-1}p\sigma^2$$
$$-2cr(n-m)(r-m-2)p'(X'X)^{-1}E(B^{-1})(X'X)^{-1}p\sigma^4$$
$$+c^2[r(n-m)+2]r(n-m)p'(X'X)^{-1}E(B^{-1})(X'X)^{-1}p\sigma^4.$$

$$(7.1.34)$$

For optimum c the conditional MSE is

$$p'Mp = rp'(X'X)^{-1}p\sigma^2$$
$$-\frac{(r-m-2)^2 r(n-m)}{(r(n-m)+2)}p'(X'X)^{-1}E(B^{-1})(X'X)^{-1}p\sigma^4. \quad (7.1.35)$$

Exercise 7.1.10. Show that optimum c is given by Equation (7.1.9). Hence (7.1.35) follows from (7.1.34).

Exercise 7.1.11. Show that for the non-full rank case the MSE of the estimable parametric functions is given by

$$p'Mp = rp'(X'X)^{+}p\sigma^2$$
$$-\frac{(r-s-2)^2 r(n-s)}{(r(n-s)+2)}p'(X'X)^{+}E(UU'BUU')^{+}(X'X)^{+}p\sigma^4.$$

$$(7.1.36)$$

Exercise 7.1.12. Show that

$$E_\beta\left[B^{-1}\sum_{i=1}^{r}(b_i - b_\bullet)(b_i - \beta_i)'\right] = (r-m-2)E[B^{-1}](X'X)^{-1}.$$

7.1.4 The Ridge Type Estimator

This section will show how to produce a generalized ridge estimator of C.R. Rao with uniformly smaller MSE than LS. Recall that one form of the BE for the full rank model is given by

$$p'\hat{\beta}_i = p'(X'X + \sigma^2 F^{-1})^{-1}(X'Y_i + \sigma^2 F^{-1}\theta). \quad (7.1.37)$$

Let $G = \sigma^2 F^{-1}$ and $\theta = 0$. Then

$$p'\hat{\beta}_i = p'(X'X + G)^{-1}X'Y_i, \quad 1 \leq i \leq r. \tag{7.1.38}$$

This is the generalized ridge regression estimator of C.R.Rao (1975). When $G = kI$ with k a positive constant (7.1.38) is the ordinary ridge regression estimator of Hoerl and Kennard (1970). When $G = UKU'$ with K a positive definite diagonal matrix (7.1.38) becomes the generalized ridge regression estimator of Hoerl and Kennard (1970). For their generalized ridge estimator as was explained in Chapter III Hoerl and Kennard obtain an expression for theoretically optimum K. Analogously for r linear models with $r > s$ a theoretically optimum G may be obtained for the ridge estimator of C.R. Rao.

To determine theoretically optimum G minimize

$$\begin{aligned} M &= \sum_{i=1}^{r} E_\beta (L'Y_i - p'\beta_i)(L'Y_i - p'\beta_i)' \\ &= L'[r\sigma^2 I + X\sum_{i=1}^{r} \beta_i\beta_i'X']L - p'\sum_{i=1}^{r}\beta_i\beta_i'X'L \\ &\quad - L'\sum_{i=1}^{r} X\beta_i\beta_i'p + p'\sum_{i=1}^{r}\beta_i\beta_i'p. \end{aligned} \tag{7.1.39}$$

by obtaining the vector of partial derivatives and setting them equal to zero. Hence

$$p'\hat{\beta}_i = p'\left(\sum_{i=1}^{r}\beta_i\beta_i'X'\right)\left[r\sigma^2 I + X\sum_{i=1}^{r}\beta_i\beta_i'X'\right]^{-1} Y_i. \tag{7.1.40}$$

The derivation of Equation (7.1.40) holds true for both the full and the non-full rank case. When $r > s$ Equation (7.1.40) may be shown to be algebraically equivalent to

$$p'\hat{\beta}_i = p'\left(X'X + r\sigma^2\left(\sum_{i=1}^{r}\beta_i\beta_i'\right)^+\right)^+ X'Y_i. \tag{7.1.41}$$

Thus, optimal G is

$$G = r\sigma^2\left(\sum_{i=}^{r}\beta_i\beta_i'\right)^+. \tag{7.1.42}$$

Two estimates of optimal G are

$$\hat{G} = r\sigma_*^2 \left(\sum_{i=1}^{r} b_i b_i' \right)^+ \tag{7.1.43}$$

and

$$\hat{G} = \left[\frac{c \sum_{i=1}^{r} b_i b_i'}{\sigma_*^2} - (X'X)^+ \right]^+ \frac{1}{\sigma_*^2}. \tag{7.1.44}$$

The estimate in (7.1.43) is obtained by substitution of unbiased estimators of b_i and σ^2 into (7.1.42). The estimate in (7.1.44) is constructed so that if:

1. a form that is algebraically equivalent to (7.1.40) and analogous to (7.1.5) is obtained;

2. the estimate is substituted into the expression obtained in 1 the EBE of C.R.Rao is obtained with $\theta = 0$ and conditional MSE uniformly smaller than that of the LS.

Exercise 7.1.13. Show that when $r > s$, $\sum_{i=1}^{r} \beta_i \beta_i'$ is of full rank, hence nonsingular. What assumptions if any must be made?

Exercise 7.1.14. Verify the algebraic equivalence of (7.1.40) and (7.1.41).

Exercise 7.1.15. Show that the estimate of optimal G does indeed lead to the estimator of C.R. Rao (1975) for $\theta = 0$ with uniformly smaller MSE than LS.

Exercise 7.1.16. Formulate the ridge estimator based on prior information when $\theta = 0$. Obtain optimal G. For which estimator of optimal G does the EBE of C.R.Rao result?

7.2 The EBE of Wind

7.2.1 The EBE

The derivation of the estimator of C.R.Rao assumed that the prior dispersion for the BE was completely arbitrary. The EBE was obtained assuming that the prior was completely unknown. The derivation of the Wind estimator will assume that the prior distribution has a structure similar to the design matrix. The diagonal matrix in the SVD of the prior will be assumed to be unknown in the derivation of the EBE of Wind. The prior dispersion will take the form

$$F = UDU'$$ (7.2.1)

where D is the diagonal matrix with elements $\omega_i{}^2/\lambda_i$. For the case where all the $\omega_i{}^2$ are equal the dispersion (7.2.1) reduces to that of Dempster (1973) considered for one linear model in Chapter IV and to be considered for r linear models in Section 7.3. Unlike Section 7.1, this section will consider the non-full rank case in the text, because the generalization from the full rank model to the non-full rank model in this case is not as straightforward.

The EBE of Wind will be formulated first for the reparametization of a non-full rank model to one of full rank. A transformation of coordinates will then be used to obtain the estimator for the non-full rank case.

To reparametize

$$Y_i = X\beta_i + \varepsilon_i, \ 1 \le i \le r$$ (7.2.2)

to a full rank model rewrite it as

$$Y_i = XU\gamma_i + \varepsilon_i, \ 1 \le i \le r.$$ (7.2.3)

where $\gamma_i = U'\beta_i$. Since $U'X'XU = \Lambda$, the LS is $g_i = \Lambda^{-1}U'X'Y_i$. Thus, from the prior assumptions on β_i

$$E(\gamma_i) = U'\theta = \phi \text{ and } D(\gamma_i) = D.$$ (7.2.4)

From (7.1.3) the BE is for d such that $p = Ud$, i.e., estimable parametric functions,

$$p'\hat{\beta}_i = d'g_i - d'\Lambda^{-1}(D + \sigma^2\Lambda^{-1})^{-1}(g_i - \phi)\sigma^2$$ (7.2.5)

Componentwise

$$\hat{\gamma}_{ij} = g_{ij} - \frac{\sigma^2}{\sigma^2 + \omega_j^2}(g_{ij} - \phi_j)\sigma^2$$ (7.2.6)

where ϕ has the components ϕ_j .

The formulation of (7.2.5) and (7.2.6) made no assumptions about the mathematical form of the prior and the population distribution. Now assume that the β_i and $\epsilon_i \,|\beta_i$ are multivariate normal. Then (Exercise 7.2.1) an unbiased estimator of $d_j = \sigma^2/(\sigma^2 + \omega_j^2)$ is

$$\hat{d}_j = \frac{(r-3)\sigma_*^2}{\lambda_j m_{jj}} \qquad (7.2.7)$$

with

$$m_{jj} = \sum_{i=1}^{r} (g_{ij} - g_{*j})^2 = U_j' B U_j. \qquad (7.2.8)$$

The U_j represent the columns of U. The σ_*^2 is the same as in (7.1.6) and $g_{*j} = U_j' b_*$. The EBE is formulated by replacing functions of unknown prior parameters by sample estimates. Thus d_j is replaced by $\hat{d}_j$ and ϕ_j by g_{*j}. Let $\hat{A}$ be a diagonal matrix with non-zero elements

$$\hat{a}_{jj} = 1 - c\hat{d}_j. \qquad (7.2.9)$$

If $\hat{\beta}_i = U'\hat{\gamma}_i$ then for estimable parametric functions

$$p'\hat{\beta}_i = d'UU'\hat{\gamma}_i = d'\hat{\gamma}_i. \qquad (7.2.10)$$

From (7.2.10) the JS of Wind

$$p'\hat{\beta}_i = p'b_* + p'U\hat{A}U'(b_i - b_*) \qquad (7.2.11)$$

is obtained.

Example 7.2.1. For the data in Example 4.4.1 all of the eigenvalues of $X'X$ are 6 and the diagonal elements of B are 0.1035, 0.1526 and 0.4867. Thus the numerical values of the estimator of Wind are

$$\hat{\beta}_1 = \begin{bmatrix} 4.723 \\ 5.907 \\ 5.716 \end{bmatrix} \quad \hat{\beta}_2 = \begin{bmatrix} 4.620 \\ 6.065 \\ 5.409 \end{bmatrix} \quad \hat{\beta}_3 = \begin{bmatrix} 5.312 \\ 6.090 \\ 5.629 \end{bmatrix} \quad \hat{\beta}_4 = \begin{bmatrix} 5.140 \\ 6.116 \\ 5.499 \end{bmatrix}$$

$$\hat{\beta}_5 = \begin{bmatrix} 5.002 \\ 6.404 \\ 5.556 \end{bmatrix} \quad \hat{\beta}_6 = \begin{bmatrix} 4.999 \\ 6.024 \\ 5.583 \end{bmatrix} \quad \hat{\beta}_7 = \begin{bmatrix} 4.911 \\ 6.152 \\ 5.702 \end{bmatrix} \quad \hat{\beta}_8 = \begin{bmatrix} 4.767 \\ 6.106 \\ 5.556 \end{bmatrix}.$$

Exercise 7.2.1. Show that $\hat{d}$ is unbiased for d.

Exercise 7.2.2. Show that $\lambda_j m_{ii}/(r-1)$ is an unbiased estimator for $\sigma^2 + \omega_j^2$. Use this information to formulate an EBE of the same type as (7.2.11).

Exercise 7.2.3. Formulate the EBE of Wind when the prior mean $\theta = 0$.

Exercise 7.2.4. Let H be any positive definite matrix. Consider the prior assumptions

$$E(\beta_i) = \beta \quad D(\beta_i) = UC'^{-1}DC^{-1}U' \qquad (7.2.13)$$

where D is a positive definite diagonal sxs matrix and C is a nonsingular s x s matrix where $\Lambda = CC'$ and $H = CDC'$.
A. Prove that there exists a C where $\Lambda = C'$ and $H = CDC'$ (see C.R. Rao (1973) p.41).
B. Formulate the BE and the corresponding EBE using the prior assumptions in (7.2.13).

7.2.2 The Average MSE

The average MSE of the estimator of Wind (7.2.11) is obtained in Theorem 7.2.1 below.

Theorem 7.2.1. Consider the setup in (7.1.1) together with assumptions(7.1.3), prior mean and prior dispersion (7.2.1). Assume $r > 3$, X has rank $s \le m$ and that the distributions of   and   are MN. Then for optimum $c = r(n-s)/(r(n-s) + 2)$ the average MSE of (7.2.11) is

$$p'Mp = E\sum_{i=1}^{r} p'(\hat{\beta}_i - \beta_i)(\hat{\beta}_i - \beta_i)'p$$

$$= \mathrm{rp}'(X'X)^+ \mathrm{p}\sigma^2 - \frac{r(n-s)(r-3)}{(r(n-s)+2)} \mathrm{p}'U\Delta U'\mathrm{p}\sigma^4 \qquad (7.2.14a)$$

where the matrix Δ is diagonal with elements

$$\delta_{jj} = \frac{1}{\lambda_j(\sigma^2 + \omega_j^2)}. \qquad (7.2.14b)$$

Proof.(Outline) Some parts of this proof will merely be outlined. The reader is invited to fill in the details in the exercises below. Lemmas 7.2.1 and 7.2.2 will be used in the calculation of the MSE. The reader is invited to prove them in Exercises 7.2.1 and 7.2.2.

Lemma 7.2.1. Let x_i, $1 \le i \le r$ be independent standard normal random variables. Then

$$E\left[\frac{x_i}{\sum_{i=1}^r x_i^2}\right] = 0. \qquad (7.2.15)$$

Lemma 7.2.2. Let x_i be as in Lemma 7.2.1 above. Then

$$E\left[\frac{1}{\sum_{i=1}^r (x_i - \bar{x})^2}\right] = \frac{1}{r-3}. \qquad (7.2.16)$$

Proof of Theorem 7.2.1. Since $\mathrm{p}'\hat{\beta}_i = d'\hat{\gamma}_i$ it suffices to find the MSE of $d'\hat{\gamma}_i$. The diagonal elements of the MDE of $\hat{\gamma}_i$ are

$$e_{ii} = E\sum_{i=1}^r (g_{ij} - \gamma_{ij})^2 - 2c(r-3)E\left[\frac{\sigma_*^2 \sum_{i=1}^r (g_{ij} - g_{\cdot j})(g_{ij} - \gamma_{ij})}{\lambda_j m_{jj}}\right]$$
$$+ \frac{c^2(r-3)^2}{\lambda_j} E\left[\frac{\sigma_*^4}{m_{jj}}\right]. \qquad (7.2.17)$$

The first term consists of the diagonal elements of the MDE of the LS of γ_{ij} and is equal to σ^2/λ_j. The second term is obtained using the sufficiency principle.

Observe that $\sigma_\bullet^2$ and g_{ij} are sufficient statistics for σ^2 and γ_{ij}. Let $g'_{\cdot j} = [g_{1j}, g_{2j}, ..., g_{rj}]$. By the sufficiency principle the j'th diagonal component for the i'th model

$$
\begin{aligned}
\ell_{jj,i} &= E_{\sigma_\bullet^2, g-j} \left[\frac{\sigma_\bullet^2 (g_{ij} - g_{\cdot j})(g_{ij} - \gamma_j)}{m_{jj}} \right] \\
&= \frac{\sigma_\bullet^2 \sigma^2}{\sigma^2 + \omega_j^2} \frac{(g_{ij} - g_{\cdot j})(g_{ij} - \gamma_j^b)}{m_{jj}}
\end{aligned}
\tag{7.2.18}
$$

where the γ_{ij}^b denotes the BE of γ_{ij}. From (7.2.18)

$$
E \left[\sum_{i=1}^{r} \ell_{jj,i} \right] = \frac{\sigma^4}{\sigma^2 + \omega_j^2}.
\tag{7.2.19}
$$

Since $\lambda_j m_{jj} \sim (\sigma^2 + \omega_j^2) \chi^2(r-1)$, from Lemma 7.2.1 it follows that

$$
E \left[\frac{\sigma_\bullet^4}{\lambda_j m_{jj}} \right] = \frac{r(n-s)+2}{r(n-s)(r-3)} \cdot \frac{\sigma^4}{(\sigma^2 + \omega_j^2)\lambda_j}.
\tag{7.2.20}
$$

Substitution of (7.2.19) and (7.2.20) into (7.2.17) yields

$$
\begin{aligned}
e_{jj} &= \frac{\sigma^2}{\lambda_j} - \frac{2c(r-3)\sigma^4}{(\sigma^2 + \omega_j^2)\lambda_j} \\
&\quad + \frac{c^2(r-3)(r(n-s)+2)}{r(n-s)} \frac{\sigma^4}{(\sigma^4 + \omega_j^2)\lambda_j}.
\end{aligned}
\tag{7.2.21}
$$

The off diagonal terms are zero. To establish this:

1. Write the expression for the off diagonal terms analogous to (7.2.17).
2. Notice that for $i \neq j \neq k$ g_{ij} and g_{ik} as well as γ_{ij} and γ_{ik} are sets of independent random variables.
3. Apply the sufficiency principle to the second term of the expression written in 1.
4. Apply Lemma 7.2.2.

Thus,

$$U'MU = r\sigma^2\Lambda^{-1} - 2c(r-3)\Delta\sigma^4$$
$$+\frac{c^2(r-3)(r(n-s)+2)}{r(n-s)}\Delta\sigma^4. \tag{7.2.22}$$

Optimum c is obtained by minimizing

$$h(c) = -2c(r-3) + \frac{c^2(r-3)(r(n-s)+2)}{r(n-s)}. \tag{7.2.23}$$

To obtain (7.2.14) and thus complete the proof of the theorem:

1. Substitute optimum c into (7.2.22).
2. Pre and post multiply (7.2.22) by U and U' respectively.
3. Use the fact that for estimable parametric functions p'UU' = p'.

Exercise 7.2.5. Prove Lemma 7.2.1.

Exercise 7.2.6. Prove Lemma 7.2.2.

Exercise 7.2.7. Fill in the details of the proof that $e_{ij} = 0$.

Exercise 7.2.8. Find optimum c.

Exercise 7.2.9. Prove Theorem 7.2.2 stated below.

Theorem 7.2.2. If the hypothesis in Theorem 7.2.1 assume that $\theta = 0$ and $r > 2$. Then the MSE of the EBE of Wind (see Exercise 7.2.3) is for optimum c

$$p'Mp = p'(X'X)^+p\sigma^2 - \frac{r(n-s)(r-2)}{(r(n-s)+2)}p'U\Delta U'p. \tag{7.2.24}$$

The matrix Δ is the same as in Theorem 7.2.1.

Exercise 7.2.10. Prove: The optimum MSE of the EBE in Exercise 7.2.4 is

$$p'Mp = p'(X'X)^+ p\sigma^2 - \frac{r(n-s)(r-3)}{(r(n-s)+2)} p'UC'^{-1}NC^{-1}U'p. \quad (7.2.25)$$

The matrix N is diagonal with elements

$$n_{ii} = \frac{1}{\sigma^2 + d_i}.$$

7.2.3 The Wind Estimator as an Approximate MMSE

Like the EBE of C.R.Rao, the EBE of Wind can be shown to be an approximate MMSE. The MSE that is minimized is obtained averaging over a quadratic loss function of the form UDU' + VV' with D a non-negative definite diagonal matrix. This includes the individual MSE for the $\hat{\gamma}_i$, the total MSE of the β_i and the predictive loss function, i.e., the cases where D has one non-zero element $1, D = I$ and $D = \Lambda$ respectively. It does not include all of the estimable parametric functions or all of the loss functions.

The estimator is assumed to have the form

$$p'\hat{\beta} = p'UU'b_\bullet + p'UAU'(b_i - b_\bullet) \qquad (7.2.26)$$

with A a diagonal matrix with elements a_i . Since $U'b_i = g_i$ and $U'b_\bullet = g_\bullet$,

$$\hat{\gamma}_i = g_\bullet + A(g_i - g_\bullet) = U'\hat{\beta}. \qquad (7.2.27)$$

The MMSE is found by minimizing

$$M = \sum_{i=1}^{r} E_\beta (\hat{\beta}_i - \beta_i)'UDU'(\hat{\beta}_i - \beta_i) = \sum_{i=1}^{r} E_\beta (\hat{\gamma}_i - \gamma_i)'D(\hat{\gamma}_i - \gamma_i)$$

$$= \sum_{i=1}^{r} E_\beta (g_\bullet - \gamma_i)'D(g_\bullet - \gamma_i) + \sum_{i=1}^{r} E_\beta (g_i - g_\bullet)'AD(g_\bullet - \gamma_i)$$

$$= \sum_{i=1}^{r} E_\beta (g_\bullet - \gamma_i)'DA(g_i - g_\bullet)' + \sum_{i=1}^{r} E_\beta (g_i - g_\bullet)'ADA(g_i - g_\bullet).$$

$$(7.2.28)$$

Computation of the expectations in Equation (7.2.28) yields

$$M = \sum_{i=1}^{r} (\gamma_i - \bar{\gamma})'D(\gamma_i - \bar{\gamma}) + tr\Lambda^{-1}D\sigma^2$$

$$-2\sum_{i=1}^{r} (\gamma_i - \bar{\gamma})'AD(\gamma_i - \bar{\gamma})$$

$$+(r-1)tr\Lambda^{-1}ADA\sigma^2 + \sum_{i=1}^{r} (\gamma_i - \bar{\gamma})'ADA(\gamma_i - \bar{\gamma}).$$

(7.2.29)

The minimization of M consists of obtaining the partial derivative of M with respect to the a_j and setting them equal to zero and solving the resulting equations. Thus,

$$a_j = 1 - \frac{(r-1)\sigma^2}{\lambda_j U'_j TU_j}, \ 1 \leq j \leq s. \tag{7.2.30}$$

Let A_m be the sxs diagonal matrix with elements in (7.2.30). The MMSE takes the form of (7.2.26) with A_m in place of A. Now $\lambda_j U'_j BU_j$ is an unbiased estimator of $\lambda_j U'_j TU_j$. Thus a_j is estimated by

$$a_j = 1 - \frac{(r-1)\sigma_*^2}{\lambda_j U'_j BU_j}, \ 1 \leq j \leq s. \tag{7.2.31}$$

Let $\hat{A}$ be the s x s diagonal matrix with elements $\hat{a}_j$. The EBE of Wind is (7.2.26) with $\hat{A}$ in place of A.

Exercise 7.2.11. Verify (7.2.29) by computing the expectations in (7.2.28).

Exercise 7.2.12. Show that the MSE of the MMSE is

$$M = r\sigma^2 tr(X'X)^+ - \sum_{j=1}^{r} \frac{(r-1)^2 d_j \sigma^4}{\lambda_j U'_j TU_j}. \tag{7.2.32}$$

Exercise 7.2.13. Verify that M in (7.2.29) is minimized by a_j in (7.2.30).

Exercise 7.2.14. Verify that $\lambda_j U'_j BU_j$ is an unbiased estimator of $\lambda_j U'_j TU_j$.

Exercise 7.2.15. A. Find the MMSE of the form

$$p'\hat{\beta}_i = p'UAU'b \tag{7.2.23}$$

where A is a positive definite diagonal matrix.
B. Show that for the approximate MMSE

$$\hat{a}_j = 1 - \frac{(r-2)\sigma_*^2}{\lambda_j U_j B_0 U'_j} \tag{7.2.34}$$

with $U_j B_0 U' = \sum_{i=1}^{r} g_{ij}^2$. This approximate MMSE is the Wind estimator with shrinkage to the origin.

7.2.4 An Optimal Ridge Regression Estimator

In Section (7.1.4) for the generalized ridge regression estimator of C.R. Rao (1975) it was shown how:

1. an optimal G could be obtained;
2. an estimator of G could be produced such that the conditional MSE of the resulting EBE was uniformly smaller than LS.

The most important feature of this estimation procedure is that all r linear models were used in constructing the EBE. The goal of this section is to solve the same problem for the generalized ridge estimator of Hoerl and Kennard (1970).

Recall that the generalized ridge estimator of Hoerl and Kennard (1970) is the special case of the generalized ridge estimator of C.R. Rao (1975) with G = UKU' with K a positive definite diagonal matrix. For the estimable parametric functions, i.e., those p such that p' = d'U'

$$\begin{aligned} p'\hat{\beta} &= p'(X'X + UKU')^+ X'Y = p'(U(\Lambda + K)U')^+ UU'X'Y \\ &= p'U(\Lambda + K)^{-1}U'X'Y = d'(\Lambda + K)^{-1}\Lambda\Lambda^{-1}U'X'Y \tag{7.2.35} \\ &= d'(\Lambda + K)^{-1}\Lambda g = d'\hat{\gamma}. \end{aligned}$$

Coordinatewise

$$\hat{\gamma}_j = \frac{\lambda_j}{\lambda_j + k_j} g_i = \left(1 - \frac{k_j}{\lambda_j + k_j}\right) g_j, \ 1 \leq j \leq s. \tag{7.2.36}$$

Hoerl and Kennard (1970) showed that for a single linear model optimum $k_j = \sigma/\gamma_j$, i.e., the k_j that minimized the individual MSE of the γ_j and suggest $\hat{k}_j = \hat{\sigma}^2/g_j$ as an estimate of optimal k_j . Simulation studies by Hoerl, Kennard and Baldwin (1975) show that the ridge estimator usually has a smaller MSE than the LS. Exact and approximate expressions for the MSE are very cumbersome. When

$$\hat{k}_j = \frac{\hat{c}_j \lambda_j}{1 - \hat{c}_j} \tag{7.2.37a}$$

with

$$\hat{c}_j = \frac{r(n-s)(r-2)\sigma_*^2}{(r(n-s)+2)\lambda_j U_j B_0 U_j'} \tag{7.2.37b}$$

the generalized ridge estimator is the same as the estimator of Wind with $\theta = 0$. It will be shown in Section 7.2.5 that the conditional MSE of the ridge estimator with estimated by (7.2.37) has smaller MSE than LS averaging over quadratic loss functions of the form $UDU' + VV'$. As was pointed out in Section (7.2.3) this includes the total MSE and the MSE of prediction.

Exercise 7.2.16. Let $\phi = U'\theta$.
A. What is the form of the ridge estimator of C.R. Rao for $\theta \neq 0$?
B. Show that if $\theta \neq 0$ the generalized ridge estimator of Hoerl and Kennard takes the form

$$\hat{\gamma}_j = \phi_j + \left(1 - \frac{k_j}{\lambda_j + k_j}\right)(g_j - \phi_j), \ 1 \leq j \leq s \tag{7.2.38}$$

C. What choice of k_j produces the estimator of Wind?

Exercise 7.2.17. A. Show that for a single linear model $k_j = \sigma^2/\gamma_j^2$ minimizes

$$m_j = E(\hat{\gamma}_j - \gamma_j)^2, \ 1 \leq j \leq s.$$

B. Find k_j that minimizes

$$m_j = \sum_{i=1}^{r} E(\hat{\gamma}_{ij} - \gamma_{ij})^2$$

i.e., optimum k_j for r linear models.

7.2.5 The Conditional MSE

The conditional MSE of the estimable parametric functions is given by Theorem 7.2.2. Corollary 7.2.1 gives the MSE for averaging over quadratic loss functions with weight matrix of the form $\Theta = UDU' + VV'$. Recall that the total loss function and the predictive loss function are included in the class of quadratic loss functions considered on Corollary 7.2.1.

Theorem 7.2.2. The MSE of the estimable parametric functions of the estimator of Wind is given by

$$M = rp'(X'X)^+ p\sigma^2 - 2c(r-3)p'UTU'p\sigma^2$$
$$+ \frac{c^2(r-3)^2(r(n-s)+2)}{r(n-s)}[p'UTU'p\sigma^2 + p'UNU'p]. \qquad (7.2.39)$$

Define

$$\delta_{\cdot j} = \sum_{i=1}^{r} (\gamma_{ij} - \bar{\gamma}_{\cdot j})^2 \frac{\lambda_j}{2\sigma^2} \text{ with } \bar{\gamma}_{\cdot j} = \frac{1}{r}\sum_{i=1}^{r} \gamma_{ij.} \qquad (7.2.40)$$

The matrix T is diagonal with elements

$$t_j = \frac{1}{\lambda_j} S_{1,r-1}(\delta_{\cdot j}), \ 1 \le j \le s. \qquad (7.2.41)$$

The notation for $S_{1,r-1}(\delta_j)$ was defined in (4.5.11). The matrix N has zero diagonal elements and off diagonal elements

$$\eta_{jt} = \sum_{i=1}^{r}(\gamma_{ij} - \bar{\gamma}_{\cdot j})(\gamma_{it} - \bar{\gamma}_{\cdot t})S_{1,r+1}(\delta_{\cdot j})S_{1,r+1}(\delta_{\cdot t}) \qquad (7.2.42)$$

The proof of the theorem is aided by Lemma 7.2.3. The method of proof of Lemma 7.2.3 will be helpful in Chapter IX.

Lemma 7.2.3. Let v_{ij}, $1 \le i \le p$, $1 \le j \le p$ be independent random variables with

$$v_{ij} \sim N(\mu_{ij},1). \qquad (7.2.43)$$

Then

$$E\left[\frac{v_{ij}}{v'_{-j}v_{-j}}\right] = \mu_{ij}S_{1,p+2}(\delta_{\cdot j}) \qquad (7.2.44)$$

with

$$v_{-j} = [v_{1j}, v_{2j}, \cdots, v_{pj}] \text{ and } \delta_{\cdot j} = \sum_{i=1}^{r}\mu_{ij}^2.$$

Proof. Observe that

$$E\left[\frac{v_{ij}}{v'_{-j}v_{-j}}\right] = E\left[\frac{v_{ij} - \mu_{ij}}{v'_{-j}v_{-j}}\right] + E\left[\frac{\mu_{ij}}{v'_{-j}v_{-j}}\right]. \qquad (7.2.45)$$

From (7.6.12)

$$E\left[\frac{\mu_{ij}}{v'_{-j}v_{-j}}\right] = \mu_{ij}S_{1,p}(\delta_{\cdot j}). \qquad (7.2.46)$$

Let $dv_{-j} = \prod_{i=1}^{p} dv_{ij}$. Now

$$E\left[\frac{1}{v'_{-j}v_{-j}}\right] = \frac{1}{(2\pi)^{p/2}}\int_{R^p}\frac{1}{v'_{-j}v_{-j}}\exp\left[-\frac{1}{2}(v_{-j} - \mu_{-j})'(v_{-j} - \mu_{-j})\right]dv_{-j}$$

$$\qquad (7.2.47)$$

is a p dimensional integral over R^p. Differentiating (4.2.47) under the integral sign with respect to μ_{ij} yields

$$E\left[\frac{v_{ij} - \mu_{ij}}{v'_{-j} v_{-j}}\right] = \frac{\partial}{\partial \mu_{ij}} E\left[\frac{1}{v'_{-j} v_{-j}}\right] = -\mu_{ij} S_{1,p}(\delta_{\cdot j}) + \mu_{ij} S_{1,p+2}(\delta_{\cdot j}). \quad (7.2.48)$$

Substitute (7.2.48) into (7.2.45) to obtain (7.2.44).

The proof of Theorem 7.2.2 consists of obtaining the elements of the MDE for γ_i. The diagonal elements are

$$n_{jj} = E_\beta\left[\sum_{i=1}^r (g_{ij} - \gamma_{ij})^2\right]$$

$$-\frac{2c(r-3)\sigma^2}{\lambda_j} E_\beta\left[\frac{\sum_{i=1}^r (g_{ij} - \gamma_{\cdot j})(g_{ij} - \gamma_{ij})}{m_{jj}}\right] \quad (7.2.49)$$

$$+\frac{c^2(r-3)^2(r(n-s)+2)\sigma^4}{r(n-s)\lambda_j} E\left[\frac{1}{m_{jj}}\right]$$

with $m_{jj} = \sum_{i=1}^r (g_{ij} - g_{\cdot j})^2$. The first term of (7.2.49) is σ^2/λ_j. An orthogonal transformation using a Helmert matrix will be helpful in obtaining the remaining terms in (7.2.49). The Helmert matrix is

$$H_0 = \begin{bmatrix} \dfrac{1}{\sqrt{2}} & -\dfrac{1}{\sqrt{2}} & 0 & \cdots & 0 \\ \dfrac{1}{\sqrt{6}} & \dfrac{1}{\sqrt{6}} & -\dfrac{2}{\sqrt{6}} & \cdots & \vdots \\ \vdots & \vdots & & \cdots & \vdots \\ \dfrac{1}{\sqrt{r(r-1)}} & \dfrac{1}{\sqrt{r(r-1)}} & \dfrac{1}{\sqrt{r(r-1)}} & \cdots & \dfrac{-(r-1)}{\sqrt{r(r-1)}} \end{bmatrix}. \quad (7.2.50)$$

Let

$$\theta_j = H_0 \gamma_{-j} \text{ and } \theta_j = H_0 g_{-j}. \quad (7.2.51)$$

Then

$$m_{jj} = \sum_{i=1}^{r} \phi_{ij}^2. \tag{7.2.52}$$

Also

$$\gamma_{-j} - \gamma_{*j} = H_0'\theta_{-j} \tag{7.2.53a}$$

and

$$g_{-j} - g_{*j} = H_0'\phi_{-j}. \tag{7.2.53b}$$

Using Lemma 7.2.3

$$E\left[\frac{\sum_{i=1}^{r}(g_{ij} - g_{*j})(g_{ij} - \gamma_{ij})}{m_{jj}}\right] = E_\beta\left[\frac{\sum_{i=1}^{r-1}\phi_{ij}(\phi_{ij} - \theta_{ij})}{m_{jj}}\right]$$

$$= \frac{(r-3)\sigma^2}{\lambda_j}E\left[\frac{1}{m_{jj}}\right] \tag{7.2.54}$$

$$= (r-3)S_{1,r-1}(\delta_{\cdot j})\sigma^2.$$

Substituting (7.2.44) and (7.2.54) into (7.2.49)

$$n_{jj} = \frac{r\sigma^2}{\lambda_j} - \frac{2c(r-3)^2}{\lambda_j}S_{1,r-1}(\delta_{\cdot j})\sigma^2 + \frac{(r-3)^2(r(n-s)+2)}{\lambda_j r(n-s)}S_{1,r-1}(\delta_{\cdot j})\sigma^2. \tag{7.2.55}$$

The off diagonal terms are

$$n_{jt} = E_\beta\sum_{i=1}^{r}(g_{ij} - \gamma_{ij})(g_{it} - \gamma_{it})$$

$$- \frac{c(r-3)\sigma^2}{\lambda_j}E_\beta\left[\frac{\sum_{i=1}^{r}(g_{ij} - g_{*j})(g_{it} - \gamma_{it})}{m_{jj}}\right] \tag{7.2.56}$$

$$- \frac{c(r-3)\sigma^2}{\lambda_t}E_\beta\left[\frac{\sum_{i=1}^{r}(g_{it} - g_{*t})(g_{ij} - \gamma_{it})}{m_{tt}}\right]$$

$$+\frac{c^2(r-3)^2[r(n-s)+2]\sigma^4}{\lambda_j\lambda_t r(n-s)}E\left[\frac{\sum_{i=1}^r(g_{ij}-g_{\cdot j})(g_{it}-g_{\cdot t})}{m_{jj}m_{tt}}\right].$$

The reader is invited to show that the first three terms of (7.2.56) are zero. Now

$$\sum_{i=1}^r(g_{ij}-g_{\cdot j})(g_{it}-g_{\cdot t})=\sum_{a=1}^{r-1}\phi_{aj}\phi_{at}. \tag{7.2.57}$$

From Lemma 7.2.3 and (7.2.57)

$$E_\beta\left[\frac{\sum_{i=1}^r(g_{ij}-g_{\cdot j})(g_{it}-g_{\cdot t})}{m_{jj}m_{tt}}\right]$$

$$=\sum_{i=1}^{r-1}E_\beta\left[\frac{\phi_{aj}}{\phi'_{-j}\phi_{-j}}\right]E_\beta\left[\frac{\phi_{at}}{\phi'_{-t}\phi_{-t}}\right] \tag{7.2.58}$$

$$=\sum_{i=1}^{r-1}\frac{\theta_{aj}\theta_{at}\lambda_j\lambda_t}{\sigma^4}S_{1,r+1}(\delta_{\cdot j})S_{1,r+1}(\delta_{\cdot t})$$

$$=\frac{\lambda_j\lambda_t}{\sigma^4}\sum_{a=1}^r(\gamma_{aj}-\bar\gamma_{\cdot j})(\gamma_{aj}-\bar\gamma_{\cdot t})S_{1,r+1}(\delta_{\cdot j})S_{1,r+1}(\delta_{\cdot t}).$$

The result (7.2.39) then follows since $p'\hat\beta=d'\hat\gamma$ for estimable parametric functions.

From the expression (7.2.39) it appears that it is going to be difficult to make analytical comparisons of the MSE of the parametric functions $p'\hat\beta$ of the estimator of Wind and the LS. However the individual MSE, the total MSE and the predictive MSE all have a smaller MSE than LS. This is an easy consequence of Corollary 7.2.1.

Corollary 7.2.1. Let D be any non negative definite diagonal sxs matrix. Let s>3. Let $\Theta=UDU'$. For optimum $c=r(n-s)/(r(n-s)+2)$ the MSE is

$$m=\sum_{i=1}^r E_\beta(\hat\beta_i-\beta_i)'\Theta(\hat\beta_i-\beta_i)=r\sigma^2\mathrm{tr}(X'X)^+\Theta$$

$$-\frac{(r-3)^2r(n-s)\sigma^2}{(r(n-s)+2)}\sum_{j=1}^s\frac{d_j}{\lambda_j}S_{jj}(\delta_{\cdot j}). \tag{7.2.59}$$

The reader is invited to prove this corollary in Exercise 7.2.20. Since $V'\hat{\beta} = 0$ the result of Corollary 7.2.1 also holds for quadratic loss functions with matrix $UDU' + VV'$.

Exercise 7.2.18. A. Show that $H_0 H_0' = I$ and $H_0' H_0 = I - (1/r)J$ where J is a matrix of ones.
B. Verify (7.2.52) and (7.2.53).

Exercise 7.2.19. Establish (7.2.48).

Exercise 7.2.20. Prove Corollary 7.2.1.

Exercise 7.2.21. Show that the Wind estimator with prior mean zero has MSE averaging over the quadratic loss function with matrix $\Theta = UDU' + VV'$ is

$$M = r\sigma^2 \text{tr}(X'X)^+ \Theta - \frac{(r-2)^2 r(n-s)\sigma^2}{(r(n-s)+2)} \sum_{j=1}^{s} \frac{d_j}{\lambda_j} S_{1,r}(\delta_{\cdot j}). \qquad (7.2.60)$$

This is the MSE of the generalized ridge estimator of Hoerl and Kennard using the estimate of k_i in Equation (7.2.37).

7.3 The Estimator of Dempster

The estimator of Dempster will now be formulated for the case of r linear models. The form of the prior dispersion is the same as the one model case; however the prior mean may be completely arbitrary because all of the r linear models can be used to estimate it. Likewise the MSE minimized to obtain the MMSE, approximated by the JS of Dempster, is obtained averaging or summing over the r linear models.

7.3.1. The Empirical Bayes Formulation

The prior assumptions for the setup (7.1.1) are

$$E(\beta_i) = \theta \text{ and } D(\beta_i) = \omega^2 (X'X)^+. \qquad (7.3.1)$$

The resulting BE is

$$p'\hat{\beta}_i = p'b_i - \frac{\sigma^2}{\sigma^2 + \omega^2}(b_i - \theta), \quad 1 \leq i \leq r. \tag{7.3.2}$$

The estimators of θ and σ^2 are given by (7.1.6) and (7.1.7) respectively. The unbiased estimator of $\delta = \sigma^2 + \omega^2$ is given by

$$\hat{\delta} = \text{tr}BX'X / s(r-1) \tag{7.3.3}$$

where B is given in Equation (7.1.8). Following the methodology of the preceding sections the EBE is given by

$$p'\hat{\beta}_i = p'b_i - \frac{c(r-1)s\sigma_*^2}{\text{tr}BX'X}p'(b_i - b_*). \tag{7.3.4}$$

The constant c is such that the MSE of $p'\beta_i$ is minimized. The estimator (7.3.4) will be called the estimator of Dempster. The case where $\theta = 0$ is considered in Exercise 7.3.1 below.

Two important differences between the estimator of Wind and the estimator of Dempster are:

1. Estimation of each coordinate of $U'\beta_i$ for the Dempster estimator uses all of the other coordinates while the Wind estimator uses only the coordinate in question to estimate a given coordinate of $U'\beta_i$.

2. In general the estimator of Wind cannot be formulated for fewer than four linear models (three linear models if the prior mean is known to be zero) but the Dempster JS can be formulated for a single linear model with zero prior mean as long as s >2 and s>3 for the case of a prior mean with constant canonical coordinates.

Both the Wind and the Dempster estimator are formulated using all r linear models.

Example 7.3.1. Consider the data from Example 7.1.1. Observe that $\text{TrBX'X} = 4.457$. The numerical values of the estimator of Dempster are given by

$$\hat{\beta}_1 = \begin{bmatrix} 4.875 \\ 5.998 \\ 5.398 \end{bmatrix} \quad \hat{\beta}_2 = \begin{bmatrix} 4.846 \\ 6.084 \\ 5.816 \end{bmatrix} \quad \hat{\beta}_3 = \begin{bmatrix} 5.040 \\ 6.098 \\ 5.517 \end{bmatrix} \quad \hat{\beta}_4 = \begin{bmatrix} 4.992 \\ 6.112 \\ 5.693 \end{bmatrix}$$

$$\hat{\beta}_5 = \begin{bmatrix} 4.953 \\ 6.271 \\ 5.615 \end{bmatrix} \quad \hat{\beta}_6 = \begin{bmatrix} 4.952 \\ 6.062 \\ 5.578 \end{bmatrix} \quad \hat{\beta}_7 = \begin{bmatrix} 4.928 \\ 6.132 \\ 5.417 \end{bmatrix} \quad \hat{\beta}_8 = \begin{bmatrix} 4.889 \\ 6.107 \\ 5.615 \end{bmatrix}.$$

Remark. The data in Example 7.3.1 was simulated data; the simulations were done using Minitab. Each of the parameters were assumed to be normally distributed with means 5,6,5.5 respectively. One possible measure of goodness of estimates is their distance from the true parameter value usually unknown in practice. One distance measure is

$$d = \sum_{i=1}^{8} (\hat{\beta}_i - \beta_i)' X' X (\hat{\beta}_i - \beta_i). \qquad (7.3.5)$$

The distances are

LS	EBE of C.R.Rao	EBE of Wind	EBE of Dempster
5.542	3.197	4.555	2.317

When the true parameter values are unknown β_i may be replaced by β_* in (7.3.5). Clearly all of the EBEs appear to be better predictors than the LS, for this particular case the best is the EBE of Dempster, the next best the EBE of C.R. Rao and the poorest the EBE of Wind.

Exercise 7.3.1. Show that if the prior mean is assumed to be zero, i.e., $\theta = 0$ then the estimator of Dempster for the setup (7.1.1) is

$$p'\hat{\beta}_i = p'b_i - \frac{c(rs - 2)\sigma_*^2}{tr B_0 X' X} p'b_i \qquad (7.3.6)$$

with $B_0 = \sum_{i=1}^{r} b_i b_i'$.

Exercise 7.3.2. Show that if the prior and population distributions are assumed to be normal an unbiased estimator of $\sigma^2 / (\sigma^2 + \omega^2)$ is $((r-1)s - 2)\sigma_*^2 / tr B X' X$. Thus the JS of Dempster can be formulated with $(r-1)s - 2$ in place of $(r-1)s$.

Exercise 7.3.3. For the prior assumptions

$$E(\beta) = \theta \text{ and } D(\beta) = (\sigma^2 + \omega^2)(X'X)^+ \Sigma(X'X)^+ - \sigma^2(X'X)^+$$

obtain the BE and the EBE corresponding to (7.2.11) in the context of r linear models.

Exercise 7.3.4. Show that if the Moore Penrose inverse $(X'X)^+$ is replaced in Dempster's prior by some other generalized inverse the same JS could be derived.

Exercise 7.3.5. Replace β_i by β_* in (7.3.5). Obtain the "distances" for the LS and the three different kinds of EBEs in Example 7.3.1 and compare.

The data for the following problem was obtained by the courtesy of Dr. James Halavin, Professor Department of Mathematics and Statistics at the Rochester Institute of Technology.

Exercise 7.3.6. The following data are coded diameters of rivets measured on three shifts. Four measurements are taken on each shift. The experiment is repeated 10 times. For each data set perform the one way ANOVA and obtain the LS of $\mu + \alpha_i$, the estimator of C.R. Rao, Wind and Dempster. Estimate their distance from the "true" parameter value using the method suggested in Exercise 7.3.5.

Shift 1				Shift 6		
1	2	3		1	2	3
206	205	208		206	202	208
206	207	208		206	202	208
202	203	206		207	200	203
202	203	206		208	201	203

Shift 2				Shift 7		
1	2	3		1	2	3
210	200	199		204	203	206
211	201	200		205	203	207
201	199	202		205	205	204
203	201	204		205	205	205

	Shift 3			Shift 8	
1	2	3	1	2	3
204	207	203	207	206	202
204	207	205	207	205	202
200	206	207	207	209	202
201	207	205	207	209	200

	Shift 4			Shift 9	
1	2	3	1	2	3
206	202	205	201	204	204
204	202	206	202	204	205
204	204	204	202	201	204
204	204	206	205	202	204

	Shift 5			Shift 10	
1	2	3	1	2	3
207	206	202	204	201	202
206	207	203	204	203	203
202	203	203	205	200	203
204	203	202	201	202	202

7.3.2 The Non- Bayesian Formulation

The development of the JS of Dempster as an approximate MMSE is along the same lines as Section 4.3 with two important differences:

1. Shrinkage is toward b_*.
2. The MSE to be minimized is obtained summing over r linear models.

The optimization problem is that of finding an estimator of the form

$$p'\hat{\beta}_i = p'b_* + (1-h)p'(b_i - b_*) \tag{7.3.7}$$

where h is a scalar such that

$$M = \sum_{i=1}^{r} E_\beta (\hat{\beta}_i - \beta_i)' X'X (\hat{\beta}_i - \beta_i) \tag{7.3.8}$$

is minimized. Substituting (7.3.8) into (7.3.7)

$$M = r s\sigma^2 - 2h(r-1)s\sigma^2 + h^2 trT(X'X) \tag{7.3.9}$$

where T was defined in (7.1.30). Differentiating (7.3.9) and setting the result equal to zero yields optimal

$$h = \frac{\sigma^2 (r-1)s}{trTX'X}. \tag{7.3.10}$$

Substitution of (7.3.10) into (7.3.7) yields the MMSE estimator

$$p'\hat{\beta} = p'b_* + \left(1 - \frac{\sigma^2 (r-1)s}{trTX'X}\right)(b_i - b_*). \tag{7.3.11}$$

When the expectation is obtained conditional on β, σ_*^2 is an unbiased estimator of σ^2 and δ in Equation (7.3.3) is an unbiased estimator of trBX'X. Thus the approximate MMSE is

$$p'\hat{\beta} = p'b_* + \left(1 - \frac{c(r-1)s\sigma_*^2}{trBX'X}\right)p'(b_i - b_*) \tag{7.3.12}$$

the same result that was obtained in (7.3.5).

Exercise 7.3.7. A. Verify (7.3.8).
. B. Verify that h obtained in (7.3.9) is the minimizing h.
C. Show that $\hat{\delta}$ is an unbiased estimator of trTX'X.

Exercise 7.3.8. Find the MMSE and the approximate MMSE of the form

$$p'\hat{\beta}_i = (1-h)p'b_i$$

for the predictive loss function.

Exercise 7.3.9. A. Find the MMSE of the form

$$p'\hat{\beta}_i = p'b_* + p'(I - q\Sigma^+(X'X)(b_i - b_*)$$

that minimizes

$$L = \sum_{i=1}^{r} E(\hat{\beta}_i - \beta_i)'\Sigma(\hat{\beta}_i - \beta_i)$$

where $\Sigma = UHU' + VV'$.
B. What is the approximate MMSE?

7.3.3 The Average MSE

One might try to obtain the average MSE of the JS (7.3.5) using the techniques in the proof of Theorem 4.4.1. Some additional work has to be done to account for the shrinkage toward b_*. Perhaps the result could be obtained more elegantly if the problem of estimating the parameters for $r > 1$ linear models could be reformulated as a problem involving one special linear model so that the result of Theorem 4.4.1 could be used. This strategy will be employed to derive the MSE of the JS (7.3.5) in the proof of Theorem 7.3.1 below.

Theorem 7.3.1. If $(r-1)s > 2$, $\epsilon_i|\beta_i$ and β_i are multivariate normal then the Dempster JS (7.3.5) has average MSE

$$M = rp'(X'X)^+p\sigma^2 - \frac{r(n-s)((r-1)s-2)}{(r(n-s)+2)s}p'(X'X)^+p\frac{\sigma^4}{\sigma^2 + \omega^2}$$

(7.3.13)

for theoretically optimum c.

Proof. The proof consists of:
1. reparametization of (7.1.1) into a setup with one linear model;
2. some algebra;
3. applying Theorem 4.4.1.

Let $Y' = [Y_1', Y_2', ..., Y_r']$, $\beta = [\beta_1, \beta_2, ..., \beta_r]$ and $\epsilon = [\epsilon_1, \epsilon_2, ..., \epsilon_r]$. Let $v = (H \otimes I)\epsilon$, $\psi = (H \otimes I)\beta$ and $W = (H \otimes I)Y$. Replace (7.1.1) by the single linear model

$$W = (I_{r-1} \otimes X)\psi + v.$$

(7.3.14)

It can be shown that v and φ are independent if ε_i and β_i are independent. Also

$$E(v \mid \psi) = 0 \text{ and } D(v \mid \psi) = \sigma^2 I_{(r-1)n}. \tag{7.3.15a}$$

From the prior assumptions about β

$$E(\psi) = 0 \text{ and } D(\psi) = \omega^2 (X'X)^+. \tag{7.3.15b}$$

For estimable parametric functions $q'\varphi$ i.e. parametric functions such that $q' = e'(I \otimes U')$ for some e' the JS of $q'\varphi$ is

$$q'\psi = q'a - \frac{c\sigma_*^2((r-1)s - 2)q'a}{a'(I \otimes X'X)a} \tag{7.3.16}$$

where a is the LS estimator of φ in (7.3.14). Now $\varphi = [\varphi_1, \varphi_2, ..., \varphi_{r-1}]$ so $q'\psi = \sum_{i=1}^{r-1} q_i'\psi_i$ and the φ_i are independent. Thus, applying Theorem 4.4.1,

$$\begin{aligned} MSE(q_i'\psi_i) &= p'(X'X)^+ p\sigma^2 \\ &- \frac{r(n-s)((r-1)s - 2)}{(r(n-s)+2)(r-1)s} p'(X'X)^+ p\frac{\sigma^4}{\sigma^2 + \omega^2}. \end{aligned} \tag{7.3.17}$$

It can be shown that

$$\begin{aligned} \sum_{i=1}^{r} p'E(\hat{\beta}_i - \beta_i)(\hat{\beta}_i - \beta)' &= \sum_{i=1}^{r} p'E(\hat{\psi}_i - \psi_i)(\hat{\psi}_i - \psi_i) \\ &+ rp'E(b_. - \bar{\beta})(b_. - \bar{\beta})p. \end{aligned} \tag{7.3.18}$$

The last term of (7.3.18) is $p'(X'X)^+ p\sigma^2$. Equation (7.3.13) follows from Equation (7.3.17) and (7.3.18).

Exercise 7.3.10. What is theoretically optimum c in Theorem 7.3.1 above?

Exercise 7.3.11. A. Show that v and ϕ are independent if ε_i and β_i are independent. B. Verify (7.3.14) and (7.3.15).

Exercise 7.3.12. Verify (7.3.18).

Exercise 7.3.13. Obtain the average MSE of the Dempster JS for r linear models for the case of a zero prior mean.

Exercise 7.3.14. Show that the average MSE of the JSL (4.2.19) is for optimum c

$$p'Mp = p'(X'X)^+ p\sigma^2 - \frac{(s-3)(n-s)}{s(n-s+2)} p'(X'X)^+ p \frac{\sigma^4}{\sigma^2 + \omega^2}. \quad (7.3.19)$$

Hint. Use the reparametized model (7.2.13). This is a special case of the setup (7.1.1). Now (7.2.18) is the EBE of C.R. Rao for this case. The result then follows.

7.3.4 The Conditional MSE

The conditional MSE of the JS may be obtained by a technique analogous to that used to obtain the average MSE in Section 7.3.3. The technique involves:
1. reparametization of (7.1.1) to a single linear model;
2. application of Theorem and Corollary 7.5.1;
3. application of (7.3.18).
The conditional MSE of the estimable parametric functions of the Dempster JS (7.3.5) is given in Theorem 7.3.2. Corollary 7.3.1 gives the conditional MSE averaging over a predictive loss function. The details of the proofs will be left to the exercises.

Theorem 7.3.2. The conditional MSE of the JS (7.3.5) for estimable parametric functions is given by

$$M = rp'(X'X)^+ p - 2c[(r-1)s-2](r-1)S_{2,(r-1)s}(\delta)p'(X'X)^+ p\sigma^2$$

$$+(r-1)\left[4c[(r-1)s-2]+c^2[(r-1)s-2]^2\left(\frac{r(n-s)+2}{r(n-s)}\right)\right]$$

$$\cdot S_{2,(r-1)s}(\delta)p'(X'X)^+ p\sigma^2 + \frac{S_{2,(r-1)s}(\delta)}{2\sigma^2\delta}p'\sum_{i=1}^r (\beta_i - \bar{\beta})(\beta_i - \bar{\beta})'p$$

$$\quad (7.3.20)$$

where $S_{2,(r-1)s}(\delta)$ is defined in Equations (4.5.12) . Also

$$\delta = \frac{1}{2\sigma^2}\sum_{i=1}^r (\beta_i - \bar{\beta})'X'X(\beta_i - \bar{\beta}).$$

Corollary 7.3.1. For optimal c the conditional MSE of the JS averaging over a predictive loss function is

$$m_2 = rs\sigma^2 - \frac{[(r-1)s-2]^2 r(n-s)}{r(n-s)+2} S_{1,(r-1)s}(\delta)\sigma^2. \qquad (7.3.21)$$

Exercise 7.3.16. Prove Theorem 7.3.2 and Corollary 7.3.1.

Exercise 7.3.17. A. Find the expression for the MSE of the parametric functions for the Dempster JS when $\theta = 0$.
B. Show that the Dempster JS derived from a prior with $\theta = 0$ has conditional MSE averaging over a predictive loss function given by

$$m_1 = rs\sigma^2 - \frac{(rs-2)^2 r(n-s)}{r(n-s)+2} S_{1,rs}(\delta + \alpha)\sigma^2 \qquad (7.3.22)$$

with $\alpha = \bar{\beta}'(X'X)\bar{\beta}/2\sigma^2$.

An interesting comparison is that of m_1 and m_2. When $\alpha = 0$, $m_1 \leq m_2$. But m_1 is an increasing function of α for fixed δ. Thus for large enough α $m_1 \geq m_2$. Values of $m_1/rs\sigma^2$ and $m_2/rs\sigma^2$ are given in Table 7.3.1 for a few values of δ and r and s. The value of α/rs where the estimators cross is also given for a few cases.
The reader may complete the third part of the table by adding the crossing values where $m_1 = m_2$ for the case where $r = 10$, $s = 5$ by doing Exercise 7.3.18 if he/she wishes.

Exercise 7.3.18. For the case where $r = 10$, $s = 5$ use a computer software package to find the value of α where the two Dempster JS have the same MSE for the values of δ given in Table 4.3.1.

Exercise 7.3.19. Verify that for fixed δ in (7.3.22) m_1 is an increasing function of α.

Exercise 7.3.20. Show that for the JSL in Chapter IV

$$E_\beta(\hat{\beta}-\beta)'X'X(\hat{\beta}-\beta) = s\sigma^2 - \frac{(s-3)^2(n-s)}{(n-s+2)} S_{1,s-2}(\delta)\sigma^2 \qquad (7.3.23)$$

with $\delta = (\beta-\bar{\beta})X'X(\beta-\bar{\beta})/2\sigma^2$ where $\bar{\beta} = U\Lambda^{-1/2}\bar{\alpha}1_s$.

Table 7.3.1

$r = 6, s = 1$	$\delta/6$	0.2	0.6	1.0
$\alpha/6$	$m_2/6\sigma^2$	0.676	0.827	0.887
0		0.336	0.730	0.815
0.5	$m_1/6\sigma^2$	0.757	0.829	0.868
1.0		0.841	0.876	0.898
crossing	$\alpha/6$	0.247	0.490	0.778

$r = 6, s = 3$	$\delta/18$	0.2	0.6	1.0
$\alpha/18$	$m_2/18\sigma^2$	0.519	0.718	0.802
0		0.371	0.610	0.719
0.2		0.517	0.673	0.754
0.4	$m_1/18\sigma^2$	0.610	0.719	0.781
0.6		0.673	0.754	0.802
crossing	$\alpha/18$	0.202	0.398	0.598

$r = 10, s = 5$	$\delta/50$	0.2	0.6	1.0
$\alpha/50$	$m_2/50\sigma^2$	0.407	0.637	0.739
0		0.316	0.569	0.687
0.2		0.471	0.636	0.723
0.4	$m_1/50\sigma^2$	0.569	0.686	0.753
0.6		0.636	0.723	0.777
crossing	$\alpha/50$			

7.4 A Generalization of the EBE

The general mathematical form for the estimators of C.R. Rao, Wind and Dempster was

$$p'\hat{\beta}_i = p'b_i - p'M(b_i - b_*),$$
(7.4.1)

where the form of the matrix M depended on the particular estimator being studied. This suggests that the estimators considered so are special cases of a general form of EBE. This general form of EBE together with a formula for its average MSE will now be presented.

7.4.1 The EBE

Let $K = UQU'$ where Q is a matrix function of b only and an unbiased estimator of $U'FU + \sigma^2 \Lambda^{-1}$. Let C be a matrix of constants. The general form of the EBE is

$$p'\hat{\beta}_i = p'b_i - p'C(X'X)^+ K^+ (b_i - b_\bullet). \qquad (7.4.2)$$

The estimators of C.R. Rao, Wind and Dempster are obtained as special cases of (7.4.2) by specifying the form of Q and letting $C = cI$. When

$$Q = \frac{U'BU}{r-1} \qquad (7.4.3)$$

the estimator of C. R. Rao (7.1.9) results. The estimator of Wind results if Q is a diagonal matrix with elements

$$q_{ij} = \frac{\lambda_j U_j' B U_j}{r-1}. \qquad (7.4.4)$$

To obtain Dempster's estimator let Q be a diagonal matrix with

$$q_{ij} = \frac{tr(X'X)B}{(r-1)s\lambda_j}. \qquad (7.4.5)$$

Exercise 7.4.1 furnishes examples of estimators where the level of prior knowledge available is intermediate between that used to obtain the estimators of C.R. Rao and Wind and Wind and Dempster.

Exercise 7.4.1. Let $s_1, s_2, \ldots, s_t$ be a sequence of positive numbers where $1 \le s_i \le s$ and $s = \sum_{j=1}^{t} s_j$. Let $U = [U_1, U_2, \ldots, U_t]$ where U_i is of dimension $m \times s_i$. Consider a prior with dispersion of the form

$$F = \sum_{j=1}^{t} U_j A U_j' .$$ (7.4.6)

Let

$$Q = \begin{bmatrix} Q_1 & 0 & \cdots & 0 \\ 0 & Q_2 & \cdots & 0 \\ \vdots & \vdots & \vdots & \vdots \\ 0 & 0 & \cdots & Q_t \end{bmatrix}$$ (7.4.7)

with

$$Q_j = \frac{1}{r-1} U_j' B U_j.$$ (7.4.8)

Let C be a diagonal matrix with elements c_j .
A. Obtain the expression for the EBE.
B. Show that when $s_j = s$, $t = 1$ the EBE reduces to that of C.R.Rao.
C. Show that when $s_j = 1$, $t = s$, $A_j = \omega_j^2/\lambda_j$ the estimator takes the form of Wind's estimator.
D. Suppose that some but not all the ω_j in part C are equal. Tell how to construct an EBE intermediate between that of Wind and Dempster. The Wind estimator should result if all the ω_j are distinct; the Dempster estimator should result if all the ω_j are equal.

7.4.2 The MSE

A general formula for the average MSE of the EBE (7.4.2) will be obtained below. The formulae for the MSEs of the estimators of C.R.Rao, Wind and Dempster may then be obtained as special cases by substitution of the prescribed form of K in each case. The MSE of (7.4.2) is obtained after calculating the expectations in

$$p'Mp = \sum_{i=1}^{r} p'E(b_i - \beta_i)(b_i - \beta_i)'p$$

$$-p'C(X'X)^+ E\left[K^+ \sum_{i=1}^{r}(b_i - b_\bullet)(b_i - \beta_i)'\right]p\sigma^4$$

$$-p'E\left[\sum_{i=1}^{r}(b_i - b_\bullet)(b_i - \beta_i)'K^+\right](X'X)^+ C'p\sigma^4 \qquad (7.4.9)$$

$$+\frac{[r(n-s)+2]}{r(n-s)}p'C(X'X)^+ E(K^+BK^+)(X'X)^+ C'p\sigma^4.$$

The first term of (7.4.9) is the MSE of the LS. The second and the third term are obtained by application of the sufficiency principle. The third term is the transpose of the second term. Observe that $b'=[b_1',b_2', ..., b_r']$ is a sufficient statistic for $\beta' = [\beta_1', \beta_2', ..., \beta_r']$ and b_i is a sufficient statistic for β_i. From the sufficiency principle

$$E_{b_i}(\beta_i) = b_i - \sigma^2(X'X)^+ A^+(b_i - \theta) \qquad (7.4.10)$$

where $A = UU'FUU' + \sigma^2(X'X)^+$. Now $b_i - \theta = b_i - b_\bullet + b_\bullet - \theta$. Thus, since $\sum_{i=1}^{r}(b_i - b_\bullet) = 0$,

$$\sum_{i=1}^{r}(b_i - b_\bullet)(b_i - \theta)' = \sum_{i=1}^{r}(b_i - b_\bullet)(b_i - b_\bullet)'$$

$$= \sum_{i=1}^{r}(b_i - b_\bullet)(b_\bullet - \theta)' = B. \qquad (7.4.11)$$

From (7.4.10) and (7.4.11)

$$p'E_b\left[K^+ \sum_{i=1}^{r}(b_i - b_\bullet)(b_i - \beta_i)'\right]p$$

$$= p'K^+ \sum_{i=1}^{r}(b_i - b_\bullet)(b_i - E_b(\beta_i))'p \qquad (7.4.12)$$

$$= p'K^+BA^+(X'X)^+ p\sigma^2.$$

For any random matrix R, $EE_b[R] = E[R]$. Thus, taking expectations of both sides of (7.4.12)

$$p'E\left[K^+ \sum_{i=1}^{r}(b_i - b_\bullet)(b_i - \beta_i)'\right]p$$

$$= p'E[K^+B]A^+(X'X)^+ p\sigma^2. \qquad (7.4.13)$$

Let

$$E_1 = E[K^+B] \tag{7.4.14a}$$

and

$$E_2 = E[K^+BK^+] . \tag{7.4.14b}$$

Substitution of (7.3.13) and (7.3.14) into (7.4.9) gives the MSE

$$p'Mp = rp'(X'X)^+p\sigma^2 - p'C(X'X)^+E_1A^+(X'X)^+p\sigma^4$$
$$-p'(X'X)^+A^+E_1(X'X)^+C'p\sigma^4 \tag{7.4.15}$$
$$+p'C(X'X)^+E_2(X'X)^+C'p.$$

An expression for optimum C will be obtained in (7.4.18). It will be valid as long as $A^+E_1E_2^+$ is independent of the population and the prior parameters. Whether this holds true for a particular estimator can be determined by calculation of E_1 and E_2, provided that these calculations are tractable. For the estimators considered so far optimum C may be obtained from (7.4.18). To obtain optimum C rewrite (7.4.15) as

$$p'Mp = rp'(X'X)^+p\sigma^2$$
$$-\frac{r(n-s)}{r(n-s)+2}p'(X'X)^+A^+E_1E_2^{+}E_1A^+(X'X)^+p\sigma^4 + p'\Omega\Omega'p. \tag{7.4.16}$$

with

$$\Omega = \alpha^{1/2}C(X'X)^+E_2^{1/2} - (X'X)^+A^+ + E_1(E_2^+)^{1/2}\alpha^{-1/2} \tag{7.4.17}$$

where $\alpha = r(n-s)/(r(n-s)+2)$. Now $\Omega = 0$ when

$$C = \frac{r(n-s)}{r(n-s)+2}(X'X)^+A^+E_1E_2^+(X'X) + V\Gamma V'. \tag{7.4.18}$$

The MSE is independent of Γ so for practical purposes let $\Gamma = 0$.

Exercise 7.4.2. A. Let Q be as given in Equations (7.4.3)-(7.4.5). Obtain optimum C and show that the MSEs of the estimators of C.R. Rao, Wind and Dempster result.
B. Let Q be as in Exercise 7.4.1. Obtain the optimum C diagonal matrix and the MSEs of the corresponding intermediate estimators.

7.5 Comparing the Efficiency of the Estimators

Two measures of efficiency, the relative savings loss (RSL) and the relative loss (RL) will be taken up in this section. The RSL compares the performance of the EBE with that of the BE. The RL compares the performance of the approximate minimum mean square error estimator with that of the minimum mean square error estimator. Thus, the performance of different approximately optimal estimators may be compared with each other and optimal estimators from both the Bayesian and the frequentist point of view.

7.5.1 The RSL

The RSL is a measure of the efficiency of the EBE as compared with the BE; i.e., a measure of efficiency of an approximately optimal estimator as compared with an optimal estimator. The RSL is the ratio of the difference between the average MSE of the EBE and the BE to the difference between the average MSE of the difference between the MSE of the LS and the BE. For the setup with r linear models and the estimators considered in this chapter the RSL depends on:

1. the number of linear models;
2. the number of observations;
3. the number of prior parameters estimated in formulating the EBE.

When more prior parameters are estimated the RSL is usually larger. The RSL is between zero and one; a smaller RSL means a more efficient estimator. Since the MSEs are obtained averaging over the prior distribution the RSL is independent of the estimated parameters. For the estimators in this chapter the RSL is also independent of the prior parameters; recall that this was not the case for the positive part estimators discussed in Chapter VI.

The defining equation for the RSL is

$$RSL = \frac{MSE(p'\hat{\beta}_{eb}) - MSE(p'\hat{\beta}_b)}{MSE(p'b) - MSE(p'\hat{\beta}_b)}.$$ (7.5.1)

A general formula for the RSL for the estimators in Section 7.4 is

$$RSL = 1 - \frac{r(n-s)}{[r(n-s)+2]} \cdot \frac{p'(X'X)^+ A^+ E_1 E_2^+ E_1 A^+ (X'X)^+ p}{p'(X'X)^+ A^+ (X'X)^+ p}.$$ (7.5.2)

For the estimators of C.R. Rao, Wind and Dempster (7.5.2) may be specialized to

$$RSL = 1 - \frac{r(n-s)}{r(n-s)+2} c$$ (7.5.3)

For the EBE of C.R.Rao

$$RSL = 1 - \frac{(n-s)(r-s-2)}{r(n-s)+2}.$$ (7.5.4)

For the EBE of Wind

$$RSL = 1 - \frac{(n-s)(r-3)}{r(n-s)+2}.$$ (7.5.5)

For the EBE of Dempster

$$RSL = 1 - \frac{(n-s)[(r-1)s-2]}{[r(n-s)+2]s}.$$ (7.5.6)

The Dempster estimator has a smaller RSL than the Wind estimator which in turn has a smaller RSL than the estimator of C.R.Rao. This is to be expected because fewer prior parameters are estimated for the EBE with smaller RSL. The number of prior parameters in the dispersion that are estimated in each case are:

1. for the estimator of C.R.Rao $s(s + 1)/2$;
2. for the estimator of Wind s ;
3. for the estimator of Dempster 1.

Exercise 7.5.1. Obtain the RSL for
A. The estimators of C.R.Rao, Wind and Dempster for zero prior mean; i.e., $\theta = 0$.
B. The generalized JS (see Exercise 7.3.3).
C. The estimators intermediate between those of Dempster and Wind.
D. The estimators intermediate between those of Wind and C.R. Rao.

Exercise 7.5.2. Observe that $MSE(p'\hat{\beta}_b)$ is independent of the prior mean θ. Let $p'\hat{\beta}_{eb0}$ denote the EBE when $\theta = 0$. Define

$$eff_\theta = \frac{MSE(p'\hat{\beta}_{eb0}) - MSE(p'\hat{\beta}_b)}{MSE(p'\hat{\beta}_{eb}) - MSE(p'\hat{\beta}_b)}.$$

A. Show that

$$eff_\theta = \frac{RSL(p'\hat{\beta}_{eb0})}{RSL(p'\hat{\beta}_{eb})}.$$

B. Find eff_θ for the estimators of C.R. Rao, Wind and Dempster.

7.5.2 The RL

As was pointed out in Section 4.7 the RL measures the efficiency of an approximate MMSE compared to an MMSE; i.e., the efficiency of the "almost" optimal estimator as compared with the optimal estimator. The MSEs considered are conditional on β , i.e., not averaging over the prior distribution. In this section the RL will be considered for the estimators of Wind and Dempster. The MSEs will be:

1. conditional on β ;
2. obtained summing over r linear models;
3. for the Dempster estimator averaged over a predictive loss function;
4. for the Wind estimator the individual MSE of the estimate of the parameter in a linear model reparametized to one of full rank with independent LS estimates, i.e., canonical orthogonal coordinates.

For the case of a non-zero prior mean the definition of the RL will be a modification of that of Section 4.7. The MMSE will be replaced by an approximate MMSE where b* is an estimate of the prior mean. This approximate MMSE will be of the form

$$p'\hat{\beta}_i = p'b_* + cp'(b - b_*), \ 1 \le i \le r \tag{7.5.7}$$

and will be denoted by MMSEM.

This definition will insure that the RL > 0. Without the modification RL < 0 is possible (see Exercise 7.5.7). The defining equation of the RL is

$$RL = \frac{MSE(JS) - MSE(MMSEM)}{MSE(LS) - MSE(MMSEM)}. \tag{7.5.8}$$

The MSE of (7.5.7), averaging over the predictive loss function and conditional on β, is after dividing by $rs\sigma^2$

$$\frac{m}{rs\sigma^2} = \frac{1}{r} + \frac{c^2(r-1)}{r} + \frac{2\delta}{rs}(1-c)^2 \tag{7.5.9a}$$

with

$$\delta = \frac{1}{2\sigma^2} \sum_{i=1}^{r} (\beta_i - \bar{\beta})'X'X(\beta_i - \bar{\beta}). \tag{7.5.9b}$$

The minimizing optimal c, obtained by differentiating (7.5.9a) and setting the result equal to zero, is

$$c = \frac{2\delta}{s(r-1) + 2\delta}. \tag{7.5.10}$$

Substituting (7.5.10) into (7.5.9) the MMSEM has MSE

$$\frac{m}{rs\sigma^2} = 1 - \frac{(r-1)^2 s}{r((r-1)s + 2\delta)}. \tag{7.5.11}$$

The Dempster JS has RL

$$RL(\delta) = 1 - \frac{(n-s)((r-1)s-2)^2}{(r(n-s)+2)(r-1)^2 s^2}[(r-1)s + 2\delta]S_{1,(r-1)s}(\delta). \quad (7.5.12)$$

When the prior mean is zero the MMSE may be used. For this case the Dempster JS has RL

$$RL(\delta, \alpha) = 1 - \frac{r(n-s)(rs-2)^2}{(r(n-s)+2)(rs)^2}[rs + 2(\delta+\alpha)]S_{1,rs}(\delta+\alpha) \quad (7.5.13)$$

with $\alpha = \dfrac{r}{2\sigma^2}\overline{\beta}'(X'X)\overline{\beta}.$

Exercise 7.5.3. Verify (7.5.9), (7.5.12) and (7.5.13).

Many of the properties of the RL are similar to those of the one model case. In most cases the proofs are similar to those in (4.7) and are left as exercises. Exercise 7.5.4 provides the reader with an opportunity to study the behavior of the RL in the context of r linear models.

Exercise 7.5.4. A. Show that

$$RL(0) = 1 - \frac{((r-1)s-2)(n-s)r}{s[r(n-s)+2](r-1)} = \frac{r}{r-1}RSL - \frac{1}{r-1}. \quad (7.5.14)$$

B. Show that $RL(\delta)$ is an increasing function and that

$$RL(0) < RL(\delta) < RL(\infty) = \lim_{\delta \to \infty} RL(\delta). \quad (7.5.15)$$

C. Show that

$$\lim_{\delta \to \infty}[(r-1)s + 2\delta]S_{1,(r-1)s}(\delta) = 1. \quad (7.5.16a)$$

D. Show that (7.5.16a) implies

$$RL(\infty) = 1 - \frac{r(n-s)((r-1)s-2)^2}{(r(n-s)+2)(r-1)^2 s^2}. \quad (7.5.16b)$$

The properties of the RL in Exercise 7.5.4 above are illustrated by the values given in Table 7.5.1 and the graphs in Figure 7.5.1.

In Section 4.7 Theorem 4.7.2 gave regions where the conditional MSE of the

JS was smaller than that of the BE and where the BE had a smaller MSE than the JS. Theorem 7.5.1 gives the analogous result in the context of r linear models with a non-zero prior mean.

Theorem 7.5.1. 1. If

$$\max(0, \frac{(r-1)s}{2}\left(\frac{c-\sqrt{RL(0)}}{1-c}\right) < \delta < \frac{(r-1)s}{2}\left(\frac{c+\sqrt{RL(\infty)}}{1-c}\right) \quad (7.5.17)$$

then the BE (7.5.7) has a smaller MSE than the JS.
2. If

$$\delta > \frac{(r-1)s}{2}\left(\frac{c+\sqrt{RL(\infty)}}{1-c}\right) \quad (7.5.18)$$

then the JS has a smaller MSE than the BE (7.5.7).

The proof of Theorem 7.5.1 depends on establishing that the lower and the upper bounds of the CMSE/rsσ^2 are

$$\ell = 1 - \frac{(r-1)^2[1-RL(0)]}{[(r-1)+2\delta/s]r} \quad (7.5.19)$$

and

$$u = 1 - \frac{(r-1)^2[1-RL(\infty)]}{[(r-1)+2\delta/s]r} \quad (7.5.20)$$

respectively. The proof is along similar lines as that of Theorem 4.7.2 and is left to the reader.

Recall the result of Rao and Shinozaki that was used in Chapter IV where for a Poisson random variable r with $E(r) = b$

$$(q+2b)^{-1} \le E\{(q+2r)^{-1}\} \le (q+2b-2)^{-1}. \quad (7.5.21)$$

Letting $b = \delta$, $q = (r-1)s - 2$ and $r = k$ the risk of the JS given in (7.3.21) is bounded by

$$1 - \frac{r(n-s)((r-1)s-2)^2}{(r(n-s)+2)rs((r-1)s-4+2\delta)}$$

$$\leq \frac{m_{jsc}}{rs\sigma^2} \leq 1 - \frac{r(n-s)((r-1)s-2)^2}{(r(n-s)+2)rs((r-1)s-2+2\delta)} \tag{7.5.22}$$

Comparing the right and the left hand sides of Inequality to (7.5.9a) and solving the resulting inequalities for δ yields Theorem 7.5.2.

Theorem 7.5.2. 1 If n or r is large in comparison with s and

$$1 + \frac{(r-1)sc}{2(1-c)} - \frac{\sqrt{((r-1)s+c-2)c}}{1-c} < \delta$$

$$< 1 + \frac{(r-1)sc}{2(1-c)} + \frac{\sqrt{((r-1)s+c-2)c}}{1-c} \tag{7.5.23}$$

the BE has a smaller risk than the JS

2. If n is large compared to r or s , $\sqrt{2/(r-1)s} < c$ and either

$$\delta < \frac{1}{2} + \frac{(r-1)sc}{2(1-c)} - \frac{\sqrt{(c+1)(c+2(r-1)s-2)}}{2(1-c)} \tag{7.5.24a}$$

or

$$\delta > \frac{1}{2} + \frac{(r-1)sc}{2(1-c)} + \frac{\sqrt{(c+1)(c+2(r-1)s-2)}}{2(1-c)} \tag{7.5.24b}$$

the BE has a larger risk than the JS.

Notice that the form of this result is the same as that of Theorem 4.7.4 with $(r-1)s$ replacing s.

Exercise 7.5.5. A. Establish (7.5.19) and (7.5.20).
B. Prove Theorem 7.5.1.

Table 7.5.2 illustrates Theorem 7.5.1. For some different numerical values of r,s and c Table 7.5.2 gives:

1. the regions in (7.5.17) and (7.5.18);
2. the actual values where the MSE of the BE and the JS are equal, i.e., the crossing points.

The crossing points were obtained by Newton's method using the software package Mathematica.

Unlike Theorem 4.7.2, Theorem 7.5.1 compared the JS with a "modified" BE, i.e., one where the prior mean was estimated from the data. The JS could have been compared with a BE with known prior mean of the form

$$p'\hat{\beta}_i = p'\theta + cp'(b_i - \theta) \tag{7.5.25}$$

with conditional MSE/$rs\sigma^2$

$$M = \frac{2(\delta + \alpha)}{rs}(1 - c)^2 + c^2 \tag{7.5.26}$$

where $\alpha = r(\bar{\beta} - \theta)'X'X(\bar{\beta} - \theta)/2\sigma^2$. The MSE (7.5.25) is less than the MSE (7.5.8) provided that

$$\alpha \leq \frac{s(1 - c)}{2(1 + c)}. \tag{7.5.27}$$

Conditions for the BE (7.5.25) to have a smaller or larger MSE than the JS may be obtained by comparison of (7.5.26) with the upper and lower bounds of the JS given in (7.5.19) and (7..5.20).

Exercise 7.5.6 A. Establish (7.5.26) and (7.5.27).
B. Obtain conditions for the MSE of the BE to be
  (1) smaller than that of the JS;
  (2) larger than that of the JS.

The RL using (7.5.25) in place of (7.5.7) is

$$RL = 1 - \frac{r(n-s)[(r-1)s - 2]^2[rs + 2(\delta + \alpha)]}{[r(n-s) + 2](rs)^2} S_{1,(r-1)s}(\delta + \alpha). \quad (7.5.28)$$

The reader is invited to verify (7.5.28) in Exercise 7.5.7.

Exercise 7.5.7. Verify (7.5.24) and give an example to show that RL< 0 is possible.

Table 7.51
Some Values of the RL

r = 4	δ/rs				
	0	1.5	3.5	5.5	Infinity
s = 1	0.667	0.844	0.872	0.878	0.889
s = 3	0.229	0.340	0.369	0.378	0.395
r = 7					
s = 1	0.333	0.483	0.521	0.533	0.555
s = 5	0.067	0.105	0.117	0.121	0.129
r = 10					
s = 4	0.056	0.087	0.097	0.101	0.108

Table 7.5.2a illustrates Theorem 7.5.1 by giving the crossing values and some regions obtained where the JS dominates the BE or vice versa. Table 7.5.2b illustrates Theorem 7.5.2b by also giving regions where one estimator dominates the other. In general the regions given by Theorem 7.5.2 are larger than those given by

Theorem 7.5.1. Some examples are depicted graphically in Figure 7.5.1.
For smaller values of c Theorem 7.5.1 gives a better approximation to the lower crossing value; for larger values of c Theorem 7.5.2 gives the better approximation. Theorem 7.5.2 gives a better approximation for the larger crossing value. Used together the endpoints of the intervals make good starting points when using the FindRoot command of Mathematica or the f solve command of Maple to approximate the crossing values numerically.

Table 7.5.2a

$r = 6, s = 1$	c	Crossing Values	MSE(BE) < MSE(JS) Theorem 7.5.1
	0.2	2.773	$\delta < 2.6$
	0.5	6.172	$\delta < 5.66$
	0.8	1.709,19.722	$2.094 < \delta < 17.906$

$r = 6, s = 5$	c	Crossing Values	MSE(BE) < MSE(JS)
	0.2	7.838	$\delta < 7.544$
	0.5	5.159,20.771	$5.428 < \delta < 19.57$
	0.8	28.654,72.94	$32.32 < \delta < 67.67$

$r = 10, s = 3$	c	Crossing Values	MSE(BE) < MSE(JS)
	0.2	8.266	$\delta < 7.97$
	0.5	5.849,22.09	$6.15 < \delta < 20.85$
	0.8	31.74,77.84	$35.63 < \delta < 72.37$

$r = 6, s = 1$	c	MSE(JS) < MSE(BE)
	0.2	$\delta > 4.06$
	0.5.	$\delta > 6.5$
	0.8	$\delta > 40$

$r = 6, s = 5$	c	MSE(JS) < MSE(BE)
	0.2	$\delta > 8.64$
	0.5.	$\delta < 2.67, \delta > 22.33$
	0.8	$\delta < 25.43, \delta > 74.59$

$r = 10, s = 3$	c	MSE(JS) < MSE(BE)
	0.2	$\delta > 9.76$
	0.5.	$\delta < 3.29, \delta > 23.71$
	0.8	$\delta < 28.47, \delta > 79.52$

Table 7.5.2b

c	MSE(BE) < MSE(JS)	MSE(JS) < MSE(BE)
0.2	$0.625 < \delta < 2.625$	$\delta > 2.962$
0.5	$0.854 < \delta < 6.146$	$\delta > 6.354$
0.8	$2.282 < \delta < 19.718$	$\delta < 1.1325, \delta > 19.867$

r = 6, s = 1

c	MSE(BE) < MSE(JS)	MSE(JS) < MSE(BE)
0.2	$1.431 < \delta < 6.818$	$\delta > 8.329$
0.5	$6.644 < \delta < 20.356$	$\delta < 4.559, \delta > 21.441$
0.8	$29.183 < \delta < 72.817$	$\delta < 27.311, \delta > 73.689$

r = 6, s = 5

c	MSE(BE) < MSE(JS)	MSE(JS) < MSE(BE)
0.2	$1.568 < \delta < 7.181$	$\delta > 8.774$
0.5	$7.359 < \delta < 21.641$	$\delta < 5.211, \delta > 22.789$
0.8	$32.284 < \delta < 77.716$	$\delta < 30.360, \delta > 78.640$

r = 10, s = 3

Exercise 7.5.8 . Using Mathematica, Maple a TI 82, 85 or 92 calculator generate the values in Table 7.5.2 for r = 5, s = 2, 4 and r =12, s =6. Some code that can be used follows.

Mathematica

```
a[x_,r_,s_]=1/(Exp[x] 2 x^((r-1) s/2-1));
b[x_,r_,s_]=Integrate[Exp[y] y^((r-1)s/2-2),{y,0,x}];
.S[x_,r_,s_]=a[x,r,s] b[x,r,s];
c[r_,s_]=((r-1) s-2)^2/(r s);
m[x_,r_,s_]=1-c[r,s] S[x,r,s];
mb[x_,c_,r_,s_]=1/r+((r-1)/r) c^2+(2 x/(r s)) (1-c)^2;
FindRoot[m[x,6,5]==mb[x,.5,6,5],{x,5}]
a1[r_,s_,x_]=1-(((r-1) s-2)^2/(r s ((r-1) s-2+2 x)));
a2[r_,s_,x_]=1-((r-1) s-2)^2/(r s ((r-1) s-4+2 x));
```

Maple

```
> a:=(x,r,s)->1/(exp(x)*2*x^((r-1)*(s/2)-1)):
> b:=(x,r,s)->int(exp(y)*y^((r-1)*s/2-2),y=0..x):
> S:=(x,r,s)->a(x,r,s)*b(x,r,s):
> d:=(r,s)->((r-1)*s-2)^2/(r*s):
> m:=(x,r,s)->1-d(r,s)*S(x,r,s):
> mb:=(x,c,r,s)->1/r+((r-1)/r)*c^2+(2*x/(r*s))*(1-c)^2:
>fsolve(m(x,6,5)=mb(x,.5,6,5),x=5..5.428);
>a1:=(r,s,x)->1-(((r-1)*s-2)^2/((r*s)*((r-1)*s-2+2*x))):
>a2:=(r,s,x)->1-(((r-1)*s-2)^2/((r*s)*((r-1)*s-4+2*x))):
```

The find root and the fsolve commands are provided as an example. The reader should provide appropriate starting values using the result of Theorem 7.5.1 as a guide.

The TI-82 program below will request values of r, s and c, graph the two functions and then request two guess values for where the graphs of the BE and the JS intersect. The reader may guess these values either by looking at the graph or using Theorem 7.5.1 or Theorem 7.5.2. The program then approximates the intersection points. If there is only one crossing value the program may be stopped after it is found by pressing the "on" button. The program runs very slowly.

TI 82 or 85

JSBEGR   • Program

```
:Prompt R,S,C
 :"fnInt(e^(Z)*Z^((R-1)*S/2-2),Z,0,X)"→Y₁
 :"1/(2*X^((R-1)*S/2-1)*e^(X))"→Y₂
 :((R-1)*S-2)^2/(S*R)→F
 :FnOff
 :"1-(Y₁*Y₂*F)"→Y₃
 :"1/R+((R-1)/R)*C^2+(2*X/(R*S))*(1-C)^2"→Y₄
 :.01→Xmin
 :Disp "Xmax?"
 :Input Xmax
 :5→Xscl
 :0→Ymin
```

```
:1→Ymax
:.1→Yscl
:DispGraph
:Trace
:StorePic Pic5
:Prompt G
:solve(Y₃-Y₄,X,G)→M
:Disp M
:Prompt H
:solve(Y₃-Y₄,X,H)→P
:Disp P
```

The TI-92 Script below allows the reader to explore the intersection points and to see where the JS has a smaller risk than the BE. After putting the script into the calculator the reader is encouraged to graph $m(x,r,s)$ and $mb(x,c,r,s)$ simultaneously and use the intersection command to find out where the two graphs cross. The solve or the n solve command may also be used.

TI 92

```
C:Define a(x,r,s)=1/(e^(x)*2*x^((r-1)*s/2-1))
C:Define b(x,r,s)=nInt(e^(y)*y^((r-1)*s/2-2),y,0,x)
C:Define f(x,r,s)=a(x,r,s)*b(x,r,s)
C:Define d(r,s)=((r-1)*s-2)^2/(r*s)
C:Define m(x,r,s)=1-d(r,s)*f(x,r,s)
C:Define mb(x,c,r,s)=1/r+((r-1)/r)*c^2+(2*x/(r*s))*(1-c)^2
```
The following two commands can be used together with solve to obtain starting values using Theorem 7.5.2.
```
C:Define a1(r,s,x)=1-((r-1)*s-2)^2/(r*s*((r-1)*s-2+2*x))
C:Define a2(r,s,x)=1-((r-1)*s-2)^2/(r*s*((r-1)*s-4+2*x))
```

Exercise 7.5.9. Obtain conditions on c in terms of r, s, δ for the endpoints in Theorem 7.5.2 to more closely approximate the crossing points of the BE and JS than Theorem 7.5.1.

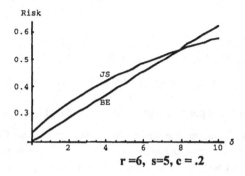

$$r = 6, \ s = 5, \ c = .2$$

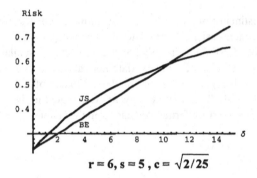

$$r = 6, \ s = 5, \ c = \sqrt{2/25}$$

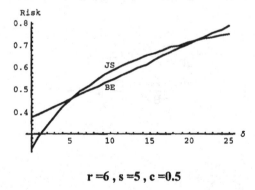

$$r = 6, \ s = 5, \ c = 0.5$$

Figure 7.5.1

7.6 The Positive Parts

This section will give a sketch of how where possible:

1. the positive parts that were taken up in Chapter VI can also be formulated for r linear models;
2. how the efficiencies of the positive parts may be calculated and studied.

The results of Chapter VI can be extended easily for the estimators of Dempster and Wind and the estimators intermediate between them. How this can be done will be taken up in Sections 7.6.1 and 7.6.2. A method of defining the positive part for the problem of estimating r multivariate normal distributions is given by Efron and Morris (1972). This method will be extended to the estimators of C.R. Rao in Section 7.6.1.

The form of the estimator of Dempster for r linear models is very similar to that for one linear model so it is easier to formulate the positive parts for r models. The components of the Wind estimator in orthogonal coordinates is similar to the form of a JS for estimating the mean of a multivariate normal distribution in Chapter II. The analog of the positive part estimator for the estimator of C.R. Rao is quite different from that of the JS. That is why they are formulated in the reverse of the order the original estimators were formulated earlier in the chapter.

7.6.1 The Estimator of Dempster

The generalized positive part may be formulated for r models by analogy with the formulation for the one model case given by Equation 6.1.11. The Dempster JS may be written in the form

$$
\begin{aligned}
p'\hat{\beta}_i &= p'b_i - \frac{c((r-1)s-2)\sigma_*^2}{\text{tr}BX'X}p'(b_i - b_.) \\
&= p'b_. + \left(1 - \frac{c((r-1)s)-2)\sigma_*^2}{\text{tr}BX'X}\right)p'(b_i - b_.).
\end{aligned}
\tag{7.6.1}
$$

Let

$$
w_r = \frac{c(r-1)s\sigma_*^2}{\text{tr}BX'X}.
\tag{7.6.2}
$$

Then the second form of the estimator in (7.6.1) may be written

$$p'\beta_i = p'b_* + (1 - w_r)p'(b_i - b_*).\tag{7.6.3}$$

By analogy with the estimator in Equation (6.1.11) the generalized positive part takes the form

$$\hat{\beta}_i = b_*$$

$$+[(1 - w_r)I_{[0,1)}(w_r) + a_1(1 - w_r)I_{[1,\alpha)}(w_r) + a_2 I_{[\alpha,\infty)}(w_r)](b_i - b_*).$$

$$\tag{7.6.4}$$

The positive parts PP_0 - PP_4 are formulated using the same choices of a_1 and a_2 and α as in the estimators of Chapter VI.

The results of Chapter VI may now be used to obtain the MSE. Using the reparametization to a setup with one linear model w_r takes the form

$$w_r = \frac{c\sigma_*^2((r-1)s - 2)}{a'(I \otimes X'X)a}.\tag{7.6.5}$$

For optimum c the MSE averaged over the loss function with matrix $I \otimes X'X$ is the same as the predictive MSE summed over the r linear models. Thus, the average and conditional MSE will take the same form as the average and conditional MSEs obtained in Chapter VI with $(r - 1)s$ replacing s. The details are left to the reader. The reader may also conduct similar numerical studies to those in Chapter VI if he/she so wishes. He/she is invited to do so in Exercises 7.6.1 and 7.6.2.

Exercise 7.6.1. A. Obtain the general form of the MSE with and without averaging over the prior for the generalized positive part estimator given by 7.6.4.
B. Give the form of the positive parts PP_0 - PP_4.
C. Give the form of the average and conditional MSE for each of the estimators PP_1- PP_4.

Exercise 7.6.2. For r = 10 and s = 3 conduct a numerical study similar to that of Chapter VI.

Exercise 7.6.3. Suppose that the prior mean $\theta = 0$. Give the form of the positive parts and their average and conditional MSEs.

Exercise 7.6.4. For the data of Example 7.1.1 find the numerical values of the positive parts PP_0 - PP_4. Find and compare the distances to the "true value" (6, 5, 5.5).

7.6.2 The Estimator of Wind

The positive parts of the estimator of Wind may be formulated by restricting the contraction factors for each of the coordinates of the estimator in orthogonal coordinates. Let

$$w_{r,j} = c\hat{d}_j = \frac{c(r-3)\sigma_\bullet^2}{\lambda_j m_{jj}}. \tag{7.6.6}$$

Recall that $\hat{d}_j$ was defined in (7.2.7) and m_{jj} was defined by (7.2.8). Let $\tilde{A}$ be a diagonal matrix with non-zero elements

$$\tilde{a}_{jj} = [(1-w_{r,j})I_{[0,1)}(w_{r,j}) + a_1(1-w_{r,j})I_{[1,\alpha)}(w_{r,j}) + a_2 I_{[\alpha,\infty)}(w_{r,j})]. \tag{7.6.7}$$

By comparison with (7.2.11) the general positive part of the JS of Wind then takes the form

$$p'\hat{\beta}_i = p'b_\bullet + p'U\tilde{A}U'(b_i - b_\bullet). \tag{7.6.8}$$

The positive parts PP_0–PP_4 may then be defined the by appropriately choosing a_1 and a_2 as done previously.

Exercise 7.6.5. A. Give the form of the positive parts PP_0–PP_4 for the estimator of Wind.
B. Find the numerical estimates for the data of Example 7.1.1.

Exercise 7.6.6. Give the form of the positive parts for the estimator of Wind when the prior mean $\theta = 0$.

The MSE may be obtained averaging over the loss function with matrix UDU' where D is a positive definite diagonal matrix. First find the individual MSEs of the j'th coordinate of the positive part $\hat{\gamma}$ summed over the r linear models, that is,

$$m_j = \sum_{i=1}^{r} E(\hat{\gamma}_{ij} - \gamma)^2, 1 \le j \le s. \tag{7.6.9}$$

The result should be the same as that for the corresponding positive part of the JS of Dempster for one linear model with $r - 3$ replacing $s - 2$ and $\delta_{.j} = \sum_{i=1}^{r} \lambda_j \gamma_j^2 / 2\sigma^2$ replacing δ. The MSE averaging over the quadratic loss function could then be calculated by finding the weighted sum

$$m = \sum_{j=1}^{s} d_j m_j \qquad (7.6.10)$$

where the d_j s are the non-zero elements of D. This technique will apply to both the conditional and the average MSEs.

7.6.3 The Estimator of C.R. Rao

Four things will be done in this section. These include:

1. giving some definitions about distance functions in metric spaces that are used in Efron and Morris (1972) to formulate positive parts for estimating the mean of a multivariate normal distribution using a matrix for the contraction factor;
2. explaining how for estimating the mean of a multivariate normal distribution the positive part may be formulated;
3. showing that the RSL and hence the average MSE is smaller than that of the MLE;
4. extending the results to the corresponding ones for the estimator of C.R. Rao.

The terminology defined here is available in most textbooks on real analysis.

Definition 7.6.1. A metric space is a set E with a given distance on E. A distance on a set E is a mapping of $E \times E$ into the set R of real numbers that has the following properties:

1. The function $d(x, y)$ is non-negative, i.e., $d(x, y) \geq 0$ for any pair of elements x, y in E.
2. The relation $d(x, y) = 0$ is equivalent to $x = y$.
3. The function $d(x,y)$ is symmetric, i.e., for any pair of elements of E, $d(x, y) = d(y,x)$.
4. For any three elements x, y, z of E the triangle inequality holds, i.e., $d(x, z) \leq d(x,y) + d(y,z)$.

A set A is compact if for every $\varepsilon > 0$ there is a finite union of sets $\overset{n}{\underset{i=1}{\cup}} B_i$ with the property that for any two points $x,y \in B_i$, $\underset{x \in B_i, y \in B_i}{\sup} d(x,y) < \varepsilon$ that contains A.[1]

Some metric spaces are also vector spaces, e.g., normed linear spaces. For such metric spaces a set a set A is convex if for any two points $x,y \in A$, $ax + (1-a)y \in A$ where a is any real number in the unit interval, i.e., $0 \le a \le 1$.

Efron and Morris (1972) use these ideas in their definition of the positive part. First assume $x_i \sim N_p(\theta_i, I)$ for $i = 1, 2,..., r$.

Definition 7.6.2 . Let $\mathfrak{R}$ be the set of all pxp symmetric matrices B that

1. are positive definite;
2. are such that $I - B$ is non-negative definite, put more formally

$$\mathfrak{R} = \{B : 0 < B \le I\}.$$

The set $\mathfrak{R}$ is a compact convex subset of the Euclidean space whose dimension is $p(p + 1)/2$ where each pxp symmetric matrix is represented by its entries above the main diagonal. Let S be any positive definite matrix. A distance function between matrices may be defined by

$$d(B,C) = \text{tr}\{(B - C)S(B - C).\tag{7.6.11}$$

If S is positive definite and $\hat{B}$ is any symmetric matrix the matrix $\hat{B}_s$ that is closest to $\hat{B}$ is such that

$$d(\hat{B}_s, \hat{B}) = \underset{B \in \mathfrak{R}}{\inf} d(B, \hat{B}).\tag{7.6.12}$$

Then, an estimator of the form $\hat{\delta}_i = (I - \hat{B})x_i$ has positive part in the form $\hat{\delta}_i = (I - \hat{B}_s)x_i$.

Exercise 7.6.7. Show that the distance function defined in (7.6.11) satisfies the properties of a metric given in Definition 7.7.1.

---

[1] This definition only applies to metric spaces not general toplogical spaces. It is good enough for the present purpose.

Exercise 7.6.8. Show that for $p = 1$ the positive rule as defined in Definition 7.6.2 for the JS estimator for the mean of a multivariate normal distribution is the classical positive part.

In Exercise 7.1.6 the reader was asked to find the Bayes estimator and the empirical Bayes estimator. The form of the BE was

$$\hat{\delta}^{b}_{i} = (I - (A + I)^{-1})x_i, \ 1 \le i \le r \tag{7.6.13}$$

and the EBE was

$$\delta^{eb}_{i} = (I - (r - p - 1)S^{-1})x_i, \ 1 \le i \le r. \tag{7.6.14}$$

where if V is the pxr data matrix $V = (x_1, x_2, ... x_r)$, $S = VV'$. To find the positive part of (7.6.14) first find the singular value decomposition of S, i.e. $S = PDP'$ where D is the diagonal matrix with elements d_i. Then let $S_+ = PD_+P'$ where D_+ is a diagonal matrix with elements $\max\{(r - p - 1), d_i\}$. The positive part estimator then becomes

$$\hat{\delta}^{eb+}_{i} = (I - (r - p - 1)S_+^{-1})x_i, \ 1 \le i \le r. \tag{7.6.15}$$

Exercise 7.6.9. Show that the positive part estimator in (7.6.15) satisfies (7.6.12).

Now recall from (7.1.8) that for the full rank case the estimator of C.R. Rao took the form

$$\begin{aligned}
p'\hat{\beta}_i &= p'b_i - cW(X'X)^{-1}B^{-1}(b_i - b_\bullet) \\
&= p'b_\bullet + p'(I - cW(X'X)^{-1}B^{-1})(b_i - b_\bullet) \\
&= d'g_\bullet + d'(I - cW\Lambda^{-1}U'B^{-1}U)(g_i - g_\bullet).
\end{aligned} \tag{7.6.16}$$

C.R. Rao p.41 has the following result. Let A and C be real mxm symmetric matrices of which C is positive definite. Then there exists a matrix R such that $A = R^{-1'}\Delta R^{-1}$ and $C = (R^{-1})'R^{-1}$ where Δ is a diagonal matrix. In Equation (7.6.16) above let $C = \Lambda$ and A be $U'B^{-1}U$. Then $\Lambda^{-1}U'B^{-1}U = R\Delta R^{-1}$. The positive part may then be formulated by replacing the matrix Δ with a diagonal matrix where the elements $\delta_i \ge 1/cW$ are replaced by $1/cW$ and the other elements remain the same. Call this matrix H. The positive part then takes the form

$$d'\hat{\gamma}_i = d'g_* + d'(I - cWH)(g_i - g_*)$$
$$= p'b_* + p'(I - cWUHU')(b_i - b_*) \qquad (7.6.17)$$
$$= p'b_i - cWUHU'(b_i - b_*).$$

It is not easy to compare the MSE properties of (7.616) and (7.6.17). However the following example suggests that the positive part estimator may do better than the estimator of C.R.Rao. Further investigation is needed.

Example 7.6.1 The Positive Part of the Estimator of C.R. Rao. Using the data in Example 7.1.1 the singular value decomposition of $UBU'\Lambda = QDQ'$ where

$$Q' = \begin{bmatrix} -0.07489 & 0.17023 & 0.98255 \\ -0.53248 & -0.83991 & 0.10493 \\ -0.84312 & 0.51534 & -0.15355 \end{bmatrix}$$

and

$$D = \begin{bmatrix} 3.00268 & 0 & 0 \\ 0 & 1.00561 & 0 \\ 0 & 0 & 0.44923 \end{bmatrix}$$

Since $cW = 1.07393$,

$$H_+ = \begin{bmatrix} 3.00268 & 0 & 0 \\ 0 & 1.07393 & 0 \\ 0 & 0 & 1.07393 \end{bmatrix},$$

$$QH_+^{-1}Q' = \begin{bmatrix} 0.92780 & 0.00763 & 0.04401 \\ 0.00763 & 0.91383 & -0.10004 \\ 0.04401 & -0.10002 & 0.35372 \end{bmatrix}$$

and the positive part estimators are

$$\hat{\beta}_1 = \begin{bmatrix} 4.917 \\ 6.149 \\ 5.815 \end{bmatrix} \quad \hat{\beta}_2 = \begin{bmatrix} 4.956 \\ 6.060 \\ 5.301 \end{bmatrix} \quad \hat{\beta}_3 = \begin{bmatrix} 4.928 \\ 6.123 \\ 5.668 \end{bmatrix} \quad \hat{\beta}_4 = \begin{bmatrix} 4.944 \\ 6.086 \\ 5.454 \end{bmatrix}$$

$$\hat{\beta}_5 = \begin{bmatrix} 4.940 \\ 6.100 \\ 5.510 \end{bmatrix} \quad \hat{\beta}_6 = \begin{bmatrix} 4.933 \\ 6.111 \\ 5.596 \end{bmatrix} \quad \hat{\beta}_7 = \begin{bmatrix} 4.920 \\ 6.141 \\ 5.769 \end{bmatrix} \quad \hat{\beta}_8 = \begin{bmatrix} 4.938 \\ 6.101 \\ 5.537 \end{bmatrix}$$

Using the measure of distance in Example 7.3.1 the positive part reduces the distance from the true values substantially. Consider the comparisons in Table 7.6.1 below.

Table 7.6.1

Estimator	Least Square	Empirical Bayes	Positive Part
Distance	5.542	3.197	2.324

Exercise 7.6.10. For the "true" value of the β parameter $(4, 7, 5)$ do a simulation of the data in Example 7.1.1. Then find the numerical estimates for the estimators of C.R. Rao, Wind , Dempster and their positive parts.

Exercise 7.6.11. How would you formulate the positive parts for estimators intermediate between those of Rao and Wind or intermediate between those of Wind and Dempster?

7.7 Summary

1. A setup with r linear models was considered.
2. Three basic James-Stein type estimators (Rao, Wind and Dempster) were obtained as empirical Bayes estimators and approximate minimum mean square error estimators.
3. The MSEs of the James-Stein type estimators were evaluated with and without averaging over the prior distribution.
4. Some estimators intermediate between those of Rao and Wind and Wind and Dempster were proposed; their average and conditional MSEs were evaluated.

5. A general form of EBE was proposed that included those of Rao, Wind and Dempster as special cases. A formula for the average MSE was obtained.

6. The conditional and average MSEs were compared with that of the LS.

7. The RSL and the RL are extended to the setup with r linear models; their properties are studied.

8. The conditional MSE of the JS and the BE are compared.

9. It is shown how the work on the positive parts in Chapter VI can be extended to the case of r linear models.

Chapter VIII

The Precision of Individual Estimators

8.0 Introduction

Although the JS has a uniformly smaller MSE than the MLE averaging over certain compound loss functions the individual components do not, in general have this property. The properties of the MSEs for the individual components, the IMSEs, can be quite different than those obtained averaging over a compound loss function, the CMSEs. The objective of this chapter is to compare the properties of these two kinds of MSEs for different setups involving univariate linear models.

The IMSE of the JS of the mean of a MN is obtained in Section 8.1. Necessary and sufficient conditions are then derived for the IMSE of the JS to be less than that of the MLE. The geometric form for the regions of the parameter space where this occurs is quite complicated. Some numerical illustrations of the results are included.

The case of a single linear model is considered in Section 8.2. Instead of the IMSE the MSE of a linear combination of the regression parameters is considered. The IMSE may be obtained as a special case. Recall that in previous chapters the formulae for the EBEs contained a constant c with a range of values where the EBE had a smaller CMSE than the LS and a theoretically optimum value that minimized it. The problem is considered relative to the MSE of linear combinations from two standpoints:

1. For what values of c is this MSE less than that of the LS?

2. For optimum c how does the value of this MSE compare with that of the LS?

Section 8.3 deals with the analogous questions for the case of r linear models a richer problem because:

1. There are more estimators to be considered.

2. More measures of efficiency suggest themselves including:

  a. the individual MSE for one linear model, the $(IMSE)_l$;

  b. the individual MSE for r linear models, the (IMSE)r;

  c. the MSE for a compound loss function for one model, the $(CMSE)_l$;

  d. the MSE for a compound loss function for r models, the (CMSE)r.

Section 8.4 provides an introduction to limited translation estimators. The limited translation estimators improve the individual risk of the JS without compromising the properties of the compound risk too much.

A summary is given in Section 8.5.

8.1 The Mean of a Multivariate Normal Distribution

C.R. Rao and Shinozaki 1978 observe that although the compound risk of the JS is uniformly less than that of the MLE, i.e.,

$$E(\hat{\theta} - \theta)'(\hat{\theta} - \theta) \le E(X - \theta)'(X - \theta) \qquad (8.1.1)$$

for all θ , the individual risk

$$E(\hat{\theta}_i - \theta_i)^2 \le E(X_i - \theta_i)^2, \ i = 1,2,...,p. \qquad (8.1.2)$$

is less only for certain portions of the parameter space.

The parameter θ is a vector with components θ_i.

The portions of the parameter space where (8.1.2) holds true is the content of Theorem 8.1.1. A similar result was proved for the estimable parametric functions in a linear regression model in Theorem 4.5.2 . The reader might like to think about how that result can be specialized to that of Theorem 8.1.1. The proof given here does not depend on Theorem 4.5.2. Two useful definitions were given by (4.5.11) and (4.5.12). They are restated here using the notation of Theorem 8.1.1 below. They are

$$S_{1,p}(\lambda) = \sum_{k=0}^{\infty} \frac{e^{-\lambda}\lambda^k}{k!(2k+p-2)} \tag{8.1.3a}$$

and

$$S_{2,p}(\lambda) = \sum_{k=0}^{\infty} \frac{e^{-\lambda}\lambda^k}{k!(2k+p)(2k+p-2)} \tag{8.1.3b}$$

where $\lambda = \theta'\theta/2$.

Theorem 8.1.1. A necessary and sufficient condition for the MSE of the individual components of the JS (the IMSE) to dominate that of the MLE is that the θ_i satisfy any of the following the equivalent inequalities

$$\frac{\theta_i^2}{\lambda} \le \frac{2S_{1,p}(\lambda) - (p+2)S_{2,p}(\lambda)}{\lambda(p+2)S_{2,p+2}(\lambda)}, \tag{8.1.4a}$$

$$\frac{\theta_i^2}{\lambda} \le \frac{4}{p+2} + \frac{(p-2.)S_{2,p}(\lambda)}{(p+2)\lambda S_{2,p+2}(\lambda)} \tag{8.1.4b}$$

and letting $\alpha_i = \theta_i^2 / \sum_{i=1}^{p} \theta_i^2$

$$\alpha_i \le \frac{2}{p+2} + \frac{(p-2.)S_{2,p}(\lambda)}{2(p+2)\lambda S_{2,p+2}(\lambda)}. \tag{8.1.4c}$$

Proof. From Equation (2.6.2) for $1 \le i \le p$ it follows that

$$m_i = E(\hat{\theta}_i - \theta_i)^2$$
$$= E(x_i - \theta_i)^2 - 2(p-2)E\left[\frac{(x_i - \theta_i)x_i}{x'x}\right]$$
$$+ (p-2)^2 E\left[\frac{x_i^2}{(x'x)^2}\right]$$

$$= 1 - 2(p-2)E\left[\frac{1}{x'x}\right]$$
$$+ [4(p-2) + (p-2)^2]E\left[\frac{x_i^2}{(x'x)^2}\right].$$

(8.1.5)

Recall that

$$E\left[\frac{1}{x'x}\right] = S_{1,p}(\lambda).$$

(8.1.6)

It can be shown that (see Exercise 8.1.1)

$$E\left[\frac{x_i^2}{(x'x)^2}\right] = \theta_i^2 S_{2,p+2}(\lambda) + S_{2,p}(\lambda).$$

(8.1.7)

Thus,

$$m_i = 1 - 2(p-2)S_{1,p}(\lambda) + (p-2)(p+2)[\theta_i^2 S_{2,p+2}(\lambda) + S_{2,p}(\lambda)] \le 1$$

(8.1.8)

iff (8.1.4a) holds true. The inequalities (8.1.4a) and (8.1.4b) are equivalent because (see Exercise 8.1.1)

$$S_{1,p}(\lambda) = pS_{2,p}(\lambda) + 2\lambda S_{2,p+2}(\lambda).$$

(8.1.9)

Notice that the second term of inequality (8.1.4b) and (8.1.4c) is positive. A sufficient condition for the i'th component of the individual risk of the JS to be less than the MLE is given by Corollary 8.1.1 below.

Corollary 8.1.1. If

$$\frac{\theta_i^2}{\lambda} \le \frac{4}{p+2}$$

(8.1.10a)

or

$$\alpha_i \le \frac{2}{p+2}$$

(8.1.10b)

then the individual component of the JS dominates that of the MLE, i.e., $m_i \leq 1$.

Intuitively the IMSE of the JS dominates that of the MLE if the square of individual parameter being estimated is small with respect to the total sum of squares of all of the parameters. Some upper bounds on α_i where the IMSE of the JS is less than that of the MLE are given for a few different values of p and $\theta'\theta$ in Table 8.1.1 below.

Table 8.1.1

$\theta'\theta$	0.4	1.2	2	4	12	20
p						
4	1	1	0.758	0.516	0.374	0.354
8	1	1	0.670	0.422	0.264	0.235

For those values of $\theta'\theta$ where the right hand side of inequality (8.1.4c) is greater than 1 the number 1 is entered in the table because $\alpha_i \leq 1$ for all values of θ_i. Note that on hyperspheres in the parameter space where the radius is small the IMSE of the JS is uniformly smaller than that of the MLE. This is illustrated in the plots in Figure 8.1.1.

Exercise 8.1.1 . Verify that the two forms of (8.1.4) are indeed equivalent and that the ratio in the second term of (8.1.4b) is positive.

Exercise 8.1.2. Verify (8.1.7) and (8.1.9).

Exercise 8.1.3. Show that the compound risk of the JS is the average of the individual risks, i.e.,

$$R = \frac{1}{p}\sum_{i=1}^{p} m_i.$$

Exercise 8.1.4. Show that $S_{2,p}(\lambda)/\lambda S_{2,p+2}(\lambda)$ is a decreasing function of λ.

Exercise 8.1.5 . Show that $S_{2,p}(\lambda)$ has the integral representation

$$S_{2,p}(\lambda) = \frac{1}{4\lambda^{p/2}e^{\lambda}}\int_0^\lambda \int_0^y e^x x^{p/2-2} dx dy.$$

Exercise 8.1.6. Use Mathematica, Maple or a TI 92 if available to generate Table 8.1.1 for p = 6 or p = 10. Some code follows.

Mathematica

```
a[p_,x_]=Integrate[Exp[z] z^(p/2-2),{y,0,x},{z,0,y}];
b[p_,x_]=1/(4 x^(p/2) Exp[x]);
c[p_,x_]=a[p,x] b[p,x];
al[p_,x_] =(2/(p+2)+((p-2)/(2 (p+2)))
(c[p,x]/(x c[p+2,x])));
```

Maple

```
> a:=(p,x)->int(int(exp(z)*(z^(p/2-2)),z=0..y),y=0..x);
> b:=(p,x)->1/(4*(x^(p/2))*exp(x));
> c:=(p,x)->a(p,x)*b(p,x);
> al:=(p,x)->2/(p+2)
+((p-2)/(2*(p+2)*x))*(c(p,x)/c(p+2,x));
```

TI-92

```
C:Define a(p,н)=∫(∫(e^z*z^(p/2-2),z,0,y),y,0,н)
C:Define b(p,н)=1/(4*н^(p/2)*e^н)
C:Define c(p,н)=a(p,н)*b(p,н)
C:Define al(p,н)=2/(p+2)+((p-2)/(2(p+2)н))(c(p,н)/c(p+2,н))
```

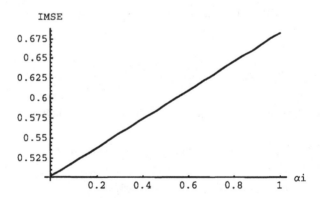

Figure 8.1.1 a IMSE when λ = 0.2

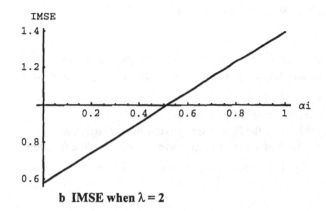

b IMSE when λ = 2

Notice that on the hypersphere $\theta'\theta = 0.4$, i.e., when $\lambda = 0.2$ as in Figure 8.1.1a the IMSE of the JS is uniformly smaller than 1 the risk of the MLE. This is not the case for Figure 8.1.1b where $\theta'\theta = 4$ or $\lambda = 2$. A three dimensional illustration is given by Figure 8.1.2 below. In this picture $\lambda = \theta'\theta/2$ while $\lambda_i = \theta_i^2/2$. Notice that for larger values of λ the risk is greater than one once λ_i is larger than some value.

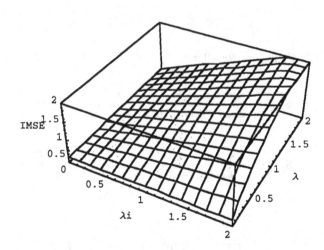

Figure 8.1.2

8.2 The Single Linear Model

The conditional MSE of the JS of the linear combinations of the regression parameters, i.e., $p'\beta$ obtained in Chapter IV may be compared with:

1. the MSE of the LS;
2. the conditional MSE of the corresponding BE;
3. the MSE of the JS averaging over a prior distribution;
4. the MSE of the JS averaging over a predictive loss function.

The objective of this section is to make some of these comparisons.

8.2.1 Comparison with the LS

From Equation (4.5.2) and the proof of Theorem 4.5.2

$$\text{MSE}(p'\hat{\beta}) = p'(X'X)^+ p\sigma^2 - 2c(s-2)p'(X'X)^+ pS_{1,s}(\delta)\sigma^2$$

$$+ [4c(s-2) + c^2(s-2)^2 \left[\frac{(n-s)+2}{(n-s)}\right]] \qquad (8.2.1)$$

$$\left[p'\beta\beta'pS_{2,s+2}(\delta) + \sigma^2 p'(X'X)^+ pS_{2,s}(\delta)\right]$$

where $\delta = \beta'(X'X)\beta/2\sigma^2$. Recall that

$$S_{1,s}(\lambda) = \sum_{k=0}^{\infty} \frac{e^{-\lambda}\lambda^k}{k!(2k+s-2)} = \frac{1}{2e^\delta \delta^{s/2-1}} \int_0^\delta e^x x^{s/2-2} dx, \qquad (8.2.2a)$$

$$S_{2,s+2}(\delta) = \sum_{k=0}^{\infty} \frac{e^{-\delta}\delta^k}{k!(2k+s)(2k+s+2)} = \frac{1}{4e^\delta \delta^{s/2+1}} \int_0^\delta \int_0^x e^y y^{s/2-1} dy dx, \qquad (8.2.2b)$$

and

$$S_{2,s}(\delta) = \sum_{k=0}^{\infty} \frac{e^{-\delta}\delta^k}{k!(2k+s)(2k+s-2)} = \frac{1}{4e^\delta \delta^{s/2}} \int_0^\delta \int_0^x e^y y^{s/2-2} dy dx. \qquad (8.2.2c)$$

A range of positive values for the parameter c where the parametric functions of the JS has a smaller MSE than the LS only exists for certain portions of the parameter space. To obtain this range of values observe that in order for the JS to have a smaller MSE than the LS the inequality

$$c < \left[\frac{p'(X'X)^+ p S_{1,s}(\delta)}{2(p'\beta\beta'p / \beta'X'X\beta) S_{2,s+2}(\delta)\delta + p'(X'X)^+ p S_{2,s}(\delta)} - 2 \right] \qquad (8.2.3)$$

must hold true. For a positive value of c to exist the inequality

$$\frac{p'\beta\beta'p}{\beta'X'X\beta} \le H(\delta) p'(X'X)^+ p \qquad (8.2.4a)$$

with

$$H(\delta) = \frac{S_{1,s}(\delta) - 2S_{2,s}(\delta)}{4\delta S_{2,s+2}(\delta)} \qquad (8.2.4b)$$

holds true. Some values of $H(\delta)$ are given in Table 8.2.1.

Conditions for the IMSE of the JS to be smaller than that of the LS may be obtained by specializing the conditions obtained in Chapter IV for the MSE of the parametric functions to be smaller than that of the LS. Recall that the MSE averaging over the predictive loss function was optimum for $c = (n - s)/((n - s)+2)$. For this value of c substituted into Equation (8.2.1) a necessary and sufficient condition for the estimable parametric functions of the JS to have a smaller MSE than the LS takes the form of the inequality

$$\frac{p'\beta\beta'p}{\beta'X'X\beta} \le \frac{p'(X'X)^+ p}{2\delta S_{2,s}(\delta)} \left[\frac{2S_{1,s}(\delta)}{s+2} - S_{2,s}(\delta) \right]. \qquad (8.2.5)$$

The condition (8.2.5) may be specialized to that for the individual components of $\gamma = U'\beta$ to have a smaller MSE than the LS by setting the i'th component of d in the estimability equation $p' = d'U'$ equal to one. Let $\delta_i = \lambda_i \gamma_i^2 / 2\sigma^2$. Then:

1. From (8.2.3) there is a range of values of the parameter c where the IMSE of the JS is smaller than that of the LS if and only if

$$\frac{\delta_i}{\delta} \le \frac{1}{4\delta S_{2,s+2}(\delta)} (S_{1,s}(\delta) - 2S_{2,s}(\delta)). \qquad (8.2.6)$$

2. When $c = (n-s)/((n-s)+2)$ the necessary and sufficient condition for the   IMSE
of the JS to be smaller than that of the LS takes the form

$$\frac{\delta_i}{\delta} \le \frac{1}{2\delta S_{2,s+2}(\delta)}\left[\frac{2S_{1,s}(\delta)}{s+2} - S_{2,s}(\delta)\right].$$

(8.2.7)

Since

$$sS_{2,s}(\delta) + 2\delta S_{2,s+2}(\delta) = S_{1,s}(\delta),$$

(8.2.8)

alternative forms of inequalities (8.2.6) and (8.2.7) take the form

$$\frac{\delta_i}{\delta} \le \frac{1}{2} + \frac{(s-2)S_{2,s}(\delta)}{2\delta S_{2,s+2}(\delta)}$$

(8.2.9)

and

$$\frac{\delta_i}{\delta} \le \frac{2}{s+2} + \frac{(s-2)S_{2,s}(\delta)}{2(s+2)\delta S_{2,s+2}(\delta)}.$$

(8.2.10)

In Exercise 8.2.1 the reader is invited to show that $S_{2,s+2}(\delta)/\delta S_{2,s}(\delta)$ is a positive
decreasing function of δ with limit zero. When $c = (n-s)/(n-s)+2)$ a sufficient
condition for the JS to have a smaller IMSE than the LS is given by

$$\frac{\delta_i}{\delta} \le \frac{2}{s+2}.$$

(8.2.11)

When

$$\frac{\delta_i}{\delta} \le \frac{1}{2}$$

(8.2.12)

there exist values of c where the JS has a smaller IMSE than the LS.

Table 8.2.1 contains some least upper bounds on δ_i/δ where there exists a c
such that the IMSE of the JS is smaller than that of the LS. Table 8.2.2 contains
some least upper bounds on δ_i/δ where the IMSE of the JS is smaller than that of
the LS when $c_{opt} = (n-s)/((n-s)+2)$. The right hand sides of the inequalities in (8.2.7)-

(8.2.10) must be less than one; when this is not the case for Inequalities (8.2.9) and (8.2.10) the IMSE of the JS always does better for c_{opt}; there always is a range of c where the JS does better than LS if the right hand side of inequalities (8.2.7) and (8.2.8) are greater than one.

Table 8.2.1

s/δ	6	10	30	50	∞	$\delta_i/\delta = 1$
6	0.724	0.621	0.536	0.521	0.500	3.10
10	0.913	0.734	0.571	0.542	0.500	5.06

Table 8.2.2

s/δ	2	10	30	50	∞	$\delta_i/\delta = 1$
6	0.459	0.280	0.259	0.255	0.250	0.627
10	0.397	0.206	0.178	0.174	0.167	0.584

Exercise 8.2.1. Show that

A. $\lim\limits_{\delta \to \infty} \dfrac{S_{1,s}(\delta)}{2\delta S_{2,s+2}(\delta)} = 1$

B. The expression $S_{1,s}(\delta)/S_{2,s+2}(\delta)$ is an increasing function of δ.

C. $\lim\limits_{\delta \to \infty} \dfrac{S_{2,s}(\delta)}{\delta S_{2,s+2}(\delta)} = 0.$

Exercise 8.2.2. Verify the equivalence of the infinite series and the iterated integrals in Equation (8.2.2).

Exercise 8.2.3. Using Mathematica, Maple or the TI 92 calculator generate Tables 8.2.1 and 8.2.2 for s = 4, 10. Some code will be given.

Mathematica

```
a[s_,d_]=1/(2 Exp[d] d^(s/2-1));
I1[s_,d_]=Integrate[Exp[y] y^(s/2-2),{y,0,d}];
s1[s_,d_]=a[s,d] I1[s,d];
b[s_,d_]=1/(4 Exp[d] d^(s/2+1));
I2[s_,d_]=Integrate[Exp[y] y^(s/2-1),{x,0,d},{y,0,x}];
s2[s_,d_]= I2[s,d] b[s,d];
```

```
c[s_,d_]=1/(4 Exp[d] d^(s/2));
I3[s_,d_]=Integrate[Exp[y] y^(s/2-2),{x,0,d},{y,0,x}];
s3[s_,d_]=c[s,d] I3[s,d];
* Use to generate Table 8.2.1
f[s_,d_]=1/2+(s-2) s3[s,d]/(4 d s2[s,d]);
*Use to generate Table 8.2.2
g[s_,d_]=2/(s+2)+((s-2) s3[s,d])/(2 (s+2) s2[s,d] d);
```

Maple

```
>i1:=(s,d)->int(exp(y)*y^(s/2-2),y=0..d):
>s1:=(s,d)->a(s,d)*i1(s,d):
>b:=(s,d)->1/(4*exp(d)*d^(s/2+1)):
>i2:=(s,d)->int(int(exp(y)*y^(s/2-1),y=0..x),x=0..d):
>s2:=(s,d)->i2(s,d)*b(s,d):
>c:=(s,d)->1/(4*exp(d)*d^(s/2)):
>i3:=(s,d)->int(int(exp(y)*y^(s/2-2),y=0..x),x=0..d):
>s3:=(s,d)->c(s,d)*i3(s,d):
>f:=(s,d)->1/2+(s-2)*s3(s,d)/(4*d*s2(s,d)):
>g:=(s,d)->2/(s+2)+((s-2)*s3(s,d))/(2*(s+2)*s2(s,d)*d):
```

TI 92

```
C:Define a(s,d)=1/((2*e^(d))*d^(s/2-1))
C:Define i1(s,d)=∫(e^(y)*y^(s/2-2),y,0,d)
C:Define s1(s,d)=a(s,d)*i1(s,d)
C:Define b(s,d)=1/((4*e^(d))*d^(s/2+1))
C:Define i2(s,d)=∫(∫(e^(y)*y^(s/2-1),y,0,н),н,0,d)
C:Define s2(s,d)=i2(s,d)*b(s,d)
C:Define c(s,d)=1/((4*e^(d))*d^(s/2))
C:Define i3(s,d)=∫(∫(e^(y)*y^(s/2-2),y,0,н),н,0,d)
C:Define s3(s,d)=c(s,d)*i3(s,d)
C:Define f(s,d)=1/2+(s-2)*s3(s,d)/(4*d*s2(s,d))
C:Define g(s,d)=2/(s+2)+((s-2)*s3(s,d))/(2*(s+2)*s2(s,d)*d)
```

From the Tables 8.2.1 and 8.2.2 or the results of Exercise 8.2.3 the reader will notice that for the computed numerical values that as δ or s increases:

1. The width of the range of values of δ_i/δ where there exists a positive c such that the parametric functions of the JS has a smaller MSE than the LS gets narrower but is bounded below by 0.5.

2. The width of the interval on δ_j where the corresponding IMSE of the JS is less than the LS gets narrower.

When $c = (n-s)/(n-s+2)$ for the i'th component of the JS a form of the IMSE is given by

$$
\begin{aligned}
\frac{m_i}{\sigma^2} &= 1 - \frac{(n-s)(s-2)^2}{(n-s+2)s} S_{1,s}(\delta) \\
&+ \frac{(n-s)(s-2)(s+2)}{(n-s+2)s}(2s\delta_i - 2\delta)S_{2,s+2}(\delta).
\end{aligned}
\tag{8.2.13}
$$

Comparing (8.2.13) with the IMSE of the LS divided by σ^2 another necessary and sufficient condition for the IMSE of the JS to be less than that of the LS is that given δ, δ_i satisfies the inequality

$$
(\delta_i - \overline{\delta}) \le \frac{(s-2)S_{1,s}(\delta)}{2s(s+2)S_{2,s+2}(\delta)}
\tag{8.2.14}
$$

where $\overline{\delta} = \sum_{i=1}^{s} \delta_i / s.$

Table 8.2.3 gives some numerical values of the right hand side of inequality (8.2.14).

Table 8.2.3

s/d	0	0.1	0.2	0.3	0.4	0.5	0.6
4	0.5	0.5001	0.5003	0.5007	0.5013	0.5021	0.5030
8	0.5	0.5042	0.5085	0.5128	0.5172	0.5216	0.5261

When $s = 4$, $\delta \ge 0.672$ and when $s = 8$ and $\delta \ge 0.6013$ the right hand side of (8.2.10) is greater than one meaning that the sufficient condition is always fulfilled.

Exercise 8.2.4. A. Show that the regions in (8.2.14) and (8.2.10) are equivalent.
B. Generate the numbers in Table 8.2.3 for $s = 6$ and $s = 10$ using the code in Exercise 8.2.3 with the additional commands.
Mathematica: `h[s_,d_]=(s-2) s1[s,d]/((2 s (s+2) s2[s,d]));`
Maple: `> h:=(s,d)->(s-2)*s1(s,d)/((2*s*(s+2)*s2(s,d))):`
TI 92 `C:Define h(s,d)=(s-2)*s1(s,d)/((2*s*(s+2)*s2(s,d)))`

Exercise 8.2.5. Show that the sum of the components given in Equation (8.2.13) sum to the CMSE of the JS.

A simple comparison of the IMSE with the CMSE may be obtained by doing Exercise 8.2.6.

Exercise 8.2.6. Show that $m_i \leq m/s$ if and only if $\delta_i \leq \delta/s$. Interpret the result.

8.2.2 Comparison of the Conditional and the Average MSE

Some interesting comparisons of the JS and the BE that is obtained for Dempster's prior may be investigated. These include:

1. comparison of the conditional MSE of the parametric functions of the JS with the average MSE;
2. the same kind of comparison as in 1 above for the IMSE.

The conditional CMSE of the JS averaged over a predictive loss function is always less than that of the LS for optimum c. The average CMSE of the LS for estimable parametric functions is the same as the conditional CMSE and is always greater than the average MSE of the BE and the EBE. The CMSE of the JS is not always less than the average MSE. The reader may show in Exercise 8.2.7 that the CMSE of the JS is less than its MSE averaging over Dempster's prior if and only if

$$\frac{\sigma^2}{\sigma^2 + \omega^2} \leq (s-2)S_{1,s}(\delta). \tag{8.2.15}$$

The right hand side of (8.2.15) is a decreasing function of δ with limit zero as $\delta \to \infty$. Thus, for a specific value of $k = \sigma^2/\omega^2$ there exists a δ_0 such that if $\delta \leq \delta_0$ the CMSE of the JS is less than the average MSE. These values of δ_0 for some values of s and k are illustrated in Table 7.24.

Table 7.2.4

s/k	0.01	0.1	1	2	5	10	100
4	101	11	1.59	0.874	0.376	0.194	0.0199
8	301	30.9	3.54	1.85	0.772	0.393	0.0399

Exercise 8.2.7. A. Show that the conditional CMSE of the JS is smaller than the average MSE if and only if Inequality (8.2.15) holds.

B. Show that the right hand side of (8.2.15) is a decreasing function of δ tending to zero as δ gets large implying the existence of a δ_0 such that if $\delta \leq \delta_0$ the conditional CMSE of the JS is less than the average MSE.

C. Obtain the values of δ_0 in Table 7.2.4 for $s = 6, 10$ and the given values of k.

Exercise 8.2.8. A. Find a necessary and sufficient condition for the conditional CMSE of the JS to be smaller than that of the corresponding BE averaged over Dempster's prior.

B. Make a table similar to Table 8.2.4 with values of δ_0 such that if $\delta \leq \delta_0$ the CMSE of the JS is less than the average MSE of the BE.

Comparisons may also be made between the conditional and the average MSE of the parametric functions of the JS . A comparison is made in Theorem 7.2.1.

Theorem 8.2.1. Let

$$a(s,\delta) = \frac{1}{(s+2)S_{2,s+2}(\delta)}\left[2S_{1,s}(\delta) - (s+2)S_{2,s}(\delta) - \frac{\sigma^2}{s(\sigma^2 + \omega^2)}\right].$$

$$(8.2.16a)$$

If $a > 0$ then a necessary and sufficient condition for the conditional MSE of the JS to be smaller than the average MSE is that β lies in the region

$$\beta'(X'X)\beta \leq a(s,\delta)\sigma^2 \qquad (8.2.16b)$$

or equivalently

$$\delta \leq \frac{1}{2}a(s,\delta). \qquad (8.2.16c)$$

Substituting the right hand side of Equation (8.2.8) into (8.2.16c) an equivalent form is given by

$$\delta \leq \frac{1}{2sS_{2,s+2}(\delta)}\left[(s-2)S_{2,s}(\delta) - \frac{k}{s(k+1)}\right]. \qquad (8.2.16d)$$

For any given value of k it can be shown that $(s-2)S_{2,s}(\delta)-k/s(k+1)-2s\delta S_{2,s+2}(\delta)$ is a decreasing function of δ. Hence there exists a δ_0 such that if $\delta \le \delta_0$ Inequality (8.2.16d) holds true. Some values of δ_0 are given in Table 8.2.5.

Table 8.2.5

s/k	0.01	0.1	1	2	5	10	100
4	0.662	0.590	0.284	0.181	0.087	0.046	0.005
8	0.594	0.536	0.272	0.176	0.086	0.046	0.005

The following simple corollary gives a necessary and sufficient condition for the IMSE of the JS without averaging over the prior to be less than the average IMSE.

Corollary 8.2.1. The IMSE of the JS without averaging over the prior is less than the IMSE of the JS averaging over the prior iff

$$\frac{\delta_i}{\delta} \le \frac{2}{s+2} + \frac{1}{2\delta(s+2)S_{2,s+2}(\delta)}\left[(s-2)S_{2,s}(\delta) - \frac{k}{s(1+k)}\right]. \qquad (8.2.17)$$

Since $S_{2,s}(\delta)$ is a decreasing function of δ with limit zero as $\delta \to \infty$ there exists a δ_0 such that if $\delta \le \delta_0$ then $(s-2)S_{2,s}(\delta)-k/s(1+k) \ge 0$. This results in the following sufficient condition for the conditional IMSE of the JS to be smaller than its average IMSE.

Corollary 8.2.2. Let δ_0 be such that if $\delta \le \delta_0$ then $(s-2)S_{2,s}(\delta)-k/s(1+k) \ge 0$. If

$$\frac{\delta_i}{\delta} \le \frac{2}{s+2} \qquad (8.2.18)$$

then the conditional IMSE of the JS is smaller than the average IMSE. The bounds for the sufficient condition are given in Table 8.2.6.

Table 8.2.6

s/k	0.01	0.1	1	2	5	10	100
4	0.333	0.333	0.333	0.333	0.277	0.144	0.015
8	0.200	0.200	0.200	0.200	0.200	0.200	0.025

Let $A = \min\{d_0, 2/(s+2)\}$. For $s = 4$, $A = 1/3$ if $k < 4.082$ and d_0 otherwise. Likewise for $s = 8$ $A = 1/5$ if $K < 12.13$ and d_0 otherwise.

Exercise 8.2.9. A. Prove Theorem 8.2.1 and its Corollary 8.2.1.
B. Show that $(s-2)S_{2,s}(\delta) - k/s(k+1) - 2s\delta S_{2,s+2}(\delta)$ is a decreasing function of δ with limit zero at infinity. Thus, for a specific value of $k = \sigma^2/\omega^2$. there exists δ_0 such that if $\delta \leq \delta_0$ $a > 0$.
C. Find δ_0 for $s = 6, 10$ and $k = \sigma^2/\omega^2$.

Exercise 8.2.10. A. Establish Corollary 8.2.2.
B. Find the entries in Table 8.2.5 for $s = 6, 10$.
C. Find the values of A for $s = 6, 10$.
   The following additional computer code will help you do Exercises 8.2.9 and 8.2.10. Use it in conjunction with the code given for Exercise 8.2.3.

Mathematica

```
h1[s_,k_,d_]=1/((s+2) s2[s,d]);
h2[s_,k_,d_]= (2 s1[s,d]-(s+2) s3[s,d]-k/(s (1+k)));
h[s_,k_,d_]=h1[s,k,d] h2[s,k,d];
l[s_,k_,d_]=(s-2) s3[s,d]-k/((k+1) s);
p[s_,k_,d_]=l[s,k,d]/(2*s*s2[s,d]);
pd[s_,k_,d_]=p[s,k,d]-d;
```

Maple

```
> h1:=(s,k,d)->1/((s+2)*s2(s,d)):
> h2:=(s,k,d)->(2*s1(s,d)-(s+2)*s3(s,d)-k/(s*(1+k))):
> h:=(s,k,d)->h1(s,k,d)*h2(s,k,d):
> l:=(s,k,d)->(s-2)*s3(s,d)-k/((k+1)*s):
> p:=(s,k,d)->l(s,k,d)/(2*s*s2(s,d)):
> pd:=(s,k,d)->p(s,k,d)-d:
```

TI 92

```
C:Define h1(s,k,d)=1/((s+2)*s2(s,d))
C:Define h2(s,k,d)=((2*s1(s,d)-(s+2)*s3(s,d)-k/(s*(1+k))))
C:Define h(s,k,d)=h1(s,k,d)*h2(s,k,d)
C:Define l(s,k,d)=(s-2)*s3(s,d)-k/((k+1)*s)
```

```
C:Define p(s,k,d)=I(s,k,d)/(2*s*s2(s,d))
C:Define q(s,k,d)=p(s,k,d)-d
```

The author found that for many of the cases when using the TI 92 he got better results by first graphing the function and then finding the zeros than by using the nSolve command giving bounds on the roots.

8.3 The Case of r Linear Models

The study of the efficiencies of JS type estimators for setups with r linear models is a very rich problem because four different kinds of related MSEs can be considered. These include:

1. the MSE averaging over both the compound loss function and the r linear models $(CMSE)_r$;
2. the MSE averaging over the compound loss function for each individual linear model $(CMSE)_l$;
3. the individual MSE averaged over the r linear models $(IMSE)_r$;
4. the individual MSE for each individual linear model $(IMSE)_l$.

The objective of this section is to compare and contrast the properties of these four measures of efficiency for different JS type estimators. The evaluation of the $(CMSE)_r$ and its properties has already been considered in Chapter VII. For the sake of completeness recall that the $(CMSE)_r$ is given by

$$m_{..} = \sum_{i=1}^{r} E_\beta (\hat{\beta}_i - \beta_i)'X'X(\hat{\beta}_i - \beta_i). \tag{8.3.1}$$

The $(CMSE)_l$ consists of the individual summands of (8.3.1), that is

$$m_{i.} = E_\beta (\hat{\beta}_i - \beta_i)'X'X(\hat{\beta}_i - \beta_i) = \sum_{j=1}^{s} E_\gamma (\hat{\gamma}_{ij} - \gamma_{ij})\lambda_j. \tag{8.3.2}$$

The $(IMSE)_l$ consists of the individual summands of the final expression in (8.3.2) divided by the λ_j , i.e.,

$$m_{ij} = E_\gamma (\hat{\gamma}_{ij} - \gamma_{ij})^2. \tag{8.3.3}$$

The $(IMSE)r$ is obtained by summation of the (IMSE) over the r linear models, i.e.,

$$m_{\cdot j} = \sum_{i=1}^{r} E_\gamma (\hat{\gamma}_{ij} - \gamma_{ij})^2. \tag{8.3.4}$$

The relationship among the various MSEs is explained by the equation

$$m_{\cdots} = \sum_{i=1}^{r}\sum_{j=1}^{s} m_{ij}\lambda_j = \sum_{i=1}^{r} m_{i\cdot} = \sum_{j=1}^{s} m_{\cdot j}\lambda_j. \tag{8.3.5}$$

The properties of the estimators of Wind and Dempster will now be considered for these different kinds of MSE.

8.3.1 The Estimator of Wind

In Chapter VII it was observed that the (IMSE)$_r$ and the (CMSE)$_r$ of the Wind estimator are both smaller than that of the LS. This is what would be expected because:

1. The MMSE that is approximated by the Wind estimator was derived by minimizing all if the components of the (IMSE)$_r$.
2. The (CMSE)$_r$ is the sum of the components of the (IMSE)$_r$ (see Equation 8.3.5).

When the mean of an MN was estimated the compound risk (CR) of the JS was uniformly smaller than that of the MLE but the individual risk (IR) was not. By analogy there is strong reason to doubt that the CMSE or the IMSE of the JS would not be uniformly smaller than that of the LS. The (IMSE)$_r$ of the Wind estimator was obtained in Equation (7.2.55) and is less than that of the LS provided that

$$c < \frac{2r(n-s)}{r(n-s)+2} \tag{8.3.6}$$

For optimum c the (CMSE)$_r$ was given by Equation (7.2.59). The result for the predictive loss function is obtained when D is the matrix of eigenvalues of $X'X$. The IMSE consists of the individual model components of the IMSE. It is given in Theorem 8.3.1 below.

Theorem 8.3.1. The (IMSE)$_l$ of the Wind estimator (7.2.11) is given by

$$m_{ij} = \frac{\sigma^2}{\lambda_j} - \frac{2c(r-3)\sigma^2}{\lambda_j}\left[\frac{(r-1)}{r}S_{1,r-1}(\delta_{\cdot j})\right]$$

$$+\frac{\sigma^2}{\lambda_j}\left[4c(r-3)+\frac{c^2(r-3)^2(r(n-s)+2)}{r(n-s)}\right]$$

$$\left[\frac{(\gamma_{ij}-\gamma_{\cdot j})^2 S_{2,r+1}(\delta_{\cdot j})}{\sigma^2}+\frac{r-1}{r}S_{2,r-1}(\delta_{\cdot j})\right] \tag{8.3.7}$$

where $\delta_{ij} = (\gamma_{ij}-\gamma_{\cdot j})^2/2\sigma^2$ and $\delta_{\cdot j} = \sum_{i=1}^{r}\delta_{ij} = \sum_{i=1}^{r}(\gamma_{ij}-\gamma_{\cdot j})^2/2\sigma^2$. Also

$$S_{1,r-1}(\delta_{\cdot j}) = \sum_{k=0}^{\infty}\frac{e^{-\delta_{\cdot j}}\delta_{\cdot j}^k}{k!(r+2k-3)}, \tag{8.3.8a}$$

$$S_{2,r+1}(\delta_{\cdot j}) = \sum_{k=0}^{\infty}\frac{e^{-\delta_{\cdot j}}\delta_{\cdot j}^k}{k!(2k+r-1)(2k+r+1)} \tag{8.3.8b}$$

and

$$S_{2,r-1}(\delta_{\cdot j}) = \sum_{k=0}^{\infty}\frac{e^{-\delta_{\cdot j}}\delta_{\cdot j}^k}{k!(2k+r-1)(2k+r-3)}. \tag{8.3.8c}$$

The proof of Theorem 8.3.1 requires Lemmas 8.3.1 and 8.3.2 below. The reader is invited to prove these results in Exercises 8.3.1 and 8.3.2.

Lemma 8.3.1. Let x_i, $1 \leq i \leq r$ be independent $N(\mu_i, 1)$ random variables. Let $m_1 = \sum_{i=1}^{r}x_i^2$. Then

$$E\left[\frac{(x_i-\mu_i)x_i}{m_1}\right] = 2E\left[\frac{x_i^2}{m_1^2}\right]-E\left[\frac{1}{m_1}\right]. \tag{8.3.9}$$

Lemma 8.3.2. Let x_i, $1 \leq i \leq r$ be independent $N(\mu_i, 1)$ random variables. Let $m_2 = \sum_{i=1}^{r}(x_i-\bar{x})^2$. Then

$$E\left[\frac{(x_i-\mu_i)(x_i-\bar{x})}{m_2}\right] = \frac{r-1}{r}E\left[\frac{1}{m_2}\right]-2E\left[\frac{(x_i-\bar{x})^2}{m_2^2}\right]. \tag{8.3.10}$$

Outline of the proof of Theorem 8.3.1. Notice that

$$
\begin{aligned}
m_{ij} = E[g_{ij} - \gamma_{ij}]^2 &- \frac{2c(r-3)\sigma^2}{\lambda_j} E\left[\frac{(g_{ij} - g_{\cdot j})(g_{ij} - \gamma_{ij})}{m_{jj}}\right] \\
&+ \frac{c^2(r-3)^2[r(n-s)+2]}{r(n-s)}\sigma^4 E\left[\frac{(g_{ij} - g_{\cdot j})^2}{\lambda_j^2 m_{jj}^2}\right].
\end{aligned}
\tag{8.3.11}
$$

Since $g_{ij} \sim N(\gamma_{ij}, \sigma^2/\lambda_j)$. The following transformations give normal random variables with variance one so that Lemma 8.3.2 can be applied.. Let $\xi_{ij} = (\lambda_j^{1/2}/\sigma)\gamma_{ij}$ and $f_{ij} = (\lambda_j^{1/2}/\sigma)g_{ij}$. Thus,

$$
\begin{aligned}
E\left[\frac{(g_{ij} - g_{\cdot j})(g_{ij} - \gamma_{ij})}{m_{jj}}\right] &= E\left[\frac{(f_{ij} - f_{\cdot j})(f_{ij} - \xi_{ij})}{n_{jj}}\right] \\
&= \left[\frac{(r-1)}{r}E\left[\frac{1}{n_{jj}}\right] - 2E\left[\frac{(f_{ij} - f_{\cdot j})^2}{n_{jj}^2}\right]\right] \\
&= \frac{\lambda_j}{\sigma^2}\left[\frac{(r-1)}{r}E\left[\frac{1}{m_{jj}}\right] - 2E\left[\frac{(g_{ij} - g_{\cdot j})^2}{m_{jj}^2}\right]\right]
\end{aligned}
\tag{8.3.12}
$$

where $n_{jj} = \sum_{i=1}^{r}(f_{ij} - f_{\cdot j})^2$. Let $\psi_{ij} = H_0 g_{ij}$ where H_0 was defined in Equation (7.2.50). It follows that $g_{ij} - g_{\cdot j} = H_0'\psi_{ij}$ and that

$$
\begin{aligned}
E\left[\frac{(g_{ij} - g_{\cdot j})^2}{m_{jj}^2}\right] &= \sum_{a=1}^{r-1} c_{ia}^2 E\left[\frac{\psi_{aj}^2}{[\sum_{a=1}^{r-1}\psi_{aj}^2]^2}\right] + \sum_{a \neq a'} c_{ba} c_{ba'} E\left[\frac{\psi_{aj}\psi_{a'j}}{[\sum_{a=1}^{r-1}\psi_{aj}^2]^2}\right] \\
&= \frac{\lambda_j}{\sigma^2} E\left[\frac{2k(\gamma_{ij} - \gamma_{\cdot j})^2/2\delta_j\sigma^2 + (r-1)/r}{(2k+r-1)(2k+r-3)}\right] \\
&= \frac{\lambda_j}{\sigma^4}(\gamma_{ij} - \gamma_{\cdot j})^2 S_{2,r+1}(\delta_{\cdot j}) + \frac{\lambda_j}{\sigma^2}\frac{(r-1)}{r} S_{2,r-1}(\delta_{\cdot j}).
\end{aligned}
\tag{8.3.13}
$$

The result follows after substitution of (8.3.12) and (8.3.13) into (8.3.11)

Exercise 8.3.1. Prove Lemmas 8.3.1 and 8.3.2. Integration by parts should prove to be useful.

Exercise 8.3.2. Fill in the details in the proof of Theorem 7.3.1.

Exercise 8.3.3. Show that the(IMSE)$_1$ of the Wind estimator of Exercise 7.3.3 is

$$
\begin{aligned}
m_{ij} = {} & \frac{\sigma^2}{\lambda_j} - \frac{2c(r-2)\sigma^2}{\lambda_j}\left[S_{1,r}(\varepsilon_{\cdot j})\right] \\
& + \frac{\sigma^2}{\lambda_j}\left[4c(r-2) + \frac{c^2(r-2)^2(r(n-s)+2)}{r(n-s)}\right] \\
& \left[\frac{\gamma_{ij}^2}{\sigma^2}S_{2,r+2}(\varepsilon_{\cdot j}) + S_{2,r}(\varepsilon_{\cdot j})\right]
\end{aligned}
\tag{8.3.14}
$$

where $\varepsilon_{\cdot j} = \gamma_{ij}^2/2\sigma^2$.

The (CMSE)$_1$ is the sum of the (IMSE)$_1$ over all s regression parameters weighted by the eigenvalues of the $X'X$ matrix. Thus,

Corollary 8.3.1. The (CMSE)$_1$ of the Wind estimator (7.2.11) is given by

$$
\begin{aligned}
m_{i\cdot} = {} & \sigma^2 s - 2c(r-3)\sigma^2\left[\frac{(r-1)}{r}\sum_{j=1}^{s}S_{1,r-1}(\delta_{\cdot j})\right] \\
& + \sigma^2\left[4c(r-3) + \frac{c^2(r-3)^2(r(n-s)+2)}{r(n-s)}\right] \\
& \left[\sum_{j=1}^{s}\frac{(\gamma_{ij}-\gamma_{\cdot j})^2}{\sigma^2}S_{2,r+1}(\delta_{\cdot j}) + \frac{r-1}{r}\sum_{j=1}^{s}S_{2,r-1}(\delta_{\cdot j})\right]
\end{aligned}
\tag{8.3.15}
$$

Exercise 8.3.4. What is the (CMSE)$_1$ of the Wind estimator in Exercise 7.2.3.

Exercise 8.3.5. Show that (CMSE)$_r$ of the Wind estimator is obtained when the (IMSE)$_1$ is summed over both indices i and j .

With the help of Theorem 8.3.1, Corollary 8.3.1 above and Corollaries 8.3.2 and 8.33 below comparisons will be made of the various MSEs of the Wind estimator and the JS. Two questions concerning these MSE that have already been considered for the single model case include:

1. For what range of values of c is m_{ij} or m_j uniformly less than the corresponding component of the LS?

2. For what values of the δ_{ij} and δ_j and hence the β parameters is the $(IMSE)_1$ or $(CMSE)_1$ less than that of the LS?

These questions are addressed for the case of r linear models by the two corollaries of Theorem (8.3.1) below.

Corollary 7.3.2. 1. Let

$$cw_j(\delta_{\cdot j}, \gamma_{ij}, r, s, n)$$

$$= \frac{2r(n-s)}{[(r(n-s)+2](r-3)}\left[\frac{(r-1)S_{1,r-1}(\delta_{\cdot j})}{r(\gamma_{ij}-\gamma_{\cdot j})^2 S_{2,r+1}(\delta_{\cdot j})/\sigma^2 + (r-1)S_{2,r-1}(\delta_{\cdot j})} - 2\right]$$

$$(8.3.16)$$

If

$$c < cw_j(\delta_{\cdot j}, \gamma_{ij}, r, s, n) \tag{8.3.17}$$

then the $(IMSE)_1$ of the j'th component of the JS is less than the $(IMSE)_1$ of the corresponding component of the LS. Furthermore if

$$c < \min_{1 \le j \le s} cw_j(\delta_{\cdot j}, \gamma_{ij}, r, s, n) \tag{8.3.18}$$

the $(IMSE)_1$ for all the components of the JS are less than that of the LS.

2. If

$$\frac{(r-1)S_{1,r-1}(\delta_{\cdot j})}{2r(\gamma_{ij}-\gamma_{\cdot j})^2 S_{2,r+1}(\delta_{\cdot j}) + (r-1)S_{2,r-1}(\delta_{\cdot j})} > 2 \tag{8.3.19}$$

or equivalently

$$\frac{(\gamma_{ij}-\gamma_{\cdot j})^2}{2\sigma^2} < \frac{(r-1)[S_{1,r-1}(\delta_{\cdot j}) - 2S_{2,r-1}(\delta_{\cdot j})]}{4rS_{2,r+1}(\delta_{\cdot j})} \tag{8.3.20}$$

such a positive c exists.

The bounds on δ_{ij} are given on for the existence of the constant c for different values of $\delta_{\cdot j}$ in Table 8.3.1 below. If in the table $\delta_{ij} \geq \delta$ there is always a value of c because $\delta_{ij} / \delta_{\cdot j}$ is at most one. Such values are marked with an asterisk.

Table 8.3.1

$s/\delta_{\cdot j}$	0.5	2.0	3.5	5.0	6.5	8.0
5	1.305 *	1.678	2.126	2.630	3.170	3.731
7	1.847 *	2.292 *	2.796	3.342	3.917	4.512
9	2.374 *	2.866 *	3.405	3.978	4.575	5.189

Exercise 8.3.6. Show that

$$S_{1,r-1}(\delta_{\cdot j}) - 2S_{2,r+1}(\delta_{\cdot j}) > 0.$$

Corollary 8.3.3. For optimum $c = r(n-s)/(r(n-s)+2)$ the Wind estimator has a smaller $(IMSE)_1$ than the LS iff

$$\delta_{ij} \leq \frac{(r-1)}{2r(r+1)S_{2,r+1}(\delta_{\cdot j})} (2S_{1,r-1}(\delta_{\cdot j}) - (r+1)S_{2,r-1}(\delta_{\cdot j}))$$

(8.3.21)

where $\delta_{ij} = (\gamma_{ij} - \gamma_{\ast})^2 / 2\sigma^2$.

Since

$$S_{1,r-1}(\delta_{\cdot j}) = 2\delta_{\cdot j}S_{2,r+1}(\delta_{\cdot j}) + (r-1)S_{2,r-1}(\delta_{\cdot j}),$$

(8.3.22)

Inequality (8.3.21) is equivalent to

$$\delta_{ij} \leq \frac{(r-1)}{r(r+1)}\left[2\delta_{\cdot j} + \frac{(r-3)S_{2,r-1}(\delta_{\cdot j})}{2S_{2,r+1}(\delta_{\cdot j})}\right]$$

(8.3.23)

The sufficient condition given by Corollary 8.3.4 then follows.

Corollary 8.3.4. If

$$\frac{\delta_{ij}}{\delta} \leq \frac{2(r-1)}{r(r+1)}$$

(8.3.24)

then the Wind estimator has a smaller $(IMSE)_1$ than the LS.

Exercise 8.3.7. Show that if the $(IMSE)_1$ of the JS is less than that of the LS for a particular component then

$$\delta_{ij} \leq \frac{(r-1)S_{1,r-1}(\delta_{\cdot j})}{r(r+1)S_{2,r+1}(\delta_{\cdot j})} \qquad (8.3.25)$$

Some bounds on δ_{ij} for different values of $\delta_{\cdot j}$ are given where the $(IMSE)_1$ of the JS is smaller than that of the LS in Table 8.3.2 a. Numbers are starred for the same reason as in Table 8.3.1. Tables 8.3.2 b and c give the bounds on δ_{ij} obtained for the sufficient and the necessary conditions respectively.

Table 8.3.2 a

$s/\delta_{\cdot j}$	0.5	2.0	3.5	5.0	6.5	8.0
5	0.502 *	0.826	1.175	1.543	1.923	2.310
7	0.515 *	0.787	1.074	1.371	1.676	1.985
9	0.519 *	0.751	0.992	1.240	1.493	1.749

8.3.2 b

$s/\delta_{\cdot j}$	0.5	2.0	3.5	5.0	6.5	8.0
5	0.133	0.533	0.933	1.333	1.733	2.133
7	0.107	0.429	0.750	1.071	1.393	1.714
9	0.089	0.356	0.622	0.889	1.156	1.422

8.3.2 c

$s/\delta_{\cdot j}$	0.5	2.0	3.5	5.0	6.5	8.0
5	1.607	1.704	1.901	2.173	2.493	2.842
7	1.332	1.505	1.722	1.971	2.241	2.527
9	1.236	1.410	1.609	1.825	2.055	2.294

Exercise 8.3.8. Verify (8.3.23) and (8.3.24).

Exercise 8.3.9. Formulate and prove results that are analogous to Corollaries 7.3.1 and 7.3.2 for the Wind estimator that was obtained when the prior mean was assumed to be zero.

The reader is invited to verify the following results comparing the $(CMSE)_1$ of the Wind estimator to the LS in Exercise 8.3.10.

Exercise 8.3.10. Show that

1. If

$$c < \frac{2r(n-s)}{[r(n-s)+2](r-3)]} \left[\frac{(r-1)\sum_{j=1}^{s} S_{1,r-1}(\delta_{\cdot j})}{\sum_{j=1}^{s} 2r\delta_{ij}S_{2,r+1}(\delta_{\cdot j}) + \sum_{j=1}^{s} S_{2,r-1}(\delta_{\cdot j})} - 2 \right]$$

(8.3.26)

then the $(CMSE)_1$ of the JS of Wind is less than that of the LS.

2. If for each j for a particular i

$$\delta_{ij} < \frac{(r-1)}{4rS_{2,r+1}(\delta_{\cdot j})} (S_{1,r-1}(\delta_{\cdot j}) - 2S_{2,r+1}(\delta_{\cdot j}))$$

(8.3.27)

then such a c exists.

3. For optimum $c = r(n-s)/(r(n-s)+2)$ the $(CMSE)_1$ of the JS of Wind is smaller than that of the LS if for each j

$$\delta_{ij} \le \frac{(r-1)}{2r(r+1)S_{2,r+1}(\delta_{\cdot j})} \left[2S_{1,r-1}(\delta_{\cdot j}) - (r+1)S_{2,r-1}(\delta_{\cdot j}) \right].$$

(8.3.28)

4. A necessary but not sufficient condition for the $(CMSE)_1$ of the JS of Wind to be smaller than that of the LS is that for each j

$$\delta_{ij} \le \frac{(r-1)S_{1,r-1}(\delta_{\cdot j})}{r(r+1)S_{2,r+1}(\delta_{\cdot j})}$$

(8.3.29)

and a sufficient but not necessary condition is that for each j

$$\frac{\delta_i}{\delta} \le \frac{2(r-1)}{(r+1)r}.$$ (8.3.30)

Exercise 8.3.12. Obtain the bounds in Table 8.3.1 and 8.3.2 for r = 6, 8, 10, 20 for δ = 0.5 + 0.5k , k =1, 2, ..., 10. Some code follows.

Mathematica

```
a[r_,d_]=1/(2 Exp[d] d^((r-1)/2-1));
I1[r_,d_]=Integrate[Exp[y] y^((r-1)/2-2),{y,0,d}];
s1[r_,d_]=a[r,d] I1[r,d];
b[r_,d_]=1/(4 Exp[d] d^((r-1)/2+1));
I2[r_,d_]=Integrate[Exp[y] y^((r-1)/2-
1),{x,0,d},{y,0,x}];
s2[r_,d_]= I2[r,d] b[r,d];
c[r_,d_]=1/(4 Exp[d] d^((r-1)/2));
I3[r_,d_]=Integrate[Exp[y] y^((r-1)/2-
2),{x,0,d},{y,0,x}];
s3[r_,d_]=c[r,d] I3[r,d];
Bound for the existence of c.
cj[r_,d_]=((r-1) (s1[r,d]-2 s3[r,d]))/(4 r s2[r,d]);
Bound for necessary and sufficient condition.
dj[r_,d_]=((r-1) (2 s1[r,d]-(r+1) s3[r,d]))/
(2 r (r+1) s2[r,d]);
Bound for sufficient condition.
e[r_,d_]=(2 (r-1)/(r (r+1))) d
Bound for necessary condition.
fj[r_,d_]=(r-1) s1[r,d]/(r (r+1) s2[r,d]);
```

Maple

```
> a:=(r,d)->1/(2*exp(d)*d^((r-1)/2-1)):
> i1:=(r,d)->int(exp(y)*y^((r-1)/2-2),y=0..d):
> s1:=(r,d)->a(r,d)*i1(r,d):
> b:=(r,d)->1/(4*exp(d)*d^((r-1)/2+1)):
> i2:=(r,d)->int(int(exp(y)*y^((r-1)/2-
1),y=0..x),x=0..d):
s2:=(r,d)->i2(r,d)*b(r,d):
c:=(r,d)->1/(4*exp(d)*d^((r-1)/2)):
> i3:=(r,d)->int(int(exp(y)*y^((r-1)/2-
2),y=0..x),x=0..d):
> s3:=(r,d)->i3(r,d)*c(r,d):
Bound for the existence of c
> cj:=(r,d)->((r-1)*(s1(r,d)-2*s3(r,d)))/(4*r*s2(r,d)):
Bound for necessary and sufficient condition.
```

```
> dj:=(r,d)->((r-1)*(2*s1(r,d)-
(r+1)*s3(r,d)))/(2*r*(r+1)*s2(r,d)):
```
Bound for sufficient condition
```
> e:=(r,d)->(2*(r-1)/(r*(r+1)))*d:
```
Bound for necessary condition.
```
> fj:=(r,d)->(r-1)*s1(r,d)/(r*(r+1)*s2(r,d)):
```

TI 92

```
C:Define a(r,d)=1/((2*e^(d))*d^((r-1)/2-1))
C:Define i1(r,d)=∫(e^(y)*y^((r-1)/2-2),y,0,d)
C:Define s1(r,d)=a(r,d)*i1(r,d)
C:Define b(r,d)=1/((4*e^(d))*d^((r-1)/2+1))
C:Define i2(r,d)=∫(∫(e^(y)*y^((r-1)/2-1),y,0,ж),ж,0,d)
C:Define s2(r,d)=i2(r,d)*b(r,d)
C:Define c(r,d)=1/((4*e^(d))*d^((r-1)/2))
C:Define i3(r,d)=∫(∫(e^(y)*y^((r-1)/2-2),y,0,ж),ж,0,d)
C:Define s3(r,d)=c(r,d)*i3(r,d)
```
Bound for the existence of c
```
C:Define cj(r,d)=((r-1)*(s1(r,d)-2*s3(r,d)))/(4*r*s2(r,d))
```
Bound for necessary and sufficient condition
```
C:Define dj(r,d)=((r-1)*(2*s1(r,d)-
(r+1)*s3(r,d))/(2*r*(r+1)*s2(r,d)))
```
Bound for sufficient condition
```
C:Define e(r,d)=(2*(r-1)/(r*(r+1)))*d
```
Bound for necessary condition.
```
C:Define fj(r,d)=(r-1)*s1(r,d)/(r*(r+1)*s2(r,d))
```

8.3.2 The Estimator of Dempster

Unlike the estimator of Wind that uses only one component from each of the r linear models to estimate each regression component the estimator of Dempster uses all of the components. Thus one might expect regions of optimality over the LS that are somewhat different with the same basic form. The different MSEs will be obtained together with a few numerical comparisons. The questions asked and the kind of conditions obtained will be similar to those in Section 8.3.1.

The (IMSE)$_1$ of the Dempster estimator (7.3.4a) is similar to that obtained for the Wind estimator with $(r-1)s$ and $(r-1)s-2$ replacing $r-1$ and in a number of places. It can be obtained by appropriately specializing (7.3.20). Theorem 8.3.2 gives the $(IMSE)_1$, the $(CMSE)_1$ and the $(IMSE)_r$.

Theorem 8.3.2. 1. The $(IMSE)_1$ of the Dempster JS is given by

$$
\begin{aligned}
m_{ij} &= \frac{\sigma^2}{\lambda_j} - \frac{2c((r-1)s-2)\sigma^2}{\lambda_j}\left[\frac{(r-1)}{r}S_{1,(r-1)s}(\delta)\right] \\
&+ \frac{\sigma^2}{\lambda_j}\left[4c((r-1)s-2)+\frac{c^2((r-1)s-2)^2(r(n-s)+2)}{r(n-s)}\right] \\
&\quad\left[\frac{(\gamma_{ij}-\gamma_{\cdot j})^2\lambda_j S_{2,(r-1)s+2}(\delta)}{\sigma^2} + \frac{r-1}{r}S_{2,(r-1)s}(\delta)\right]
\end{aligned}
\tag{8.3.31}
$$

2. The $(CMSE)_1$ is given by

$$
\begin{aligned}
m_{j\cdot} &= \sigma^2 s - 2cs((r-1)s-2)\sigma^2\left[\frac{(r-1)}{r}S_{1,(r-1)s}(\delta)\right] \\
&+\sigma^2\left[4c((r-1)s-2)+\frac{c^2((r-1)s-2)^2(r(n-s)+2)}{r(n-s)}\right] \\
&\quad\left[\frac{\sum_{j=1}^{s}(\gamma_{ij}-\gamma_{\cdot j})^2\lambda_j S_{2,(r-1)s+2}(\delta)}{\sigma^2} + \frac{(r-1)s}{r}S_{2,(r-1)s}(\delta)\right]
\end{aligned}
\tag{8.3.32}
$$

3.The $(IMSE)_r$ is given by

$$
\begin{aligned}
m_{ij} &= \frac{r\sigma^2}{\lambda_j} - \frac{2c((r-1)s-2)\sigma^2}{\lambda_j}\left[(r-1)S_{1,(r-1)s}(\delta)\right] \\
&+ \frac{\sigma^2}{\lambda_j}\left[4c((r-1)s-2)+\frac{c^2((r-1)s-2)^2(r(n-s)+2)}{r(n-s)}\right] \\
&\quad\left[\frac{\lambda_j\sum_{i=1}^{r}(\gamma_{ij}-\gamma_{\cdot j})^2 S_{2,(r-1)s+2}(\delta)}{\sigma^2} + (r-1)S_{2,(r-1)s}(\delta)\right].
\end{aligned}
\tag{8.3.33}
$$

Unlike the estimator of Wind when $c = r(n-s)/(r(n-s)+2)$ the $(IMSE)_r$ of the estimator of Dempster is not uniformly less than that of the LS.

Similar conditions to those obtained for the estimator of Wind may be derived for the estimator of Dempster where there exists a c where the $(IMSE)_1$, $(CMSE)_r$ and $(IMSE)_r$ are smaller than that of the LS. Furthermore when c is optimal for the $(CMSE)_r$ inequalities for the other MSEs to be smaller than LS may be obtained. These are given in the series of corollaries of Theorem 8.3.2 that follow.

Corollary 8.3.5. Let

$$cd_j(\delta,\gamma_{ij},r,s,n)$$

$$= \frac{2r(n-s)}{[(r(n-s)+2\}(r-3)} \cdot$$

$$\left[\frac{(r-1)S_{1,r(-1)s}(\delta)}{r\lambda_j(\gamma_{ij}-\gamma_{\cdot j})^2 S_{2,(r+1)s+2}(\delta)/\sigma^2 + (r-1)S_{2,(r-)sl}(\delta)} - 2 \right] \tag{8.3.34a}$$

For those components where

$$c < cd_j(\delta,\gamma_{ij},r,s,n) \tag{8.3.34b}$$

the $(IMSE)_1$ of the JS of Dempster is smaller than the $(IMSE)_1$ of the LS. Such a positive c exists provided that

$$\frac{\lambda_j(\gamma_{ij}-\gamma_{\cdot j})^2}{2\sigma^2} < \frac{(r-1)[S_{1,(r-1)s}(\delta) - 2S_{2,(r-1)s}(\delta)]}{4rS_{2,(r-1)s+2}(\delta)}. \tag{8.3.34c}$$

For the value of c where the $(CMSE)_r$ is minimized, that is, $c = r(n-s)/(r(n-s)+2)$. Corollary (8.3.6) gives necessary and sufficient conditions for the JS of Dempster to have a smaller $(IMSE)_1$ than the LS.

Corollary 8.3.6. The $(IMSE)_1$ of the JS of Dempster is smaller than the $(IMSE)_1$ of the LS iff

$$\frac{(\gamma_{ij}-\gamma_{\cdot j})^2\lambda_j}{2\sigma^2} \le \frac{(r-1)(2S_{1,(r-1)s}(\delta) - ((r-1)s+2)S_{2,(r-1)s})}{2r((r-1)s+2)S_{2,(r-1)s+2}(\delta)} \tag{8.3.35}$$

Since

$$S_{1,(r-1)s}(\delta) = 2\delta S_{2,(r-1)s+2}(\delta) + (r-1)sS_{2,(r-1)s}(\delta), \tag{8.3.36}$$

Inequality (8.3.36) is equivalent to

$$\frac{\lambda_j(\gamma_{ij}-\gamma_{\cdot j})^2}{2\sigma^2} \le \frac{(r-1)}{r((r-1)s+2)}\left[2\delta + \frac{((r-1)s-2)S_{2,(r-1)s}(\delta)}{2S_{2,(r-1)s+2}(\delta)}\right]$$

(8.3.37)

resulting in the sufficient condition given in Corollary 8.3.7.

Corollary 8.3.7. If

$$\frac{\lambda_j(\gamma_{ij}-\gamma_{\cdot j})^2}{2\sigma^2} \le \frac{2(r-1)\delta}{r((r-1)s+2)}$$

(8.3.38)

then the $(IMSE)_1$ of the JS of Dempster is smaller than that of the LS.

A necessary condition is given in Corollary 8.3.8.

Corollary 8.3.8. If the $(IMSE)_1$ of the JS is smaller than that if the LS then

$$\frac{\lambda_j(\gamma_{ij}-\gamma_{\cdot j})^2}{2\sigma^2} \le \frac{(r-1)S_{1,(r-1)s}(\delta)}{r((r-1)s+2)S_{2,(r-1)s+2}(\delta)}.$$

(8.3.39)

The above conditions are similar to those obtained for the for the Wind estimator with two important differences:

1. The right hand side of all the inequalities is a function of the same variable δ instead of δ_j.

2. For smaller λ_j the region where the $(IMSE)_1$ of the JS is smaller than that of the LS is larger.

The second difference is important because multicollinear data is frequently characterized by the $X'X$ matrix having small eigenvalues. Thus for multicollinear data the Dempster JS might be preferable over the Wind estimator.

The bounds for $r = 5, 9$ and $s = 1, 2, 3$ are given in Table 8.3.3 below. Table 8.3.3a gives the bounds on δ_{ij} where there exists a c such that the $(IMSE)_1$ of the Dempster JS is smaller than that of the LS (Corollary 8.3.5). For $c = r(n - s)/(r(n- s)+2)$ Table 8.3.3 b gives the bounds for the necessary and sufficient condition for the $(IMSE)_1$ of the Dempster JS to be smaller than that of the LS (Corollary 8.3.6). Table 8.3.3 c gives the bounds for the sufficient condition in Corollary 8.3.7 and the bounds for the necessary condition in Corollary 8.3.8 are given in Table 8.3.3d. Notice that the bounds are the same as those for the Wind estimator when s = 1.

In computing the values of these various bounds using Mathematica and Maple on a Macintosh Peforma 6115 and a IICX the author ran into numerical difficulties for cases where $(r-1)s > 30$ because the denominator in the fractions are very close to zero. The reader is hereby cautioned about this.

Table 8.3.3 a

r	s/δ	0.5	2.0	3.5	5.0	6.5	8.0
5	1	1.305	1.678	2.126	2.630	3.170	3.731
	2	2.136	2.580	3.064	3.580	4.117	4.671
	3	2.952	3.430	3.935	4.461	5.002	5.555
9	1	2.374	2.866	3.405	3.978	4.575	5.189
	2	4.179	4.735	5.311	5.904	6.511	7.127
	3	6.020	6.553	7.149	7.756	8.371	8.993

8.3.3b

r	s/δ	0.5	2.0	3.5	5.0	6.5	8.0
5	1	0.502	0.826	1.175	1.543	1.923	2.310
	2	0.467	0.676	0.893	1.116	1.343	1.574
	3	0.450	0.604	0.762	0.923	1.086	1.251
9	1	0.519	0.751	0.992	1.240	1.493	1.749
	2	0.489	0.625	0.763	0.903	1.044	1.187
	3	0.480	0.573	0.670	0.768	0.866	0.965

8.3.3c

r	s/δ	0.5	2.0	3.5	5.0	6.5	8.0
5	1	0.133	0.533	0.933	1.333	1.733	2.133
	2	0.080	0.320	0.560	0.800	1.040	1.280
	3	0.057	0.229	0.400	0.571	0.743	0.914
9	1	0.089	0.356	0.622	0.889	1.156	1.422
	2	0.049	0.198	0.346	0.494	0.642	0.790
	3	0.034	0.137	0.239	0.342	0.444	0.547

8.3.3d

r	s/δ	0.5	2.0	3.5	5.0	6.5	8.0
5	1	1.607	1.704	1.901	2.173	2.493	2.842
	2	1.113	1.269	1.448	1.643	1.849	2.064
	3	1.001	1.130	1.269	1.415	1.567	1.722
9	1	1.236	1.410	1.609	1.825	2.055	2.294
	2	1.054	1.174	1.299	1.429	1.562	1.697
	3	1.007	1.087	1.178	1.271	1.365	1.460

Exercise 8.3.13. Obtain the numerical value of the bounds in Table 8.3.3 for r = 6,7,8, s =1, 2.3
for δ = 0.5 + 0.5k , k =1, 2, ..., 10. Some code follows.

Mathematica

```
a[r_,s_,d_]=1/(2 Exp[d] d^(((r-1) s)/2-1));
I1[r_,s_,d_]=Integrate[Exp[y] y^(((r-1) s)/2-2),
{y,0,d}];
s1[r_,s_,d_]=a[r,s,d] I1[r,s,d];
b[r_,s_,d_]=1/(4 Exp[d] d^(((r-1) s)/2+1));
I2[r_,s_,d_]=Integrate[Exp[y] y^(((r-1) s)/2-1),
{x,0,d},{y,0,x}];
s2[r_,s_,d_]= I2[r,s,d] b[r,s,d];
c[r_,s_,d_]=1/(4 Exp[d] d^(((r-1) s)/2));
I3[r_,s_,d_]=Integrate[Exp[y] y^(((r-1) s)/2-2),{x,0,d},
{y,0,x}];
s3[r_,s_,d_]=c[r,s,d] I3[r,s,d];
```
Bound for the existence of c.
```
cj[r_,s_,d_]=((r-1)*(s1[r,s,d]-2 s3[r,s,d]))/(4 r
s2[r,s,d]);
```
Bound for necessary and sufficient condition.
```
dj[r_,s_,d_]=((((r-1)  )(2 s1[r,s,d]-((r-1) s+2)*
s3[r,s,d]))/(2 r ((r-1) s+2) s2[r,s,d]));
```
Bound for sufficient condition
```
e[r_,s_,d_]=(2 (r-1))/(r ((r-1) s+2)) d;
```
Bound for necessary condition
```
fj[r_,s_,d_]=(r-1) s1[r,s,d]/(r ((r-1)s+2) s2[r,s,d]);
```

Maple

```
> a:=(r,s,d)->1/(2*exp(d)*d^((r-1)*s/2-1)):
>i1:=(r,s,d)->int(exp(y)*y^((r-1)*s/2-2),y=0..d):
> s1:=(r,s,d)->a(r,s,d)*i1(r,s,d):
```

```
> b:=(r,s,d)->1/(4*exp(d)*d^((r-1)*s/2+1)):
>i2:=(r,s,d)->int(int(exp(y)*y^((r-1)*s/2-
1),y=0..x),x=0..d):
>s2:=(r,s,d)->i2(r,s,d)*b(r,s,d):
>c:=(r,s,d)->1/(4*exp(d)*d^((r-1)*s/2)):
>i3:=(r,s,d)->int(int(exp(y)*y^((r-1)*s/2-
2),y=0..x),x=0..d):
> s3:=(r,s,d)->i3(r,s,d)*c(r,s,d):
```
Bound for the existence of c.
```
>cj:=(r,s,d)->((r-1)*(s1(r,s,d)-
2*s3(r,s,d)))/(4*r*s2(r,s,d)):
```
Bound for necessary and sufficient condition.
```
>dj:=(r,s,d)->((((r-1))*(2*s1(r,s,d)-((r-
1)*s+2)*s3(r,s,d)))/(2*r*((r-1)*s+2)*s2(r,s,d))):
```
Bound for sufficient condition
```
> e:=(r,s,d)->(2*(r-1)/(r*((r-1)*s+2)))*d;
```
Bound for necessary conditon
```
> fj:=(r,s,d)->(r-1)*s1(r,s,d)/(r*((r-1)*s+2)*s2(r,s,d)):
```

TI 92

```
C:Define a(r,s,d)=1/((2*e^(d))*d^((r-1)*s/2-1))
C:Define i1(r,s,d)=∫(e^(y)*y^((r-1)*s/2-2),y,0,d)
C:Define s1(r,s,d)=a(r,s,d)*i1(r,s,d)
C:Define b(r,s,d)=1/((4*e^(d))*d^((r-1)*s/2+1))
C:Define i2(r,s,d)=∫(∫(e^(y)*y^((r-1)*s/2-1),y,0,ж),ж,0,d)
C:Define s2(r,s,d)=i2(r,s,d)*b(r,s,d)
C:Define c(r,s,d)=1/((4*e^(d))*d^((r-1)*s/2))
C:Define i3(r,s,d)=∫(∫(e^(y)*y^((r-1)*s/2-2),y,0,ж),ж,0,d)
C:Define s3(r,s,d)=c(r,s,d)*i3(r,s,d)
```
Bound for the existence of c.
```
C:Define cj(r,s,d)=((r-1)*(s1(r,s,d)-2*s3(r,s,d)))/(4*r*s2(r,s,d))
```
Bound for necessary and sufficient condition.
```
C:Define dj(r,s,d)=((r-1)*(2*s1(r,s,d)-((r-
1)*s+2)*s3(r,s,d))/(2*r*((r-1)*s+2)*s2(r,s,d)))
```
Bound for sufficient condition
```
C:Define e(r,s,d)=2*(r-1)*d/(r*((r-1)*s+2))
```
Bound for necessary conditon
```
C:Define fj(r,s,d)=(r-1)*s1(r,s,d)/(r*((r-1)*s+2)*s2(r,s,d))
```

The reader is invited to establish properties of the $(CMSE)_1$ and the $(IMSE)_r$ that are analogous to the properties of the $(IMSE)_1$ that were given in Corollaries 8.3.5-8.3.8 in Exercise 8.3.14. and Exercise 8.3.15.

Exercise 8.3.14. Establish the following properties of the $(CMSE)_1$ as easy consequences of (8.3.32).
A.(1) If

$$c < \frac{2r(n-s)}{r(n-s)+2} \left[\frac{(r-1)sS_{1,(r-1)s}(\delta)}{2r\sum_{j=1}^{s}\lambda_j(\gamma_{ij}-\gamma_{\bullet j})^2/2\sigma^2+(r-1)sS_{2,(r-1)s}(\delta)} - 2 \right]$$

$$(8.3.40)$$

then the $(CMSE)_1$ of the Dempster JS is smaller than the $(CMSE)_1$ of the LS.
(2) If

$$\frac{\sum_{j=1}^{s}\lambda_j(\gamma_{ij}-\gamma_{\bullet j})^2}{2s\sigma^2} < \frac{(r-1)(S_{1,(r-1)s}(\delta)-2S_{2,(r-1)s}(\delta)}{4rS_{2,(r-1)s+2}(\delta)}$$

$$(8.3.41)$$

a positive c exists such that if (8.3.40) is satisfied the $(CMSE)_1$ of the Dempster JS is smaller than that of the LS.
B. A necessary and sufficient condition for the $(CMSE)_1$ of the Dempster JS to be smaller than the $(CMSE)_1$ of the LS is that

$$\frac{\sum_{j=1}^{s}\lambda_j(\gamma_{ij}-\gamma_{\bullet j})^2}{2s\sigma^2} \le \frac{(r-1)(2S_{1,(r-1)s}(\delta)-((r-1)s+2)S_{2,(r-1)s}(\delta)}{2r((r-1)s+2)S_{2,(r-1)s+2}(\delta)}$$

$$(8.3.42)$$

C. If

$$\frac{\sum_{j=1}^{s}\lambda_j(\gamma_{ij}-\gamma_{\bullet j})^2}{2s\sigma^2} \le \frac{2(r-1)\delta}{r((r-1)s+2)}$$

$$(8.3.43)$$

then the $(CMSE)_1$ of the JS of Dempster is smaller than that of the LS.

D. If the $(CMSE)_1$ of the JS is smaller than that if the LS then

$$\frac{\sum_{j=1}^{s} \lambda_j (\gamma_{ij} - \gamma_{\bullet j})^2}{2s\sigma^2} \leq \frac{(r-1)S_{1,(r-1)s}(\delta)}{r((r-1)s+2)S_{2,(r-1)s+2}(\delta)}. \tag{8.3.44}$$

Exercise 8.3.15. Establish the following properties of the $(IMSE)_r$ from Equation (8.3.33).
A.(1) If

$$c < \frac{2r(n-s)}{(r(n-s)+2)} \left[\frac{(r-1)S_{1,(r-1)s}(\delta)}{2\lambda_j \sum_{i=1}^{r} (\gamma_{ij} - \gamma_{\bullet j})^2 / 2\sigma^2 + (r-1)S_{2,(r-1)s}(\delta)} - 2 \right] \tag{8.3.45}$$

then the $(IMSE)_r$ of the Dempster JS is smaller than that of the LS.

(2) If

$$\frac{\lambda_j \sum_{i=1}^{r} (\gamma_{ij} - \gamma_{\bullet j})^2}{2r\sigma^2} < \frac{(r-1)(S_{1,(r-1)s}(\delta) - 2S_{2,(r-1)s}(\delta)}{4rS_{2,(r-1)s+2}(\delta)} \tag{8.3.46}$$

then a positive c exists where (8.3.45) is satisfied.

B. A necessary and sufficient condition for the $(IMSE)_r$ of the Dempster JS to be smaller than that of the LS is that the inequality

$$\frac{\lambda_j \sum_{i=1}^{r} (\gamma_{ij} - \gamma_{\bullet j})^2}{2r\sigma^2} \leq \frac{(r-1)(2S_{1,(r-1)s}(\delta) - ((r-1)s+2)S_{2,(r-1)s}(\delta)}{2r((r-1)s+2)S_{2,(r-1)s+2}(\delta)} \tag{8.3.47}$$

be satisfied.

C. If

$$\frac{\lambda_j \sum_{i=1}^{r} (\gamma_{ij} - \gamma_{\cdot j})^2}{2r\sigma^2} \leq \frac{2(r-1)\delta}{r((r-1)s+2)} \tag{8.3.48}$$

then the $(IMSE)_r$ of the Dempster JS is smaller than that of the LS.
D. If the $(IMSE)_r$ of the JS is smaller than that if the LS then

$$\frac{\lambda_j \sum_{i=1}^{r} (\gamma_{ij} - \gamma_{\cdot j})^2}{2r\sigma^2} \leq \frac{(r-1)S_{1,(r-1)s}(\delta)}{r((r-1)s+2)S_{2,(r-1)s+2}(\delta)}. \tag{8.3.49}$$

Notice that when an appropriate weighted average of $(\gamma_{ij} - \gamma_{\cdot j})^2$ is used the bounds for the $(IMSE)_1$, the $(CMSE)_r$ and the $(IMSE)_r$ are the same.

If the above inequalities are divided by δ for the Dempster estimator or δ_j for the Wind estimator bounds can be obtained for the ratio of the individual parameter or the average sum of parameters averaged over s parameters or r models to the δ parameters. The bounds for the sufficient condition can then be shown to be the limit for large δ of those of the necessary and the necessary and sufficient condition for the $(IMSE)_1$,$(CMSE)_1$ and $(IMSE)_r$ to be smaller for the Dempster or Wind JS than the LS. This is the content of Theorem 8.3.3.

Theorem 8.3.3. Assume that $(r-1)s > 2$. Let

$$g(\delta,r,s) = \frac{(r-1)(2S_{1,(r-1)s}(\delta) - ((r-1)s+2)S_{2,(r-1)s}(\delta))}{2r((r-1)s+2)S_{2,(r-1)s+2}(\delta)}, \tag{8.3.50}$$

$$a(r,s) = \frac{2(r-1)}{r((r-1)s+2)} \tag{8.3.51}$$

and

$$b(\delta,r,s) = \frac{(r-1)S_{1,(r-1)s}(\delta)}{r((r-1)s+2)\delta S_{2,(r-1)s+2}(\delta)}. \tag{8.3.52}$$

Then

$$a(r,s) \leq g(\delta,r,s) \leq b(\delta,r,s), \tag{8.3.53}$$

$$\lim_{\delta \to \infty} g(\delta, r, s) = a(r, s) \tag{8.3.54}$$

and

$$\lim_{\delta \to \infty} b(\delta, r, s) = a(r, s). \tag{8.3.55}$$

Proof. To establish (8.3.52) notice that from (8.3.36) that

$$2S_{1,(r-1)s}(\delta) - ((r-1)s + 2)S_{2,(r-1)s}(\delta)$$
$$= 4\delta S_{2,(r-1)s+2}(\delta) + ((r-1)s - 2)S_{2,(r-1)s}(\delta) \tag{8.3.56}$$
$$\geq 4\delta S_{2,(r-1)s+2}(\delta)$$

and that

$$2S_{1,(r-1)s}(\delta) - ((r-1)s + 2)S_{2,(r-1)s}(\delta) \leq 2S_{1,(r-1)s}(\delta). \tag{8.3.57}$$

The inequality (8.3.52) then follows.

   The limiting relationships in (8.3.54) and (8.3.55) will follow once it is established that

$$\lim_{\delta \to \infty} \frac{S_{1,(r-1)s}(\delta)}{\delta S_{2,(r-1)s+2}(\delta)} = 2 \tag{8.3.58}$$

and

$$\lim_{\delta \to \infty} \frac{S_{2,(r-1)s+2}(\delta)}{\delta S_{2,(r-1)s}(\delta)} = 0. \tag{8.3.59}$$

Observe that

$$S_{1,(r-1)s}(\delta) = \frac{1}{2e^{\delta}\delta^{(r-1)s/2}} \int_0^{\delta} e^x x^{(r-1)s/2 - 2} dx, \tag{8.3.60a}$$

$$S_{2,(r-1)s+2}(\delta) = \frac{1}{4e^{\delta}\delta^{s/2+1}} \int_0^{\delta} \int_0^x e^y y^{(r-1)s/2 - 1} dy dx \tag{8.3.60b}$$

and

$$S_{2,(r-1)s}(\delta) = \frac{1}{4e^{\delta}\delta^{(r-1)s/2}}\int_0^{\delta}\int_0^x e^y y^{\,(r-1)s/2-2}dydx.$$ (8.3.60c)

Thus, by a double application of L'Hospital's rule and the Fundamental Theorem of Integral Calculus

$$\lim_{\delta\to\infty}\frac{S_{1,(r-1)s}(\delta)}{\delta S_{2,(r-1)s+2}(\delta)}$$

$$=\lim_{\delta\to\infty}\frac{2\delta\int_0^{\delta}e^x x^{(r-1)s/2-2}dx}{\int_0^{\delta}\int_0^x y^{\,(r-1)s/2-1}e^y dydx}$$

$$=\lim_{\delta\to\infty}\frac{2\delta^{(r-1)s/2-1}e^{\delta}+2\int_0^{\delta}e^x x^{(r-1)s/2-2}dx}{\int_0^{\delta}y^{\,(r-1)s/2-1}e^y dy}$$

$$=\lim_{\delta\to\infty}\frac{2e^{\delta}\delta^{(r-1)s/2-1}+((r-1)s-2)e^{\delta}\delta^{(r-1)s/2-2}+2e^{\delta}\delta^{(r-1)s/2-2}}{e^{\delta}\delta^{(r-1)s/2-1}}$$

$$=\lim_{\delta\to\infty}2+\frac{(r-1)s}{\delta}=2.$$
(8.3.61)

This establishes (8.3.58). To establish (8.3.59) observe that

$$\lim_{\delta\to\infty}\frac{S_{2,(r-1)s}(\delta)}{\delta S_{2,(r-1)s+2}(\delta)}$$

$$=\lim_{\delta\to\infty}\frac{\int_0^{\delta}\int_0^x y^{\,(r-1)s/2-2}e^y dydx}{\int_0^{\delta}\int_0^x y^{\,(r-1)s/2-1}e^y dydx}$$

$$=\lim_{\delta\to\infty}\frac{\delta^{(r-1)s/2-2}e^{\delta}}{\delta^{(r-1)s/2-1}e^{\delta}}=\lim_{\delta\to\infty}\frac{1}{\delta}=0.$$
(8.3.62)

Again a double application of L'Hospital's rule is used.

Table 8.3.4 illustrates the limiting behavior of g(δ, r, s) and b(δ, r, s) for the case where r = 5 and s =3. Figure 8.3.3 illustrates the Theorem.

Table 8.3.4

δ	1	10	100	1000	10000
a(r,s)	0.114	0.114	0.114	0.114	0.114
g(δ,r,s)	0.501	0.147	0.117	0.115	0.114
b(δ,r,s)	1.043	0.193	0.121	0.115	0.114

Exercise 8.3.16. Make a table that is similar to Table 8.3.4 for r = 6, s = 1, 2 , 3.

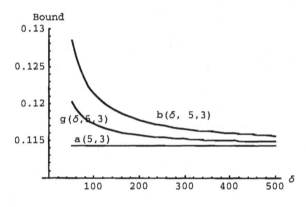

Figure 8.3.1

8.4 The Limited Translation Estimators

It has been seen that although the compound risk of the JS estimators is smaller than that of the LS the individual risk of the JS may not be. Efron and Morris (1972b) describe compromises between the JS estimator and the MLE. that have smaller individual risks than the JS and slightly larger compound risks. These compromise estimators limit translation away from the MLE.

This section will consist of an elaboration of some of the ideas in Efron and Morris (1972). This deals mainly with the problem of estimating the mean of a multivariate normal distribution. The author believes that the ideas of this section can be extended to the models considered in this book and hopes to pursue it as a future project.

A limited translation estimator is a compromise between the JS and the MLE. The basic form of the limited translation estimator for the JS is given by

$$\hat{\theta}_i^D = \left(1 - \frac{(p-2)}{\sum\limits_{i=1}^{p} x_i^2} \rho_D \left(\frac{(p-2)x_i^2}{\sum\limits_{i=1}^{p} x_i^2} \right) \right) x_i, \quad 1 \le i \le p \tag{8.4.1a}$$

where

$$\rho_D \left(\frac{(p-2)x_i^2}{\sum\limits_{i=1}^{p} x_i^2} \right) = \min \left(1, \frac{D}{|x_i| \sqrt{\frac{(p-2)}{\sum\limits_{i=1}^{p} x_i^2}}} \right). \tag{8.4.1b}$$

provided that $0 \le D \le \sqrt{p-2}$. If $D = 0$ then (8.4.1a) reduces to the MLE. If $D = \sqrt{p-2}$ then (8.4.1a) reduces to the JS. Otherwise

$$\hat{\theta}_i^D = \left(1 - \frac{D}{|x_i|} \sqrt{\frac{p-2}{\sum\limits_{i=1}^{p} x_i^2}} \right) x_i, \quad 1 \le i \le p. \tag{8.4.1c}$$

Efron and Morris call ρ a relevance function.

The computation of the Bayes and the conditional individual and compound risk functions will now be outlined. To do this a modified relative savings loss of the individual components is defined. The RSLM for the j'th component is defined by

$$s_p(\rho)_j = \frac{(\text{IMSEMLE})_j - (\text{IMSELT})_j}{(\text{IMSEML})_j - (\text{IMSEJS})_j} \tag{8.4.2}$$

Since the MLE has IMSE 1, and for the JS

$$(\text{IMSE})_j = 1 - \frac{(p-2)}{p(1+A)}, \qquad (8.4.3)$$

it follows that for the limited translation estimator with relevance function ρ

$$(\text{IMSELT})_j = 1 - s_p(\rho)\frac{(p-2)}{p(1+A)}. \qquad (8.4.4)$$

Efron and Morris show that $s_p(\rho)_j$ is independent of A and j and is given by the formula in (8.4.5 below. From now on the subscript j will not be used. Now

$$s_p(\rho) = 1 - E(1 - \rho((p-2))W_p))^2 \qquad (8.4.5)$$

where W_p is a Beta(3/2, $(p-1)/2$) random variable.

Exercise 8.4.1. A. How does the RSL defined by Equation (8.4.2) differ from that defined previously?
B. Express the IMSELT in terms of the previously defined RSL.

Exercise 8.4.2. Establish (8.4.5).

Notice that the average MSE of the limited translation estimators are always smaller than that of the MLE. This will not be the case for the conditional MSE. Table 8.4.1 gives the Bayes risk for some different values of A and D. The graph for $p = 4$ is given by Figure 8.4.1. Notice that when D is somewhat close to $\sqrt{p-2}$ the average individual risk of the limited translation estimator is very close to that of the JS; that is why the graph of the limited translation estimator is not plotted for values of $D > 0.5\sqrt{2}$.

Table 8.4.1

D/A	0	0.50	1.00	1.50	2.00
0(MLE)	1.0000	1.0000	1.0000	1.0000	1.0000
$0.25\sqrt{2}$	0.6874	0.7916	0.8437	0.8749	0.8958
$0.50\sqrt{2}$	0.5464	0.6976	0.7732	0.8186	0.8488
$0.75\sqrt{2}$	0.5042	0.6695	0.7521	0.8017	0.8347
$\sqrt{2}$(JS)	0.5000	0.6667	0.7500	0.8000	0.8333

$$p = 4$$

D/A	0	0.50	1.00	1.50	2.00
0(MLE)	1.0000	1.0000	1.0000	1.0000	1.0000
$0.25\sqrt{6}$	0.4932	0.6262	0.7196	0.7757	0.8131
$0.50\sqrt{6}$	0.2747	0.5165	0.6374	0.7099	0.7582
$0.75\sqrt{6}$	0.2507	0.5004	0.6253	0.7003	0.7502
$\sqrt{6}$(JS)	0.2500	0.5000	0.6250	0.7000	0.7500

$$p = 8$$

D/A	0	0.50	1.00	1.50	2.00
0(MLE)	1.0000	1.0000	1.0000	1.0000	1.0000
$0.25\sqrt{10}$	0.3181	0.5454	0.6590	0.7272	0.7727
$0.50\sqrt{10}$	0.1778	0.4518	0.5889	0.6711	0.7259
$0.75\sqrt{10}$	0.1668	0.4445	0.5834	0.6667	0.7223
$\sqrt{10}$(JS)	0.1667	0.4444	0.5833	0.6667	0.7222

$$p = 12$$

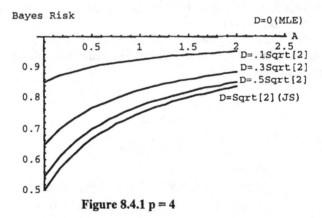

Figure 8.4.1 p = 4

The IMSE of the JS was obtained in Section 8.1 Equation (8.1.8). Thus, the IMSE of the limited translation error takes the form

$$IMSE(LT) = s_p(\rho)IMSE(JS) + 1 - s_p(\rho). \qquad (8.4.6)$$

The IMSE of the i'th component JS is a maximum for a particular value of δ when all of the coordinates of a parameter θ are zero except the i'th parameter, i.e., $\lambda = \theta_i^2/2$. Thus, for $1 \le i \le p$,

$$\max IMSE(JS)_i = 1 - 2(p-2)S_{1,p}(\theta_i^2/2)$$
$$+ (p-2)(p+2)[\theta_i^2 S_{2,p+2}(\theta_i^2/2) + S_{2,p}(\theta_i^2/2)],$$
$$(8.4.7)$$

and thus,

$$\max IMSE(LT) = s_p(\rho)\max IMSE(JS) + 1 - s_p(\rho). \qquad (8.4.8)$$

Now when $IMSE(JS) < 1$, $IMSE(LT)$ is a decreasing function of $s_p(\rho)$ and the JS dominates the limited translation estimators and the MLE. Likewise if $IMSE(JS) > 1$, $IMSE(LT)$ is an increasing function of $s_p(\rho)$ and the MLE and the limited translation estimators do better with respect to the MLE. This is illustrated by Table 8.4.2 for $p = 8$ and Figure 8.4.2 below for $p = 4$.

Exercise 8.4.3. A. Show that $f(a,x) = ax + 1 - a$ is an increasing function of a if $x > 1$ and a decreasing function of a if $x < 1$.
B. Use the result in A to show if the JS has maximum IMSE less than one it is the dominant estimator with respect to this measure; otherwise the MLE is the dominant estimator.

Table 8.4.2 a

a/θ_i	0	0.25	0.50	0.75	1
0(MLE)	1	1	1	1	1
0.25	0.75	0.824	0.858	0.909	0.973
0.50	0.50	0.648	0.715	0.818	0.946
0.75	0.25	0.472	0.573	0.727	0.920
1(JS)	0	0.296	0.430	0.636	0.893

Individual Risk < 1

8.4.2b

a/θ_i	2	4	6	8	10
0(MLE)	1	1	1	1	1
0.25	1.236	1.342	1.234	1.154	1.106
0.50	1.472	1.685	1.468	1.307	1.211
0.75	1.708	2.027	1.702	1.461	1.317
1(JS)	1.944	2.370	1.935	1.614	1.422

Individual Risk >1

For the portion of the parameter space where the maximum (worst case scenario) individual component risks of the limited translation estimators exceed that of the MLE they are smaller than that of the JS. The compound risk of the JS is uniformly smaller than that of the limited translation estimators. However for larger values of D the sacrifice is not that great. This is illustrated by Table 8.4.3 and Figure 8.4.3.

The compound risk of the JS is given by

$$m = 1 - s_p(\rho)\frac{(p-2)^2}{p}\sum_{k=0}^{\infty}\frac{e^{-\lambda}\lambda^k}{k!(p+2k-2)} \qquad (8.4.9)$$

where $\lambda = \theta'\theta/2$.

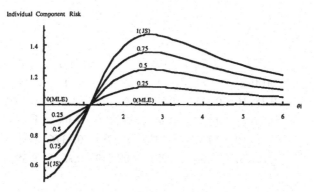

Figure 8.4.2 p = 4

Table 8.4.3

D/λ	0	0.5	1	1.5	2
0(MLE)	1	1	1	1	1
$0.25\sqrt{2}$	0.6874	0.7540	0.8024	0.8381	0.8648
$0.50\sqrt{2}$	0.5464	0.6431	0.7133	0.7651	0.8039
$0.75\sqrt{2}$	0.5042	0.6098	0.6866	0.7432	0.7856
$\sqrt{2}$(JS)	0.5000	0.6065	0.6839	0.7410	0.7838

$$p = 4$$

D/λ	0	0.5	1	1.5	2
0(MLE)	1	1	1	1	1
$0.25\sqrt{6}$	0.4392	0.5029	0.5555	0.5994	0.6363
$0.50\sqrt{6}$	0.2747	0.3570	0.4250	0.4818	0.5296
$0.75\sqrt{6}$	0.2507	0.3357	0.4060	0.4646	0.5141
$\sqrt{6}$(JS)	0.2500	0.3351	0.4055	0.4642	0.5136

$$p = 8$$

D/λ	0	0.5	1	1.5	2
0(MLE)	1	1	1	1	1
$0.25\sqrt{10}$	0.3181	0.3711	0.4173	0.4579	0.4937
$0.50\sqrt{10}$	0.1778	0.2417	0.2974	0.3463	0.3895
$0.75\sqrt{10}$	0.1668	0.2315	0.2880	0.3376	0.3813
$\sqrt{10}$(JS)	0.1667	0.2314	0.2879	0.3375	0.3813

$$p = 12$$

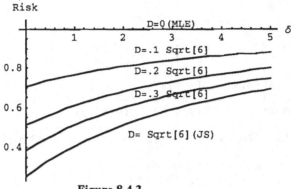

Figure 8.4.3

The limited translation estimators provide a way to substantially reduce the component risk of an estimator while not sacrificing a great deal from the overall risk. The limited translation rules:

1. have a smaller maximum component MSE than the JS when the component risk exceeds one;

2. a compound MSE less than the MLE that compares very favorably with the JS for large enough values of D.

Exercise 8.4.4. Find and compare the numerical values of the Bayes risk, the maximum individual risk and the compound risks of the MLE, the JS and the limited translation estimators for $D = .1\sqrt{p-2}, .3\sqrt{p-2}, .5\sqrt{p-2}, 1$ and $p = 3, 5, 7$. Use one of the computer programs that follow.

Mathematica

```
f[w_,p_]=(Gamma[p/2+1]/(Gamma[3/2]*Gamma[(p-1)/2]))*
(w^(1/2))*(1-w)^((p-3)/2);
g[D_,p_]=(Integrate[f[w,p],{w,0,D^2/(p-2)}]
+Integrate[f[w,p] D/Sqrt[(p-2) w],{w,D^2/(p-2),1}]);
h[D_,p_]=(Integrate[f[w,p],{w,0,D^2/(p-2)}]
+Integrate[f[w,p] D^2/((p-2) w),{w,D^2/(p-2),1}]);
Savings Loss
s[D_,p_]=2 g[D,p]-h[D,p];
Bayes Risk (LT)
br[A_,D_,p_]=1-((p-2)/p) s[D,p] (1/(A+1));
aj[p_,x_]=Exp[-x]/(2 x^(p/2-1));
a[A_,p_]=(p-2)/(p (1+A));
bj[p_,x_]=Integrate[Exp[y] y^(p/2-2),{y,0,x}];
cj[p_,x_]=aj[p,x] bj[p,x];
```

Compound Risk JS
```
mjs[p_,x_]=1-(p-2)^2/p cj[p,x];
```
Compound Risk LTJS
```
ltjs[D_,p_,x_]=s[D,p] mjs[p,x]+1-s[D,p];
al[p_,x_]=Integrate[Exp[z] z^(p/2-2),{y,0,x},{z,0,y}];
b1[p_,x_]=1/(4 x^(p/2) Exp[x]);
c1[p_,x_]=al[p,x] b1[p,x];
ms[a_,p_,x_]=(1- 2 (p-2) (p c1[p,x]+2 x c1[p+2,x])
+(p-2) (p+2) (2 a c1[p+2,x]+c1[p,x]));
```
Maximum Risk JS
```
ax[p_,x_]=1-((p-2)^2) c1[p,x]+p (p-2) 2 x c1[p+2,x];
sup[p_,th_]=ax[p, ((th)^2)/2];
```
Maximum Risk Limited Translation (JS)
```
lt[r_,p_,th_]=r sup[p,th]+1-r;
```


Maple

```
>f:=(w,p)->(1/Beta(3/2, (p-1)/2))*((w^(1/2))*(1-w)^((p-
3)/2)):
>g:=(D,p)->int(f(w,p),w=0..D^2/(p-
2))+int(f(w,p)*D/sqrt((p-2)*w),w=D^2/(p-2)..1):
h:=(D,p)->int(f(w,p),w=0..D^2/(p-2))+int(f(w,p)*D^2/((p-
2)*w),w=(D^2/(p-2))..1):
```
Savings Loss
```
> s:=(D,p)->2*g(D,p)-h(D,p):
> a:=(A,p)->(p-2)/(p*(1+A)):
```
Bayes Risk LT
```
> br:=(A,D,p)->1-((p-2)/p)*s(D,p)*(1/(A+1)):
> aj:=(p,x)->exp(-x)/(2*x^(p/2-1)):
> bj:=(p,x)->int(exp(y)*y^(p/2-2),y=0..x):
> cj:=(p,x)->aj(p,x)*bj(p,x):
```
JS (Compound Risk)
```
> mjs:=(p,x)->1-((p-2)^2/p)*cj(p,x):
```
LTJS (Compound Risk)
```
> ltjs:=(D,p,x)->s(D,p)*mjs(p,x)+1-s(D,p):
> al:=(p,x)->int(int(exp(z)*z^(p/2-2),z=0..y),y=0..x):
> b1:=(p,x)->1/(4*x^(p/2)*exp(x)):
> c1:=(p,x)->al(p,x)*b1(p,x):
```
Maximum Individual Risk JS
```
> ax:=(p,x)->1-((p-2)^2)*c1(p,x)+p*(p-2)*2*x*c1(p+2,x):
```

Maximum Individual Risk JS
```
> sup:=(p,th)->ax(p,((th)^2)/2):
```
Maximum Individual Risk LT
```
> lt:=(r,p,th)->r*sup(p,th)+1-r:
```

TI 92

```
C:Define be(c,d)=∫(н^((c-1))*(1-н)^(d-1),н,0,1)
C:Define f(w,p)=(1/be(3/2,(p-1)/2))*((w^(1/2)))*(1-w)^((p-3)/2)
C:Define g(D,p)=∫(f(w,p),w,0,D^2/(p-2))+∫(f(w,p)*D/√((p-2)*w),w,D^2/(p-2),1)
C:Define h(D,p)=∫(f(w,p),w,0,D^2/(p-2))+∫(f(w,p)*D^2/((p-2)*w),w,(D^2/(p-2)),1)
```

Savings Loss
```
C:Define s(D,p)=2*g(D,p)-h(D,p)
C:Define bb(A,p)=(p-2)/(p*(1+A))
```

Bayes Risk
```
C:Define br(A,D,p)=1-bb(A,p)*s(D,p)
C:Define aj(p,н)=e^(-н)/(2*н^(p/2-1))
C:Define bj(p,н)=∫(e^(y)*y^(p/2-2),y,0,н)
C:Define cj(p,н)=aj(p,н)*bj(p,н)
C:Define mjs(p,н)=1-((p-2)^2/p)*cj(p,н)
```

Compound Risk (LT)
```
C:Define ltjs(D,p,н)=s(D,p)*mjs(p,н)+1-s(D,p)
C:Define a1(p,н)=∫(∫(e^z*z^(p/2-2),z,0,y),y,0,н)
C:Define b1(p,н)=1/(4*н^(p/2)*e^н)
C:Define d1(p,н)=a1(p,н)*b1(p,н)
```

Maximum individual component risk (JS)
```
C:Define ma(p,н)=1-(p-2)^2*d1(p,н)+p*(p-2)*2*н*d1(p+2,н)
C:Define su(p,th)=ma(p,th^2/2)
```

Maximum individual component risk (LT)
```
C:Define lt(t,p,th)=t*su(p,th)+1-t
```

8.5 Summary

The properties of the individual risks of James-Stein type estimators were studied in this chapter. Conditions were given for individual risks of JS estimators to be smaller than the MLE or the LS. The cases that were considered included:

1. estimating the mean of a multivariate normal distribution;
2. the single linear model;
3. r linear models.

The JS of Dempster and Wind was considered. For r models the individual risks were considered for the individual models and averaging over the r linear models. Some numerical studies were done and some Computer Algebra Programs are included for the reader who wishes to carry these studies further.

An introduction to the limited translation estimators was presented.

Chapter IX

The Multivariate Linear Model

9.0 Introduction

The use of a multivariate model is appropriate when the response variable in an experiment has several different characteristics that might be related in some way, e.g., height, weight, blood pressure, glucose level. The purpose of this chapter is to extend as much as possible the results of Chapters IV and VII to the multivariate model. Thus, the objectives include:

1. formulation of the JS for a multivariate model by both the EB and the pure frequentist approach;
2. formulation of EBEs for the simultaneous estimation of parameters of r multivariate linear models;
3. where possible, evaluation of the exact MSEs of the EBEs with and without averaging over the prior distribution;
4. approximation of the MSE for some of the cases where it is not possible to obtain exact results.

The multivariate model may be formulated as an extension of the univariate model. Section 9.1 explains how this is done, gives the form of the LS estimator and shows how the concept of estimable parametric functions may be generalized to the multivariate case.

The BE are derived in Section 9.2. Exercise 9.2.2 invites the reader to show that in the context of a multivariate model the ridge estimators are special cases of the BE thus generalizing the corresponding result for the univariate case.

The variables in a multivariate model are correlated. When the variables are uncorrelated the setup reduces to that of p models considered in the previous chapter. When the dispersion matrix is known the generalization of the various EBEs to the multivariate case is straightforward and the MSEs are easy generalizations of those obtained for the univariate case.

For the case of an unknown dispersion matrix the exact MSEs cannot always be obtained. More work is required than in the univariate case to obtain the exact MSEs for those cases where the computation is possible. Examples of cases where it is possible to obtain exact MSEs for an unknown dispersion matrix include;

1. the Wind estimator for one linear model (See Section 9.3);
2. the estimator of C.R. Rao for r models (See Section 9.4.).

The Wind estimator cannot be formulated for a single univariate model. The reason that it can be formulated for a single multivariate model with p > 2 variables is that all of the p variables may be used to simultaneously estimate the unknown prior parameters. Although Sclove (1971) formulated a similar JS to the one to be considered here, he did not give an EB or approximate MMSE formulation. Furthermore, he did not obtain an explicit formula for the MSE. However, he did prove that his estimator had a smaller conditional MSE than the LS. The formulation of the Wind estimator by both the Bayesian and the non-Bayesian approach together with formulae for the MSE for the single model case is given in Section 9.3.

Practical problems whose solutions call for the simultaneous estimation of parameters include:

1. a replicated experiment using multivariate models;
2. the construction of a selection index for choosing individuals with high intrinsic genetic values.

C.R. Rao (1977) formulates simultaneous estimators of vector parameters that represent the genetic worth of an individual. The EBEs of C.R. Rao together with the evaluation of their efficiency for the case of r multivariate models is the subject of Section 9.4.

In addition to computing the exact MSEs mentioned above an example of how to compute an approximate MSE using a Taylor series approximation for the dispersion of the errors will be given and comments about the goodness of the approximation will be made. The example to be considered in Section 9.5 is the estimator of Dempster for one linear model. The basic principle can be applied to other estimators. An indication of how some of the ideas in Chapters VI, VII and VIII might be extended to the multivariate case will be given in the exercises. Some of the results that might be obtained in solving these exercises would constitute answers to some questions that are open as far as the author knows.

9.1 The Multivariate Linear Model

As has already been mentioned the multivariate model is appropriate when $p > 1$ possibly related characteristics are being observed together as a response variable. When the univariate model is employed it is only possible to observe one characteristic at a time.

C.R. Rao (1973) shows how to formulate the multivariate model as a convenient generalization of the univariate model. The multivariate linear model takes the form

$$Y = (I \otimes X)\beta + \varepsilon \tag{9.1.1}$$

where X is an nxm matrix of rank $s \leq m$, β is an mp-dimensional vector of parameters with m-dimensional components β_i, $1 \leq i \leq p$, ε is an mp dimensional error vector and $Y = [Y_1, Y_2, ..., Y_p]$ with Y_j an n dimensional observation vector for a particular attribute being studied.

An important feature of a multivariate model is that although the individual observations may be independent the characteristics being studied may indeed be related. This leads to the assumption that the error terms have mean zero and dispersion $\Sigma \otimes I$, i.e.,

$$E(\varepsilon \mid \beta) = 0 \text{ and } D(\varepsilon \mid \beta) = \Sigma \otimes I. \tag{9.1.2}$$

The matrix Σ is a pxp dispersion matrix. When $p = 1$, the setup (9.1.1) reduces to that of a single univariate linear model. When $\Sigma = \sigma^2 I$ the setup reduces to that of p uncorrelated (independent in the case of normal error terms) linear models.

Since the multivariate model is formulated as an extension of the univariate model together with the assumption about the error terms in (9.1.2), the multivariate LS is a weighted least square estimator; i.e.,

$$b = [(I \otimes X)'(\Sigma^{-1} \otimes I)(I \otimes X)]^+ (\Sigma^{-1} \otimes I)Y$$
$$= (I \otimes (X'X)^+ X')Y. \tag{9.1.3}$$

A bivariate linear model is illustrated in Example 5.1.1.

Example 5.1.1. Performance of Students at the End of a Course as Compared with Performance at the Beginning.

The data below is concerned with scores on examinations in an undergraduate Complex Variables course taught by the author. The author's institution, Rochester Institute of Technology (RIT) is on the quarter system; each quarter has ten weeks of classes and one week for final examinations. There were four hour exams administered two weeks apart and a final examination. The independent variables X_1 and X_2 represent the percentage scores on the first two hour examinations. The response variable observed is the percentage score on the last hour examination and the final examination Y_1 and Y_2 respectively.

X_1	X_2	Y_1	Y_2
64	89	88	77
67	93	86	82
97	87	95	88
47	55	28	51
94	99	91	97
100	93	91	91
78	97	99	97
78	75	91	90
58	62	45	64
75	79	52	67

The bivariate model is

$$\begin{bmatrix} Y_1 \\ Y_2 \end{bmatrix} = \begin{bmatrix} X & 0 \\ 0 & X \end{bmatrix} \begin{bmatrix} \beta_{01} \\ \beta_{11} \\ \beta_{21} \\ \beta_{02} \\ \beta_{12} \\ \beta_{22} \end{bmatrix} + \varepsilon$$

with the 10 x 3 X matrix having the first column consisting of ones and the second and third column consisting of the ten entries under X_1 and X_2. The Y_1 and Y_2 consist of the entries in the columns in the data above. The least square estimator consists of the column vector

$$b = \begin{bmatrix} -46.71 \\ 0.313 \\ 1.202 \\ 4.513 \\ 0.356 \\ 0.589 \end{bmatrix}$$

The Σ matrix is estimated by

$$\hat{\Sigma} = \begin{bmatrix} 1233.89 & 551.907 \\ 551.907 & 398.286 \end{bmatrix}.$$

The correlation coefficient between Y_1 and Y_2 is 0.787. Thus, it appears that there is some kind of linear relationship between performance on the last hour test and the final examination. Also, in the univariate regression of the last hour examination on the first two 78% of the variation is accounted for; Likewise for the regression of the final exam on the first two hour examinations 81% of the variation is accounted for. Thus it appears that a students performance at the end of the quarter is related to how he or she does at the beginning.

Exercise 9.1.1. A. Show that if A is a nonsingular matrix then

$$(A \otimes B)^+ = A^{-1} \otimes B^+.$$

Where was this fact used in the derivation of (9.1.3)?

Exercise 9.1.2. Suppose W is an nxp matrix of multivariate observations, β is an mxp matrix of parameters and ε is an nxp dimensional error matrix, X is as in the model (9.1.1). Give an alternate formulation of the multivariate model. (See for example Morrison (1967).)

Exercise 9.1.3. Recall that the SVD of

$$X = \begin{bmatrix} S' & T' \end{bmatrix} \begin{bmatrix} \Lambda^{1/2} & 0 \\ 0 & 0 \end{bmatrix} \begin{bmatrix} U \\ V \end{bmatrix} = S' \Lambda^{1/2} U'.$$

A. Show that $I \otimes X$ has SVD

$$I \otimes X = (I \otimes S')(I \otimes \Lambda^{1/2})(I \otimes U').$$

B. For the multivariate model, a parametric function $p'\beta$ is estimable iff one of the following hold true:

1. $p'(I \otimes V) = 0$;
2. $p' = d'(I \otimes U')$ for some sp dimensional vector d;
3. $p' = t'(I \otimes X)$ for some vector t.

Show that 1, 2 and 3 are equivalent definitions of estimability.
C. Show that in a multivariate model the LS is independent of the generalized inverse for estimable parametric functions.

Exercise 9.1.4. The following data represents student's examination scores in an undergraduate Calculus based Probability course taught by the author. The response variable Y_1 is the final examination score; the response variable Y_2 is the average of all hour examinations, homework and the final examination and X is the grade on the first hour examination.

X	71	56	48	62	80	57	86	73	68	38	75	60
Y_1	64	86	66	89	95	68	66	88	54	78	95	56
Y_2	70	82	64	90	90	66	72	85	59	74	97	66

Formulate the appropriate multivariate model and find the LS estimators. Estimate Σ.

9.2 The BE

Recall that the two basic steps in deriving an EBE are:

1. obtain the BE;
2. substitute sample estimates for unknown prior parameters.

This section is concerned with the first step, deriving the BE.

A prior distribution will be assumed that has a dispersion matrix with a structure similar to that of the dispersion of the error terms. Recall that the error vector was assumed to have a dispersion of the form $\Sigma \otimes I$. Thus, by analogy with the univariate model it will be assumed that the β parameters of a prior distribution with mean and dispersion of the form

$$E(\beta) = \theta \text{ and } D(\beta) = \Sigma \otimes F. \tag{9.2.1}$$

The resulting form of the BE is an easy generalization of the results obtained for the univariate model. From (7.1.4) with $\Sigma \otimes I$ in place of Σ and $\Sigma \otimes F$ in place of F the BE is

$$p'\hat{\beta}_i = p'\theta + p'(\Sigma \otimes F)(I \otimes X').$$
$$[(I \otimes X)(\Sigma \otimes F)(I \otimes X') + \Sigma \otimes I]^{-1}(Y - (I \otimes X)\theta). \tag{9.2.2}$$

Letting $p' = [p_1', p_2',..., p_p']$, $Y' = [Y_1', Y_2',...,Y_p']$ and $\theta' = [\theta_1', \theta_2',..., \theta_p']$,

$$p'\hat{\beta}_i = p_i'\theta_i + p_i'FX(XFX' + I)^{-1}(Y - X\theta_i) \tag{9.2.3}$$

The exercises below show how to extend some results from the univariate to the multivariate case.

Exercise 9.2.1. Equivalent definitions of estimability for multivariate models. Prove that the following two conditions for estimability of are equivalent.
1. For each i, $1 \leq i \leq p$ there exists a d_i such that

$$p_i' = d_i'U' \tag{9.2.4}$$

2. There exists a d such that

$$p' = d'(I \otimes U'). \tag{9.2.5}$$

Exercise 9.2.2. Generalized ridge regression estimator of C.R. Rao. Show that when $\theta = 0$ and $F = URU'$ with R a positive definite matrix

$$p_i'\hat{\beta}_i = p_i'(X'X + F^+)^+ X'Y_i, \ 1 \le i \le p. \tag{9.2.6}$$

For each of the following choices of F assuming $\theta = 0$ obtain the corresponding ridge type estimator.

A. $F = (1/k)UU'$; (ordinary ridge estimator)

B. $F = (1/k)(X'X)^+$;(contraction estimator)

C. $F = U K^{-1}U'$, K a positive definite diagonal matrix(the generalized Hoerl and Kennard Ridge estimator)

Exercise 9.2.3. The average MSE of the BE

A. Show that for the estimable parametric functions three forms of the average MSE of the BE are

$$m_1 = p'(\Sigma^{-1} \otimes X'X + \Sigma^{-1} \otimes (UU'FUU')^+)^+ p, \tag{9.2.7a}$$

$$m_2 = p'\Sigma \otimes (X'X)^+ p$$
$$-p'\Sigma \otimes (X'X)^+ (\Sigma \otimes (X'X)^+ + \Sigma \otimes (UU'FUU')^+)^+ \Sigma \otimes (X'X)^+ p \tag{9.2.7b}$$

and

$$m_3 = p'(\Sigma \otimes F)p$$
$$-p'(\Sigma \otimes F)(\Sigma \otimes UU'FUU' + \Sigma \otimes (X'X)^+)^+ (\Sigma \otimes F)p. \tag{9.2.7c}$$

B. Show that the LS has average MSE

$$m_0 = p'(\Sigma \otimes (X'X)^+ p, \tag{9.2.7d}$$

hence, the BE has smaller average MSE than the LS.

Exercise 9.2.4. Comparison of the conditional MSE of the BE with the LS.

A. Obtain an expression for the conditional MSE of the BE for the multivariate model.

B. Prove that the conditional MSE of the BE is less than or equal to that of the LS iff

$$(\beta - \theta)'(2\Sigma \otimes UU'FUU' + \Sigma \otimes (X'X)^+)^+ (\beta - \theta) \leq 1. \qquad (9.2.8)$$

9.3 The EBE of Wind for One Linear Model

Unlike the univariate case the estimators of both Wind and Dempster may be formulated for one multivariate linear model. The reason the estimator of Wind can be formulated for the multivariate case with $p > 2$ is that all p variables play a role similar to that of the r models in the univariate case. The Wind estimator has an exact conditional and average MSE averaging over quadratic loss functions of the form $\Sigma \otimes UDU'$ with D a diagonal matrix.

The derivation of the Wind estimator by the EB and the frequentist approach and the calculation of its MSEs will be dealt with in this section. Unlike the Wind estimator the Dempster estimator does not have an exact MSE when Σ is unknown and estimated by a Wishart matrix. The computation of an approximate MSE will be dealt with in Section 9.5.

9.3.1. The Wind Estimator as an EBE

The discussion immediately above Equation 9.2.1 and the form of the prior dispersion for the univariate case would suggest that the multivariate generalization of the prior assumptions should take the form

$$E(\beta) = 0 \text{ and } D(\beta) = \Sigma \otimes UDU' \qquad (9.3.1)$$

where D is a diagonal matrix with j'th element ω_j/λ_j and the λ_j are the eigenvalues of X'X. Let $\gamma_i = U'\beta_i$ and γ_{ij} be the j'th element of γ_j . Then the BE of γ_{ij} is

$$\gamma_{ij}^b = g_{ij} - \frac{1}{1 + \omega_j^2} g_{ij}. \qquad (9.3.2)$$

When β and ε conditional on β are multivariate normal, an unbiased estimator of $d_j = 1/(1 + \omega_j^2)$ is

$$\hat{d}_j = \frac{(n - s)(p - 2)}{(n - s - p + 1)\lambda_j g_{-j}' \hat{\Sigma} g_{-j}}. \qquad (9.3.3)$$

where $\hat{\Sigma}$ is the within sum of squares matrix in multivariate analysis of variance MANOVA. From C.R. Rao (1973) or Kshirsagar (1972)

$$\frac{(n-s)}{g'_{-j}\hat{\Sigma}g_{-j}} \sim \frac{\chi^2(n-p-s+1)}{g'_{-j}\Sigma g_{-j}} \tag{9.3.4}$$

The numerator and the denominator of the right hand side of (9.3.4) are independent random variables.
Also

$$g'_{-j}\hat{\Sigma}^{-1}g_{-j} = b'(I \otimes U_j)\hat{\Sigma}^{-1}(I \otimes U'_j)b. \tag{9.3.5}$$

From the methodology already developed and employed for the univariate case the EBE of γ_{ij} takes the form

$$\hat{\gamma}_{ij} = g_{ij} - c\hat{d}_j g_{ij} \tag{9.3.6}$$

where c is a scalar that minimizes the MSE. By virtue of (9.3.5) d_j may be written in terms of b. Let

$$\hat{a}_j = 1 - d_j. \tag{9.3.7}$$

Hence,

$$\hat{\gamma}_{ij} = \hat{a}_j g_{ij} \tag{9.3.8}$$

Let A be the s x s diagonal matrix with elements a_j. Then

$$\hat{\gamma}_i = Ag_i \tag{9.3.9}$$

and

$$\hat{\gamma} = (I \otimes \hat{A})g. \tag{9.3.10}$$

Define

$$\hat{\beta} = (I \otimes U)\hat{\gamma}. \tag{9.3.11}$$

For the estimable parametric functions the estimator of Wind is

$$p'\hat{\beta} = p'(I \otimes U\hat{A}U')b. \qquad (9.3.12)$$

A variant of (9.3.12) for a known prior mean θ is

$$p'\hat{\beta} = p'\theta + p'(I \otimes U\hat{A}U')(b - \theta) \qquad (9.3.13)$$

where $\hat{a}_j$ is defined by (9.3.7) and

$$\hat{d}_j = \frac{(n-s)(p-2)}{(n-s-p+1)\lambda_j(g_{-j} - U_j'\theta)'\hat{\Sigma}^{-1}(g_{-j} - U_j'\theta)} \qquad (9.3.14)$$

The data used in Example 9.3.1 and Exercise 9.3.4 were obtained through the courtesy of Dr. James Halavin, Professor, Department of Mathematics and Statistics at the Rochester Institute of Technology.

Example 9.3.1. Characteristics of rivets coming off an assembly line.
According to specifications and prior information rivets coming off an assembly line should have the following dimensions:
Diameter x 1000: 205
Thickness x 1000:55
Depth of Cut (Score) x 1000:36.
Four measurements were taken in each of three shifts.

Shift	Diameter	Thickness	Score
1	206	56	33
1	206	54	35
1	202	58	38
1	202	58	36
2	205	54	36
2	207	54	37
2	203	57	34
2	203	57	35
3	208	55	36
3	208	55	35
3	206	54	35
3	206	54	36

The multivariate linear model is

$$Y = (I_3 \otimes X)\beta + \varepsilon$$

where

$$X = \begin{bmatrix} I_4 & 0 & 0 \\ 0 & I_4 & 0 \\ 0 & 0 & I_4 \end{bmatrix}$$

The LS estimators are

$$\begin{array}{lll} b_{11} = 204 & b_{21} = 204.5 & b_{31} = 207 \\ b_{12} = 56.5 & b_{22} = 55.5 & b_{32} = 54.5 \\ b_{13} = 35.5 & b_{23} = 35.5 & b_{33} = 35.5 \end{array}$$

Now

$$\hat{\Sigma} = \frac{1}{9} \begin{bmatrix} 31 & -19 & -5 \\ -19 & 21 & 1 \\ -5 & 1 & 19 \end{bmatrix}$$

and

$$\hat{\Sigma}^{-1} = \begin{bmatrix} 0.6963 & 0.6229 & 0.1505 \\ 0.6229 & 0.9869 & 0.1120 \\ 0.1505 & 0.1120 & 0.5074 \end{bmatrix}$$

In Section 9.4 optimum c will be obtained as

$$c = \frac{n-s-p+1}{n-s-p+3}.$$

Thus, for this case optimum $c = 7/9$. Now

$$(b_{-1} - \theta)'\hat{\Sigma}^{-1}(b_{-1} - \theta) = 1.1574$$
$$(b_{-2} - \theta)'\hat{\Sigma}^{-1}(b_{-2} - \theta) = 0.2544$$
$$(b_{-3} - \theta)'\hat{\Sigma}^{-1}(b_{-3} - \theta) = 1.6683$$

The contraction factors are

$$t_1 = 1 - \frac{1}{4(1.1474)} = 0.7840$$

$$t_2 = 1 - \frac{1}{4(0.2544)} = 0.0173$$

$$t_3 = 1 - \frac{1}{4(1.6683)} = 0.8501$$

and thus the EBE are given by

$$\hat{\beta}_{i1} = \begin{bmatrix} 205 \\ 55 \\ 36 \end{bmatrix} + 0.7840 \begin{bmatrix} -1 \\ 1.5 \\ -0.5 \end{bmatrix} = \begin{bmatrix} 204.22 \\ 56.18 \\ 35.61 \end{bmatrix}$$

$$\hat{\beta}_{i2} = \begin{bmatrix} 205 \\ 55 \\ 36 \end{bmatrix} + 0.0173 \begin{bmatrix} -0.5 \\ 0.5 \\ -0.5 \end{bmatrix} = \begin{bmatrix} 204.99 \\ 55.01 \\ 35.99 \end{bmatrix}$$

$$\hat{\beta}_{i3} = \begin{bmatrix} 205 \\ 55 \\ 36 \end{bmatrix} + 0.8501 \begin{bmatrix} 2 \\ -0.5 \\ 0.5 \end{bmatrix} = \begin{bmatrix} 206.70 \\ 54.51 \\ 35.57 \end{bmatrix}$$

Exercise 9.3.1. Fill in the details of the formulation of the BE (9.3.3).

Exercise 9.3.2. Given the prior assumptions

$$E(\beta) = 0 \text{ and } D(\beta) = \Sigma \otimes \omega^2 (X'X)^+$$

obtain the estimator of Dempster for the multivariate model.

Exercise 9.3.3. (Alternative Wind Estimator) Recall that for a positive definite matrix H there is a nonsingular sxs matrix C with $\Lambda = CC'$ and $H = CDC'$ with D a positive definite diagonal matrix. Let $Y = U'\beta$, $\eta = (I \otimes C')Y$.

A. 1. Show that a reparametized full rank model is

$$Y = (I \otimes XUC^{-1})\eta + \varepsilon \qquad (9.3.16)$$

2. For estimable parametric functions

$$p'\beta = t'\eta$$

B. Given the prior assumptions

$$E(\beta) = 0 \text{ and } D(\beta) = \Sigma \otimes UC^{-1}DC'^{-1}U' \qquad (9.3.17)$$

obtain the BE and the EBE for $t'\eta$, $d'\gamma$ and $p'\beta$.

Exercise 9.3.4. For the applications considered in Example 9.3.1 using the prior information given there and the data below:
A. Perform the ANOVA and obtain the LS estimators.
B. Find the JS estimator.
C. Suppose no prior information is available. Find the JS estimator using the grand means of the three variables as estimators of the prior mean.

Shift	Diameter	Thickness	Depth of Cut
1	207	55	29
1	206	54	28
1	202	57	30
1	203	58	30
2	206	53	31
2	207	53	31
2	203	58	28
2	203	58	29
3	202	59	28
3	203	59	28
3	203	59	29
3	202	58	30

9.3.2 The Wind Estimator as an Approximate MMSE

The objective is to obtain a MMSE of the form

$$p'\hat{\beta} = p'(I \otimes UAU')b \qquad (9.3.18)$$

where A is a diagonal matrix. Let $\Theta = UDU'$ with D a non-negative diagonal matrix. The objective is to obtain A so that

$$
\begin{aligned}
m &= E_\beta[((I \otimes UAU')b - \beta)'(\Sigma^{-1} \otimes \Theta)((I \otimes UAU')b - \beta)] \\
&= E_\beta[((I \otimes A)g - \gamma)'(\Sigma^{-1} \otimes D)((I \otimes A)g - \gamma)].
\end{aligned}
\qquad (9.3.19)
$$

is minimized. The optimum estimator, i.e., the MMSE (see Exercise 9.3.5), is

$$\hat{\gamma} = (I \otimes A_m)g \qquad (9.3.20)$$

where A_m is the diagonal matrix with elements

$$
\begin{aligned}
a_j &= 1 - \frac{p}{p + \lambda_j \gamma'_{-j} \Sigma^{-1} \gamma} \\
&= 1 - \frac{p}{p + \lambda_j \beta'(I \otimes U_j)\Sigma^{-1}(I \otimes U'_j)\beta}, \quad 1 \le j \le 5.
\end{aligned}
\qquad (9.3.21)
$$

Exercise 9.3.5. Fill in the details of the derivation of (9.3.20) and (9.3.21).

Exercise 9.3.6. Show that

$$\hat{\alpha}_j = \frac{(n - s - p - 1)}{(n - s)} b'(I \otimes U_j)\hat{\Sigma}^{-1}(I \otimes U'_j)b \qquad (9.3.22)$$

is an unbiased estimator of

$$\alpha_j = p + \lambda_j \beta'(I \otimes U_j)\Sigma^{-1}(I \otimes U'_j)\beta. \qquad (9.3.23)$$

Use this information to show how to formulate the Wind estimator (9.3.12) as an approximate MMSE.

Exercise 9.3.7. Show how to formulate the alternative Wind estimator (Exercise 9.3.3) as an approximate MMSE.

Exercise 9.3.8. Formulate the positive parts $PP_0 - PP_4$ of the JS (9.3.12).

9.3.3 The Average MSE and the RSL

The MSE averaging over the prior assumptions (9.2.1) and over the quadratic loss function with matrix $\Sigma \otimes UDU'$ will be obtained. This MSE is uniformly smaller than the LS. These facts are stated more formally and proved in Theorem 9.3.1

Theorem 9.3.1. Assume that:

1. The error vector ε conditional on β is a multivariate normal random variable.
2. The vector of parameters β is multivariate normal and satisfies the prior assumptions (9.1.4).
3. The number of variables is greater than two, i.e., $p > 2$.
4. The matrix $\Theta = UDU'$ with D a positive definite diagonal matrix. Then for optimum

$$c = \frac{n - s - p + 1}{n - s - p + 3} \tag{9.3.24}$$

the JS of Wind has MSE averaging over the prior and the quadratic loss function with matrix $\Sigma^{-1} \otimes \Theta$

$$m = E(\hat{\beta} - \beta)'(\Sigma^{-1} \otimes \Theta)(\hat{\beta} - \beta)$$
$$= ptr(X'X)^+ \Theta - \frac{(p-2)(n-s-p+1)}{(n-s-p+3)} \sum_{j=1}^{s} \frac{d_j}{\lambda_j (1 + \omega_j^2)}. \tag{9.3.25}$$

Proof. Since $\hat{\gamma}_{-j} = [\gamma_{1j}, \gamma_{2j}, ..., \gamma_{pj}]$, letting $m_{-j} = E(\hat{\gamma}_{-j} - \gamma_{-j})' \Sigma^{-1} (\hat{\gamma}_{-j} - \gamma_{-j})$,

$$E(\hat{\beta} - \beta)'(\Sigma^{-1} \otimes \Theta)(\hat{\beta} - \beta)$$
$$= E(\hat{\gamma} - \gamma)'(\Sigma^{-1} \otimes D)(\hat{\gamma} - \gamma)$$

$$= \sum_{j=1}^{s} d_j E(\hat{\gamma}_{-j} - \gamma_{-j})' \Sigma^{-1} (\hat{\gamma}_{-j} - \gamma_{-j})$$

$$= \sum_{j=1}^{s} d_j m_{-j}.$$

(9.3.26)

Then

$$m_{-j} = E(g_{-j} - \gamma_{-j})' \Sigma^{-1} (g_{-j} - \gamma_{-j})$$
$$- \frac{2c(n-s)(p-2)}{\lambda_j(n-s-p+1)} E \left[\frac{g'_j \Sigma^{-1} (g_{-j} - \gamma_{-j})}{(g'_j \hat{\Sigma}^{-1} g_{-j})^2} \right]$$
$$+ \frac{c^2 (n-s)^2 (p-2)^2}{\lambda_j^2 (n-s-p+1)} E \left[\frac{g'_j \Sigma^{-1} g_{-j}}{(g'_j \hat{\Sigma}^{-1} g_{-j})^2} \right].$$

(9.3.27)

The first term of (9.3.27) is the MSE of the LS in orthogonal coordinates, p/λ_j. The computation of the second term is accomplished using the sufficiency principle mentioned earlier and result (viib) of C.R. Rao (1973). Thus,

$$E \left[\frac{g'_j \Sigma^{-1} (g_{-j} - \gamma_{-j})}{g'_j \hat{\Sigma}^{-1} g_{-j}} \right] = \frac{(n-s-p+1)}{(n-s)(1+\omega_j^2)}.$$

(9.3.28)

Likewise

$$E \left[\frac{g'_j \Sigma^{-1} g_{-j}}{(g'_j \hat{\Sigma}^{-1} g_{-j})^2} \right] = \frac{(n-s-p+1)(n-s-p+3)}{(n-s)^2} E \left[\frac{1}{g'_j \Sigma^{-1} g_{-j}} \right]$$
$$= \frac{(n-s-p+1)(n-s-p+3)\lambda_j}{(n-s)^2 (p-2)(1+\omega_j^2)}$$

(9.3.29)

Substitution of (9.3.28) and (9.3.29) into (9.3.27) yields

$$m_{-j} = \frac{p}{\lambda_j} - \frac{2c(p-2)}{\lambda_j(1+\omega_j^2)} + \frac{c^2(p-2)(n-s-p+3)}{\lambda_j(n-s-p+1)(1+\omega_j^2)}.$$

(9.3.30)

The result (9.3.25) then follows for optimum c.

Exercise 9.3.9. A. Fill in the details of the computation of (9.3.28) and (9.3.29).

B. Show that the MSE of the EBE is less than that of the JS if

$$c < \frac{2(n-s-p+1)}{(n-s-p+3)}$$

and that optimum c is as given in Theorem 9.3.1.
C. Extend Theorem 9.3.1 to cases where

$$\Theta = UDU' + \mathbf{1}'V'$$

Exercise 9.3.10. From Exercise 9.3.7 the alternative Wind estimator is

$$p'\hat{\beta} = p'(I \otimes UC^{-1}\hat{A}C'^{-1}U')b \tag{9.3.31a}$$

where A is a diagonal matrix with elements

$$\hat{a}_j = 1 - \frac{(p-2)(n-s)c}{(n-s-p+1)b'(I \otimes UC_j)\hat{\Sigma}^{-1}(I \otimes C_j'U')b}. \tag{9.3.31b}$$

Prove the following theorem and thus verify the formula given for the average MSE given by the formula in (9.3.32).

Theorem 9.3.2. Assume that 1, 2 and 3 of the hypothesis of Theorem 9.3.1 hold true. Let $\Theta = UC\Delta C'U'$ where Δ is a positive definite diagonal matrix with elements δ_j and C was defined in Exercise 9.3.3. Then for optimum c as obtained in Theorem 9.3.1(Exercise 9.3.9) the MSE averaging over prior assumptions (9.3.16) is

$$E(\hat{\beta} - \beta)'(\Sigma^{-1} \otimes \Theta)(\hat{\beta} - \beta)$$

$$\tag{9.3.32}$$

$$= ptr\Theta - \frac{(p-2)(n-s-p+1)}{(n-s-p+3)} \sum_{j=1}^{s} \frac{\delta_j}{1+\omega_j^2}.$$

Recall the RSL was defined by the equation

$$RSL = \frac{MSE(EBE) - MSE(BE)}{MSE(LS) - MSE(BE)}. \tag{9.3.33}$$

The Wind estimator and the alternative Wind estimator both have RSL for the average MSEs computed in this section

$$RSL = 1 - \frac{(p-2)(n-s-p+1)}{p(n-s-p+3)}. \tag{9.3.34}$$

Exercise 9.3.11. Establish the formula for the RSL (9.3.34).

9.3.4 The Conditional MSE and the RL

The goal of this section is to obtain the conditional MSE of the Wind estimator. The main result is contained in Theorem 9.3.3.

Theorem 9.3.3. Assume that hypothesis 1,2 and 4 of Theorem 5.3.1 are satisfied. Then for optimum c obtained in Theorem 9.3.1 the conditional MSE of the Wind JS

$$m = E_\beta (\hat{\beta} - \beta)'(\Sigma^{-1} \otimes \Theta)(\hat{\beta} - \beta)$$

$$= p \operatorname{tr}(X'X)^+ \Theta - \frac{(p-2)^2(n-s-p+1)}{(n-s-p+3)} \sum_{j=1}^{s} \frac{d_j}{\lambda_j} S_{1,p}(\delta_{-j}) \tag{9.3.35}$$

where

$$S_{1,p}(\delta_{-j}) = \sum_{h=0}^{\infty} \frac{e^{-\delta_{-j}} \delta_{-j}^h}{h!(p+2h-2)}$$

and

$$\delta_{-j} = \frac{\gamma'_{-j} \Sigma \gamma_{-j}}{2} = \frac{\lambda_j}{2} \beta'(I \otimes U_j)\Sigma^{-1}(I \otimes U'_j)\beta.$$

The main steps of the proof will be given; the details of the proof will be left to the exercises.

Outline of Proof. Equation (9.3.26) also holds true for expectations conditional on β with m_j the same as in Equation (9.3.27). By integration by parts

$$E_\beta\left[\frac{g'_j\Sigma^{-1}(g_{-j}-\gamma_{-j})}{g'_j\hat{\Sigma}^{-1}g_j}\right] = \frac{(p-2)(n-s-p+1)}{(n-s)}S_{1,p}(\delta_{-j}). \qquad (9.3.36)$$

Furthermore

$$E_\beta\left[\frac{g'_j\Sigma^{-1}g_j}{[g'_j\hat{\Sigma}^{-1}g_j]^2}\right] = \frac{(n-s-p+1)(n-s-p+3)}{(n-s)^2}\lambda_j S_{1,p}(\delta_{-j}). \quad (9.3.37)$$

Substituting these conditional expectations into Equation (9.3.27) after summation over j,

$$m = p\sum_{j=1}^{s}\frac{d_j}{\lambda_j} - 2c(p-2)^2\sum_{j=1}^{s}\frac{d_j}{\lambda_j}S_{1,p}(\delta_{-j})$$
$$+\frac{c^2(p-2)^2(n-s-p+3)}{(n-s-p+1)}\sum_{j=1}^{s}\frac{d_j}{\lambda_j}S_{1,p}(\delta_{-j}). \qquad (9.3.38)$$

Optimum c may be obtained by the usual Calculus argument.

Exercise 9.3.12. A. Fill in the details of the calculation of the expectations in (9.3.36) and (9.3.37).
B. Show that optimum c is the same as in Theorem 9.3.1.

Exercise 9.3.13 Assume that:
1. The number of variables is p > 2.
2. The quadratic loss function has matrix is $\Sigma\otimes P$ where $P = UC\Theta C'U'$, Θ is any positive definite diagonal matrix and C was defined in Exercise 9.3.3. Show that for optimum c obtained in Theorem 9.3.1,

$$E_\beta(\hat{\beta}-\beta)'(\Sigma^{-1}\otimes P)(\hat{\beta}-\beta) = ptr\Theta$$
$$-\frac{(p-2)^2(n-s-p+1)}{(n-s-p+3)}\sum_{j=1}^{s}\theta_j S_{1,p}(\delta_{-j})$$

$$(9.3.39)$$

where

$$\delta_{-j} = \frac{\beta'(I\otimes UC_j)\Sigma^{-1}(I\otimes C'_jU')\beta}{2}.$$

Exercise 9.3.14. Show that for the prior assumptions (9.3.1)

$$E[S_{1,p}(\delta_{-j})] = \frac{1}{p-2}.$$

Exercise 9.3.15. A. Show that the JS of Wind has

$$RL = 1 - \frac{(p-2)^2(n-s-p+1)}{(n-s-p+3)f} \sum_{j=1}^{s} \frac{d_j}{\lambda_j} S_{1,j}(\delta_{-j}) \qquad (9.3.40)$$

with

$$f = \sum_{j=1}^{s} \frac{d_j}{p + 2\delta_{-j}/\lambda_j}.$$

B. To what extent can the properties of the RL that were obtained in Chapter IV and Chapter VII be extended to the multivariate model.

Exercise 9.3.16. A. To what extent can the conditions obtained in Chapter IV and Chapter VII for the conditional MSE of a JS to be smaller than the corresponding BE be extended to the estimator of Wind in the multivariate case?
B. If possible obtain ranges for different values of n and p for the EBE of Wind to have a smaller conditional MSE than the BE for both the univariate and the multivariate case.

Exercise 9.3.17. Find a suitable estimator of the K matrix so that the Wind estimator is a special case of the generalized ridge regression estimator with uniformly smaller conditional and average MSE than the JS for loss functions of the form considered above.

Exercise 9.3.18. A. Obtain the average and the conditional MSEs of $PP_0 - PP_4$ formulated in Exercise 9.3.8. (The derivation of Hotelling's T^2 on pp. 541-542 of C.R. Rao (1973) should prove useful.)

B. Do numerical studies of the efficiencies of these estimators similar to those done in Chapter VI.

9.4 The EBE for r Linear Models

9.4.1 The Estimator of C.R. Rao

The estimator of C.R. Rao will now be generalized to a setup with r multivariate linear models. The setup consists of the r models

$$Y_i = (I \otimes X)\beta_i + \varepsilon_i, \quad 1 \leq i \leq r, \, r > sp + 2 \tag{9.4.1}$$

where Y_i is a vector of m p variate observations, X is a known nxm matrix of rank s $\leq$ m, β is an mp dimensional vector of parameters and the error terms ε are np dimensional vectors. Assume that the prior assumptions on β_i take the form

$$E(\beta_i) = \theta \text{ and } D(\beta_i) = H. \tag{9.4.2}$$

Also assume that the distribution of the error terms conditional on β_i have mean and dispersion

$$E(\varepsilon_i \mid \beta_i) = 0 \quad D(\varepsilon_i \mid \beta_i) = \Sigma \otimes I. \tag{9.4.3}$$

respectively where H is an mp x mp non-negative definite matrix and β_i is an mp dimensional vector of parameters. Assumptions (9.4.2) and (9.4.3) are just multivariate generalizations of those used in Chapter VII for the univariate model. By analogy with the univariate case, the linear BE for known θ, Σ and H

$$\begin{aligned}
p'\beta_i &= p'\theta + p'H(I \otimes X')[(I \otimes X)H(I \otimes X') + \Sigma \otimes I]^{-1} \\
&\quad (Y_i - (I \otimes X)\theta) \\
&= p'b_i - p'\Sigma \otimes (X'X)^+[(I \otimes UU')H(I \otimes UU') + \Sigma \otimes I]^+ \\
&\quad (Y_i - (I \otimes X)\theta).
\end{aligned} \tag{9.4.4}$$

For estimable parametric functions

$$p'\hat{\beta}_i = d'\hat{\gamma}_i$$
$$= d'g_i - d'(\Sigma \otimes \Lambda^{-1})[(I \otimes U')H(I \otimes U) + \Sigma \otimes I]^{-1}(g_i - (I \otimes U')\theta).$$

$$(9.4.5)$$

When β, Σ and H are unknown, by the methodology used several times before, unbiased estimators are substituted into the BE to obtain the EBE. These unbiased estimators for θ, Σ and H are respectively

$$rg_. = \sum_{i=1}^{r} g_i, \qquad (9.4.6)$$

$$W = r(n-s)\hat{\Sigma} = \sum_{i=1}^{r} W_i, \qquad (9.4.7)$$

where the W_i are the within sum of squares matrices of the individual models and

$$M = (r-1)[(I \otimes U)H_.(I \otimes U') + \hat{\Sigma}^{-1} \otimes \Lambda^{-1}]$$
$$= \sum_{i=1}^{r} (g_i - g_.)(g_i - g_.)'. \qquad (9.4.8)$$

Substitution of (9.4.6)-(9.4.8) into (9.4.5) yields the EBE

$$d'\hat{\gamma}_i = d'g_i - d'(E \otimes I)(W \otimes \Lambda^{-1}M^{-1})(g_i - g_.)$$
$$= p'b_i - p'(EW \otimes (X'X)^+)R^+(b_i - b_.) \qquad (9.4.9)$$

where

$$R = (I \otimes U)(I \otimes U')B(I \otimes U)(I \otimes U'), \qquad (9.4.10a)$$

$$B = \sum_{i=1}^{r} (b_i - b_.)(b_i - b_.)' \qquad (9.4.10b)$$

and E is a diagonal matrix.

Example 9.4.1. Consider the experiment described in Example 9.3.1 concerning measurements of characteristics of rivets coming off an assembly line. Below the results of ten replicates of the experiment taken from data obtained through the courtesy of Dr. James Halavin RIT Department of Mathematics and Statistics. The objective is to obtain simultaneous estimates of the parameters for ten multivariate models.

Replicate 1

Shift	Diameter	Thickness	Score
1	210	68	34
1	205	68	34
1	206	72	34
1	208	73	35
2	208	70	34
2	208	71	36
2	208	71	34
2	208	69	34
3	208	71	35
3	208	70	35
3	206	71	33
3	204	70	34

Replicate 2

Shift	Diameter	Thickness	Score
1	203	76	34
1	201	76	34
1	201	76	33
1	205	76	35
2	204	74	34
2	205	74	34
2	205	75	33
2	201	75	34
3	201	78	34
3	207	78	33
3	207	71	32
3	208	71	32

Replicate 3

Shift	Diameter	Thickness	Score
1	208	71	33
1	205	68	34
1	206	74	33
1	204	72	32
2	204	75	34
2	205	75	34
2	205	74	31
2	204	75	31
3	204	74	29
3	205	76	30
3	205	74	28
3	207	73	28

Replicate 4

Shift	Diameter	Thickness	Score
1	207	70	30
1	205	71	32
1	204	70	30
1	207	71	30
2	208	69	32
2	204	69	30
2	205	69	30
2	202	70	30
3	202	72	30
3	206	73	31
3	207	70	29
3	203	69	30

Replicate 5

Shift	Diameter	Thickness	Score
1	203	74	30
1	206	74	30
1	207	71	32
1	205	71	30
2	206	71	29
2	202	72	29
2	202	77	32
2	202	76	32
3	205	76	34
3	204	76	33
3	204	72	31
3	205	72	31

Replicate 6

Shift	Diameter	Thickness	Score
1	205	76	32
1	205	75	33
1	205	73	32
1	205	73	33
2	205	73	32
2	205	74	32
2	203	73	32
2	203	75	32
3	204	72	32
3	204	74	32
3	205	75	33
3	205	74	32

Replicate 7

Shift	Diameter	Thickness	Score
1	204	75	31
1	204	75	33
1	205	74	33
1	205	75	32
2	206	74	29
2	206	74	28
2	294	76	30
2	204	75	31
3	203	76	31
3	203	76	30
3	204	73	31
3	204	73	31

Replicate 8

Shift	Diameter	Thickness	Score
1	204	75	31
1	204	74	32
1	204	74	29
1	204	74	31
2	206	73	30
2	205	73	30
2	204	74	30
2	205	74	31
3	205	75	32
3	205	75	32
3	205	74	32
3	206	74	32

Replicate 9

Shift	Diameter	Thickness	Score
1	204	74	30
1	204	72	30
1	204	74	28
1	204	74	29
2	203	75	30
2	203	75	29
2	204	74	32
2	203	74	31
3	204	75	31
3	204	73	31
3	205	74	30
3	206	75	31

Replicate 10

Shift	Diameter	Thickness	Score
1	202	76	34
1	202	76	34
1	204	75	32
1	204	75	33
2	205	76	30
2	200	77	30
2	201	77	31
2	200	79	31
3	208	71	31
3	208	71	31
3	207	71	30
3	208	70	30

Consider the variables Thickness and Score. Consider the bivariate linear models

$$Y_i = (I_2 \otimes I_3 \otimes 1_4)\beta_i + \varepsilon_i, \ 1 \le i \le 10$$

where

$$\beta_i' = [\beta_{i11}', \quad \beta_{i12}', \quad \beta_{i13}', \quad \beta_{i21}', \quad \beta_{i22}', \quad \beta_{i23}'] \ .$$

The LS are[1]

$$
\tilde{b}' = \begin{bmatrix} b_1' \\ b_2' \\ b_3' \\ b_4' \\ b_5' \\ b_6' \\ b_7' \\ b_8' \\ b_9' \\ b_{10}' \end{bmatrix} = \begin{bmatrix}
70.25 & 70.25 & 70.50 & 35.25 & 34.50 & 34.25 \\
76.00 & 74.50 & 74.50 & 34.00 & 33.75 & 33.00 \\
71.25 & 74.75 & 74.25 & 33.00 & 32.50 & 28.75 \\
70.57 & 69.25 & 71.00 & 30.50 & 30.50 & 30.00 \\
72.50 & 74.00 & 74.00 & 30.50 & 30.50 & 32.25 \\
74.25 & 73.75 & 73.75 & 32.50 & 32.00 & 32.25 \\
74.75 & 74.75 & 74.50 & 32.25 & 29.50 & 30.75 \\
74.25 & 73.50 & 74.50 & 30.75 & 30.25 & 32.00 \\
73.50 & 74.50 & 74.25 & 29.25 & 30.50 & 30.75 \\
75.50 & 77.25 & 70.75 & 33.25 & 30.50 & 30.50
\end{bmatrix} .
$$

Now

$$b_*' = \begin{bmatrix} 73.28 & 73.65 & 73.20 & 32.03 & 31.45 & 31.45 \end{bmatrix},$$

---

[1] The LS b and the EBE $\hat{\beta}$ is a 60 dimensional column vector . Each row of the matrix $\tilde{b}'$ and the matrix $\tilde{\hat{\beta}}'$ is the portion of the column vector for the i'th model. The reason for using this somewhat awkward notation is to be able to fit all 60 elements of the vector into a smaller space.

$$B = \begin{bmatrix} 37.67 & 31.78 & 13.42 & 3.637 & -7.617 & 2.536 \\ 31.78 & 47.90 & 14.76 & 3.900 & -9.550 & -9.300 \\ 13.42 & 14.76 & 26.35 & -7.300 & -5.275 & -2.713 \\ 3.637 & 3.900 & -7.300 & 25.56 & 18.08 & 7.950 \\ -7.617 & -9.550 & -5.275 & 18.08 & 24.85 & 12.60 \\ 2.536 & -9.300 & -2.713 & 7.950 & 12.60 & 23.10 \end{bmatrix}$$

and

$$B^{-1} = \begin{bmatrix} 0.1037 & -0.0560 & -0.0259 & -0.0405 & 0.0635 & -0.0576 \\ -0.0560 & 0.0721 & -0.0131 & -0.0214 & 0.0035 & 0.0391 \\ -0.0259 & -0.0131 & 0.0652 & 0.0497 & -0.0404 & 0.0102 \\ -0.0405 & -0.0214 & 0.0297 & 0.1452 & -0.1261 & 0.0205 \\ 0.0635 & 0.0035 & -0.0404 & -0.1261 & 0.1761 & -0.0630 \\ 0.0576 & 0.0391 & 0.0102 & 0.0205 & -0.0630 & 0.0939 \end{bmatrix}.$$

Let

$$c = \frac{r - sp - 2}{r(n - s) + 1}.$$

The motivation for this particular choice of c is given by the result of Exercise 9.4.5. Since $r = 10$, $s = 3$ and $p = 2$ it follows that $c = 2/91$. Also

$$W_1 = \begin{bmatrix} 24.5 & 3.75 \\ 3.78 & 6.50 \end{bmatrix} \quad W_2 = \begin{bmatrix} 50 & 6.5 \\ 6.5 & 4.75 \end{bmatrix} \quad W_3 = \begin{bmatrix} 24.25 & 0.75 \\ 0.75 & 13.75 \end{bmatrix}$$

$$W_4 = \begin{bmatrix} 11.75 & 3.50 \\ 3.50 & 8.00 \end{bmatrix} \quad W_5 = \begin{bmatrix} 51 & 22 \\ 22 & 18.75 \end{bmatrix} \quad W_6 = \begin{bmatrix} 4.25 & 0.75 \\ 0.75 & 1.75 \end{bmatrix}$$

$$W_7 = \begin{bmatrix} 12.5 & 0.25 \\ 0.25 & 8.5 \end{bmatrix} \quad W_8 = \begin{bmatrix} 2.75 & 0.75 \\ 0.75 & 6.50 \end{bmatrix} \quad W_9 = \begin{bmatrix} 6.75 & -3.25 \\ -3.25 & 8.50 \end{bmatrix}$$

$$W_{10} = \begin{bmatrix} 6.50 & 3.50 \\ 3.50 & 4.75 \end{bmatrix}.$$

Thus,

$$W = \sum_{i=1}^{10} W_i = \begin{bmatrix} 204.25 & 38.50 \\ 38.50 & 80.75 \end{bmatrix}.$$

The EBE of C.R. Rao is

$$\tilde{\beta}' = \begin{bmatrix} 70.36 & 70.22 & 70.51 & 34.25 & 34.50 & 37.20 \\ 75.81 & 74.55 & 74.49 & 33.69 & 33.69 & 33.03 \\ 71.37 & 74.67 & 74.18 & 32.46 & 32.46 & 28.79 \\ 70.46 & 69.43 & 71.11 & 30.53 & 30.53 & 30.08 \\ 72.69 & 73.89 & 73.92 & 30.54 & 30.52 & 32.17 \\ 74.19 & 73.78 & 73.73 & 32.49 & 32.00 & 32.24 \\ 74.74 & 74.88 & 74.36 & 32.15 & 29.64 & 30.71 \\ 74.22 & 73.56 & 74.43 & 30.74 & 30.29 & 31.97 \\ 73.51 & 74.39 & 74.32 & 29.36 & 30.40 & 30.78 \\ 75.46 & 77.14 & 70.95 & 33.25 & 30.48 & 30.54 \end{bmatrix}.$$

Exercise 9.4.1. Verify (9.4.5) and the equality in (9.4.9).

Exercise 9.4.2. Show how to formulate (9.4.9) as an approximate MMSE.

Exercise 9.4.3. Find the EBE for the variables (Diameter, Thickness), (Diameter, Score). Is it possible to find the EBE for all three variables with only ten replications? Explain.

Exercise 9.4.4. Is it possible to formulate the positive part of the estimator of C.R. Rao for the multivariate case? If so how would you go about it?

9.4.2 The Average MSE of the EBE

The computation of the average MSE of the EBE of C.R. Rao for the multivariate case presents some new difficulties because of:

1. the nonexistence of a theoretically optimum c;
2. the dispersion Σ is estimated by a matrix of random variables.

The main result is given in Theorem 9.4.1. The MSE formula presented there specializes to that obtained earlier for the univariate model when p=1.

Theorem 9.4.1. If for the setup (9.4.1):

1. the r models are independent and r>sp + 2;
2. the distribution of β and ε conditional on β is multivariate normal;
3. assumptions (9.4.2) and (9.4.3) hold true;
4. the parametric functions $p'\beta$ are estimable;

then the average MSE of the multivariate EBE (7.4.9) is

$$
\begin{aligned}
m = \sum_{i=1}^{r} p'E(\hat{\beta}_i - \beta_i)(\hat{\beta} - \beta_i) &= rp'(\Sigma \otimes (X'X)^+)p \\
&- r(n-s)p'(\xi \otimes I)\Gamma p - r(n-s)p'\Gamma(\xi \otimes I)p \\
&+ \frac{r(n-s)}{r-sp-2}p'(\xi\Sigma\xi \otimes U\Delta U')p.
\end{aligned}
\tag{9.4.11}
$$

The matrix

$$
\begin{aligned}
\Gamma = (\Sigma \otimes (X'X)^+)[(I \otimes UU')H(I \otimes UU' \\
+ \Sigma \otimes (X'X)^+]'(\Sigma \otimes (X'X)^+).
\end{aligned}
\tag{9.4.12a}
$$

The sxs matrix Δ is the sum of the p diagonal s x s blocks of the matrix

$$
\Omega = (I \otimes \Lambda^{-1})[(I \otimes U')H(I \otimes U') + \Sigma \otimes \Lambda^{-1}]'(\Sigma \otimes \Lambda^{-1}).
\tag{9.4.12b}
$$

Proof. For the estimable parametric functions

$$m = d'(\Sigma \otimes \Lambda^{-1})d$$

$$-d'(\xi \otimes I)E[(W \otimes \Lambda^{-1})M^{-1}\sum_{i=1}^{r}(g_i - g_.)(g_i - \gamma_i)']d$$

$$-d'E[\sum_{i=1}^{r}(g_i - \gamma_i)(g_i - g_.)'M^{-1}(W \otimes \Lambda^{-1})(\xi \otimes I)]d$$ (9.4.13)

$$+d'(\xi \otimes I)E[(W \otimes \Lambda^{-1})M^{-1}(W \otimes \Lambda^{-1})](\xi \otimes I)d.$$

The computation of the second and the third term of (9.4.13) is similar to that of the univariate case.

Computation of the fourth term of (9.4.13) requires some work that was not needed in the univariate case because of the matrix W. Observe that

$$d'E[(W \otimes \Lambda^{-1})M^{-1}(W \otimes \Lambda^{-1})]d = d'EE_w[(W \otimes \Lambda^{-1})M(W \otimes \Lambda^{-1})]$$

(9.4.14)

Since M and W are independent

$$E_w[(W \otimes \Lambda^{-1})M^{-1}(W \otimes \Lambda^{-1})] = (W \otimes \Lambda^{-1})E(M^{-1})(W \otimes \Lambda^{-1}).$$ (9.4.15)

Now M is an sp variate Wishart distribution with $r - 1$ degrees of freedom and dispersion matrix

$$A = (I \otimes U)H(I \otimes U') + \Sigma \otimes \Lambda^{-1},$$ (9.4.16)

i.e.,

$$M \sim W_{sp}(r - 1, A).$$ (9.4.17)

Thus,

$$E(M^{-1}) = \frac{1}{r - sp - 2}A^{-1}.$$ (9.4.18)

From (9.4.16) and (9.4.18)

$$E_w[(W \otimes \Lambda^{-1})M^{-1}(W \otimes \Lambda^{-1})]$$
$$= (W \otimes \Lambda^{-1})E(M^{-1})(W \otimes \Lambda^{-1}) \qquad (9.4.19)$$
$$= \frac{1}{r - sp - 2}(W \otimes \Lambda^{-1})A^{-1}(W \otimes \Lambda^{-1}).$$

Now

$$(W \otimes \Lambda^{-1})A^{-1}(W \otimes \Lambda^{-1}) = (W \otimes I)(I \otimes \Lambda^{-1})A^{-1}(I \otimes \Lambda^{-1})(W \otimes I)$$
$$(9.4.20)$$

Let

$$D^{-1} = (I \otimes \Lambda^{-1})A^{-1}(I \otimes \Lambda^{-1}) \qquad (9.4.21)$$

and

$$T = (W \otimes I)D^{-1}(W \otimes I). \qquad (9.4.22)$$

Let T_{ij} be the s x s block in the i'th row and the j'th column of the p x p matrix of s x s blocks of T. Then

$$T_{ij} = \sum_{l=1}^{p}\sum_{k=1}^{p} w_{lk}I_s D^{kl}I_s w_{lj} \qquad (9.4.23)$$

where D^{kl} is an sxs block in the k'th row and the l'th column of the pxp matrix of sxs blocks of D^{-1}. From (5.4.23) and Equation 13, p.161 Anderson (1958)

$$E(T_{ij}) = \sum_{k=1}^{p}\sum_{l=1}^{p} r(n - s)^2 \sigma_{lk}ID^{kl}I\sigma_{lj}$$
$$+ \sum_{k=1}^{p}\sum_{l=1}^{p} r(n - s)^2 \sigma_{il}ID^{kl}I\sigma_{kj} \qquad (9.4.24)$$
$$+ \sum_{k=1}^{p}\sum_{l=1}^{p} r(n - s)^2 \sigma_{ij}ID^{kl}I\sigma_{kl}$$

since

$$W \sim W_p(r(n - 1), \Sigma). \qquad (9.4.25)$$

From (9.4.24)

$$E(T) = [r(n-s)+1]r(n-s)(\Sigma \otimes I)D^{-1}(\Sigma \otimes I)$$
$$+r(n-s)\Sigma \otimes \sum_{i=1}^{p}(D^{-1} \otimes (\Sigma \otimes I)_{ll}) \tag{9.4.26}$$

where $D^{-1}(\Sigma \otimes I)_{ll}$ is the l'th diagonal sxs block of $D^{-1}(\Sigma \otimes I)$. From (9.4.26) and (9.4.19) the fourth term of (9.4.13) is

$$d'E[(W \otimes \Lambda^{-1})M^{-1}(W \otimes \Lambda^{-1})]d$$
$$= \frac{(r(n-s)+1)r(n-s)}{r-sp+2}d'(\Sigma \otimes \Lambda^{-1})A^{-1}(\Sigma \otimes \Lambda^{-1})d. \tag{9.4.27}$$

Equation (9.4.11) now follows by estimability.

Exercise 9.4.5. A. Obtain the first three terms of (9.4.13).
B. Show how to obtain the last two terms of (9.4.11) from (9.4.27).

Exercise 9.4.6. Show that when

$$c = \frac{r-sp-2}{r(n-s)+1} \tag{9.4.28}$$

the first three terms of (9.4.11) is minimized and the resulting MSE is

$$m = p'\Sigma \otimes (X'X)^{+}p - \frac{(r-sp-2)r(n-s)}{r(n-s)+1}p'Tp$$
$$+\frac{r(n-s)(r-sp-2)}{[r(n-s)+1]^2}p'(\Sigma \otimes U\Delta U')p. \tag{9.4.29}$$

Exercise 9.4.7. Formulate the EBE of C.R. for the case where the prior mean is zero and obtain its MSEs.

9.5 The Estimator of Dempster (An Example of an Approximate MSE)

As has already been pointed out it is not always possible to find the exact MSE for an EBE of the parameters of a multivariate model. For the case of the seemingly unrelated regression equations model Srivistava (1970) obtained an approximate formula for the conditional MSE. The technique will be applied here to show how to obtain the approximate average MSE of the Dempster estimator for a MANOVA model. The method of computation will be illustrated for a single linear model but can be applied to other cases.

Consider a single multivariate linear model together with the prior assumptions in (9.3.14). The reader who has solved Exercises 9.3.1 and 9.3.2 will observe that for estimable parametric functions the EBE is

$$p'\hat{\beta} = p'b - \frac{1}{1+\omega^2}p'b \qquad (9.5.1)$$

and the EBE of Dempster is

$$p'\hat{\beta} = p'b - \frac{cp'b}{b'(\hat{\Sigma}^{-1} \otimes X'X)b}. \qquad (9.5.2)$$

In Exercise 9.5.3 the reader is invited to obtain the average MSE of (9.5.2) for the case where Σ is known. This is a relatively straightforward problem in the light of what has already been done. Given certain conditions explained below the result of Exercise 9.5.3 may be used to obtain the MSE for the case of unknown Σ.

Definition 9.5.1 that is used by Srivistava (1970) will help to quantify the level of goodness of the approximate MSEs.

Definition 9.5.1. A function G(t) is $O_p(t^\alpha)$, i.e., of the same order of magnitude in probability as t^α if given any two constants ε and η however small, there exists a T depending on ε and η such that

$$P\left[\frac{G(t)}{t^\alpha}\right] > 1 - \eta \text{ for all } t > T.$$

The original definition was formulated for sequences by Mann and Wald (1943). It is given in Definition 9.5.2 below and it is equivalent to Definition 9.5.1 for sequences, i.e., functions on the integers.

Definition 9.5.2. (Mann and Wald (1943)). The sequence $X_N = O_p(f(N))$ if for each $\varepsilon > 0$ there exists an $A_\varepsilon > 0$ such that $P[|X_N| \leq A_\varepsilon f(N)] \geq 1 - \varepsilon$ for all values of N.

Exercise 9.5.1. Show that for sequences Definitions 9.5.1 and 9.5.2 are equivalent.

Exercise 9.5.2. Let $X_1, X_2,..., X_n \sim N(\mu, \sigma^2)$. Let $s_n^2 = \sum_{i=1}^n (x_i - \bar{x})^2 /(n-1)$. This is the usual unbiased estimator of the variance σ^2. Show that this estimator

A. is consistent for σ^2;

B. is $O_p(n^\alpha)$ if $\alpha \geq -1/2$.

The MSE of the EBE (9.5.2) may be approximated to the same order as $\Delta = \hat{\Sigma} - \Sigma$ or a higher order. The approximation will be obtained here to the same order as Δ assuming $\Delta \sim O_p(n^{-1/2})$. With this in mind observe that

$$E[p'(\hat{\beta} - \beta)(\hat{\beta} - \beta)'p]$$

$$= E[p'(b - \beta)(b - \beta)'p] - 2cp'E\left[\frac{b(b - \beta)'}{b'(\hat{\Sigma}^{-1} \otimes X'X)b}\right]p \qquad (9.5.3)$$

$$+ c^2 p'E\left[\frac{bb'}{[b'(\hat{\Sigma}^{-1} \otimes X'X)b]^2}\right]p.$$

The matrix $\hat{\Sigma}^{-1}$ will be approximated by functions of Σ up to $O_p(n^{-1/2})$ by obtaining a few terms of the Taylor expansion of $(\Sigma + \Delta)^{-1}$. Thus,

$$\hat{\Sigma}^{-1} = (\Sigma + \Delta)^{-1}$$

$$= \Sigma^{-1/2}(I + \Sigma^{-1/2}\Delta\Sigma^{-1/2})^{-1}\Sigma^{-1/2}$$

$$= \Sigma^{-1/2}(I - \Sigma^{-1/2}\Delta\Sigma^{-1/2})\Sigma^{-1/2} + O_p(n^{-1})$$

$$= \Sigma^{-1} - \Sigma^{-1}\Delta\Sigma^{-1} + O_p(n^{-1}). \qquad (9.5.4)$$

The remaining terms of (9.5.4) are $O_p(n^{-1})$ because the matrix Δ appears at least twice. Let $C = \Sigma^{-1}\Delta\Sigma^{-1} - O_p(n^{-1})$. Observe that C is $O_p(n^{-1/2})$. Then

$$b'(\hat{\Sigma}^{-1} \otimes X'X)b = b'(\Sigma^{-1} \otimes X'X)b - b'(C \otimes X'X)b \qquad (9.5.5)$$

and

$$\frac{1}{b'(\hat{\Sigma}^{-1} \otimes X'X)b}$$

$$= \frac{1}{b'(\Sigma^{-1} \otimes X'X)b\left[1 - b'(C \otimes X'X)b/b'(\Sigma^{-1} \otimes X'X)b\right]}$$

$$= \frac{1}{b'(\Sigma^{-1} \otimes X'X)b} + O_p(n^{-1/2}). \tag{9.5.6}$$

Furthermore

$$\frac{1}{[b'(\Sigma^{-1} \otimes X'X)b]^2} = \frac{1}{b'(\Sigma^{-1} \otimes X'X)b} + O_p(n^{-1/2}). \tag{9.5.7}$$

Equations (9.5.6) and (9.5.7) may be used in conjunction with the MSE for known Σ to approximate the MSE to $O_p(n^{-1/2})$ for unknown Σ. The approximation may be refined to a higher order by computing more terms in the series in (9.5.4)-(9.5.7) and using the results together with the MSE formula for known Σ.

Exercise 9.5.3. A. Assume Σ is known. Find the average and the conditional MSE for the estimator of Dempster, i.e.,

$$m = E(\hat{\beta} - \beta)'(\Sigma^{-1} \otimes X'X)(\hat{\beta} - \beta).$$

B. Assume $\Delta = \hat{\Sigma} - \Sigma + O_p(n^{-1/2})$. Obtain the approximate and the average MSE to $O_p(n^{-1})$.

C. For the results of A and B as appropriate give exact or approximate expressions for the RSL and the RL.

Exercise 9.5.4. Formulate the estimators of Wind and Dempster for r multivariate linear models. Use the methods of this section to obtain the approximate MSEs.

Exercise 9.5.5. Formulate the positive parts PP_0 - PP_4 for the JS estimator of Dempster. Is it possible to show that the average and conditional MSEs of the positive parts is smaller than that of the JS?

9.6 Summary

1. The formulation of the multivariate linear model (MANOVA) along the lines of C.R. Rao (1973) was reviewed.

2. The BEs and the EBEs were formulated for the multivariate generalization of the priors of C.R. Rao, Wind and Dempster.

3. An exact expression for the average and conditional MSE of the Wind estimator of the parameters of a single MANOVA model was obtained.

4. The results of C.R. Rao for r linear models were generalized to setups with r multivariate models.

5. To provide an example of a large sample approximation to an MSE, it was shown how the approximate MSE could be obtained for the Dempster JS for the single linear model case.

6. An indication of how some of the results in Chapters VI, VII and VIII can be extended to the multivariate linear model was given.

Chapter X

Other Linear Model Setups

10.0 Introduction

The estimators and the results about their efficiencies that were examined in Chapters I – IX can be expanded to many other different linear model setups for a wide variety of applications in the physical, engineering, economic and social sciences. Three different kinds of linear model setups will be considered in this chapter. These include:

1. the Seemingly Unrelated Regression Model;
2. the simultaneous estimation of the parameters for r linear models when the design matrices and the error variances are different;
3. the Kalman filter.

Chapter VII concentrated on the problem of the simultaneous estimation of parameters for r uncorrelated linear models. Two other important problems are:

1. the estimation of the parameters in a $(r + 1)$'st linear model using the $(r + 1)$'st observation and the past observations of the response variable in the r linear models;

531

2. the prediction of the response variable in the $(r + 1)$'st model assuming the observations for the first r models are available.

This chapter will also include a brief introduction to these problems.

The SURE model is a generalization of both the setup with r linear models considered in Chapter VII and the MANOVA model that was considered in Chapter IX. Unlike the setup with r linear models and the MANOVA model the design matrices for each variable are different. Unlike the setup in Chapter VII the error terms of the individual models are contemporaneously correlated. This kind of setup is very important to econometric applications. For example, one might like to use two linear regression models to compare the effect of allocation of resources by two similar firms on some measure of productivity. It is easy to see that the error terms of such models will be. contemporaneously correlated. The effects of such contemporaneous correlation on the form and the efficiency of the LS, the JS and the ridge regression estimator will be studied in Section 10.1.

C.R. Rao (1975) obtains the linear Bayes and the empirical Bayes estimators of parameters of r linear regression models for the case where the variances of the error terms and the design matrices in the individual models are different. This situation could arise if an experiment is replicated in not quite the same way due to practical considerations or errors on the part of the technicians taking the data. For example some of the replications might have different missing values and the variability of the error terms might be different due to circumstances beyond the experimenters control. Section 10.2 includes a brief description of the estimators for this type of circumstance.

Instead of estimating the parameters of r linear models using all of the models one can estimate the parameters of the $(r+1)$'st model using the $(r+1)$'st set of observations and the other r linear models. Estimation of the parameters of the other r linear models is no longer of interest; one is interested only in estimating the current individual parameters. A discussion of estimators of Rao, Wind and Dempster for the current individual and their efficiencies is also given in Section 10.2.

Another problem that is considered by C.R. Rao (1975) is that where there are $r+1$ linear models where the last observation of the $(r+1)$'st model is unavailable and has to be predicted. This problem is described in Section 10.2 and examples are given for the estimators for C.R. Rao, Wind and Dempster.

The Kalman filter is used in many diverse fields of application. These include tracking of space satellites, quality control and time series analysis in econometrics. The formulation of one and r Kalman filters is described in Section 10.3 and includes some examples that use simulated data.

10.1 The Seemingly Unrelated Regression Equation Model (SURE)

The Seemingly Unrelated Regression Equation model (SURE) is a generalization of the multivariate linear model that was studied in Chapter IX. The multivariate linear model was viewed there as p linear univariate models with correlated error terms and identical X matrices. Like the multivariate model the SURE model may also be viewed as p or to avoid confusing notations g models with correlated error terms. However the design matrices need not be equal.

The setup would be as follows.

$$
\begin{bmatrix} Y_1 \\ Y_2 \\ \cdot \\ \cdot \\ \cdot \\ Y_g \end{bmatrix} = \begin{bmatrix} X_1 & 0 & \cdot & \cdot & \cdot & 0 \\ 0 & X_2 & & & & 0 \\ & & 0 & & & \\ \cdot & \cdot & \cdot & & & \\ \cdot & \cdot & \cdot & & & \\ 0 & 0 & & & & X_g \end{bmatrix} \begin{bmatrix} \beta_1 \\ \beta_2 \\ \cdot \\ \cdot \\ \cdot \\ \beta_g \end{bmatrix} + \begin{bmatrix} \varepsilon_1 \\ \varepsilon_2 \\ \cdot \\ \cdot \\ \cdot \\ \varepsilon_g \end{bmatrix}
\qquad (10.1.1)
$$

where the X_i are $n_i \times m_i$ dimensional matrices. As was the case for a multivariate model it will be assumed that

$$
E(\varepsilon \mid \beta) = 0 \text{ and } D(\varepsilon \mid \beta) = \Sigma \otimes I. \qquad (10.1.2)
$$

When all of the design matrices are identical, i.e., all the X_i are equal the model (10.1.1) reduces to the MANOVA model that was studied in Chapter IX.

The assumptions in (10.2) allow for two different regression equations to have contemporaneously correlated error terms. There are a number of practical situations where this type of model is desirable particularly in econometric models. For example Zellner (1962) used two different regression equations to explain investment for two large corporations General Electric and Westinghouse. The dependent variables were the real gross investments of the firms and the independent variables were the value of the outstanding shares at the beginning of the period studied and the initial value of the firm's real capital stock. For this situation it seems reasonable that the error terms might be contemporaneously correlated.

The effects of contemporaneous correlation on the efficiency of the least square estimators, the ridge type estimators and the James-Stein type estimators will be taken up in Section 10.1.1, Section 10.1.2 and Section 10.1.3 respectively.

10.1.1 The Least Square Estimators

Two kinds of least square estimators may be studied, the ordinary least square estimator for each individual model and a weighted least square estimator where the setup is viewed as a single linear model. Before formulating the model for g variables an example for the case g = 2 will be considered.

Example 10.1.1. The setup (10.1) becomes

$$\begin{bmatrix} Y_1 \\ Y_2 \end{bmatrix} = \begin{bmatrix} X_1 & 0 \\ 0 & X_2 \end{bmatrix}\begin{bmatrix} \beta_1 \\ \beta_2 \end{bmatrix} + \begin{bmatrix} \varepsilon_1 \\ \varepsilon_2 \end{bmatrix}$$

It is assumed that $X_1'X_2 = 0$ and that $X_2'X_1 = 0$. Furthermore

$$\Sigma = \sigma^2 \begin{bmatrix} 1 & \rho \\ \rho & 1 \end{bmatrix}$$

where ρ is the correlation coefficient for the random variables ε_1 and ε_2. The ordinary least square estimator for each individual model is given by

$$b_i = (X_i'X_i)^{-1}X_i'Y_i , \; i = 1,2.$$

Since

$$\Sigma^{-1} = \frac{1}{\sigma^2(1-\rho^2)}\begin{bmatrix} 1 & -\rho \\ -\rho & 1 \end{bmatrix},$$

it follows that the weighted LS is given by

$$b_1^a = (X_1'X_1)^{-1}X_1'(Y_1 - \rho Y_2)$$

and

$$b_2^a = (X_2'X_2)^{-1}X_2'(Y_2 - \rho Y_1).$$

For the ordinary LS the variance is given by

$$V(b_i) = \sigma^2 (X_i'X_i)^{-1}, \; i = 1,2.$$

For the weighted LS

$$V(b_i^a) = \sigma^2 (1 - \rho^2)(X_i'X_i)^{-1}, \; i = 1,2.$$

Thus, especially when there is significant correlation between the variables, it is advantageous to use the weighted LS. This weighted LS is often referred to in the literature as the Aiken estimator.

In general the Aiken estimator takes the form

$$b^a = (X'(\Sigma^{-1} \otimes I)^{-1}X)^{-1}X'\Sigma^{-1}Y. \qquad (10.1.3)$$

Denote the elements of Σ by σ_{ij} and the elements of Σ^{-1} by σ^{ij}. When $X_i'X_j = 0$ the components of b^a are

$$b_i^a = (X_i'X_i)^{-1}X_i'Y_i + \sum_{j \neq i} \frac{\sigma^{ij}}{\sigma^{ii}}(X_i'X_i)^{-1}X_i'Y_j. \qquad (10.1.4)$$

Exercise 10.1.1. Show that if Σ is known then b_i^a is an unbiased estimator of β_t.

Exercise 10.1.2. What is the form of the variance of (10.13) and (10.14) and how does it compare to that of the OLS?

10.1.2 The Ridge Type Estimators

Assuming a zero prior mean and that F is positive definite three equivalent forms of the BE are

$$\begin{aligned} \hat{\beta} &= FX'(XFX' + \Sigma)^{-1}Y = (F^{-1} + X'\Sigma^{-1}X)^{-1}X'\Sigma^{-1}Y \\ &= b^a - (X'\Sigma^{-1}X)^{-1}(F + (X'\Sigma^{-1}X)^{-1})^{-1}b^a. \end{aligned} \qquad (10.1.5)$$

The Aiken ridge estimator is the second expression in (10.1.5). For the case where

$$F = \begin{bmatrix} F_1 & 0 & . & . & . & 0 \\ 0 & F_2 & 0 & . & . & . \\ . & . & . & . & . & . \\ . & . & . & . & . & . \\ . & . & . & . & . & 0 \\ 0 & . & . & . & . & F_g \end{bmatrix} \qquad (10.1.6)$$

and $X_i'X_j = 0$ for the i'th model the estimator takes the form

$$\hat{\beta}_i = \left(\frac{1}{\sigma^{ii}} F_i^{-1} + X_i'X_i \right)^{-1} \left(X_i'Y_i + \sum_{j \neq i} \frac{\sigma^{ij}}{\sigma^{ii}} X_j'Y_i \right) \qquad (10.1.7)$$

Let $G_i = \dfrac{1}{\sigma^{ii}} F_i^{-1}$. Then the Aiken ridge estimator (10.17) takes the form

$$\hat{\beta}_i = \left(G_i + X_i'X_i \right)^{-1} \left(X_i'Y_i + \sum_{j \neq i} \frac{\sigma^{ij}}{\sigma^{ii}} X_j'Y_i \right). \qquad (10.1.8)$$

The ridge estimator of C.R. Rao takes the form

$$\hat{\beta}_i = (X_i'X_i + G_i)^{-1} X_i'Y_i \qquad (10.1.9)$$

Example 10.1.2 The form of the Aiken ridge estimator for the two equation model studied in Example 10.1.1 is

$$\begin{bmatrix} \hat{\beta}_1 \\ \hat{\beta}_2 \end{bmatrix} = \begin{bmatrix} (X_1'X_1 + G_1)^{-1} X_1'(Y_1 - \rho Y_2) \\ (X_2'X_2 + G_2)^{-1} X_2'(Y_2 - \rho Y_1) \end{bmatrix}$$

The goal is to compare the efficiency of the Aiken ridge estimator (10.1.8) with that of the ridge estimator of C.R. Rao (10.1.9). The Aiken ridge estimator makes use of the contemporaneous correlation between the models. The ridge estimator of C.R. Rao does not.

The computation of the dispersion of the Aiken ridge estimator will be easier after it is written in the alternative form

$$\hat{\beta}_i = (G_i + X_i'X_i)^{-1}(X_i'X_i)b_i^a \quad . \tag{10.1.10}$$

The reader that did Exercise 10.1.2 obtained

$$D(b_i^a) = \frac{1}{\sigma^{ii}}(X_i'X_i)^{-1} \tag{10.1.11}$$

for the dispersion of the Aiken estimator. Using (10.1.10) and (10.1.11) the dispersion of the Aiken ridge estimator is

$$D(\hat{\beta}_i) = \frac{1}{\sigma^{ii}}(G_i + X_i'X_i)^{-1}(X_i'X_i)(G_i + X_i'X_i)^{-1}. \tag{10.1.12}$$

The reader is invited to show in Exercise 10.1.3 that the squared bias matrix is given by

$$B(\hat{\beta}_i) = (G_i + X_i'X_i)^{-1}G_i\beta_i\beta_i'G_i(G_i + X_i'X_i)^{-1} \quad . \tag{10.1.13}$$

The conditional MDE of the Aiken ridge estimator is the sum of the expressions in (10.1.12) and (10.1.13). Hence

$$MDE(\hat{\beta}_i) = (G_i + X_i'X_i)^{-1}\left(\frac{1}{\sigma^{ii}}(X_i'X_i) + G_i\beta_i\beta_i'G_i\right)(G_i + X_i'X_i)^{-1}. \tag{10.1.14}$$

The MDE averaging over the prior distribution may be obtained from (10.1.15) after noticing that

$$E[G_i\beta_i\beta_i'G_i] = \frac{1}{\sigma^{ii}}G_i. \tag{10.1.15}$$

Finding the expectation of (10.1.14) using (10.1.15) yields the average MDE ,

$$MDE(\hat{\beta}_i) = \frac{1}{\sigma^{ii}}(G_i + X_i'X_i)^{-1} \tag{10.1.16}$$

Exercise 10.1.3. Establish (10.1.13).

The corresponding expression for the dispersion, squared bias, conditional MDE and average MDE for the ridge estimator of C.R.Rao are given by

$$D(\hat{\beta}_i) = \sigma_{ii}(G_i + X_i'X_i)^{-1}(X_i'X_i)(G_i + X_i'X_i)^{-1}, \qquad (10.1.17)$$

$$B(\hat{\beta}_i) = (G_i + X_i'X_i)^{-1}G_i\beta\beta'G_i(G_i + X_i'X_i)^{-1}, \qquad (10.1.18)$$

$$MDE(\hat{\beta}_i) = (G_i + X_i'X_i)^{-1}\left(\sigma_{ii}(X_i'X_i) + G_i\beta\beta'G_i\right)(G_i + X_i'X_i)^{-1} \qquad (10.1.19)$$

and

$$MDE(\hat{\beta}_i) = \sigma_{ii}(G_i + X_i'X_i)^{-1} \qquad (10.1.20)$$

respectively.

The Aiken estimators are more efficient than the ordinary LS and the corresponding ridge estimators because

$$\frac{1}{\sigma^{ii}} < \sigma_{ii}. \qquad (10.1.21)$$

To establish (10.1.21) notice that

$$\frac{1}{\sigma^{ii}} = \frac{|\Sigma|}{|D_{ii}|} \qquad (10.1.22)$$

where $|\Sigma|$ is the determinant of Σ and $|D_{ii}|$ is the determinant of the minor of Σ obtained by deleting the i'th row and the i'th column of Σ. Observe that

$$|\Sigma| = \left|\sigma_{11} - \Sigma_{12}D_{11}\Sigma_{12}'\right|\left|D_{11}\right| \le \sigma_{11}\left|D_{11}\right| \qquad (10.1.23)$$

The inequality in (10.1.23) becomes an equality when all the elements of Σ_{12} are zero. If $i \ne 1$ exchange the i'th row with the first row of Σ and then exchange the i'th column with the first column. Call the new matrix thus formed $\tilde{\Sigma}$. Now

$$\left|\tilde{\Sigma}\right| = |\Sigma|, \left|\tilde{D}_{11}\right| = |D_{ii}| \text{ and } \tilde{\sigma}_{11} = \sigma_{ii}. \qquad (10.1.24)$$

The result (10.1.21) follows.

Remember that a positive semidefinite matrix A is larger than a positive semidefinite matrix B if A −B is positive semidefinite. The Aiken ridge estimator is more efficient than that of C.R.Rao in the sense that :

1. its dispersion is smaller;
2. its conditional and average MDE is smaller;
3. the squared bias to variance ratio p'Bp/p'Dp is larger .

Exercise 10.1.5. Obtain expressions for the MDE with and without averaging over an appropriate prior distribution for the estimators in Example 10.1.2.

Exercise 10.1.6. Give the form of the ordinary and generalized Hoerl Kennard ridge, the contraction and generalized contraction estimators as special cases of the Aiken ridge estimator.

When Σ is unknown it can be estimated by a consistent estimator S such that

$$\Delta = (S \otimes I) - (\Sigma \otimes I) \text{ is } O_p(n^{-\frac{3}{2}}). \tag{10.1.25}$$

Srivistava and Giles (1987) show that when S is the within sum of squares matrix estimator and the error vector is symmetrically distributed the estimator

$$b^a = (X'(S^{-1} \otimes I)^{-1}X)^{-1}X'S^{-1}Y \tag{10.1.26}$$

is unbiased to order $O(n^{-1})$. This means that the bias is $O(n^{-a})$ where a > 1. Thus the approximate estimator is asymptotically unbiased. They also present an asymptotic approximation for the dispersion matrix. Please see Srivistava and Giles (1987) for details.

10.1.3 A James Stein Type Estimator

A JS type estimator will now be derived as an empirical Bayes estimator when orthogonality condition $X'_i X_j = 0$ holds true. Assume that the prior mean is zero and that $F = (X'\Omega X)^{-1}$ where Ω has elements ω_{ij} . The last form of the BE in (10.1.5) becomes

$$\hat{\beta} = b^a - (X'\Sigma^{-1}X)^{-1}((X'\Omega^{-1}X)^{-1} + (X'\Sigma^{-1}X)^{-1})^{-1}b^a \quad . \tag{10.1.27}$$

Denoting the elements of Ω^{-1} by ω^{ij} the coordinates of the estimator in (10.1.27) simplify to

$$\hat{\beta}_i = b_i^a - \frac{(1/\sigma^{ii})}{(1/\sigma^{ii} + 1/\omega^{ii})} b_i^a \qquad (10.1.28)$$

Using the usual procedure of replacing unknown parameters by estimates one form of the EBE would be for $1 \leq i \leq g$

$$\hat{\beta}_i = b_i^a - \frac{c(1/s^{ii})}{b_i^a (X_i'X_i) b_i^a} b_i^a \qquad (10.1.29)$$

where s^{ii} are the elements of S^{-1}, the inverse of the matrix of consistent estimators of the elements of Σ. A large sample approximation to the mean square error of (10.1.29) may be obtained using the method outlined in Chapter VII and in Srivastava and Giles (1987).

When Σ is known the MSE of (10.1.29) may be calculated exactly. The result will be given in Example 10.1.3 for the case of the two equation model .

Exercise 10.1.7. Assume Σ is known.
A. Show that $b_i^a (X_i'X_i) b_i^a / m_i$ is an unbiased estimator of $1/\sigma^{ii} + 1/\omega^{ii}$.
B. Show that $(m_i - 2)/ b_i (X_i'X) b_i$ is an unbiased estimator of $1/(1/\sigma^{ii} + 1/\omega^{ii})$.
Following the methodology of the previous chapters the JS would take the form

$$\hat{\beta}_i = b_i^a - \frac{(m_i - 2)(1/\sigma^{ii})}{b_i^a (X_i'X_i) b_i^a} b_i^a, 1 \leq i \leq g. \qquad (10.1.30)$$

Example 10.1.3. For the two equation model $1/\sigma^{ii} = \sigma^2 (1 - \rho^2)$. The JS then takes the form

$$\hat{\beta}_i = b_i^a - \frac{(m_i - 2)\sigma^2 (1 - \rho^2)}{b_i^a (X_i'X_i) b_i^a} b_i^a, i = 1, 2. \qquad (10.1.31)$$

When $\rho = 0$ (10.1.31) takes the form of the JS for a parameter of an individual model. When $\rho = 1$ it takes the form of the Aiken estimator and the MSE is zero.

The MSE averaging over the predictive loss function for the given model is for i = 1,2

$$\text{mse}(\rho,\delta)_i = E(\hat{\beta}_i - \beta_i)'X_i'X_i(\hat{\beta}_i - \beta_i) = \sigma^2(1-\rho^2)m_i$$

$$-\sigma^2(1-\rho^2)(m_i-2)^2 \sum_{k=1}^{\infty} \frac{e^{-\chi}\chi^k}{k!(m_i+2k-2)} \qquad (10.1.32)$$

where $\chi = \delta/(1-\rho^2)$. The reader may show that for any fixed δ the partial derivative of $MSE(\rho, \delta)_i$ is negative. When $\rho = 0$ the MSE in (10.1.32) is that of the JS without taking contemporaneous correlation into account. Thus, the higher the correlation the smaller the MSE. However, it can be shown that the improvement in the risk of the estimator (10.1.31) over the Aiken estimator is not as great as that of the JS over the LS.

Table 10.1 gives a numerical comparison of the LS, the JS the Aiken estimator and the JS taking the correlation into account for selected values of δ and ρ when m_i = 8.

Table 10.1a

ρ/δ	0	0.2	0.4	0.6	0.8	1
0	0.250	0.286	0.319	0.350	0.379	0.406
0.2	0.240	0.276	0.309	0.340	0.368	0.394
0.4	0.210	0.246	0.278	0.308	0.335	0.360
0.6	0.160	0.195	0.227	0.254	0.279	0.301
0.8	0.090	0.124	0.151	0.173	0.192	0.208
1.0	0	0	0	0	0	0

Table 10.1b

ρ/δ	2	3	4	5	∞
0	0.514	0.592	0.650	0.694	1
0.2	0.500	0.576	0.632	0.674	0.96
0.4	0.459	0.527	0.575	0.611	0.84
0.6	0.384	0.436	0.472	0.497	0.64
0.8	0.257	0.283	0.299	0.309	0.36
1.0	0	0	0	0	0

For $m_i = 6$ the behavior of the risk is illustrated by a contour plot and a three dimensional plot in Figure 10.1.1 and 10.1.2 respectively.

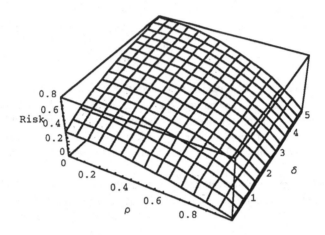

Figure 10.1.1

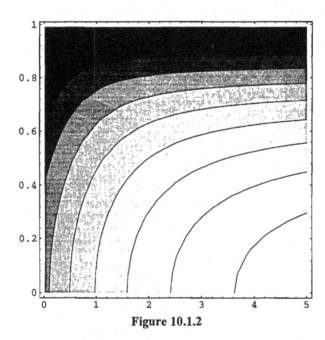

Figure 10.1.2

In Figure 10.1.2 the horizontal axis represents the values of δ and the vertical axis ρ.

Exercise 10.1.8. Show that for fixed δ the risk in (10.1.32) is a decreasing function of ρ for $0 \le \rho \le 1$.

Exercise 10.1.9. Explain why the risk of (10.1.31) given by (10.1.32) is not as great an improvement over the risk of the Aiken estimator as the JS is over the LS.

Exercise 10.1.10. Generate Tables 10.1.1 for p = 6 and p =12. Make contour and three dimensional plots for p = 4 ,8,12. Some code follows.

Mathematica

```
a[d_,s_]=Exp[-d]/(2 d^(s/2-1));
b[d_,s_]=Integrate[Exp[x] x^(s/2-2),{x,0,d}];
c[d_,s_]=a[d,s] b[d,s];
mjs[d_,s_]=1- (s-2)^2/s c[d,s];
m[h_,d_,s_]= mjs[d/(1-h^2),s];
r[h_]=1-h^2;
msur[h_,d_,s_]=r[h]*m[h,d,s];
mle[d_,h_]=1;
```

Examples of Plot Commands

```
Plot3D[msur[h,d,8],{h,0,.99},{d,.01,5},AxesLabel->
{"rho","delta","Risk"}]
ContourPlot[msur[h,d,8],{d,.01,5},{h,0,.99},
AxesLabel->{"rho","delta"}]
```

Maple

```
> a:=(d,s)->exp(-d)/(2*d^(s/2-1));
> b:=(d,s)->int((exp(x))*x^(s/2-2),x=0..d);
c:=(d,s)->a(d,s)*b(d,s);
> mjs:=(d,s)->1-((s-2)^2/s)*c(d,s);
m:=(h,d,s)->mjs(d/(1-h^2),s);
> r:=h->1-h^2;
> msur:=(h,d,s)->r(h)*m(h,d,s);
```

Examples of Plot Commands

```
> plot3d(msur(h,d,8),d=.01..5,h=0..0.99);
> with(plots):
> contourplot(msur(h,d,8),d=.01..5,h=0..0.99);
```

TI 92

```
C:Define a(ᴋ,s)=nInt(e^(y)*y^(s/2-2),y,0,ᴋ)
C:Define b(ᴋ,s)=(s-2)^2/s
C:Define c(ᴋ,s)=e^(-ᴋ)/(2*ᴋ^(s/2-1))
C:Define d(ᴋ,s)=a(ᴋ,s)*b(ᴋ,s)*c(ᴋ,s)
C:Define mj(ᴋ,s)=1-d(ᴋ,s)
C:Define r(h)=1-h^2
C:Define m(h,ᴋ,s)=mj(ᴋ/r(h),s)
C:Define ms(h,ᴋ,s)=r(h)*m(h,ᴋ,s)
```

Graphs in three dimensions may be plotted by putting the calculators into the 3d graphing mode and graphing.

TI82

```
SURE          • Program
 :Disp "S?"
 :Disp "D?"
 :Disp "H?"
 :Input S
 :Input D
 :Input H
 :1-H^2→G
 :D/G→L
 :fnInt((ᴋ^(S/2-2))*e^ᴋ,ᴋ,0,L)→A
 :1/((2*L^(S/2-1))*e^L)→B
 :(S-2)^2/S→F
 :G*(1-(A*B*F))→R
 :Disp R
```
This program will only calculate values. It will not plot graphs.

10.2 Some Aspects of the Simultaneous Estimation Problem

C.R. Rao (1975) and Chapter VII of this volume deal with the problem of simultaneously estimating the parameters in r linear models when the assumptions about each of the individual linear models are the same. For the setup with r linear models C.R. Rao also considers three other problems. They are:

1. the simultaneous estimation of the parameters in r linear models where the assumptions about the individual models are different;
2. the estimation of the parameters in a (r + 1)st linear model using the (r + 1)st response variable and the past observations of the response variable in the r linear models;
3. the prediction of the response variable in the (r+1) model assuming the observations for the first r models are available.

The objective of this section is to formulate these three problems and to consider how the methods already developed in this book may help in understanding them.

10.2.1 The Simultaneous Estimation Problem Under Different Assumptions on Linear Models

In Chapter VII the problem of simultaneously estimating the parameters for r uncorrelated linear models where the assumptions about the models were all the same was considered. The MANOVA model could be viewed in that context assuming that the X matrices were the same but the errors form the individual linear models were correlated. The SURE model assumed distinct X matrices for the g linear correlated models.

C.R. Rao (1975) considers the simultaneous estimation of parameters for r uncorrelated linear models when the X matrices are different. This formulation will now be presented. To avoid certain kinds of difficulties the full rank case will be considered. Consider r linear models

$$Y_i = X_i \beta_i + \epsilon_i, \ 1 \le i \le r \tag{10.2.1}$$

where the X_i are $n_i$xm matrices. Again it is assumed that the parameters are random variables with a given prior mean and variance. For the error terms it is assumed that

$$E(\epsilon_i \mid \beta_i) = 0 \text{ and } D(\epsilon_i \mid \beta_i) = \sigma_i^2 I. \tag{10.2.2}$$

While the assumptions about the individual models may be different the prior knowledge is still assumed to be the same. It is assumed that for the prior mean and variance

$$E(\beta_i) = \theta \text{ and } D(\beta_i) = F. \tag{10.2.3}$$

By analogy with what has already been done if σ_t^2, θ and F are known the optimum estimator (linear BE) is

$$\hat{\beta}_i = b_i - \sigma_i^2(X_i'X_i)^{-1}(F + \sigma_i^2(X_i'X_i)^{-1})^{-1}(b_i - \theta). \qquad (10.2.4)$$

The b_i are the LS estimators for the ith model. When σ_t^2, θ and F are unknown suitable estimates may be substituted into (10.2.4) to obtain empirical Bayes estimates

$$\hat{\beta}_i = b_i - \sigma_{i*}^2(X_i'X_i)^{-1}(F + \sigma_{i*}^2(X_i'X_i)^{-1})^{-1}(b_i - b_*) \qquad (10.2.5)$$

The following estimates are suggested,

$$(n_i - m)\sigma_{i*}^2 = Y_i'Y_i - Y_i'X_ib_i, \quad 1 \le i \le r, \qquad (10.2.6)$$

$$b_* = \frac{1}{r}\sum_{i=1}^{r} b_i, \qquad (10.2.7)$$

$$(r-1)(F_* + \frac{1}{k}\sum_{i=1}^{r}(X_i'X_i)^{-1}\sigma_{i*}^2) = B = \sum_{i=1}^{r}(b_i - b_*)(b_i - b_*)' \qquad (10.2.8)$$

and

$$(r-1)F_* = B - \frac{(r-1)}{r}\sum_{i=1}^{r}(X_i'X_i)^{-1}\sigma_{i*}^2. \qquad (10.2.9)$$

When the X matrices are different the estimators of Wind and Dempster cannot be formulated because the form of the prior would require that the prior dispersion matrices be different for the r linear models.

Exercise 10.2.1. What difficulties are encountered if the estimation of parameters for the non-full rank case is considered?

10.2.2 The Estimation of Parameters for the Current Individual

Consider a set of r linear models as given in Chapter VII and an (r+1)'th linear model
of the same type

$$Y_{r+1} = X\beta_{r+1} + \varepsilon_{r+1} \tag{10.2.10}$$

The objective is the estimation of the current parameter $p'\beta_{p+1}$ only using Y_{r+1} and
the past observations. The estimation of $p'\beta_i$, $1 \le i \le r$ associated with the past
observations $Y_1, Y_2, ..., Y_r$ are no longer of interest. Suppose β_{p+1} is considered to
be a random variable with

$$E(\beta_{r+1}) = \theta \text{ and } D(\beta_{r+1}) = F. \tag{10.2.11}$$

For the full rank case the BE is given by

$$p'\hat{\beta}_{r+1} = p'b_{r+1} - p'(X'X)^{-1}(F + \sigma^2(X'X)^{-1})^{-1}(b_{r+1} - \theta)\sigma^2 \tag{10.2.12}$$

and using the notations of Chapter VII the empirical Bayes estimator takes the
form

$$p'\hat{\beta}_{r+1} = p'b_{r+1} - cWp'(X'X)^{-1}B^{-1}(b_{r+1} - b_*) \tag{10.2.13}$$

with $c = (r - m - 2)/(r(n - m) + 2)$.

The MSE of the EBE can be computed easily conditional on the past observations.
If β_{r+1} is the true value than

$$\begin{aligned}
\text{MSE}(p'\hat{\beta}_{r+1}) = {} & p'(X'X)^{-1}p\sigma^2 - 2c\sigma^2Wp'(X'X)^{-1}B^{-1}(X'X)^{-1}p \\
& + c^2W^2p'(X'X)^{-1}B^{-1}(\delta\delta' + \sigma^2(X'X)^{-1})B^{-1}(X'X)^{-1}p
\end{aligned} \tag{10.2.14}$$

where $\delta = \beta_{r+1} - \beta_*$. The parametric functions of the LS ignoring the previous
observations $Y_1, Y_2,...Y_r$ has MSE $p'(X'X)^+p\sigma^2$. The following Theorem may now
be stated.

Theorem 10.2.1. Let $M = (2/cW)B - (X'X)^{-1}$.If M is positive definite then

$$\text{MSE}(p'\hat{\beta}_{r+1}) \le \text{MSE}(p'b_{r+1}) \tag{10.2.15}$$

iff

$$\delta' M^{-1} \delta \le 1. \tag{10.2.16}$$

Exercise 10.2.2. Verify (10.2.14).

Exercise 10.2.3. Give the form of the estimator (10.2.13), the MSE (10.2.14) and the generalization of Theorem 10.2.1 to the non full rank case.

Example 10.2.1 . Consider the data of Example 7.2.1. Using the first seven of the eight linear models as past data and the Y matrix from the eighth data set as the current observation the empirical Bayes estimator for β_8 will be obtained. This can then be compared with the estimate of the parameter obtained using all eight linear models.

For the seven linear models $c = 2/107$ and $W = 39.192$. Also

$$b_8 = \begin{bmatrix} 5.023 \\ 6.110 \\ 5.517 \end{bmatrix},$$

$$b_. = \begin{bmatrix} 4.922 \\ 6.082 \\ 5.590 \end{bmatrix},$$

$$B = \begin{bmatrix} 0.09483 & 0.01332 & -0.02983 \\ 0.01332 & 0.10481 & 0.00215 \\ -0.02983 & 0.00215 & 0.4818 \end{bmatrix}$$

and

$$B^{-1} = \begin{bmatrix} 10.9569 & -1.40607 & 0.68460 \\ -1.40607 & 9.72267 & -0.13045 \\ 0.68460 & -0.13045 & 2.11841 \end{bmatrix}.$$

Since $X'X = 6I$, it follows that

$$\hat{\beta}_8 = b_8 - (2/107) \cdot (39.192) \cdot (1/6) \cdot B^{-1}(b_8 - b_\bullet)$$

$$= \begin{bmatrix} 5.021 \\ 6.110 \\ 5.464 \end{bmatrix}.$$

Recall that in Example 7.1.1

$$\hat{\beta}_8 = \begin{bmatrix} 4.861 \\ 6.143 \\ 5.524 \end{bmatrix}.$$

The squared distance between these estimators as defined in the examples of Chapter VII is 0.1817.

Exercise 10.2.4. For the values given in Example 7.2.1.
A. Obtain the matrix M.
B. Obtain the singular value decomposition of M and observe that it is a positive definite matrix.
C. Reduce the coordinates of δ to orthogonal coordinates by making an orthogonal transformation using the orthogonal matrix in the SVD of M.
D. What is the form of the ellipsoid that insures that the EBE has a smaller MSE than the LS? In particular what are its axis?

A similar presentation can be given for the estimators of Dempster and Wind. For the Dempster estimator for the non-full rank case the BE is

$$p'\hat{\beta}_{r+1} = p'b_{r+1} - \frac{\sigma^2}{\sigma^2 + \omega^2}(b_{r+1} - \theta) \tag{10.2.17}$$

and the EBE is

$$p'\hat{\beta}_{r+1} = p'b_{r+1} - \frac{c((r-1)s-2)\sigma_\bullet^2}{\text{tr}BX'X} p'(b_{r+1} - b_\bullet). \tag{10.2.18}$$

The constant $c = r(n-s)/(r(n-s)+2)$. The MSE averaging over the predictive loss function takes the form

$$m = \sigma^2 s - \frac{2c((r-1)s-2)\sigma_*^2\sigma^2 s}{trBX'X} +$$
$$\frac{c^2((r-1)s-2)^2\sigma_*^4}{[trBX'X]^2}(\delta'(X'X)\delta + \sigma^2 s)$$

$$(10.2.19)$$

This MSE is smaller than that of the LS iff

$$\delta'(X'X)\delta \leq \left(\frac{2((r(n-s)+2)trBX'X}{r(n-s)((r-1)s-2)\sigma_*^2} - 1\right)\sigma^2 s. \qquad (10.2.20)$$

provided that the quantity on the right hand side of (10.2.30) is positive.

Exercise 10.2.5. Verify (10.2.19) and (10.2.20).

Example 10.2.2. The JS of Dempster for the data of Example 7.2.1 is

$$\hat{\beta}_8 = b_8 - ((16/107)(39.192)/(6(0.6814)))(b_8 - b_*)$$

$$= \begin{bmatrix} 4.878 \\ 6.070 \\ 5.649 \end{bmatrix}.$$

In Example 7.3.20

$$\hat{\beta}_8 = \begin{bmatrix} 4.889 \\ 6.107 \\ 5.615 \end{bmatrix}$$

The squared distance between these two estimators is small, 0.015876 .

Exercise 10.2.5. Obtain the numerical form of the ellipsoid in (10.2.20). if possible. The estimator of Wind takes the form

$$p'\hat{\beta}_{r+1} = p'b_\bullet + p'U\hat{A}U'(b_{r+1} - b_\bullet)$$
$$= p'b_{i+1} - p'(I - U\hat{A}U')(b_{r+1} - b_\bullet). \tag{10.2.21}$$

The matrix $\hat{A}$ depends solely on the past observations. In orthogonal coordinates the MSE of the individual components conditional on the past observations is given by

$$m_j = \frac{\sigma^2}{\lambda_j} - \frac{2c(r-3)\sigma_\bullet^2}{\lambda_j m_{jj}} \cdot \frac{\sigma^2}{\lambda_j} + \frac{c^2(r-3)^2\sigma_\bullet^4}{[\lambda_j m_{jj}]^2}\left[\frac{\sigma^2}{\lambda_j} + (\gamma_{r+1,j} - g_{\bullet j})^2\right]. \tag{10.2.22}$$

This IMSE is smaller than the LS iff

$$(\gamma_{r+1,j} - g_{\bullet j})^2 \le \frac{\sigma^2}{\lambda_j}\left[\frac{2(r(n-s)+2)\lambda_j m_{jj}}{r(n-s)(r-3)\sigma_\bullet^2} - 1\right] \tag{10.2.23}$$

provided that the quantity on the right hand side of the inequality is positive.

Example 10.2.3. For the Wind estimator recall that the diagonal elements of the matrix $\hat{A}$ take the form

$$a_j = 1 - \frac{(r-3)W}{(r(n-s)+2)\lambda_j\sum_{i=1}^{r}(g_{ij} - g_{\bullet j})^2}.$$

For the data of Example 7.2.1

$$a_1 = 1 - \frac{4(39.192)}{(107)(6)(0.09483)} = -1.5750,$$

$$a_2 = 1 - \frac{4(39.192)}{(107)(6)(0.1048)} = -1.3300$$

$$a_3 = 1 - \frac{4(39.192)}{(107)(6)(0.4818)} = 0.4931$$

and

$$\hat{\beta}_8 = \begin{bmatrix} 4.763 \\ 6.045 \\ 5.523 \end{bmatrix}.$$

In Example 7.2.1

$$\hat{\beta}_8 = \begin{bmatrix} 4.767 \\ 6.106 \\ 5.556 \end{bmatrix}.$$

The distance between these two estimators is 0.02985. In this instance the distance between the estimator of β_8 as the current individual and as the EBE where all eight models are used to estimate it is least for the estimator of Dempster, a bit larger for the estimator of Wind and largest for the estimator of C.R. Rao.

Exercise 10.2.6. Obtain the numerical form of the ellipsoid in 10.2.33. if possible.

Exercise 10.2.7. A. When $\theta = 0$ what is the form of the estimators of the current individual for the estimators of C.R. Rao, Wind and Dempster?
B. Derive conditions similar to those obtained in Theorem 10.2.1, (10.2.20) and (10.2.32).

Exercise 10.2.8. What is the form of the estimators for the current individual for priors that supply information about the dispersion intermediate between that of Dempster and Wind and intermediate between that of Wind and C. R. Rao?

10.2.3. Prediction of a Future Observation

Consider $r + 1$ models where the last component of Y_{r+1} is unavailable and hence has to be predicted. The case where the prior dispersion is known will be considered first. Consider the partitioned matrices

$$Y_{r+1} = \begin{bmatrix} Y_{r+1,1} \\ Y_{r+1,2} \end{bmatrix} \text{ and } X = \begin{bmatrix} X_1 \\ X_2 \end{bmatrix} \tag{10.2.24}$$

where $Y_{r+1,1}$ consists of the first $n-1$ components of Y_{r+1} and X_1 is an $(n-1) \times m$ matrix. Define

$$b_1 = (X_1'X_1)^{-1}X_1'Y_1. \tag{10.2.25}$$

The linear predictor of the n'th observation y_n given $Y_{r+1,1}$ is given by

$$\hat{y}_n = X_2\theta + (X_1FX_2')'(X_1FX_1' + \sigma^2 I)^{-1}(Y_1 - X_1\theta). \tag{10.2.26}$$

The prediction variance is given by

$$m = \sigma^2 + X_2'FX_2 - (X_1FX_2')'(X_1FX_1' + \sigma^2 I)^{-1}(X_1FX_2'). \tag{10.2.27}$$

Exercise 10.2.9. Verify (10.2.26) and (10.2.27).

Exercise 10.2.10. Suppose the weighted least square estimator is in place of an ordinary least square estimator used with weight matrix

$$V = \begin{bmatrix} V_1 & v \\ v & w \end{bmatrix}$$

where V_1 is an $(n-1) \times (n-1)$ matrix. Obtain results analogous to those in Equations (10.2.26) and (10.2.27).

In Chapter V in Theorem 5.5.1 alternative forms of the BE were obtained for the full rank case. Analogous alternative forms may be obtained for the regression linear predictor of y_n. In particular it can be shown that

$$\hat{y}_n = X_2 b_1 - X_2(X_1'X_1)^{-1}(F + \sigma^2(X_1'X_1)^{-1})^{-1}(b_1 - \theta)\sigma^2 \tag{10.2.28}$$

where

$$b_1 = (X_1'X_1)^{-1}X_1'Y_{r+1,1}.$$

Exercise 10.2.11. A. What are the alternative forms of the predictor of y_n analogous to the alternative forms of the BE that are obtained in Theorem 5.5.1? B. Establish the equivalence of these alternative forms.

When σ^2, θ and F are unknown the estimates based on Y_1, Y_2,...Y_r as given in (7.1.5) to (7.1.7) may be substituted into (10.2.38) to obtain the empirical Bayes predictor (EBP)

$$\hat{y}_n = X_2 b_1 - c W X_2 (X_1'X_1)^{-1} B^{-1}(b_1 - b_*).$$
(10.2.29)

The above results may also be obtained for the estimators of Dempster and Wind. Substituting Dempster's prior dispersion $(F = \omega^2 (X_1'X_1)^{-1})$ into (10.2.28)

$$\hat{y}_n = X_2 b_1 - \frac{\sigma^2}{\sigma^2 + \omega^2} X_2 b_1.$$
(10.2.30)

The prediction EBE may then be written

$$\hat{y}_n = X_2 b_1 - \frac{((r-1)m - 2)r(n-m)\hat{\sigma}^2}{(r(n-m)+2)\mathrm{tr}BX_1'X_1} X_2 (b_1 - b_*).$$
(10.2.31)

Assume that the singular value decomposition of

$$X_1'X_1 = U_1 \Lambda_1 U_1'.$$

The EBP that corresponds to the estimator of Wind takes the form

$$\hat{y}_n = X_2 b_1 - X_2 (I - U\hat{\Lambda}U')(b_1 - b_*).$$
(10.2.32)

Example 10.2.4. For the data of Example 7.1.1.

$$X_1 = \begin{bmatrix} 1_6 & 0 & 0 \\ 0 & 1_6 & 0 \\ 0 & 0 & 1_5 \end{bmatrix} \quad \text{and} \quad X_2 = [0 \ 0 \ 1].$$

Then

$$b_1 = \begin{bmatrix} 5.023 \\ 6.110 \\ 5.454 \end{bmatrix}$$

and

$$\hat{y}_n = 5.454 - (2/107)(39.192)\begin{bmatrix} 0 & 0 & 1 \end{bmatrix}\begin{bmatrix} \dfrac{1}{6} & 0 & 0 \\ 0 & \dfrac{1}{6} & 0 \\ 0 & 0 & \dfrac{1}{5} \end{bmatrix}.$$

$$\begin{bmatrix} 10.9569 & -1.40607 & 0.68460 \\ -1.40697 & 9.72267 & -0.13045 \\ 0.68460 & -0.13045 & 2.11841 \end{bmatrix}\left(\begin{bmatrix} 5.023 \\ 6.110 \\ 5.454 \end{bmatrix} - \begin{bmatrix} 4.922 \\ 6.082 \\ 5.590 \end{bmatrix}\right)$$

$$= 5.487.$$

The actual observation in the original data set was 5.83.
    To obtain the Dempster estimate observe that

$$BX_1'X_1 = \begin{bmatrix} 1.8262 & -0.2343 & 0.1369 \\ -0.2345 & 1.6205 & -0.0261 \\ 0.1141 & -0.0217 & 0.4237 \end{bmatrix}$$

and that Tr $BX_1'X_1 = 3.8703$. Then

$$y_n = 5.454 - \left(\frac{15 \cdot 1}{17 \cdot 3.87028}\right)\begin{bmatrix} 0 & 0 & 1 \end{bmatrix}\left(\begin{bmatrix} 5.023 \\ 6.110 \\ 5.454 \end{bmatrix} - \begin{bmatrix} 4.922 \\ 6.082 \\ 5.590 \end{bmatrix}\right)$$

$$= 5.485.$$

The future observation for the Wind estimator will be obtained for the data of 7.1.1. Using the expressions for the a_j in Example 10.2.3 a_1 and a_2 are the same. However

$$a_3 = 1 - \frac{4(39.192)}{(107)(6)(0.3291)} = 0.2580$$

Then

$$y_n = 5.454 - \begin{bmatrix} 0 & 0 & 1 \end{bmatrix} \begin{bmatrix} 2.8405 & 0 & 0 \\ 0 & 2.5700 & 0 \\ 0 & 0 & 0.7420 \end{bmatrix} \left(\begin{bmatrix} 5.023 \\ 6.110 \\ 5.454 \end{bmatrix} - \begin{bmatrix} 4.922 \\ 6.082 \\ 5.590 \end{bmatrix} \right)$$

$$= 5.555.$$

10.3 The Kalman Filter

The Kalman Filter (KF) has found successful application in many diverse areas. These include signal processing in aerospace tracking , underwater sonar, quality control, short term forecasting and many others. Originally the KF was formulated by Kalman (1960) and Kalman and Bucy (1961) to solve problems in optimal control theory. Meinhold and Singpurwalla (1983) explain the KF from a statisticians viewpoint as a recursive Bayes estimator. A modified form of their formulation will be presented in this section.

Section 10.3.1 will present and give examples of the KF for one linear model. The discussion will be similar to that of Gruber (1990).

Perhaps an applied situation calls for several KFs. For example there could be several stations on earth tracking the motion of a satellite. Maybe a medical researcher wants to analyze the results concerning life lengths from a replicated dose response experiment. For these types of problems it would be useful to consider r KFs. Setups involving r KFs will be described in Section 10.3.2.

10.3.1 The KF for One Linear Model

The KF will be formulated as an iterative BE where each iteration provides the prior information for the next step. The KF differs from the conventional linear model because the regression coefficients change with time. In the conventional linear model the parameters are time independent. The KF consists of a linear model together with a stochastic linear relationship (system equation) between the parameters at time t and t + 1. The estimators at time t serve as prior estimates for the estimators at time t +1 obtained as linear Bayes estimators. For t = 0 the initial prior assumptions must be specified. The KF is then formulated by the following steps:

1. The linear BE is obtained at t = 1 from the initial prior mean and variance using the system equation.
2. The updated dispersion is obtained using the system equation and the initial prior dispersion,
3. The linear BE and the updated dispersion serve as the new prior mean and dispersion.
4. The linear BE is obtained again using the new prior mean and dispersion.
5. The process is repeated.

With this in mind the mathematical formulation goes as follows:

1. Let t = 0, 1, 2,... be discrete time points. Consider a linear model

$$Y_t = X_t\beta_t + \varepsilon_t \tag{10.3.1}$$

where X_t is a fixed nxm matrix of rank $s_t \leq m$, β_t is an m dimensional vector of parameters , ε_t is an n dimensional error vector and Y_t is an n dimensional vector of observations. It will be assumed that for the error vector ε_t

$$E(\varepsilon_t \mid \beta_t) = 0 \text{ and } D(\varepsilon_t \mid \beta_t) = \sigma_t^2 L \tag{10.3.2}$$

2. Observe that the β_t are random variables. The system equation takes the form of a stochastic linear relationship between β_t and β_{t-1} ,i.e.,

$$\beta_t = M_t\beta_{t-1} + \eta_t, \tag{10.3.3}$$

where M_t is an m x m matrix. The η_t is an m dimensional error vector that satisfies

$$E(\eta_t \mid \beta_t) = 0 \text{ and } D(\eta_t \mid \beta_t) = W_t. \tag{10.3.4}$$

3. The initial prior assumptions, that is, those assumed for t = 0

$$E(\beta_0) = \theta \text{ and } D(\beta_0) = F_0.$$ (10.3.5)

4. From (10.3.3) and (10.3.5) observe that

$$E(\beta_1 \mid Y_0) = M_1\theta$$ (10.3.6)

and

$$D(\beta_1 \mid Y_0) = M_1 F_0 M_1' + W_1.$$ (10.3.7)

5. The quantities obtained in (10.3.6) and (10.3.7) are the new prior mean and dispersion. Hence, let

$$\theta_1 = M_1\theta \quad \text{and} \quad F_1 = M_1 F_0 M_1' + W_1.$$ (10.3.8)

6. The new prior assumptions in (10.3.8) may be used to obtain the linear BE by the methods of Chapter V. The linear BE takes the form

$$p'\hat{\beta}_1 = p'M_1\theta + p'F_1 X_1'(X_1 F_1 X_1' + \sigma_1^2 I)^{-1}(Y_1 - X_1 M_1\theta).$$ (10.3.9)

7. The process may now be repeated. Thus, the BE $p'\hat{\beta}_2$ may be obtained from the prior assumptions

$$\theta_2 = M_2\hat{\beta}_1 \quad \text{and} \quad F_2 = M_2 \Sigma_1 M_2' + W_2.$$ (10.3.10)

where

$$\begin{aligned}
\Sigma_1 &= E(\hat{\beta}_1 - \beta)(\hat{\beta}_1 - \beta)' \\
&= F_1 - F_1 X_1'(\sigma_1^2 I + X_1 F_1 X_1')^{-1} X_1 F_1.
\end{aligned}$$ (10.3.11)

8. The estimator $p'\hat{\beta}_t$ is defined by induction. Once $p'\hat{\beta}_{t-1}$ is obtained $p'\hat{\beta}_t$ is the BE for the prior assumptions with

$$E(\beta_t) = M_t\hat{\beta}_{t-1} \text{ and } D(\beta_t) = M_t \Sigma_{t-1} M_t' + W_t,$$ (10.3.12)

where

$$\Sigma_{t-1} = F_{t-1} - F_{t-1}X'_{t-1}(\sigma_t^2 I + X_{t-1}F_{t-1}X'_{t-1})^{-1}X_{t-1}F_{t-1}.$$ (10.3.13)

The BE takes the form

$$p'\hat{\beta}_t = p'\hat{\beta}_{t-1} + p'F_tX'_t(X_tF_tX'_t + \sigma_t^2 I)^{-1}(Y_t - X_tM_t\hat{\beta}_{t-1}).$$ (10.3.14)

This explains how the KF is an iterative BE where each iteration provides the prior information for the next step. The matrix

$$P_t = F_tX'_t(X_tF_tX'_t + \sigma_t^2 I)^{-1}$$ (10.3.15)

is called the Kalman gain.

Example 10.3.1. The Iterative Procedure for Obtaining Estimates in the KF
Assume that the linear model is

$$Y_t = \mu_t + \varepsilon_t.$$

The system equation is given by

$$\mu_t = \mu_{t-1} + \eta_t.$$

Assume that the initial prior distribution has mean 1 and variance 9. Also assume that

$$E(\varepsilon_t \mid \mu_t) = 0, \ D(\varepsilon_t \mid \mu_t) = 2$$
$$E(\eta_{t} \mid \mu_t) = 0, \ D(\eta_t \mid \mu_t) = 1.$$

It is observed that

$$Y'_1 = [1.2 \quad 1.8 \quad 1.7 \quad 2.6 \quad 2.4 \quad 2.0]$$

$$Y'_2 = \begin{bmatrix} 2.4 & 2.8 & 3.0 & 3.1 & 3.5 & 3.6 \end{bmatrix}.$$

The estimates of the parameters μ_1 and μ_2 are obtained by following the steps on (10.3.1) to (10.3.14). Since $E(\mu_0)=1$ and $D(\mu_0)=9$ it follows that

$$E(\beta_1|Y_0) = 1 \text{ and } D(\beta_1|Y_0) = 9 + 1 = 10.$$

Thus, the new prior mean and dispersion is $\theta_1 = 1$ and $F_1 = 10$. The estimate of μ_1 is

$$\hat{\mu}_1 = 1 + 10 \cdot 1_6'(2I + 1_6 \cdot 10 \cdot 1_6')^{-1} \left(\begin{bmatrix} 1.2 \\ 1.8 \\ 1.7 \\ 2.6 \\ 2.4 \\ 2.0 \end{bmatrix} - 1_6 \cdot 1 \right) = 1.91935.$$

Now

$$\Sigma_1 = 10 - 10 \cdot 1_6'(2I + 1_6 \cdot 10 \cdot 1_6')^{-1} 1_6 \cdot 10 = 0.332581$$

and

$$F_2 = .332581 + 1 = 1.332851.$$

Thus,

$$\hat{\mu}_2 = 1.91935 + 1.332851 \cdot 1_6' \cdot$$

$$(2I + 1_6 \cdot 1.332851 \cdot 1_6')^{-1} \left(\begin{bmatrix} 2.4 \\ 2.8 \\ 3.0 \\ 3.1 \\ 3.5 \\ 3.6 \end{bmatrix} - 1_6 \cdot 1.332851 \right) = 2.83714$$

Example 10.3.2. A Tracking Filter for Linear Motion. Consider a system that tracks the position of a particle for linear motion. The state of the system is

$$\begin{bmatrix} s \\ v \end{bmatrix} = \begin{bmatrix} \text{position} \\ \text{velocity} \end{bmatrix}$$

The linear model for measurements at discrete points in time is

$$Y_t = \begin{bmatrix} 1 & 0 \end{bmatrix} \begin{bmatrix} s_t \\ v_t \end{bmatrix} + \varepsilon_t$$

where ε_t is normally distributed with variance 1.[1] The system equation is

$$\begin{bmatrix} s_t \\ v_t \end{bmatrix} = \begin{bmatrix} 1 & 1 \\ 0 & 1 \end{bmatrix} \begin{bmatrix} s_{t-1} \\ v_{t-1} \end{bmatrix} + \begin{bmatrix} (1/3)\eta_t \\ 1\eta_t \end{bmatrix}$$

where η_t is normally distributed with variance 2. Assume that the error terms in the linear model and the system equation are independent. The initial value is

$$\begin{bmatrix} s_0 \\ v_0 \end{bmatrix} = \begin{bmatrix} 0 \\ 5 \end{bmatrix}.$$

The parameters of the KF will be obtained using simulated data. By using Minitab a simulation of the first ten observations of the error terms for the system equation and the linear model were obtained. This information together with the initial value of the position and velocity was used to obtain simulated measurements of the position at time t Y_t for $t = 1...10$. The values obtained for the Y's were

Table 10.3.1

Y_1	Y_2	Y_3	Y_4	Y_5
4.593	9.265	15.704	20.888	29.191

Y_6	Y_7	Y_8	Y_9	Y_{10}
31.080	35.853	41.376	46.628	53.614

The initial prior assumptions are

$$E(\beta_0) = \begin{bmatrix} 0 \\ 5 \end{bmatrix} \text{ and } D(\beta_0) = \begin{bmatrix} 1 & 0 \\ 0 & 1 \end{bmatrix}.$$

Also

---

[1] When the error term of the linear model is normally distributed the disturbance is called white noise. Otherwise the term colored noise is used.

$$W_t = \begin{bmatrix} \frac{2}{9} & \frac{2}{3} \\ \frac{2}{3} & 2 \end{bmatrix}.$$

The first five KF estimates are given as

Table 10.3.2

$\hat{\beta}_1$	$\hat{\beta}_2$	$\hat{\beta}_3$	$\hat{\beta}_4$	$\hat{\beta}_5$
$\begin{bmatrix} 4.6783 \\ 4.7161 \end{bmatrix}$	$\begin{bmatrix} 9.2901 \\ 4.6385 \end{bmatrix}$	$\begin{bmatrix} 15.3543 \\ 5.7298 \end{bmatrix}$	$\begin{bmatrix} 20.9256 \\ 5.6081 \end{bmatrix}$	$\begin{bmatrix} 28.6812 \\ 7.2544 \end{bmatrix}$

Exercise 10.3.1 Suppose that $Y_3' = \begin{bmatrix} 2.8 & 3.5 & 3.6 & 3.7 & 3.8 & 4.0 \end{bmatrix}$. Find $\hat{\mu}_3$.

Exercise 10.3.2. For the data in Example 10.3.2
A. Find the next five estimates of $\hat{\beta}_t$.
B. What are the updated prior means and dispersions.
C. Suppose that the initial prior dispersion is the same as the matrix W_t. Find the estimates $\hat{\beta}_t$. Compare your results to those in part A and Example 10.3.2.

Exercise 10.3.3. A. Obtain alternative forms of the BE using the methods of Theorems 5.5.1 and 5.5.2.
B. Obtain alternative forms of the average MSE analogous to those obtained in Equations (5.7.5),(5.7.6) and (5.7.8).

Exercise 10.3.4. Consider two KFs where the initial prior information is of the form

$$E(\beta_0) = \theta_{01,} \ D(\beta_0) = F_{01}$$

$$E(\beta_0) = \theta_{02,} \ D(\beta_0) = F_{02} \tag{10.3.16}$$

Let $p'\hat{\beta}_{t1}$ and $p'\hat{\beta}_{t2}$ be the estimators derived at the t'th stage. If $F_{01} \le F_{02}$ prove that averaging over the prior distribution

$$MSE(p'\hat{\beta}_{t1}) \le MSE(p'\hat{\beta}_{t2}) . \tag{10.3.17}$$

Exercise 10.3.5. Let Y_t be the simple transform for the number of defectives observed in a sample at time t. The parameter θ_{1t} represents the true defective index of the process. The observation equation is

$$Y_t = \theta_{1t} + \varepsilon_t.$$

The system equation is

$$\theta_t = \theta_{t-1} + w_t.$$

Assume that $\theta_0 = 1$, $F_0 = 1$, $\sigma_\varepsilon^2 = 1$, $W_t = 1$. Give the BE $\hat{\theta}_t$ for stages t = 1 to t = 4 in terms of θ_{t-1}.

There are a number of other ways to obtain the KF. In addition to being obtained as a linear BE it can also be obtained:

1. as a mixed estimator using weighted LS;
2. as a minimax estimator;
3. as a generalized ridge estimator.

The reader is given the opportunity to show this in Exercises 10.3.5 and 10.3.6.

Assume that F_0 is positive definite. To formulate the KF as a mixed estimator for the full rank case consider the augmented linear model

$$\begin{bmatrix} Y_t \\ M_t\hat{\beta}_{t-1} \end{bmatrix} = \begin{bmatrix} X_t \\ I \end{bmatrix}\beta_t + \begin{bmatrix} \varepsilon_t \\ \eta_t \end{bmatrix} \qquad (10.3.18)$$

together with the assumptions

$$E(\varepsilon_t) = 0, \quad D(\varepsilon_t) = \sigma_t^2 I$$
$$E(\eta_t) = 0, \quad D(\eta_t) = \tau_t^2 F_t. \qquad (10.3.19)$$

The KF is then the weighted LS from (10.3.18) using the assumptions from (10.3.19). It takes the form

$$p'\hat{\beta}_t = (\tau_t^2 X_t'X_t + \sigma_t^2 F_t^{-1})^{-1}(\tau_t^2 X_t'Y_t + \sigma_t^2 F_t^{-1}M_t\hat{\beta}_{t-1}). \qquad (10.3.20)$$

A matrix R_t can be found where $F_t^{-1} = R_t'R_t$. Let $r_t = R_t M_t \hat{\beta}_t$. The matrix F_t is defined by induction using Equations (10.3.12) and (10.3.13). An example for the non-full rank case is given in Exercise 10.3.4.

Exercise 10.3.6. Assume that X_t are time independent, i.e., $X_t = X$ and that $M_t = I$. Assume that X is on non-full rank. Using the augmented linear model analogous to (10.3.18) find the estimator analogous to (10.3.20). Under what conditions is this estimator a recursive BE?

Exercise 10.3.7. The following sequence of steps will allow the reader to derive the minimax version of the KF. See Gruber (1990) for more details. Consider the linear model (10.3.1) together with the system equation (10.3.3).
A. Let θ and F_0 be the initial values of an m dimensional vector and an mxm positive definite matrix. Let

$$\Omega_1 = \left\{ \beta : (\beta - M_1\theta)(\beta - M_1\theta)' \leq F_1 \right\}. \tag{10.3.21}$$

For a linear estimator of the form

$$p'\hat{\beta}_1 = p'\theta + L_1'(Y_1 - X\theta) \tag{10.3.22}$$

find the maximum risk on Ω_1.
B. Find L_1 so that the maximum risk found in A is minimized.
C. From the system equation show that

$$\theta_1 = M_1\theta \quad \text{and} \quad F_1 = M_1 F_0 M_1' + W_1 \tag{10.3.23}$$

D. Find the optimum minimax estimator for

$$\theta_2 = M_2\hat{\beta}_1 \text{ and } F_2 = M_2\Sigma_1 M_2' + W_2$$

where on Ω_1

$$\Sigma_1 = \max E_\beta (\hat{\beta}_1 - \beta)(\hat{\beta}_1 - \beta)'. \tag{10.3.24}$$

E. At the t'th step let

$$\Omega_t = \left\{ \beta_t : (\beta_t - M_t\hat{\beta}_{t-1})(\beta_t - M_t\hat{\beta}_{t-1})' \leq F_t \right\} \tag{10.3.25}$$

where

$$F_t = M_t \Sigma_{t-1} M'_t + W_t$$

and on Ω_{t-1}

$$\Sigma_{t-1} = \max E_\beta (\hat{\beta}_{t-1} - \beta_{t-1})(\hat{\beta}_{t-1} - \beta_{t-1})'. \qquad (10.3.26)$$

Find the maximum risk of the linear estimator of the form

$$p'\hat{\beta}_t = p'\hat{\beta}_{t-1} + L'_t(Y_t - X_t\hat{\beta}_{t-1}) \qquad (10.3.27)$$

on Ω_t and then show that the optimum minimax estimator takes the form (10.3.14).

Exercise 10.3.8. This exercise will enable the reader to derive the KF as a generalized ridge estimator.
A. Using a Lagrange multiplier argument minimize

$$D_1 = (B_1 - M_1\theta)'H_1(B_1 - M_1\theta) \qquad (10.3.28)$$

subject to

$$(B_1 - b_1)X'_1X_1(B_1 - b_1) = \Phi_0, \qquad (10.3.29)$$

where b_1 is the LS for the first stage. Obtain

$$\hat{\beta}_1 = (H_1 + \lambda_t X'_1 X_1)^+ (H_1\theta + \lambda_t X'_1 X_1 b_1) \qquad (10.3.30)$$

B. On

$$S_1 = \left\{ \beta_1 : (\beta_1 - M_1\theta_0)(\beta_1 - M_1\theta)' \leq H_1 \right\} \qquad (10.3.31)$$

let

$$p'\Sigma_1 p = Var(p'\hat{\beta}_1) + MaxBias(p'\hat{\beta}_1). \qquad (10.3.32)$$

Now

$$H_2 = (M_2\Sigma_1^+ M'_2 + W_1)^+. \qquad (10.3.33)$$

Using $M_1\hat{\beta}_1$ in place of θ and the rest of the items with subscript 2 in place of 1 derive the analogue of (10.3.30) for the second stage.

C. For

$$H_t = (M_t \Sigma_{t-1}^+ M_t' + W_t)^+. \tag{10.3.34}$$

where on

$$S_{t-1} = \left\{ \beta_{t-1} : (\beta_{t-1} - M_{t-1}\hat{\beta}_{t-2})(\beta_{t-1} - M_{t-1}\hat{\beta}_{t-2})' \le H_{t-1} \right\}, \tag{10.3.35}$$

$$p'\Sigma_{t-1}p = \text{Var}(p'\hat{\beta}_{t-1}) + \text{MaxBias}(p'\hat{\beta}_{t-1}) \tag{10.3.36}$$

show that the minimization of

$$D_t = (B_t - M_t\hat{\beta}_{t-1})'H_1(B_t - M_t\hat{\beta}_{t-1}) \tag{10.3.37}$$

subject to

$$(B_t - b_t)X_t'X_t(B_t - b_t) = \Phi_t, \tag{10.3.38}$$

where b_t is the LS for the t'th stage, yields

$$\hat{\beta}_t = (H_t + \lambda_1 X_1'X_1)^+ (H_t\theta + \lambda_1 X_1'X_1 b_t). \tag{10.3.39}$$

The ridge estimators of Hoerl and Kennard and the contraction estimator of Mayer and Willke may be formulated for a half KF for the case where the X matrices are constant and $M_t = I$. In a half KF the prior mean is not updated but the prior dispersion is. In the case of the ridge and contraction estimator for one linear model the prior mean is assumed to be zero at each stage. Thus, consider the linear model

$$Y_t = X\beta_t + \varepsilon_t \tag{10.3.40}$$

with the system equation

$$\beta_t = \beta_{t-1} + \eta_t. \tag{10.3.41}$$

Assume that

$$E(\varepsilon_t \mid \beta_t) = 0 \text{ and } D(\varepsilon_t \mid \beta_t) = \sigma_t^2 L \qquad (10.3.42)$$

The ridge estimators of Hoerl and Kennard were derived as Bayes estimators with prior dispersion UDU' where D is a diagonal matrix. The updated priors will have dispersion matrices in this form if

$$E(\eta_t \mid \beta_t) = 0 \text{ and } D(\eta_t \mid \beta_t) = U\Delta_t^{-1}U'. \qquad (10.3.43)$$

The initial prior assumptions are

$$E(\beta_0) = 0 \text{ and } D(\beta_0) = UD_0^{-1}U'\sigma_0^2 = F_0 \qquad (10.3.44)$$

where D_0 is a positive definite diagonal matrix. Employing the system equation

$$E(\beta_1 \mid Y_0) = 0 \text{ and } D(\beta_1 \mid Y_0) = U(D_0^{-1} + \Delta_1^{-1})U' \qquad (10.3.45)$$

implying that

$$D_1^{-1} = D_0^{-1} + \Delta_1^{-1}. \qquad (10.3.46)$$

Likewise it can be shown that

$$D_t^{-1} = D_{t-1}^{-1} + U'\Sigma_{t-1}U. \qquad (10.3.47)$$

At stage t the ridge estimator takes the form

$$p'\hat{\beta}_t = p'(X'X + \sigma_t^2 UD_t U')^+ X'Y_t. \qquad (10.3.48)$$

The diagonal elements may be estimated iteratively by

$$\frac{1}{d_{ti}} = \frac{\sigma_{t-1}^2}{\lambda_i + d_{t-1i}} + \frac{1}{\delta_{ti}} = \frac{\sigma_{t-1}^2 \delta_{ti} + \lambda_i + d_{t-1i}}{\delta_{ti}(\lambda_i + d_{t-1i})}. \qquad (10.3.49)$$

Solving for d_{ti} yields

$$d_{ti} = \frac{\delta_{ti}(\lambda_i + d_{t-1i})}{\sigma_{t-1}^2 \delta_{ti} + \lambda_i + d_{t-1i}}. \qquad (10.3.50)$$

For constant $\delta_{ti} = \delta$ and σ^2 the steady state solution consists of the positive root of the quadratic equation

$$x^2 + (\sigma^2\delta + \lambda_i - \delta)x + \delta\lambda_i = 0, \tag{10.3.51}$$

$$x = \frac{-(\lambda_i + \delta(\sigma^2 - \lambda_i)) + \sqrt{4\delta\lambda_i + (\lambda_i + \delta(\sigma^2 - \lambda_i))^2}}{2}. \tag{10.3.52}$$

Exercise 10.3.9. A. Show that the root in (10.3.52) is positive.
B. Find the other root of (10.3.51) and show that it is negative.

If the initial estimator is an ordinary ridge estimator, i.e., $D_0 = kI$ the estimator at the t'th stage need not be an ordinary ridge estimator. One way to obtain an ordinary ridge estimator at each stage is to find the harmonic mean of the d_{ti} from the equation

$$\frac{1}{d_t} = \sum_{i=1}^{s} \frac{1}{d_{ti}} \tag{10.3.53}$$

where s is the rank of the X matrix and use d_t for the next iteration.

Exercise 10.3.10. Suggest some other ways to have an ordinary ridge regression estimator at each stage for the KF based on Equations 3.4.7-3.4.9.

The analogues of the contraction estimators of Mayer and Willke may be obtained by letting

$$W_t = \frac{\sigma_t^2}{\delta_t}(X'X)^+ \tag{10.3.54}$$

and

$$F_0 = \frac{\sigma^2}{k_0}(X'X)^+. \tag{10.3.55}$$

Then

$$\Sigma_t = \sigma_t^2(k_t(X'X) + (X'X))^+ = \frac{\sigma_t^2}{1+k_t}(X'X)^+ \ . \tag{10.3.56}$$

Consequently,

$$F_t = \left[\frac{\sigma_t^2}{\delta_t} + \frac{\sigma_{t-1}^2}{k_{t-1}+1}\right](X'X)^+. \tag{10.3.57}$$

Now

$$\frac{\sigma_t^2}{k_t} = \left[\frac{\sigma_t^2}{\delta_t} + \frac{\sigma_{t-1}^2}{k_{t-1}+1}\right] \ . \tag{10.3.58}$$

Solving Equation (10.3.58) for k_t yields

$$k_t = \frac{\sigma_t^2\delta_t(1+k_{t-1})}{\sigma_t^2 k_{t-1} + \sigma_{t-1}^2\delta_t + \sigma_t^2}. \tag{10.3.59}$$

At stage t the contraction estimator is

$$p'\hat{\beta}_t = \frac{1}{1+k}p'b_t = \left(1-\frac{1}{1+k}\right)p'b_t. \tag{10.3.60}$$

The JS as an EBE is given by

$$p'\hat{\beta}_t = \left(1-\frac{(n-s)(s-2)\hat{\sigma}^2}{(n-s+2)b_t(X'X)b_t}\right)p'b_t. \tag{10.3.61}$$

Exercise 10.3.11. Give the details of the derivation of (10.3.61) from an empirical Bayes point of view.

Exercise 10.3.12. Find the form of the steady state solution to Equation (10.3.59) for constant σ and δ.

10.3.2 The KF for r Linear Models

The problem of simultaneous estimation of the parameters for r KFs will be considered. There will be some important differences in the procedure from that of Section 10.3.1. Suppose the Kalman filter is formulated in the same way as in 10.3.1? The mean of the posterior distribution depends on the observations. Hence at each stage the updated prior distribution will be different for each linear model. The prior dispersion would be the same. The optimization problem of finding the recursive linear BE should be formulated the same way at each stage. This will be accomplished by taking the new prior mean to be the average of the posterior mean for all of the r linear models. This will result in a minor modification of the computation of the dispersions that will be explained below.

With this in mind the mathematical formulation is as follows:

1. For discrete time points $t = 0, 1, 2,...$ consider r linear models

$$Y_{ti} = X_t \beta_{ti} + \varepsilon_{ti}, \ 1 \le i \le r \tag{10.3.62}$$

where again X_t is a fixed nxm matrix of rank $s_t \le m$. For each i β_{ti} is an m dimensional vector and Y_{ti} is an n dimensional vector of observations. For each of the error vectors ε_{ti}

$$E(\varepsilon_{ti} \mid \beta_{ti}) = 0 \text{ and } D(\varepsilon_{ti} \mid \beta_{ti}) = \sigma_t^2 L \tag{10.3.63}$$

2. With each linear model there is a system equation in the form of a stochastic linear relationship between β_n and $\beta_{t-1,i}$, i.e.,

$$\beta_{ti} = M_t \beta_{t-1,i} + \eta_{ti}, \tag{10.3.64}$$

with M_t an mxm matrix. For each i the η_{ti} is an m dimensional error vector where

$$E(\eta_{ti} \mid \beta_{ti}) = 0 \text{ and } D(\eta_{ti} \mid \beta_{ti}) = W_t. \tag{10.3.65}$$

3. Assume again that the initial prior assumptions are

$$E(\beta_0) = \theta \text{ and } D(\beta_0) = F_0. \tag{10.3.66}$$

4. As was the case before

$$\theta_1 = E(\beta_1 \mid Y_{0i}) = M_1\theta \text{ and } F_1 = D(\beta_1 \mid Y_{0i}) = M_1F_0M_1 + W_1. \quad (10.3.67)$$

5. Using the prior assumptions in (10.3.67) the linear BE at stage 1 takes the form

$$p'\hat{\beta}_{1i} = p'M_1\theta + p'F_1X_1'(X_1F_1X_1' + \sigma_1^2I)^{-1}(Y_{1i} - X_1M_1\theta), \ 1 \le i \le r.$$
$$(10.3.68)$$

Observe that the components of the estimator are apt to be different because the observed Y values would be for each linear model.

6. The new prior mean would be

$$\theta_2 = M_2\bar{\hat{\beta}}_1 \qquad\qquad (10.3.69)$$

where $\bar{\hat{\beta}}_1 = \sum_{i=1}^{r}\beta_{1i}/r$. The new prior dispersion would be

$$F_2 = M_2\Sigma_1M_2' + W_2 \qquad\qquad (10.3.70)$$

where

$$\begin{aligned}\Sigma_1 &= E(\hat{\beta}_{1i} - \beta_i)(\hat{\beta}_{1i} - \beta_i)' \\ &= F_1 - F_1X_1'(\sigma_1^2I + X_1F_1X_1')^{-1}X_1F_1.\end{aligned} \qquad (1.3.71)$$

The dispersion matrix Σ_1 is the same for each of the r linear models.

7. The estimator at the t'th stage is again defined by induction. Thus, once $p'\hat{\beta}_{t-1,i}$ is obtained $p'\hat{\beta}_{ti}$ is the BE for the prior assumptions with

$$E(\beta_{ti}) = M_t\bar{\hat{\beta}}_{t-1} \text{ and } D(\beta_{ti}) = M_t\Sigma_{t-1}M_t' + W_t \qquad (1.3.72)$$

where again

$$\Sigma_{t-1} = F_{t-1} - F_{t-1}X_{t-1}'(\sigma_t^2I + X_{t-1}F_{t-1}X_{t-1})^{-1}X_{t-1}F_{t-1}. \qquad (10.3.73)$$

The BE takes the form

$$p'\hat{\beta}_{ti} = p'\bar{\beta}_t + p'F_tX_t'(X_tF_tX_t' + \sigma_t^2I)^{-1}(Y_t - X_tM_t\bar{\hat{\beta}}_{t-1}). \qquad (10.3.74)$$

The principal difference between this and the one model case is the pooling of the prior assumptions at each stage. For the full rank case

$$p'\hat{\beta}_{ti} = p'b_{ti} - p'(X_t'X_t)^{-1}(F_t + \sigma_t^2(X_t'X_t)^{-1})^{-1}(b_{ti} - M_t\bar{\hat{\beta}}_t). \qquad (10.3.75)$$

Suppose the initial prior mean and dispersion cannot be determined? Formulae for the BE may be written in terms of θ and F_0. At each stage θ_t and F_t can be estimated from the data. The estimator of C.R. Rao may then be formulated at the t'th stage the same way as for the original setup with r linear models. The estimators of Wind and Dempster may be formulated for the case where the X matrices are the same at each stage and the W_t take the form in (10.3.43). The MSEs with and without averaging over the prior information are obtained the same way provided that the expectations at stage t are conditional on the parameters at stage $t-1$.

Exercise 10.3.13. A. Obtain the alternative forms of the BE (10.3.74) for the full and the non full rank case.
B. Under what conditions can ridge and contraction estimators be formulated? What is the explicit form of the estimators?

Exercise 10.3.14. A. Give the form of the estimators of Wind, Dempster and C.R. Rao for the setup with r KFs.
B. Obtain the MSEs with and without averaging over the prior at the t'th stage conditional on the parameters at stage $t-1$.

Example 10.3.3. Satellite Tracking. Suppose r weather satellites are launched. For each one of them let θ_{ti} represent the position and the speed at time t with respect to the center of the earth. The position and the speed of the satellites are unknown parameters that cannot be measured directly. Tracking stations around the earth obtain measurements of the distance to the satellite and the accompanying angles. For each satellite these are measured by the observation vector Y_{ti}. Geometric principles are used to map Y_{ti} into θ_{ti} and are incorporated into the design matrix X_t. The ε_t represents the measurement error. The resulting models are

$$Y_{ti} = X_t\theta_{ti} + \varepsilon_i, \quad 1 \le i \le r. \qquad (10.3.76)$$

The position and the speed change according to the laws of physics concerned with orbiting bodies is prescribed by the M_t in the system equation. The error terms η_{ti} allow for deviations from these laws due to various phenomena, for example the non-uniformity of the earth's magnetic field. The resulting system equations are

$$\theta_{ti} = M_t\theta_{t-1,i} + \eta_{ti}, \ 1 \le i \le r .\qquad(10.3.77)$$

Example 10.3.4. Navigation. The idea for this example was obtained from Lewis (1986) and Gruber (1990). Suppose a convoy of five ships move east at 10 knots. The ships all have identical performance characteristics, e.g., five PT boats. Each ship's velocity is assumed to be constant except for the effect of wind gusts and wave actions. An estimate of each ships position s_{ti} and velocity v_{ti} is required. Two cases will be considered:

1. The initial prior distribution of the ships is normal with mean $s_0 = 0$ and variance $\sigma_0^2 = 3$. The initial velocity also has a normal prior distribution with mean 10 ($s_0 = 10$) and its variance is 3. The ships all start from the same place the same way. The objective is to find the Bayes estimate of position s_{ti} and velocity v_{ti} for each of the ships.

2. The initial prior distributions of the ships is normal with unknown parameters. However it is assumed that the ships all start from the same place the same way. This might be the case if a ship from the US Navy is tracking a convoy of enemy ships. Now the goal is to find the empirical Bayes estimate of position s_{ti} and velocity v_{ti} for each of the ships.

The system dynamics and the measurement process will be modeled. For the simulated data the Bayes case, then the empirical Bayes case will be considered. During hour k the i'th ship moves with velocity v_{ki} knots. Thus, its position changes according to the equations

$$S_{k+1,i} = S_{k,i} + V_{k,i}.$$

As a result of the unknown effect of wind and waves

$$v_{k+1,i} = v_k + w_{ki}.$$

It will be assumed that w_k is normally distributed. Navigational fixes are taken at k = 1,2,3,.... Position is determined to within an error covariance of 3 but no direct information is available about velocity.

Let

$$\beta_{ki} = \begin{bmatrix} s_{ki} \\ d_{ki} \end{bmatrix}.$$

The output equation is

$$y_{ki} = \begin{bmatrix} 1 & 0 \end{bmatrix} \beta_{ki} + \varepsilon_{ki} \text{ with } \varepsilon_{ki} \sim N(0,3).$$

Only positions are observed and ε_{ki} is the error of measurement. The dynamic equations are

$$\beta_{k+1,i} = \begin{bmatrix} 1 & 1 \\ 0 & 1 \end{bmatrix} \beta_k + \begin{bmatrix} 0 \\ 1 \end{bmatrix} w_{ki}$$

where $w_k \sim N(0,2)$.

A simulation similar to that of Example 10.3.3 was performed for each of the five models. Table 10.3.3 has the simulated observed Y values for each model for t=1,2,3,4,5.

Table 10.3.3

			Model		
Time	1	2	3	4	5
1	9.5820	7.6205	8.2413	9.2880	10.1886
2	18.2505	20.9539	19.6727	19.9826	15.5760
3	37.8663	26.4570	29.1016	26.5720	27.5725
4	52.1449	45.9744	41.4906	37.7006	40.5091
5	65.0850	59.3837	50.7120	48.5055	54.6150

For the initial prior distribution it is assumed that

$$\theta_0 = \begin{bmatrix} 0 \\ 10 \end{bmatrix} \text{ and } F_0 = \begin{bmatrix} 3 & 0 \\ 0 & 3 \end{bmatrix}$$

and

$$W = \begin{bmatrix} 0 & 0 \\ 0 & 2 \end{bmatrix}.$$

Then

$$\theta_1 = M\theta_0 = \begin{bmatrix} 1 & 1 \\ 0 & 1 \end{bmatrix}\begin{bmatrix} 0 \\ 10 \end{bmatrix} = \begin{bmatrix} 10 \\ 10 \end{bmatrix}$$

and

$$F_1 = MF_0M' + W = \begin{bmatrix} 1 & 1 \\ 0 & 1 \end{bmatrix}\begin{bmatrix} 3 & 0 \\ 0 & 3 \end{bmatrix}\begin{bmatrix} 1 & 0 \\ 1 & 1 \end{bmatrix} + \begin{bmatrix} 0 & 0 \\ 0 & 2 \end{bmatrix} = \begin{bmatrix} 6 & 3 \\ 3 & 5 \end{bmatrix}.$$

For example the BE for at t = 1 for model 1 and model 2 is

$$\hat{\beta}_{11} = \theta_1 + F_1 X'(XF_1X' + \sigma^2 I)^{-1}(Y_{11} - X\theta_1)$$

$$= \begin{bmatrix} 10 \\ 10 \end{bmatrix} + \begin{bmatrix} 6 & 3 \\ 3 & 5 \end{bmatrix}\begin{bmatrix} 1 \\ 0 \end{bmatrix}\left(\begin{bmatrix} 1 & 0 \end{bmatrix}\begin{bmatrix} 6 & 3 \\ 3 & 5 \end{bmatrix}\begin{bmatrix} 1 \\ 0 \end{bmatrix} + 3\right)^{-1}.$$

$$\left(9.5820 - \begin{bmatrix} 1 & 0 \end{bmatrix}\begin{bmatrix} 10 \\ 10 \end{bmatrix}\right)$$

$$= \begin{bmatrix} 9.7213 \\ 9.8607 \end{bmatrix}$$

and

$$\hat{\beta}_{12} = \theta_1 + F_1 X'(XF_1X' + \sigma^2 I)^{-1}(Y_{12} - X\theta_1)$$

$$= \begin{bmatrix} 10 \\ 10 \end{bmatrix} + \begin{bmatrix} 6 & 3 \\ 3 & 5 \end{bmatrix}\begin{bmatrix} 1 \\ 0 \end{bmatrix}\left(\begin{bmatrix} 1 & 0 \end{bmatrix}\begin{bmatrix} 6 & 3 \\ 3 & 5 \end{bmatrix}\begin{bmatrix} 1 \\ 0 \end{bmatrix} + 3\right)^{-1}.$$

$$\left(7.6205 - \begin{bmatrix} 1 & 0 \end{bmatrix}\begin{bmatrix} 10 \\ 10 \end{bmatrix}\right)$$

$$= \begin{bmatrix} 8.4137 \\ 9.2068 \end{bmatrix}.$$

Similarly

$$\hat{\beta}_{13} = \begin{bmatrix} 8.7609 \\ 9.3804 \end{bmatrix}, \hat{\beta}_{14} = \begin{bmatrix} 9.5254 \\ 9.76278 \end{bmatrix} \text{ and } \hat{\beta}_{15} = \begin{bmatrix} 10.1258 \\ 10.0629 \end{bmatrix}.$$

Then the mean of the $\hat{\beta}_{1j}, j = 1,2,3,4,5$ is

$$\bar{\hat{\beta}}_1 = \begin{bmatrix} 9.3094 \\ 9.6547 \end{bmatrix}$$

and

$$\theta_2 = M\bar{\hat{\beta}}_1 = \begin{bmatrix} 18.9641 \\ 9.6547 \end{bmatrix}.$$

Now

$$\Sigma_1 = F_1 - F_1 X' (XF_1 X' + \sigma^2 I)^{-1} XF_1$$

$$= \begin{bmatrix} 6 & 3 \\ 3 & 5 \end{bmatrix} - \begin{bmatrix} 6 & 3 \\ 3 & 5 \end{bmatrix}\begin{bmatrix} 1 \\ 0 \end{bmatrix}\left([1 \quad 0]\begin{bmatrix} 6 & 3 \\ 3 & 5 \end{bmatrix}\begin{bmatrix} 1 \\ 0 \end{bmatrix} + (3)\right)^{-1}[1 \quad 0]\begin{bmatrix} 6 & 3 \\ 3 & 5 \end{bmatrix}$$

$$= \begin{bmatrix} 2 & 1 \\ 1 & 4 \end{bmatrix}$$

and

$$F_2 = M\Sigma_1 M' + W = \begin{bmatrix} 1 & 1 \\ 0 & 1 \end{bmatrix}\begin{bmatrix} 2 & 1 \\ 1 & 4 \end{bmatrix}\begin{bmatrix} 1 & 0 \\ 1 & 1 \end{bmatrix} + \begin{bmatrix} 0 & 0 \\ 0 & 2 \end{bmatrix} = \begin{bmatrix} 8 & 5 \\ 5 & 6 \end{bmatrix}.$$

By substitution

$$\hat{\beta}_{21} = \begin{bmatrix} 18.4451 \\ 9.6547 \end{bmatrix} \quad \hat{\beta}_{22} = \begin{bmatrix} 20.4112 \\ 10.5592 \end{bmatrix} \quad \hat{\beta}_{23} = \begin{bmatrix} 19.4794 \\ 9.9768 \end{bmatrix}$$

$$\hat{\beta}_{24} = \begin{bmatrix} 19.7048 \\ 10.1177 \end{bmatrix} \quad \hat{\beta}_{25} = \begin{bmatrix} 17.1161 \\ 8.7307 \end{bmatrix}.$$

To simulate the data it was assumed that $\sigma^2 = 3$. Suppose the initial prior mean and dispersion is unknown. For this particular data set it can be shown that the LS

$$b_{ij} = y_{ij}$$

The estimable parametric functions estimate the position only. The EBE of the position at each stage then takes the form

$$\hat{\beta}_{ij} = \overline{y}_{i\cdot} + \left(1 - \frac{3(2)}{\sum\limits_{j=1}^{5} (y_{ij} - \overline{y}_{i\cdot})^2} \right) (y_{ij} - \overline{y}_{i\cdot}), \ 1 \le j \le 5$$

The estimates for the first three stages are given in Table 10.3.4 below.

Table 10.3.4

Stage	Model				
	1	2	3	4	5
1	8.7523	9.4246	9.2461	8.8531	8.5444
2	19.1054	18.1788	18.6179	18.5117	20.022
3	26.6511	30.5616	29.6552	30.5222	30.2793

Exercise 10.3.15. A. Show for Example 10.3. that

$$\theta_3 = \begin{bmatrix} 28.7743 \\ 9.7429 \end{bmatrix} \text{ and } F_3 = \begin{bmatrix} 8.6364 & 5.0909 \\ 5.0909 & 5.7273 \end{bmatrix}.$$

B. Use the result in A to show that

$$\hat{\beta}_{31} = \begin{bmatrix} 35.5223 \\ 13.7207 \end{bmatrix}, \ \hat{\beta}_{32} = \begin{bmatrix} 27.0544 \\ 8.7291 \end{bmatrix}, \hat{\beta}_{33} = \begin{bmatrix} 29.0172 \\ 9.8861 \end{bmatrix}$$

$$\hat{\beta}_{34} = \begin{bmatrix} 34.4207 \\ 13.0714 \end{bmatrix} \text{ and } \hat{\beta}_{35} = \begin{bmatrix} 27.8823 \\ 9.2172 \end{bmatrix}.$$

C. Find θ_4 and F_4 and find the KF estimates for t = 4.

Exercise 10.3.16. For Example 10.3

A. Show that the LS estimators for position are the observations.

B. Show that the estimable parametric functions contain the position vector only.

C. Using the methodology of this book explain how the EBE was obtained.

D. Obtain the EBE for the fourth and fifth stage.

10.4 Summary

In this Chapter:

1. The effects of contemporaneous correlation on the least square, ridge and James-Stein estimator were studied for the seemingly unrelated regression equation model.

2. The problems of estimation of the parameters for r linear models with different design matrices and error matrices, estimating the current individual and predicting an unknown observation by Bayes and empirical Bayes methods were introduced.

3. Setups with one and with r Kalman filters were formulated and examples of recursive estimation of the parameters of these models were given.

Chapter XI

Summary and Conclusion

11.0 Introduction

The amount of research that has been done on shrinkage estimators in general and the James-Stein and ridge regression estimator in particular during the last thirty years cannot be summarized in a single volume. A great deal of work remains to be done. The topics that were presented in this volume were chosen because of their importance to the specialized area of this book and to statistics in general. Of course the contents of any book are dictated by the interests and prejudices of the author and what he/she feels comfortable writing about. In this regard this book is no exception.

This chapter will:

1. summarize what was covered in the book;
2. explain why different topics were chosen and taken up where they were;
3. account for any omissions that the author is aware of;
4. come to some general conclusions relative to the results of the research that was presented;
5. suggest new directions for further research to be done by the author or other workers in the field.

Each of Chapters I-X will be dealt with in a separate subsection of Section 11.1, Sections 11.1. - 11.1.10. Some general conclusions about the subject matter of the entire book will be presented in Section 11.2. Few equations will be used in this chapter.

11.1 Overview of the Chapters

11.1.1 Chapter I Introduction

After a general overview of the chapter's contents the James-Stein and the ridge regression estimator are introduced. It is seen that the James-Stein estimator uses all of the coordinates in the parameter space to simultaneously estimate the individual coordinates. Surprisingly this actually produces a better estimator than the MLE when at least three variables are involved because its risk averaged over all coordinates of the parameter space is smaller. The form of the estimator and its risk is presented. It is seen that the risk of the JS is smaller than that of the MLE when there are three or more variables. For data taken from the US Weather Almanac it is seen that the sum of the squares of the deviations for the JS estimator of the differences between the average temperatures for July 1990 and the all time in ten cities was smaller than that of the MLE.

The Least Square Estimator is derived and shown to be a poor choice for multicollinear data. The ordinary ridge estimator is derived by the method of Hoerl and Kennard (1970). The variability of the ridge and the LS estimator is compared for a regression of the research expenditures of the United States on that of other countries. The data is highly multicollinear. It is shown that the variability of the LS for this data is considerably larger than that of the ridge estimator.

The intent of the historical survey is to review enough of the main references in the field to give the interested reader many entry points into the literature. A number of the topics mentioned in the historical survey are not taken up later in the book, for example Pitman Efficiency and empirical Bayes confidence intervals. These are important and interesting topics but it was decided based on the author's experience and background to limit the mathematical development to point estimation and measures of efficiency based on the mean square error. However the interested reader can pursue these topics using the given references. No doubt there are references that could have been included that were not. Chances are, however that most of these references are contained in the bibliographies of the articles that are referenced.

Since the list of references runs twenty seven pages, it might be worthwhile to mention some of the basic ones again. The author began his research during the winter of 1976 as a graduate student at the University of Rochester under the direction of P.R.S.S. Rao. The work began by reading C.R. Rao (1975) and the main references in that article. These included Hoerl and Kennard (1970), Efron and Morris (1971a,1973), Stein (1956) and James and Stein (1961). Some other useful references for beginners include C.R. Rao (1973) Chapter 4, Toutenburg (1982), Vinod and Ullah(1981) and C.R. Rao and Toutenburg (1995).

Two kinds of estimators considered throughout the book are empirical Bayes estimators and approximate minimum mean square error estimators. Two examples of empirical Bayes estimators are presented. They are:

1. the well known example due to Robbins, the father of empirical Bayes;
2. an example of estimation of variance components due to C.R.Rao.

An example of an approximate minimum mean square error estimator that is also an EBE is given.

11.1.2 Chapter II The Stein Paradox

In order to make the book accessible to readers with different levels of preparation some basic decision theory concepts are summarized. These include maximum likelihood estimation, the general decision model, admissibility, the frequentist risk and the Bayes risk. Three different heuristic arguments are given to support the fact that one would expect the MLE to be inadmissible. These are:

1. Stein's original argument;
2. an argument based on elementary geometry;
3. an argument based on the LS principle.

The James-Stein estimator is formulated as an empirical Bayes estimator and as an approximate MSE. Since for both formulations the Stein estimator approximates an optimal estimator with smaller risk than the MLE it stands to reason that the JS would also have this property. By obtaining an explicit formula it is shown that frequentist and the Bayes risk of the JS is uniformly smaller than that of the MLE hence proving that the MLE is inadmissible.

Although the JS is used to prove the inadmissibility of the MLE it too is inadmissible. This is demonstrated by observing that the positive part estimator has uniformly smaller MSE than the JS.

11.1.3 Chapter III The Ridge Estimators of Hoerl and Kennard

An exposition of the basic information about ridge regression estimators found in the literature was given. It was shown how the derivation of the ordinary ridge regression estimator by Hoerl and Kennard (1970) taken up in Chapter I could be generalized to include different kinds of ridge and contraction estimators. Three different measures of efficiency are studied :

1. the MSE of the parametric functions of the regression parameters;
2. the total MSE (TMSE);
3. the individual MSE of regression parameters after a reparametization of the regression model where the design matrix is orthogonal.

In general for a given value of the ridge parameter or matrix these MSEs are less than that of the LS only for a range of the β parameters. For the MSE of the parametric functions an ellipsoid in the parameter space can be found where the ridge type estimator has a smaller MSE than the LS. It is shown that there exists a value of the ridge parameter where the ordinary ridge estimator of Hoerl and Kennard has a smaller TMSE than the LS for a given value of the β parameter. The individual MSEs are used to obtain values of the elements of the matrix of ridge parameters where the generalized ridge estimator has minimum IMSE.

A large number of Monte Carlo simulation studies have been done about ridge type estimators. In general it has been found from these that ridge estimators are closer to the true parameter values than ordinary least square estimators. Ridge estimators usually perform best when the data is highly multicollinear. The results of some of these simulation studies is discussed.

Some of the different methods of picking the ridge parameter are discussed. The ridge parameters are determined either by estimation of a function of different parameters from the data or by graphical methods. The method that appears to be the most popular is the ridge trace, that is a plot of the regression coefficients vs. the ridge parameter values. An appropriate k is guessed upon at the point where the parameter values appear to become stable. This technique and relies very heavily on the judgment of the data analyst.

It was decided not to include large sample approximations to the MSE, bias and variance. For this reason the consistency and large sample properties of double K class estimators is not included.

11.1.4 Chapter IV James-Stein Estimators for a Single Linear Model

James-Stein type estimators are developed for the parameters of a single linear model as special cases of empirical Bayes estimators and as approximate minimum mean square error estimators. For the sake of generality the estimators are developed for the non- full rank linear model. A brief review of least square estimators, generalized inverses and estimable parametric functions is presented.

The linear BE is derived for the special case of Dempster's prior.[1] The prior dispersion is proportional to the $X'X$ matrix where X is the design matrix. The JS EBE is obtained by substituting estimates of functions of the unknown parameters. The parametric functions of the JS estimator have smaller MSE averaging over the prior distribution than the LS for all values of the prior dispersion. However the parametric functions of the JS have a smaller MSE than the LS without averaging over the prior only for a range of values of the β parameters. When averaged over a the predictive loss function, i.e., the quadratic loss function with matrix $X'X$. the MSE is uniformly smaller than that of the LS. Conditions are given on the quadratic loss functions and the estimable parametric functions for the JS to have a uniformly smaller MSE than the LS. Regions for the β parameters where the JS is has a smaller MSE than LS are also given.

The Stein estimator may also be derived as an approximate minimum mean square error estimator. The minimum mean square error estimator (MMSE) is the constant multiple of the LS estimator that minimizes the mean square error of prediction. The constant is a function of the unknown parameters. The approximate MMSE is obtained by replacing the minimizing constant by a function of the sample estimators.

Two important measures of efficiency are defined:

1. the Relative Savings Loss (RSL) due to Efron and Morris;
2. the Relative Loss (RL) due to the author.

The RSL is a measure of efficiency of the EBE compared with the BE using the Bayes risk. The RL is a measure of efficiency of the approximate MMSE compared with the MMSE using the frequentist risk. The RSL is constant while the RL is found to be a bounded increasing function of $\delta=\beta'X'X\beta/\sigma^2$ with the RSL as a lower bound and a function of the RSL as an upper bound.

The MSE of the BE averaged over the prior distribution is always less than that of the EBE. When the MSE of these estimators was considered without averaging over the prior distribution it was found that the MSE of prediction of the BE was smaller than that of the EBE only for a range of values of the β parameters. The MSE of the EBE is uniformly less than that of the LS while as δ tends to infinity

---

[1] Dempster's prior is also referred to as Zellner's g prior in the literature.

the BE blows up and the JS is clearly smaller. However analytical conditions are obtained where the BE and the JS cross twice and the risk of the JS is smaller than that of the LS for values of δ near zero. Thus, the behavior of the conditional and the average MSE of the JS and the BE are quite different.

The risk and the crossing values of the JS and the BE are determined for a number of special cases using the software package Mathematica. Since only a few cases could be considered the program was provided. The code should work for all versions of Mathematica on all platforms. The computations were performed on a Macintosh Performa 6115. For Maple users the program was rewritten and tested using Maple. Also included is a program for the TI 82 and a script for the TI 92. Computer algebra systems like Mathematica and Maple show a great amount of promise as important tools for statistical research.

11.1.5 Chapter V Ridge Estimators from Different Points of View

There are many different kinds of optimization problems that may be solved from both a Bayesian and a frequentist viewpoint that result in ridge type estimators. This chapter was intended to be a survey of these methods.

The chapter begins with a review of LS estimators for the non-full rank case. This includes a review of estimability of parametric functions, generalized inverses and a proof of the Gauss Markov Theorem.

Ridge type estimators are obtained by solving the appropriate optimization problem as:

1. a constrained least square estimator that results from the solution of the LS problem with a constraint;
2. by generalizing the original argument of Hoerl and Kennard;
3. as a minimax estimator;
4. as a special case of the Bayes estimator.

Using a considerable amount of linear algebra several different equivalent forms of the BE were derived. For the full rank case these equivalencies hold true for all parametric functions. For the non-full rank case they only hold true for estimable parametric functions or given certain conditions on the prior mean and dispersion. These equivalent forms are used to:

1. see when the solutions to the various optimization problems yield equivalent estimators;
2. obtain ridge type estimators as special cases of mixed estimators, linear Bayes estimators and minimax estimators;

3. obtain equivalent forms of the MSE of the estimators with and without averaging over a prior distribution;

4. formulate empirical Bayes estimators in Chapter VII.

The MSE of the different estimators with and without averaging over the prior distribution are obtained and compared. Averaging over a prior distribution the BE always has a smaller MSE than the LS. More precise prior information, i.e., a smaller dispersion matrix (recall that matrix A is smaller than matrix B if B − A was non-negative definite) means that the BE has a smaller MSE.

The problem is not as simple for the comparisons of the MSE without averaging over the prior. The BE and ridge regression estimators are found to be smaller than the LS for a range of the parameter values. This range usually takes the form of an ellipsoid in the parameter space. It is very hard to compare the conditional MSE of two BEs in general. Under certain special conditions the ordinary and generalized ridge regression estimators and the contraction estimators can be compared with respect to the MSE, the variance and the bias.

The jackknifed ridge regression estimator is presented. It is observed that although it does have a smaller bias than the usual ridge estimator its variance generally is larger and its signal to noise ratio smaller.

A project for the future would be reformulating of some of the ideas of this chapter and Gruber (1990) in terms of the theory of estimating functions (see for example Godambe (1991).)

11.1.6 Chapter VI Improving the James Stein Estimator: The Positive Parts

Five different ways to truncate the JS are considered leading to the formulation of five different positive part estimators PP_0.-PP_4. Numerical comparisons of the size and the properties of the Bayes and the frequentist risk are made for the problem of estimating the mean of a multivariate normal distribution and for the single linear model. Similar comparisons for r linear models and the multivariate linear model still have to be done. All of the positive part estimators have a smaller risk than the JS but no one positive part estimator does better than another for all prior dispersions or for the entire range of the parameter space. However it appears that the two best positive parts of the five are the classical positive part PP_0 for smaller values of the parameters and PP_2 for larger values.

Since the numerical comparisons could only be presented for a few cases due to lack of space the reader is invited to do them for other cases. For this purpose Mathematica, Maple, TI 82 programs and TI 92 scripts are given in the exercises. The computations and graphs were done using Mathematica 2.2. The programs were rewritten and tested in Maple and on the graphing calculators. The given Mathematica code should work for all versions of Mathematica.

For a long time it was known that the classical positive part of the JS was inadmissible but an explicit estimator with uniformly smaller risk was not available. Shao and Strawderman (1994) found an estimator with uniformly smaller risk. Some of the work of Shao and Strawderman is summarized. It would be worthwhile to make some numerical comparisons of risk of the estimators of Shao and Strawderman with PP_0-PP_4 to see how much of an improvement actually is obtained.

11.7 Chapter VII The Simultaneous Estimation Problem

The problem of simultaneous estimation of the parameters for r uncorrelated linear regression models turns out to be a very rich problem because many different empirical Bayes estimators can be formulated depending on the amount of knowledge available about the prior distribution. The three main kinds of estimators that are studied are those of C.R.Rao, Wind and Dempster. For the estimator of C.R. Rao the prior dispersion is completely unknown; for the estimator of Wind the diagonal matrix of its singular value decomposition is unknown; the prior used to obtain the Dempster estimator is an unknown constant multiple of the Moore Penrose inverse of the $X'X$ matrix. The estimator of Dempster was studied for the single linear model in Chapter IV and the results obtained there are extended to r linear models. The estimators of Wind and C.R. Rao could not be formulated for a single univariate model.

For all three kinds of estimators the parametric functions have a smaller MSE than the LS when averaged over their respective priors. The parametric functions of the estimator of C.R. Rao have a uniformly smaller MSE than the LS without averaging over the prior distribution. The estimators of Wind and Dempster do not have this property. However the MSE averaged over a quadratic loss function whose matrix has the same structure as the prior dispersion does have a uniformly smaller MSE than the LS. It would be nice if a general theorem along these lines could be stated and proved.

Expressions for the MSE of ridge estimators when the biasing parameter k is stochastic are generally very complicated. For ridge estimators that are algebraically equivalent to those of Wind and C.R. Rao type with prior mean zero estimates of the biasing matrix K are such that the form of the MSE for matrix, individual, total and predictive MSE takes on a relatively simple form and is uniformly less than that of the LS.

For the estimator of Dempster the results about the analytical and numerical comparison of the MSE with the BE in Chapter IV are extended to r linear models and some computer programs are included for the reader who wishes to obtain numerical results not obtained here. An outline of the structure of the positive parts and their MSE for r linear models is undertaken. Additional research needs to be done to get a more complete picture.

11.1.8 Chapter VIII The Precision of Individual Estimators

Although the compound risk of the JS is uniformly smaller than that of the LS or the MLE the individual MSE (IMSE) usually does not have that property. The conditional MSE of the individual components or the parametric functions is less than that of the LS or the MLE only for a range of the β parameters. The geometric structure of this range is very complicated.

For the problem of estimating the mean of a multivariate normal distribution and for the single linear model necessary and sufficient conditions on the parameters are obtained for the IMSE and for the risk of the parametric functions to have smaller MSE than the LS. For some special cases numerical values of the bounds on the functions of the parameters are obtained using Mathematica. Programs are provided in Mathematica, Maple and for the TI 82 and the TI 92 for the reader interested in obtaining these bounds for values not considered in the text and for verifying the results obtained there.

The case of r linear models turns out to be a very rich problem because the estimators of C.R. Rao, Dempster, Wind and their variants need to be investigated. Four kinds of MSE are considered:

1. the $(IMSE)_1$, i.e., the MSE of the individual components for each individual linear model;
2. the $(IMSE)_r$, i.e., the sum of the r $(IMSE)_1$ for each of the r linear model;
3. the $(CMSE)_1$, i.e., the risk averaged over a compound loss function of the components for each individual linear model;
4. the $(CMSE)_r$, the compound risk, i.e. the $(CMSE)_1$ summed over all r linear models.

For the estimator of Wind and that of C.R. Rao the $(IMSE)_r$ is uniformly smaller than that of the LS. It was observed earlier that the $(CMSE)_r$ is smaller than that of the LS for all three kinds of estimators when the predictive loss function is used. For the estimator of Wind and Dempster conditions are obtained for the different MSEs to be smaller than that of the LS, numerical estimates are given for special cases and computer programs were supplied. Results still need to be obtained for the $(IMSE)_1$ and the $(CMSE)_1$ for the estimators of C.R. Rao. This problem should also be investigated for the multivariate linear model and the seemingly unrelated regression equation model considered in Chapter IX and X. Work also needs to be done along these lines for the positive part estimators.

The limited translation estimators represent a compromise between the JS and the MLE that has a smaller IMSE and a slightly larger CMSE. An introduction to them is given with numerical illustrations of how they compare with each other, the usual JS and the MLE. Some MSEs are computed. The problem is considered for estimating the mean of a multivariate normal distribution; further investigations are needed for the different linear model setups.

11.1.9 Chapter IX The Multivariate Linear Model

The formulation of the multivariate linear model the MANOVA along the lines of C.R. Rao (1973) is reviewed. The linear BE is formulated with the ridge and contraction estimators as special cases.

The estimators of C.R. Rao, Wind and Dempster are considered in specific linear model contexts. The estimator of C.R. Rao is considered for r linear models . The MSE of the parametric functions is obtained averaging over the prior distribution. When the variance matrix of the error terms is unknown it turns out to be quite messy. There no longer exists an optimal constant that guarantees that the estimator is uniformly smaller than the LS.

The analogue of the estimator of Wind is considered for the case of a single linear model. This is possible for the multivariate model since functions of the different variables can be used as simultaneous estimates. For this case it is possible to obtain the average and conditional MSE averaging over a form of the predictive loss function in a form where exact comparisons can be made with the LS.

For the most part it was decided to limit the discussion to situations where large sample approximation was not necessary. However for purposes of illustration the Dempster estimator is formulated for the multivariate linear models and a large sample approximation to its average MSE was obtained.

Some of the questions asked in the previous chapters may also be asked about the estimators studied here for the MANOVA model. These include:

1. formulation of the positive parts and obtaining their Bayes and empirical Bayes estimator;
2. the study of the individual risk function in comparison with the compound risk function;

3. extending the concepts of relative loss and relative savings loss and the comparison of the BE and the EBE.

These questions are not discussed in the text but instead are formulated as exercises. The problems posed by these exercises are open questions to the best of the author's knowledge.

11.1.10 Chapter X Other Linear Model Setups

This chapter extends the ideas considered in Chapters I-IX to some new situations that are not considered there. Three main items are dealt with:

1. the seemingly unrelated regression equation (SURE) model;
2. some variations of the simultaneous estimation problem for r linear models;
3. the Kalman Filter.

The seemingly unrelated equation model is formulated as a generalization of the MANOVA model where the X matrix associated with each variable need not be the same. The least square, ridge and James Stein estimator are studied from two points of view :

1. as estimators that take into account the contemporaneous correlation between the variables ;
2. as ordinary least square, ridge or James-Stein estimators that neglect the existence of contemporaneous correlation.

As one might expect the estimators that take contemporaneous correlation into account are more efficient. Examples are given to illustrate this improvement in efficiency for multivariate models with two variables. Numerical values of the risk of the JS with and without considering contemporaneous correlation are given. As was done in the previous chapters computer algebra code is provided for readers who want to explore further.

Three variations of the simultaneous estimation problem considered in Chapter VII are;

1. the simultaneous estimation of the parameters in r linear models where the assumptions about the individual models are different;
2. the estimation of the parameters in a $(r + 1)$ st linear model using the $(r + 1)$ st response variable and the past observations of the response variable in the r linear models;

3. the prediction of the response variable in the $(r+1)$'st model assuming the observations for the first r models are available.

One problem connected with the simultaneous estimation of parameters in r linear models where assumptions about individual models are different is that of finding a reasonable approximation to the MSE. The estimator and the predictor mentioned in items 2 and 3 above are obtained using the estimators of Wind, C.R. Rao and Dempster.

The Kalman filter besides being very important to many diverse fields of application is an important application of the linear Bayes estimator. The Kalman filter is formulated for one and r linear models. By using simulated observations the generation of the iterated BE is illustrated.

11.2 Conclusion

Two important ways to improve estimators by shrinking were considered in this volume, the use of ridge type estimators and the use of Stein rule estimators. From the Bayesian viewpoint these estimators are always formulated to have smaller Bayes risk than the least square estimator. The efficiency of the estimators from the frequentist point of view is always improved as compared with that of the LS or MLE for at least a range of values of the parameter space. Ridge type estimators are generally more precise estimators for highly multicollinear data.

Improving estimators by shrinking and the study of James-Stein and ridge regression estimators are indeed areas where a great deal of research has been done and remains to be done.

References

Adkins, L.C. (1992). "Finite Sample Moments of a Bootstrap Estimator of the James-Stein Rule." Econometric Reviews, 11(2), 173-193.

Adkins, L.C. and Hill, R.C. (1989). "Risk Characteristics of a Stein-like Estimator for the Probit Regression Model." Economics Letters, 30, 19-26.

Adkins, L.C. and Hill, R.C. (1990). "An Improved Confidence Ellipsoid for the Linear Regression Model." Journal of Statistical Computation and Simulation, 36, 9-18.

Adkins, L.C., Hill, R .C. (1996). "Using Prior Information in the Probit Model: Empirical Risks of Bayes, Empirical Bayes and Stein Estimators." Bayesian Analysis in Statistics and Econometrics. Berry D.A., Chaloner, K.M. and Geweke, J.K. eds. John Wiley and Sons. New York.

Akdeniz, F. and Kaciranlar, S. (1995). "On the Almost Unbiased Generalized Liu Estimator and Unbiased Estimation of the Bias and MSE." Communications in Statistics Theory and Methods, 24(7), 1789-1797.

Alam, K. (1973). "A Family of Admissible, Minimax Estimators of the Mean of a Multivariate Normal Distribution." Annals of Statistics, 1(3), 517-525.

Allen, D.M. (1971). "Mean Square Error of Prediction as a Criterion for Selecting Variables." Technometrics 13, 469-475.

Allen, D.M. (1974). "The Relationship between Variable Selection and Data Augmentation and a Method for Prediction." Technometrics, 16, 125-127.

Ali, M.A. (1989). "On a Class of Shrinkage Estimators of the Vector of Regression Coefficients." Communications in Statistics-Theory and Methods, 18(12), 4491-4500.

Ali, M.A.(1991). "Impacts of auxiliary information on outliers in regression." Communications in Statistics –Theory and Methods, 20, 3271-3281 .

Anderson, T.W. (1958). An Introduction to Multivariate Statistical Analysis. New York: John Wiley and Sons, Inc.

Andrews, D.F. , Herzberg, A.M. (1985). Data: a Collection of Problems from Many Fields for the Student and Research Worker. Springer Verlag. New York.

Bacon, R.W. and Hausman, J.A. (1974). "The Relationship Between Ridge Regression and the Minimum Mean Squared Error Estimator of Chipman." Oxford Bulletin of Economics and Statistics 36, 115-124.

Bair, F.(ed.)(1992). The Weather Almanac. Gale Research Inc., Detroit.

Baksalary, J.K. (1984). "Comparing Stochastically Restricted Estimators in a Linear Regression Model." Biometrics Journal , 26, 555-557.

Baksalary, J.K. and Kala, R. (1981). "Linear Transformations Preserving Best Unbiased Estimators in a General Gauss- Markov Model." The Annals of Statistics, 8, 913-916.

Baksalary, J.K. and Kala, R. (1983). "Partial Orderings between Matrices One of which is of Rank One." Bulletin of the Polish Academy of Science, Mathematics, 31, 5-7.

Baksalary, J.K. and Markiewicz, A. (1990). "Admissible Linear Estimators of an Arbitrary Vector of Parametric Functions in the General Gauss-Markov Model." Journal of Statistical Planning and Inference, 26,161-171.

Baksalary, J.K., Markiewicz, A. and Rao, C.R. (1995). "Admissible Linear Estimation in the General Gauss-Markov Model with respect to an Arbitrary Quadratic Loss Function." Journal of Statistical Planning and Inference, 44(3), 341-348.

Baksalary, J.K. and Pordzik, P.R. (1993). "Preliminary Test Estimation of a Vector of Parametric Functions in the General Gauss Markov Model." Journal of Statistical Planning and Inference, 3 6, 227-240.

Bannerjee, K.S. and Karr, R.N. (1971). "A Comment on Ridge Regression, Biased Estimation for Non-Orthogonal Problems." Technometrics, 13(4), 895-898.

Baranchik, A.J. (1964). Multiple Regression and Estimation of the Mean of a Multivariate Normal Distribution. Technical Report 51. Department of Statistics, Stanford University.

Baranchik, A.J. (1970). "A Family of Minimax Estimators ,of the Mean of a Multivariate Normal Distribution." Annals of Mathematical Statistics 41, 642-645.

Baranchik,A.J.(1973). "Inadmissibility of Maximum Likelihood Estimators in Some Multiple Regression Problems with Three or More Independent Variables." Annals of Statistics, 1, 312-322.

Barry, D. (1995). "A Bayesian Analysis for a Class of Penalised Likelihood Estimates." Communications in Statistics - Theory and Methods, 24(4), 1057-1071.

Bar-Shalom, Y. and Fortmann, T.E. (1988). Tracking and Data Association. Academic Press, Boston.

Baye, M.R. and Parker, D.F. (1984). "Combining Ridge and Principal Component Regression." Communications is Statistics. Theory and Methods, 13(2), 197-205.

Becker, R.A., Chambers, J.M. and Wilks, A.R. (1988). The New S Language. Bell Telephone Laboratories, Murray Hill.

Belsey, D.A., Kuh, E. and Welsch, R.E. (1980). Regression Diagnostics. John Wiley, New York.

Beran, R. (1987). "Prepivoting to Reduce Error Level of Confidence Sets." Biometrika, 74, 457-468.

Beran, R. (1995). "Stein Confidence Sets and the Bootstrap." Statistica Sinica, 5, 109-127.

Berger, J.O. (1976a). " Admissible Minimax Estimation of a Multivariate Normal Mean with Arbitrary Quadratic loss." Annals of Statistics, 4, 223-226.

Berger, J.O. (1976b). "Minimax Estimation of a Multivariate Normal Mean Under Arbitrary Quadratic Loss.". Journal of Multivariate Analysis, 6, 256-264.

Berger, J.O. (1985). Statistical Decision Theory and Bayesian Analysis. Springer Verlag, New York.

Berger, J.O. and Berliner, L.M. (1986). "Robust Bayes and Empirical Bayes Analysis with ε Contaminated Priors." The Annals of Statistics, 14(2), 461- 466.

Berger, J.O. and Bock, M.E. (1976). "Eliminating Singularities of Stein-type Estimators of Location Vectors." Journal of Multivariate Analysis, 52, 338-351.

Berndt, E. (1991). The Practice of Econometrics: Classic and Contemporary. Addison Wesley, Reading, Massachusetts.

Berry, J.C. (1994). "Improving the James-Stein Estimator using the Stein Variance Estimator." Statistics and Probability Letters, 20, 241-245.

Bilodeau, M. and Kariya, T. (1989). "Minimax Estimators in the Normal MANOVA Model." Journal of Multivariate Analysis, 28, 260-270.

Birkes, D. and Dodge, Y. (1993). Alternative Methods of Regression. John Wiley, New York.

Blair, Frank (editor) (1992). The Weather Almanac. Sixth Edition. Gale Research Inc., Detroit.

Bock, M.E.(1975). "Minimax Estimators of the Mean of a Multivariate Normal Distribution." Annals of Statistics, 3, 209-218.

Bock, M.E., Yancey, T.A. and Judge G.G. (1973). "The Statistical Implication of Preliminary Test Estimators in Regression." Journal of the American Statistical Association, 68, 109-116.

Brandwein, A.C .and Strawderman, W.E. (1990). "Stein Estimation: The Spherically Symmetric Case." Statistical Science, 5(3), 356-369.

Brandwein, A.C. and Strawderman, W.E. (1991). "Generalizations of James-Stein Estimators under Spherical Symmetry." Annals of Statistics, 19, 1639-1650.

Brandwein, A.C., Ralescu, S. and Strawderman, W.C. (1993). "Shrinkage Estimators of the Location Parameter for Certain Spherically Symmetric Distributions." Annals Institute of Statistical Mathematics, 45(3), 551-565.

Breiman, L. (1995). "Better Subset Regression Using the Nonnegative Garrote." Technometrics, 37(4), 373-384.

Brown, K.G. (1978). "On Ridge Estimators in Rank Deficient Models." Communications in Statistics---Theory and Methods, A7(2), 187-192.

Brown, L. (1971). "Admissible Estimators, Recurrent Diffusion and Insoluble Boundary Value Problems." Annals of Mathematical Statistics, 42, 855-903.

Brown, P. J. (1993). Measurement, Regression and Calibration. Clarendon Press, United Kingdom.

Brownstone, D. (1990). "Bootstrapping Improved Estimators for Linear Regression Models." Journal of Econometrics, 44, 171-187.

Bunke, O. (1975). "Minimax Linear, Ridge and Shrunken Estimators for Linear Parameters." Mathematiche Operationsforschung und Statistik, 6, 697-701.

Bunke, H. and Bunke, O. (1986). Statistical Inference in Linear Models. John Wiley and Sons, New York.

Campbell, S.L. and Meyer, C.D. (1991). Generalized Inverses of Linear Transformations. Dover Publications, Inc., New York.

Carter, G.M. and Rolph, J.E. (1974). "Empirical Bayes Methods Applied to Estimating Fire Alarm Probabilities." Journal of the American Statistical Association, 69, 880-885.

Carter, R.A.L. (1981). "Improved Stein-Rule Estimator for Regression Problems." Journal of Econometrics, 17, 113-123, and Erratum, Journal of Econometrics, 17, 393-394.

Carter, R.A.L. (1984). "Double k-Class Shrinkage Estimators in Multiple Regression." Journal of Quantitative Economics, 1, 27-47.

Carter, R.A.L., Srivastava, M.S., Srivastava, V.K. and Ullah, A. (1990). "Unbiased Estimation of the MSE Matrix of Stein-Rule Estimators, Confidence Ellipsoids and Hypothesis Testing." Econometric Theory, 6, 63–74.

Casella, G. (1985). "An Introduction to Empirical Bayes Data Analysis." American Statistician, 39, 83- 87

Casella, G. and Hwang, J.T. (1983). "Empirical Bayes Confidence Sets for the Mean of a Multivariate Normal Distribution." Journal of the American Statistical Association, 78, 688-698.

Cellier, D. and Fourdrinier, D. (1995a). "Shrinkage Positive Rule Estimators for Spherically Symmetric Distributions." Journal of Multivariate Analysis, 53, 194-209.

Cellier, D. and Fourdrinier, D. (1995b). "Shrinkage Estimators Under Spherical Symmetry for the General Linear Model." Journal of Multivariate Analysis, 52(2), 338-351.

Cellier, D., Fourdrinier, D., and Robert, C. (1989a). "Controlled Shrinkage Estimators (A Class of Estimators Better than the Least Squares Estimator with Respect to a General Quadratic Loss, for Normal Observations)." Statistics, 20(1), 13-22.

Cellier, D., Fourdrinier, D., and Robert, C. (1989b). "Robust Shrinkage Estimators of the Location Parameter for Elliptically Symmetric Distributions." Journal of Multivariate Analysis, 29, 39-52.

Chalton, D.O. and Troskie, C.G. (1992). "Identification of Outlying and Influential Data with Biased Estimation: A Simulation Study." Communications in Statistics-Simulation, 21(2), 607-626.

Chang. Y.T. (1995). "Two-Stage James-Stein Estimators of the Mean Based on Prior Knowledge." Communications in Statistics-Theory and Methods, 24(9), 2211-2227.

Chaturvedi, A. ,Van Hoa, T. and Shulka, G. (1993). "Performance of the Stein-Rule Estimators when the Disturbances are Misspecified as Spherical." The Economic Studies Quarterly, 44(2), 97-107.

Chawla, J.S. (1988). "A Note on General Ridge Estimator." Communications in Statistics–Theory and Methods, 17(3), 739 -744.

Chawla, J.S.(1990). "A Note on Ridge Regression." Statistics and Probability Letters, 9, 343-345.

Chen, J. and Hwang (1988). "Improved Set Estimators for the Coefficients of a Linear Model when the Error Distribution is Spherically Symmetric with Unknown Variances." Canadian Journal of Statistics, 16, 293-299.

Chi, X.W. and Judge, G. (1985). "On Assessing the Precision of Stein's Estimator." Economic Letters, 18, 143-148.

Copas, J.B. (1983). "Regression Prediction and Shrinkage(with discussion)." Journal of the Royal Statistical Society B, 45(3), 311-354.

Cramer, H. (1946). Mathematical Methods of Statistics. Princeton University Press, Princeton, New Jersey.

Craven, P. and Wahba, G. (1979) . "Smoothing Noisy Data with Spline Functions: Estimating the Correct Degree of Smoothing by the Method of Generalized Cross Validation." Numerical Mathematics, 31, 377-403.

Crivelli, A., Firinguetti, L., Montano, R. and Munoz, M. (1995). "Confidence Intervals in Ridge Regression by Bootstrapping the Dependent Variable: A Simulation Study ." Communications in Statistics- Simulation, 24(3), 631-632.

Crouse, R.H. and Holzworth, R.J. (1988). "Combining Prior Information with Ridge Regression." American Statistical Association: 1988 Proceeding of the Business and Economic Statistics Section, 302–307.

Crouse, R.H. , Jin, C. and Hanumara, R.C. (1995). "Unbiased Ridge Estimation with Prior Information and Ridge Trace." Communications in Statistics-Theory and Methods, 24(9), 2341-2354.

Daniel, C. and Wood, F. (1971). Fitting Equations to Data. John Wiley and Sons. New York.

De Jong, S. (1995). "PLS Shrinks." Journal of Chemonmetrics, 9, 323-326.

Delaney, N.J. and Chatterjee, S. (1986). "Use of the Bootstrap and Cross Validation in Ridge Regression." Journal of Business and Economic Statistics, 4(2), 255-262.

Dempster, A.P. (1973). "Alternatives to Least Squares in Multiple Regression." Multivariate Statistical Inference, D.G. Kabe and R.P. Gupta ed., Amsterdam: North Holland, 25-40.

Dempster, A.P., Schatzoff, M. and Wermuth, N. (1977). "A Simulation Study of Alternatives to Ordinary Least Squares." Journal of the American Statistical Association, 72, 77--91.

Dhillion, U.S., Shilling, J.D., and Sirmans, C.F. (1987). "Choosing Between Fixed and Adjustable Rate Mortgages." Journal Money Credit Banking, 19, 260-267.

Draper, N. R. Van Nostrand, R. Craig (1979). "Ridge Regression and James-Stein Estimation: Review and Comments." Technometrics, 21, 451-466.

Draper, N.R. and Smith, H. (1981). Applied Regression Analysis. (second edition), John Wiley and Sons. New York.

Drygas, II, (1983). "Sufficiency and Completeness in the General Gauss-Markov Model." Sankhya, Series A, 45, 88-98.

Duncan, D.B. and Horn, S.D. (1972). "Linear Dynamic Recursive Estimation from the Viewpoint of Regression Analysis." Journal of the American Statistical Association , 38, 815-821.

Economic Report of the President (1988). Transmitted to the Congress in February 1988 Together with The Annual Report of the Council Of Economic Advisers. United States Government Printing Office, Washington: 1988.

Edlund, P. (1989). "On Identification of Transfer Function Models by Biased Regression Methods." Journal of Statistical Computation and Simulation, 31, 131-148.

Efron, B. (1979). "Bootstrap Methods: Another Look at the Jackknife." Annals of Statistics, 7, 1-26.

Efron, B. (1980). The Jackknife, Bootstrap and Other Resampling Plans. SIAM. Philadelphia, Pa.

Efron, B. (1987). "Better Bootstrap Confidence Intervals (with discussion)." Journal of the American Statistical Association, 82, 171-200.

Efron, B. and Gong, G. (1983). " A Leisurely Look at the Jackknife, Bootstrap and other Resampling Plans." The American Statistician, 37, 36-45.

Efron, B. and Morris, C. (1971). "Limiting the Risk of Bayes and Empirical Bayes Estimators, Part I: The Bayes Case." Journal of the American Statistical Association, 66, 807-815.

Efron, B. and Morris, C. (1972a). "Empirical Bayes on Vector Observations: An Extension of Stein's Method." Biometrika, 59, 335-347.

Efron, B. and Morris, C. (1972b). "Limiting the Risk of Bayes and Empirical Bayes Estimators, Part I: The Empirical Bayes Case." Journal of the American Statistical Association, 67, 130-139.

Efron, B. and Morris, C. (1973). "Stein's Estimation Rule and its Competitors." Journal of the American Statistical Association, 65, 117-130.

Efron, B. and Morris, C. (1976a). "Families of Minimax Estimators of the Mean of a Multivariate Normal Distribution." Annals of Statistics, 4, 11-21.

Efron, B. and Morris, C. (1976b). "Multivariate Empirical Bayes and Estimation of Covariance Matrices." Annals of Statistics, 4, 22-32.

Efron, B. and Morris, C. (1977). "Stein's Paradox in Statistics." Scientific American, 236(5), 119-127.

Efroymson, M.A. (1960). "Multiple Regression Analysis." in Mathematical Methods for Digital Computers, eds. A. Ralston and H.S. Wilf. John Wiley and Sons. New York.

Egerton, M.F. and Laycock, P.J. (1982). "An Explicit Formula for the Risk of the James-Stein Estimators." Canadian Journal of Statistics, 10, 199-205.

Evans, S.N. and Stark, P.B. (1996). "Shrinkage Estimators Skorokhod's Problem and Stochastic Integration by Parts." The Annals of Statistics, 24(2), 809-815.

Farebrother, R.W. (1975). "The Minimum Mean Square Error Linear Estimator and Ridge Regression." Technometrics, 17(1), 127-128.

Farebrother, R.W. (1976). "Further Results on the Mean Square Error of Ridge Regression." Journal of the Royal Statistical Society, B. 38, 248-250.

Farebrother, R.W. (1978). "A Class of Shrinkage Estimators." Journal of the Royal Statistical Society, B. 40, 47-49.

Farebrother, R.W. (1984). "The Restricted Least Squares Estimator and Ridge Regression." Communications in Statistics--Theory and Methods, 13(2), 191-196.

Fay. R.E. and Herriot, R.A. (1979). "Estimates for Income for Small Places: An Application of James-Stein Procedures to Census Data." Journal of the American Statistical Association ,74, 269-277.

Ferguson, T.S.(1969). Mathematical Statistics: a Decision Theoretic Approach. Academic Press, New York.

Firinguetti, L. (1989). " A Simulation Study of Ridge Regression Estimators with Autocorrelated Errors." Communications in Statistics -Simulation, 18(2), 673-702.

Firinguetti, L. (1990) . "A Note on the Evaluation of Expected values of Ratios of Random Variables with an Application to Ridge Regression." Estadistica, 42, 1-16.

Flack, V.F. (1989). "Predictability Measures for Ridge Regression Models." Communications in Statistics-Theory and Methods, 18(2), 755-788.

Fomby, T.B. and Johnson, S.R. (1977). "MSE Evaluation of Ridge Estimators based on Stochastic Prior Information." Communications in Statistics, Series A, 6,1245-1258.

Fomby, T.B., Hill, R.C. and Johnson, S.R. (1984). Advanced Econometric Methods. Springer Verlag, New York.

Fomby, T.B. and Samanta, S.K. (1991). "Application of Stein Rules to Combination Forecasting." Journal of Business and Economic Statistics, 9(4), 391-407.

Fourdirnier, D. and Robert, C. (1995). "Intrinsic Losses for Empirical Bayes Estimation: A Note on Normal and Poisson Cases." Statistics and Probability Letters, 23, 35-44.

Fourdirnier, D. and Strawderman, W.E. (1996). "A Paradox Concerning Shrinkage Estimators: Should a Known Scale Parameter be Replaced by an Estimated Value in the Shrinkage Factor." Journal of Multivariate Analysis , 59, 109-140.

Fourdrinier, D. and Wells, M. T. (1995). "Estimation of a Loss Function for Spherically Symmetric Distributions in the General Linear Model." Annals of Statistics, 23(2), 571-592.

Freedman, D.A. (1981). "Bootstrapping Regression Models." Annals of Statistics, 9, 1218-1228.

Fule, E. (1995). "On Ecological Regression and Ridge Estimation." Communications in Statistics-Simulation, 24(2), 385-398.

Gana, R. (1995) "Ridge Regression Estimation of the Linear Probability Model." Journal of Applied Statistics, 22(4), 537-539.

George, E.I. (1986a). "Minimax Multiple Shrinkage Estimation." Annals of Statistics, 14, 188-205.

George, E.I. (1986b). "Combining Minimax Shrinkage Estimators." Journal of the American Statistical Association, 81, 437-445.

George, E.I. and Casella, G. (1994). "An Empirical Bayes Confidence Report." Statistica Sinica, 4, 617-638.

George, E.I. and Oman, S.D. (1996). "Multiple-Shrinkage Principal Component Regression." The Statistician , 45(1), 111-124.

Ghosh, M. (1993). " On the Γ-Minimaxity of a Class of Shrinkage Estimators." Statistics and Decisions, Supplement Issue No.3, 31-53.

Ghosh, M. and Dey, D.K. (1986). "On the Inadmissibility of Preliminary Test Estimators when the Loss involves a Complexity Cost." Annals of the Institute of Statistical Mathematics, 38, 419-427.

Ghosh, J.K., Mukerjee, R. and Sen, P.K. (1996). "Second-Order Pitman Admissibility and Pitman Closeness: The Multiparameter Case and Stein-Rule Estimators." Journal of Multivariate Analysis, 57, 52-68.

Ghosh, M., Saleh, A.K.Md. E. and Sen, P.K. (1989). "Empirical Bayes Subset Estimation in Regression Models." Statistics and Decisions, 7, 15-35.

Ghosh, M. and Shieh, G. (1991). "Empirical Bayes Minimax Estimators of Matrix Normal Means." Journal of Multivariate Analysis, 38, 306-318,

Ghosh, M. and Sinha, B. K. (1988). "Empirical and Hierarchical Bayes Competitors of Preliminary Test Estimators in Two Sample Problems." Journal of Multivariate Analysis, 27, 206-277.

Gibbons, D. G. (1981). "A Simulation Study of Some Ridge Estimators." Journal of the American Statistical Association, 76(373), 131-139.

Giles, J.A. and Giles, D.E.A. (1993). "Preliminary-Test Estimation of the Regression Scale Parameter when the Loss Function is Asymmetric." Communications in Statistics - Theory and Methods, 22(6), 1709-1733.

Giles, J.A., Giles D.E.A. and Ohtani, K. (1996). "The Exact Risks of Some Pre-Test and Stein-Type Regression Estimators Under Balanced Loss." Communications in Statistics -Theory and Methods ,25(12),2901-2924.

Girard, D.A. (1989). "A Fast 'Monte Carlo Cross Validation'." Procedures for Large Least squares Problems with Noisy Data." Numerical Mathematics, 56, 1-23.

Girard, D.A. (1991). "Asymptotic Optimality of the Fast Randomized Versions of GGV and C_L and Regularization." The Annals of Statistics, 19(4), 1950-1963.

Giri, N.C. (1996). Multivariate Statistical Analysis. Marcel Dekker, Inc., New York.

Gnot, S., Trenkler, G. and Zmyslony, R. (1995). "Nonnegative Minimum Biased Quadratic Estimation in the Linear Regression Models." Journal of Multivariate Analysis, 54, 113-125.

Godambe, V.P. ed. (1991) Estimating Functions. Clarendon Press, Oxford.

Godambe, V.P. (1994) "Linear Bayes and Optimal Estimation." Technical Report Series. Department of Statistics and Actuarial Science. University of Waterloo. STAT 94-11.

Godambe, V.P. and Kale, B.K. (1991). "Estimating Functions: an Overview." Estimating Functions edited by V.P. Godambe. Clarendon Press, Oxford.

Goldstein, M. and Smith, A.F.M. (1974). " Ridge Type Estimator for Regression Analysis." Journal of the Royal Statistical Society B, 36, 284--291.

Golub, G.H., Heath, M. and Wahba, G. (1979). "Cross-Validation as a Method for Choosing a Good Ridge Parameter." Technometrics, 21(2), 215-223.

Golub, G.H. and Van Loan C.F. (1983). Matrix Computations. Johns Hopkins University Press, Baltimore, Maryland.

Goutis, C. (1996). "Partial Least Squares Algorithm Yields Shrinkage Estimators." The Annals of Statistics, 24(2), 816-824.

Graybill, F.A. (1976).Theory and Application of the Linear Model. Wadsworth, Boston.

Greene, W.. (1990). Econometric Analysis. Macmillian, New York.

Grewal, M.S. and Andrews, A.P. (1993). Kalman Filtering Theory and Practice. Prentice Hall, Engelwood Cliffs, New Jersey.

Griffin, B. S. and Krutchkoff, R. G. (1971). "Optimal Linear Estimators: An Empirical Bayes Version with Application to the Binomial Distribution." Biometrika, 58, 195-201.

Gross, J. (1996). "Estimation using the Linear Regression Model with Incomplete Ellipsoidal Restrictions." Acta Applicandae Mathematicae, 43, 81-85.

Gross, J. (1996) "On a Class of Estimators in the General Gauss- Markov Model." Communications in Statistics -Theory and Methods, 25(2), 381-388.

Gruber, M.H.J. (1979). Empirical Bayes, James-Stein and Ridge Regression Type Estimators for Linear Models. Unpublished Ph.D. Thesis, University of Rochester.

Gruber, M.H.J. (1985). "A Comparison of Bayes Estimators and Constrained Least Square Estimators." Communications in Statistics-Theory and Methods, 14(2), 479-489.

Gruber, M.H.J. (1990). Regression Estimators A Comparative Study. Academic Press, Boston.

Gruber, M.H.J.(1991). "The Efficiency of Jack-knifed and Usual Ridge Type Estimators." Statistics and Probability Letters, 11, 49-51.

Gruber, M.H.J.(1994). "The Comparison of the Efficiency of Bayes and Empirical Bayes Estimators." Proceedings of the American Statistical Association..

Gruber, M.H.J. and Rao, P.S.R.S.(1982). "Bayes Estimators for Linear Models with Less than Full Rank." Communications in Statistics-Theory and Methods 11(1), 59-69.

Guo, Y.Y. and Pal, N. (1992). "A Sequence of Improvements over the James-Stein Estimator." Journal of Multivariate Analysis, 42, 302-317.

Hadzivukovic, S. Nikolic-Djoric,E. and Cobanovic, K. (1992). "The Choice of Perturbation Factor in Ridge Regression." Journal of Applied Statistics, 19(2), 223-230.

Haff, L.R (1978). "The Multivariate Normal Mean with Intraclass Correlated Components: Estimation of Urban Fire Alarm Probabilities." Journal of the American Statistical Association, 73, 767-774.

Haff, L.R. and Johnson, R.W. (1986). "The Superharmonic Condition for Simultaneous Estimation of Means in Exponential Families." Canadian Journal of Statistics, 14, 43-54.

Harewood, S.L. (1992). "A Stein Rule Estimator which Shrinks towards the Ridge Regression Estimator." Economics Letters, 40, 127-133.

Hartigan, J.A. (1969). "Linear Bayesian Methods." Journal of the Royal Statistical Society B, 31, 440-454.

Harville, D.H. (1976). "Extension of the Gauss-Markov Theorem to Include the Estimation of Random Effects." The Annals of Statistics, 4(2), 384--395.

Hasegawa, H. (1995). "On Risk Comparisons of Some Estimators in a Linear Regression Model under Inagaki's Loss Function and Nonnormal Error Terms." Communications in Statistics-Theory and Methods, 24(7), 1655-1685.

Hawkins, D.L. and Han, C. (1989). " A Minimum Average Risk Approach to Shrinkage Estimators of the Normal Mean." Annals of the Institute of Statistical Mathematics, 41(2), 347-363.

He'bel, P., Faivre, R. Goffinet, B. and Wallach, D. (1993). "Shrinkage Estimators Applied to Prediction of French Winter Wheat Yield". Biometrics, 49, 281-293.

Hemmerle, W.J. (1975). "An Explicit Solution for Generalized Ridge Regression." Technometrics, 17, 309-314.

Heiligers, B. and Markiewicz, A. (1996). "Linear Sufficiency and Admissibility in Restricted Linear Models." Statistics and Probability Letters, 30, 105-111.

Hill, B.M. (1994). "On Steinian Shrinkage Estimators: The Finite/Infinite Problem and Formalism in Probability and Statistics." Aspects of Uncertainty edited by R.P. Freeman and A.F.M. Smith. John Wiley and Sons. Ltd., New York.

Hill, R.C. and Judge, G.G. (1987). "Improved Prediction in the Presence of Multicollinearity." Journal of Econometrics, 35, 83-100.

Hocking, R.R. (1996). Methods and Applications of Linear Models Regression and the Analysis of Variance. John Wiley and Sons. New York.

Hoerl, A.E. (1959) "Optimum Solution of Many Variables Equations." Chemical Engineering Progress, 55(11), 69-78.

Hoerl, A.E. (1962). "Application of Ridge Analysis to Regression Problems." Chemical Engineering Progress, 58(3), 54-59.

Hoerl, A.E. and Kennard, R.W.(1970a). "Ridge Regression: Biased Estimation for Nonorthogonal Problems." Technometrics, 12, 55-67.

Hoerl, A.E. and Kennard, R.W. (1970b). "Ridge Regression: Applications to Nonorthogonal Problems." Technometrics, 12, 69-82.

Hoerl, A.E. and Kennard, R.W. (1975). "A Note on a Power Generalization of Ridge Regression." Technometrics, 17(2), 569.

Hoerl, A.E. and Kennard, R.W. (1976). "Ridge Regression : Iterative Estimation of the Biasing Parameter." Communications in Statistics, A5, 77-88.

Hoerl, A.E. and Kennard, R.W. (1990). "Ridge Regression: Degrees of Freedom in the Analysis of Variance." Communications in Statistics-Simulation, 19(4), 1485-1495.

Hoerl, A.E., Kennard, R.W., and Baldwin, K.F. (1975). "Ridge Regression: Some Simulations." Communications in Statistics: Theory and Methods, 4(2), 105-123.

Hoerl, R.W., Schuenemeyer, J.H. and Hoerl, A.E. (1986). "A Simulation of Biased Estimation and Subset Selection Regression Techniques." Technometrics, 28(4), 369-380.

Hoffmann, K.(1995). "A Note on the Kuks-Olman Estimator." Statistics, 26,185-187.

Hogg, R.V. and Craig, A.T. (1995). Introduction to Mathematical Statistics. Fifth Edition. Macmillan Publishing Company, New York.

Holzworth, R.J. (1996). "Policy Capturing with Ridge Regression." Organizational Behavior and Human Decision Processes, 68, 171-179.

Hosmane, B.S. (1988). "On a Generalized Stein Estimator of Regression Coefficients." Communications in Statistics, Theory and Methods, 17(6), 1735-1740.

Huang, H.C. and Cressie, N. (1996) "Spatio-temporal Prediction of Snow Water Equivalent Using the Kalman Filter." Computational Statistics and Data Analysis, 22, 159-175.

Hudson, H.M. (1974). "Empirical Bayes Estimation." Technical Report No. 58. Stanford University. Department of Statistics. Stanford, California.

Humak, K.M.S. (1977).   Statistische Methoden der Modellbildung, Band 1. Statistische Inferenz fur Linear Parameter. Akademie-Verlag, Berlin.

Hwang, J.T. and Casella, G. (1982). "Minimax Confidence Sets for the Mean of a Multivariate Normal Distribution." Annals of Statistics, 10, 868-881.

Hwang, J.T. and Chen, J. (1986). "Improved Confidence Sets for the Coefficients of a Linear Model with Spherically Symmetric Errors." Annals of Statistics, 14(2), 444-460.

Hwang, J.T.G. and Ullah, A. (1994). "Confidence Sets Centered at James-Stein Estimators: A Surprise Concerning the Unknown Variance Case." Journal of Econometrics, 60, 145-156.

Inagaki, N. (1977). "Two Errors in Statistical Model Fitting." Annals of the Institute of Statistical Mathematics, 29,131-152.

James, W. and Stein, C. (1961). "Estimation with Quadratic Loss." Proceedings of the Fourth Berkeley Symposium on Mathematics and Statistics. Berkeley: University of California Press 1, 361-379.

Jones, M.C. and Copas, J.B. (1986). "On the Robustness of Shrinkage Predictors in Regression to Differences between Past and Future Data." Journal of the Royal Statistical Society B, 48(2), 223-237.

Judge, G.G and Bock, M.E. (1978). The Statistical Implications of Pretest and Stein-Rule Estimators. North Holland Publishing Company, New York.

Judge, G.G., Griffiths, W.E. Hill, R.C. and Lee, T.C. (1980). The Theory and Practice of Econometrics. John Wiley and Sons, New York.

Judge, G.G. , Hill, R.C. and Bock, M.E. (1990). "An Adaptive Bayes Estimator of the Multivariate Normal Mean Under Quadratic Loss." Journal of Econometrics, 44, 189-213.

Judge G.G. and Yancy, T. (1986). Improved Methods in Econometrics. North Holland Amsterdam.

Kadane, J.B. (1970). "Testing Overidentifying Restrictions when the Disturbances are Small." Journal of the American Statistical Association, 65, 182-185.

Kadane, J.B. (1971). "Comparison of K-Class Estimators when the Disturbances are Small." Econometrica, 39, 723-738.

Kadiyala, K. (1984). "A Class of Almost Unbiased and Efficient Estimators of Regression Coefficients." Economic Letters, 16, 293-296.

Kagan, A.M., Linnik, Y.V. and Rao, C.R. (1973). Characterization Problems in Mathematical Statistics. Wiley, New York.

Kalman, D. (1996). "A Singularly Valuable Decomposition: The SVD of a Matrix." The College Mathematics Journal , 27(1), 2-23.

Kalman, R.E. (1960). "A New Approach to Linear Filtering and Prediction Problems." Journal of Basic Engineering, 82, 32-45.

Kalman, R.E. and Bucy, R.S. (1961). "New Results in Linear Filtering and Prediction." Journal of Basic Engineering, 83, 95-108.

Kariya, T., Konno, Y. and Strawderman, W.E. (1996). "Double Shrinkage Estimators in the GMANOVA Model." Journal of Multivariate Analysis, 56, 245-258.

Keating , J.P. and Czitrom, V. (1989). "A Comparison of James-Stein Regression with Least Squares in the Pitman Nearness Sense." The Journal of Computation and Simulation, 34, 1-9.

Keating, J.P. and Mason, R.L. (1988). "James-Stein Estimation from an Alternative Perspective." American Statistician, 42, 160-164.

Keating, J.P., Mason, R.L. and Sen, P.K., (1993). Pitman's Measure of Closeness: A Comparison of Statistical Estimators. Society of Industrial and Applied Mathematics, Philadelphia.

Kejian, L. (1993). "A New Class of Biased Estimate in Linear Regression." Communications in Statistics-Theory and Methods, 22(2), 393-402.

Kempthorne, P. J. (1988). "Controlling Risks under Different Loss Functions: The Compromise Decision Problem." Annals of Statistics, 16, 1594-1608 .

Ki, F. and Tsui, K. (1990). "Multiple-Shrinkage Estimators of Means in Exponential Families." The Canadian Journal of Statistics, 18(1), 31-46.

Ki, Y.F. (1992). "Multiple Shrinkage Estimators in Multiple Linear Regression." Communications in Statistics-Theory and Methods, 21(1), 111-136.

Kibria, G. (1996). "On Preliminary Test Ridge Regression Estimators for Linear Restrictions in a Regression Model with Non-Normal Disturbances." Communications in Statistics -Theory and Methods, 25(10), 2349-2369.

Kim, M. and Hill, R.C. (1995). "Shrinkage Estimation in Nonlinear Regression The Box-Cox Transformation." Journal of Econometrics, 66, 1-33.

Kim, P.T. (1987). "Recentered Confidence Sets for the Mean of a Multivariate Distribution when the Scale Parameter is Unknown." Ph.D. Thesis, Department of Mathematics , University of California at San Diego, La Jolla California.

Kleffe, J, and Rao, J. N. K. (1992). Estimation of Mean Square Error of Empirical Best Linear Unbiased Predictors under a Random Error Variance Linear Model. Journal of Multivariate Analysis, 43, 1-15.

Kshirsagar, A.M. (1972). Multivariate Analysis. Marcel Dekker, New York.

Kubokawa, T. (1991). " An Approach to Improving the James-Stein Estimator." Journal of Multivariate Analysis, 36, 121-126.

Kubokawa, T. (1994). "A Unified Approach to Improving Equivariant Estimators." Annals of Statistics, 22, 290-298.

Kubowaka, T., Saleh, A.K., Md, E. and Morita, K. (1992). Improving on MLE of Coefficient Matrix in a Growth Curve Model. Journal of Statistical Planning and Inference, 31, 169-177.

Kuks, J. and Olman, V. (1971). "Minimax Linear Estimation of Regression Coefficients." (Russian). Izvestija Akademii Nauk Estonkoi SSR, 20,480-482.

Kuks, J. and Olman, V.(1972). "Minimax Linear Estimation of Regression Coefficients II (Russian) ." Izvestija Akademii Nauk Estonkoi SSR, 21,66-72.

Kullback, S. (1959). Information Theory and Statistics. John Wiley, New York.

Kullback. S. and Liebler, R.A. (1951). "On Information Sufficiency." Annals of Mathematical Statistics, 22, 79-86.

Laird, N.M. and Louis, T.A. (1987). "Bootstrapping Empirical Bayes Estimators to Account for Sampling Variation (with Discussion)." Journal of the American Statistical Association , 82, 739-757.

Laird, N.M. and Louis, T.A. (1989). "Empirical Bayes Confidence Intervals for a Series of Related Experiments." Biometrics, 45, 481-495.

Lauter, H. (1975). "A Minimax Linear Estimator for Linear Parameters under Restrictions in form of Inequalities." Mathematiche Operationsforschung und Statistik, 6, 689-695.

Lauterbach, J. and Stahlecker, P. (1988). "Approximate Minimax Estimation in Linear Regression: A Simulation Study." Communications in Statistics, Simulations, 17(1), 209-227.

Lawless, J.F. and Wang, P. (1976). "A Simulation Study of Ridge and Other Regression Estimators." Communications in Statistics. Theory and Methods, 5, 307-323.

Le Cessie, S. and Van Houwelingen, J.C. (1992). "Ridge Estimators in Logistic Regression." Applied Statistics, 41(1), 191-201.

Lehmann E.L. (1959) Testing Statistical Hypothesis. John Wiley and Sons, New York.

Lewis, F.L. (1986). Optimal Estimation with an Introduction to Stochastic Control Theory. John Wiley and Sons, New York.

Liang, K.Y. and Liu, X.H. (1991). "Estimating Equations in Generalized Linear Models with Measurement Error." Estimating Functions edited by V.P. Godambe. Clarendon Press, Oxford.

Liang, K.Y. and Zeger, S.L. (1988). "On the Use of Concordant Pairs in Matched Case-Control Studies." Biometrics 44, 1145-1156.

Lindley, D.V. (1956). "On a Measure of Efficiency Provided by an Experiment." Annals of Mathematical Statistics, 27, 986-1005.

Lindley, D.V. (1962). "Contribution to Discussion on Paper by C. M. Stein." Journal of the Royal Statistical Society B, 35, 379-421.

Lindley, D.V. (1965).Introduction to Probability and Statistics from a Bayesian Viewpoint. Part 2. Inference. Cambridge at the University Press.

Lindley, D.V. and Smith, A.F.M. (1972). "Bayes Estimate for the Linear Model (with discussion) Part 1." Journal of the Royal Statistical Society B, 34, 1-41.

Liski, E.P. (1988). "A Test of the Mean Square Error Criterion for Linear Admissible Estimators." Communications in Statistics-Theory and Methods, 17(11), 3743--3756.

Liski, E.P. (1989). "Comparing Stochastically Restricted Linear Estimators in a Regression Model." Biometrics Journal, 31, 313-316.

Lowerre, J.M. (1974). "On the Mean Square Error of Parameter Estimates for Some Biased Estimators." Technometrics, 16(3), 461-464.

Lu, K. and Berger, J.O. (1989). "Estimated Confidence Procedures for Normal Means." Journal of Statistical Planning and Inference 23, 1-19.

Lu, W. (1993). "Estimations in the Normal regression Empirical Bayes Model." Communications in Statistics-Theory and Methods, 22(6), 1773-1794.

Mackinnon, M.J. and Puterman, M.L. (1989). "Collinearity in Generalized Linear Models." 18(9), 3463-3472.

Mann, H. B. and Wald, A. (1943). "On Stochastic Limit and Order Relationships." Annals of Mathematical Statistics, 14, 216-226.

Marchand, E. (1993). "Estimation of a Multivariate Mean with Constraints on the Norm." The Canadian Journal of Statistics, 21(4), 359-366.

Marchand, E. and Giri, N.C. (1993). "James-Stein Estimation with Constraints on the Norm." Communications is Statistics-Theory and Methods, 22(10) , 2903-2924.

Markiewicz, A. (1996). "Characterization of General Ridge Estimators." Statistics and Probability Letters , 27, 145-148.

Marquardt, D.W. (1970). "Generalized Inverses, Ridge Regression, Biased Linear Estimation, and Non-Linear Estimation." Technometrics, 12(3), 591-612.

Marquardt, D.W. and Snee, R.D. (1975). "Ridge Regression in Practice." The American Statistician , 29(1), 3-20.

Mason, R.L. and Blaylock, N.W. (1991). "Ridge Regression Estimator Comparisons using Pitman's Measure of Closeness." Communications   in Statistics–Theory and Methods, 20(11), 3629-3641.

Mason, R.L., Keating, J.P., Sen, P.K. and Blaylock, N.W. (1990). "Comparison of Linear Estimators using Pitman's Measure of Closeness." Journal of   the American Statistical Association, 85, 579-581.

Mayer, L.S. and Willke, T.A. (1973). "On Biased Estimation in Linear Models." Technometrics, 15, 497-508.

Mc Cullagh, P. and Nelder, J.A. (1983). Generalized Linear Models. Chapman and Hall, New York.

Mc Donald, G.C. and Galarneau, D.I. (1975). "A Monte Carlo Evaluation of Some Ridge Type Estimators." Journal of the American Statistical Association,   70, 407-416.

Mc Donald, G.C. and Schwing, R.C. (1973). "Instabilities of Regression Estimates Relating Air Pollution to Mortality." Technometrics, 15, 463-481.

Md, A.K., Saleh, E. and Han, C.(1990). "Shrinkage Estimation in Regression Analysis." Estadistica, 42(139), 40-63.

Meeden, G. (1972), "Some admissible empirical Bayes procedures." Annals of Mathematical Statistics , 43, 96-101.

Mehta, J.S. and Srinivasan, R. (1971). "Estimation of the Mean by Shrinkage to a Point." Journal of the American Statistical Association , 66, 86-90.

Meinhold, R.J. and Singpurwalla, N.D. (1983) "Understanding the Kalman Filter." The American Statistician, 37(2), 123-127.

Menjoge, S.S. (1981). "Some New Procedures in Simultaneous Estimation of Parameters their Compound and Individual Risks." Unpublished Ph.D. Thesis. Department of Statistics , University of Rochester, Rochester, New York.

Menjoge, S.S. (1983). "Improved Estimators with the Weighted and Compounded Loss Functions." Communications in Statistics- Theory and Methods, 12(2),   131-139.

Menjoge, S.S. (1984). "On Double k-Class Estimators of Coefficients in Linear Regression." Economics Letters, 15, 295-300.

Miller, A.J. (1990). Subset Selection in Regression. Chapman and Hall. New York.

Mittelhammer, R.C. (1984). "Restricted Least Squares, Pre Test, OLS and Stein Rule Estimators: Risk Comparisons under Model Misspecification." Journal of Econometrics, 25, 151-164.

Mood, A., Graybill, F. and Boes, D.C. (1974). Introduction to the Theory of Statistics. Third Edition .Mc Graw Hill, New York.

Morris, C.N. (1983) "Parametric Empirical Bayes Inference: Theory and Applications." Journal of the American Statistical Association, 78, 47-65.

Morrison, D.F. (1967). Multivariate Statistical Methods. McGraw Hill, New York.

Mundlak, Y. (1981). "On the Concept of Nonsignificant Functions and its Implications for Regression Analysis." Journal of Econometrics, 16, 139-150.

Nagata, Y. (1983). "Admissibility of Some Preliminary Test Estimators for the Mean of Normal Distribution." Annals of the Institute of Statistical Mathematics, 37, 365-373.

Nagata, Y. (1997). "Stein Type Confidence Interval of the Disturbance Variance in a Linear Regression Model with Multivariate Student-t Distributed Errors." Communications in Statistics–Theory and Methods, 26(2), 503-523.

Nebebe, F. and Sim, A. (1990). "The Relative Performances of Improved Ridge Estimators and an Empirical Bayes Estimator: Some Monte Carlo Results." Communications in Statistics-Theory and Methods, 19(9), 3469-3495.

Neter, J., Wasserman, W. and Kutner, M.A. (1985). Applied Linear Models. Richard D. Irvin, Homewood, Illinois.

Neymann, J. (1963). "Two Breakthroughs in the Theory of Statistical Decision Making." Estadistica Espanola, 18, 5-28.

Nickerson, D.M. (1988). "Dominance of the positive-part version of the James-Stein estimator." Statistics and Probability Letters, 7, 97-103.

Nieto, F.H. and Guerrero, V.M. (1995). "Kalman Filter for Singular and Conditional State-Space Models when the System State and the Observational Error are Correlated." Statistics and Probability Letters, 22, 303-310.

Nomura, M. (1988). "On the Almost Unbiased Ridge Regression Estimator." Communications in Statistics, Simulation, 17(3), 729-743.

Nomura, M, and Ohukubo, T. (1985). "A Note on Combining Ridge and Principal Component Regression." Communications in Statistics. Theory and Methods, 14, 2489–2493.

Noor, I. and Mehta, J.S. (1988). "Finite Sample Properties of Generalized Ridge-Type Shrinkage Estimator." Pakistan Journal of Statistics, 4(2)A, 115-127.

Noor, I. and Ahmad, N. (1995). "Exact and Approximate Relative Efficiency of Generalized Ridge-Type Shrinkage Estimator." Pakistan Journal of Statistics, 11(1), 17-22.

Noor, I. and Mehta, J.S. (1989). "A Study of Some Shrinkage Type Estimators." Communications in Statistics-Theory and Methods, 18(12), 4437-4457.

Nyquist, H. (1988). "Applications of the Jackknife Procedure in Ridge Regression." Computational Statistics and Data Analysis, 177 -183.

Obenchain, R.L. (1975). "Ridge Analysis Following a Preliminary Test of the Shrunken Hypothesis." Technometrics, 17, 431-435.

Obenchain, R.L. (1977). "Classical F-Tests and Confidence Regions for Ridge Regression." Technometrics, 19(4), 429-439.

Ohtani, K. (1986). "Distribution and Density Functions of the Feasible Generalized Ridge Regression Estimator." Communications in Statistics-Theory and Methods, 22, 2733-2746.

Ohatani, K. (1993). "A Comparison of the Stein Rule and the Positive Part Stein Rule Estimators in a Misspecified Linear Regression Model." Econometric Theory, 9, 668-679.

Ohatani, K. (1995). "Generalized Ridge Regression Estimators Under the LINEX Loss Function." Statistical Papers, 36, 99 - 110.

Ohtani, K. (1996a). "Exact Small Sample Properties of an Operational Variant of the Minimum Mean Squared Error Estimator." Communications in Statistics-Theory and Methods, 25(6), 1223-1231.

Ohtani, K. (1996b). "Further Improving the Stein Rule Estimator using the Stein Variance Estimator in a Misspecified Linear Regression Model." Statistics and Probability Letters, 29, 191-199.

Ohtani K. (1996c). "On an Adjustment of Degrees of Freedom in the Minimum Mean Squared Error Estimator." Communications in Statistics- Theory and Methods, 25(12), 3049-3058.

Ohtani, K, and Kozumi, H. (1996). "The Exact General Formulae for the Moments and the MSE Dominance of the Stein-Rule and Positive Part Stein-Rule Estimators." Journal of Econometrics, 74, 273-287.

Pal, N. and Chano, C.H. (1996). "Risk Analysis and Robustness of Four Shrinkage Estimators." Calcutta Statistical Association Bulletin, 44, 35-59.

Peddada, S.D., Nigam, A.K. and Saxena, A.K. (1989). "On the Inadmissibility of Ridge Estimator in a Linear Model." Communications in Statistics-Theory and Methods, 18(10), 3571-3585.

Peele, L. and Ryan, T.P. (1982). "Minimax Linear Estimators with Application to Ridge Regression." Technometrics, 24 (2), 157-159.

Pfeffermann, D. (1984). "On Extensions of the Gauss-Markov Theorem to the Case of Stochastic Regression Coefficients." Journal of the Royal Statistical Society, B, 46(1), 139-148.

Perron, F. (1993). " Estimation of a Mean Vector in a Two Sample Problem." Journal of Multivariate Analysis, 46, 254-261.

Phillips, P.C.B. (1984). "The Exact Distribution of the Stein-Rule Estimator." Journal of Econometrics, 25, 123-131.

Pitman, E.J.G. (1937). "The Closest Estimates of Statistical Parameters." Proceedings of the Cambridge Philosophical Society, 33, 212-222.

Pliskin, J.L. (1987). "A Ridge Type Estimator and Good Prior Means." Communications in Statistics-Theory and Methods, 16(12), 3429--3437.

Potthoff, R.F. and Roy, S.N. (1964). "A Generalized Multivariate Analysis of Variance Model Useful Especially for Growth Curve Problems." Biometrika 51, 313-326.

Precht, M. and Rao, P.S.S.N.V.P. (1985) "An Evaluation of Biased Estimators of Regression Coefficients-A Simulation Study-." Statistiche Hefte, 26, 263-285.

Press, S. J. and Rolph, J. E. (1986). "Empirical Bayes Estimation of the Mean in a Multivariate Normal Distribution." Communications in Statistics-Theory and Methods, 15, 2201- 228.

Price, M.J. (1982). "Comparisons Among Regression Estimators Under the Generalized MSE Criterion." Communications in Statistics-Theory and Methods, 11(17), 1965-1984.

Quenouille, M. (1956). "Notes on Bias in Estimation." Biometrika, 43, 353-360.

Rao, C.R.(1973). Linear Statistical Inference and its Applications, Second Edition. Wiley, London.

Rao, C.R.(1975). "Simultaneous Estimation of Parameters in Different Linear Models and Applications to Biometric Problems." Biometrics, 31, 545-554.

Rao, C.R. (1976a). "Estimation of Parameters in a Linear Model. The 1975 Wald Memorial Lectures." Annals of Statistics, 4(6), 1023-1037.

Rao, C.R. (1976b). "Characterization of Prior Distributions and Solution to a Compound Decision Problem." Annals of Statistics 4, 823-835.

Rao, C.R. (1977). "Simultaneous Estimation of Parameters, A Compound Decision Problem." Statistical Theory and Related Topics, S. Gupta and D.S. Moore ed. New York: Academic Press, 327-350.

Rao, C.R. (1994). "Some Statistical Problems in Multitarget Tracking." in Statistical Decision Theory and Related Topics. S.S. Gupta and J.O. Berger Eds. Springer Verlag, New York.

Rao, C.R. and Mitra, S.K. (1971). Generalized Inverses of Matrices and its Applications. John Wiley and Sons, New York.

Rao, C.R. and Shinozaki, N. (1978). "Precision of Individual Estimators in Simultaneous Estimation of Parameters." Biometrika, 65(1), 30-32.

Rao, C.R. and Toutenburg, H. (1995). Linear Models: Least Squares and Alternatives. Springer-Verlag, New York.

Ravishanker, N., Dey, D. and Wu, L. (1995). "Shrinkage Estimation in Time Series Using a Bootstrapped Covariance Estimate." The Journal of Statistical Computation and Simulation, 53, 259-267.

Ravishanker, N., Dey, D. and Wu, L. (1996). "Shrinkage Estimation of Contemporaneous Outliers in Concurrent Time Series." Communications in Statistics -Simulation, 25(3), 643-656.

Rayner, R.K. (1989). "Bootstrap Inversion of Edgeworth Expansions for Nonparametric Confidence Intervals." Statistics and Probability Letters, 8, 201-206.

Reinsel, G. (1984). "Estimation and Prediction in a Multivariate Random Effects Generalized Linear Model." Journal of the American Statistical Association, 79, 406-414.

Robbins, H. (1955). "An Empirical Bayes Approach to Statistics." Proceedings of the Third Berkeley Symposium on Mathematical Statistics and Probability, Berkeley: University of California Press, 1, 157-163.

Robbins, H. (1983). "Some Thoughts on Empirical Bayes estimation." Annals of Statistics, 11, 713-723.

Robert, C. (1988). "An Explicit Formula for the Risk of the Positive Part Estimator." Canadian Journal of Statistics, 16(2), 161-168.

Robert, C. (1994). The Bayesian Choice: A Decision-Theoretic Motivation. Springer-Verlag, New York.

Robert, C. and Casella, G. (1990). "Improved Confidence Sets for Spherically Symmetric Distributions." Journal of Multivariate Analysis, 32, 84-94.

Robert, C. and Saleh, A.K. Md.E. (1989). "Recentered Confidence Sets: A Review." MSI Technical Report, Cornell University, Ithaca, N.Y.

Rolph, J. E. (1976). "Choosing Shrinkage Estimators for Regression Problems." Communications in Statistics, A5, 789-802.

Rubin, D.B. (1980). "Using Empirical Bayes Techniques in the Law School Validity Studies." Journal of the American Statistical Association, 75, 801-816.

Rubinfeld, D. (1977). "Voting in Local School Elections: A Micro Analysis." Review of Economic Statistics, 59, 30-42.

Rukhin, A.L.(1995). "Admissibility: Survey of a Concept in Progress." International Statistical Review, 63(1), 95-115.

Rutherford, J. R. and Krutchkoff, R. G. (1967). "The Empirical Bayes Approach: Estimating the Prior Distribution." Biometrika, 54, 326-328.

Rutherford, J. R. and Krutchkoff, R. G. (1969a). "Some Empirical Bayes Techniques in Point Estimation." Biometrika , 56, 133-137.

Rutherford, J. R. and Krutchkoff, R. G. (1969b). "Epsilon Asymptotic Optimality of Empirical Bayes Estimators. Biometrika , 56, 220-233.

Sajjan, S.G., Basawa, I.V. (1996). "Empirical Bayes Prediction for a Mixed Linear Model with Autoregressive Errors." Statistics and Probability Letters, 29, 1-7.

Sarkar, N. (1989). "Comparisons among some Estimators in Misspecified Linear Models with Multicollinearity." Annals Institute of Statistical Mathematics, 41, 717-724.

Sarkar, N. (1996). "Mean Square Error Matrix Comparison of Some Estimators in Linear Regressions with Multicollinearity." Statistics and Probability Letters, 30, 133-138.

Saxena, A. (1984). "Bayesian Estimation of TG Ridge Model." Statistica Neerlandica, 38(4), 256-260.

Schipp, B., Trenkler, G., and Stahlecker, P. (1988). "Minimax Estimation with Additional Linear Restrictions-A Simulation Study." Communications in Statistics Simulation, 17(2), 393-406.

Schipp, B. and Toutenburg, H. (1996). "Feasible Minimax Estimators in the Simultaneous Equations Model Under Partial Restrictions." Journal of Statistical Planning and Inference, 50, 241-250.

Schmidt, P. (1976). Econometrics. Marcel Dekker, New York.

Sclove, S. L. (1968). "Improved Estimators for Coefficients in Linear Regression." Journal of The American Statistical Association, 63, 596-606.

Sclove, S.L.(1971). "Improved Estimation of Parameters in Multivariate Regression." Sankya A, 33, 61-67.

Sclove, S.L., Morris, C. and Radhakrishnan, R. (1972). "Non-Optimality of Preliminary-Test Estimators for the Mean of a Multivariate Normal Distribution." Annals of Mathematical Statistics, 43, 1481–1490.

Searle, S.R. (1971). Linear Models. John Wiley and Sons, Inc, New York.

Sen, P.K., Kubokawa, T. and Saleh, A.K. M.E. (1989). "The Stein Paradox in the Sense of the Pitman Measure of Closeness." The Annals of Statistics, 17(3), 1375-1386.

Sengupta, D. (1991). "On Shrinkage toward an Arbitrary Estimator." Statistics and Decisions, 9, 81-105.

Shannon, C. E. (1948). "A Mathematical Theory of Communication." Bell System Technical Journal, 27, 379-423.

Shao, P.Y.S. and Strawderman, W.E. (1994). "Improving on the James-Stein Positive Part Estimator." The Annals of Statistics, 22(3), 1517-1538.

Shao, P.Y.S. and Strawderman, W.E. (1995). Improving on the Positive Part of the UMVUE of a Noncentrality Parameter of a Noncentral Chi-Square Distribution." Journal of Multivariate Analysis, 52-66.

Shao, P.Y.S. and Strawderman, W.E. (1996). "Improving on the MLE of a Positive Normal Mean." Statistica Sinica, 6, 259-274.

Shiaishi, T. and Konno, Y. (1995). "On Construction of Improved Estimators in Multiple-Design Multivariate Linear Models Under General Restriction." Annals of Institute of Statistical Mathematics, 46(4), 665-674.

Shaefer, R. L. (1986). "Alternative Estimators in Logistic Regression when the Data are Collinear." Journal of Statistical Simulation, 25, 75-91.

Shaefer, R.L. , Roi, L.D. and Wolfe, R.A. (1984). "A Ridge Logistic Estimator." Communications in Statistics–Theory and Methods, 13, 99-113.

Shinozaki, N. (1974). " A Note on Estimating the Mean Vector of a Multivariate Normal Distribution with General Quadratic Loss Function." Keio Engineering Reports, 27(7), 105-112.

Shinozaki, N. (1989). "Improved Confidence Sets for the Mean of a Multivariate Normal Distribution." Annals of the Institute of Statistical Mathematics, 41(2), 331-346.

Shinozaki, N. and Chang, Y.T. (1993). "Minimaxity of Empirical Bayes Estimators of the Means of Independent Normal Variables with Unequal Variances." Communications in Statistics–Theory and Methods, 22(8), 2147-2169.

Shinozaki, N. and Chang, Y.T. (1996). "Minimaxity of Empirical Bayes Estimators Shrinking Toward the Grand Mean when Variances are Unequal." Communications in Statistics-Theory and Methods, 25(1), 183-199.

Silvapulle, M.J. (1991). "Robust Ridge Regression Based on an M Estimator." Australian Journal of Statistics, 33(3), 319-333.

Silvey, S.D.(1969). "Multicollinearity and Imprecise Estimation." Journal of the Royal Statistical Society, B, 31, 539-552.

Singh, B., Chaubey, Y.P. and Dwivedi, T.P. (1986). "An Almost Unbaised Ridge Estimator." Sankya: The Indian Journal of Statistics, B48, 342-346.

Singh, R.K. ,Pandey, S.K., Srivastava, V.K. (1994). "A Generalized Class of Shrinkage Estimators in Linear Regression when Disturbances are not Normal." Communications in Statistics-Theory and Methods, 23(7), 2029-2046.

Singh, R.S. (1977). "Improvement on Some Known Nonparametric Uniformly Consistent Estimators of Derivative of a Density." Annals of Statistics ,5, 394-399.

Singh, R.S. (1979). "Empirical Bayes Estimation in Lesbegue-Exponential Families with Rates Near the Best Possible Rate." Annals of Statistics, 7, 890-902.

Sommers, R.W. (1964). Sound Application of Regression Analysis in Chemical Engineering. Presented at the American Institute of Chemical Engineers Symposium on Avoiding Pitfalls in Engineering Applications of Statistical Methods, Memphis, Tennessee.

Soofi, E.S. (1990). "Effects of Collinearity on Information about Regression Coefficients." Journal of Econometrics, 43, 255 -274.

Srivastava. A. K. (1996). "Estimation of Linear Regression Model with Rank Deficient Observations Matrix under Linear Restrictions." Microelectronics and Reliability, 36(1), 109-110.

Srivastava, A, K. and Shulka, Praveen (1996). "Minimax Estimation in a Linear Model." Journal of Statistical Planning and Inference 50, 77-89.

Srivastava, M.S. and Bilodeau, M. (1989). "Stein Estimation under Elliptical Distributions." Journal of Multivariate Analysis. 28, 247-259.

Srivastava,V.K. (1970). "The Efficiency of Estimating Seemingly Unrelated Regression Equations." Annals of the Institute of Statistical Mathematics, 22,483-493.

Srivastava, V.K. and Agnihotri, B.S. (1980). "Estimation of Regression Models under Linear Restrictions." Biometrics Journal, 22(3), 279-280.

Srivastava, V.K. and Chaturvedi, A. (1986). "A Necessary and Sufficient Condition for the Dominance of an Improved Family of Estimators in Linear Regression Models." Economic Letters, 20, 345-349.

Srivastava,V.K. and Giles, D.E.A. (1987).Seemingly Unrelated Regression Equations Models: Estimation and Inference. Marcel Dekker Inc., New York.

Srivastava, V.K. and Giles, D.E.A. (1991). "Unbiased Estimation of the Mean Squared Error of the Feasible Generalised Ridge Regression Estimator." Communications in Statistics-Theory and Methods, 20(8), 2375-2386.

Srivastava,V.K. and Ullah, A. (1995). " Stein Rule Estimation in Models with a Lagged Dependent Variable." Communications in Statistics–Theory and Methods, 24(5), 1343-1353.

Srivastava,V.K. and Upadhyaya,S. (1978). "Large Sample Approximations in Seemingly Unrelated Regression Equations." Annals of the Institute of Statistical Mathematics, 30, 89-96.

Stahlecker, P., and Trenkler, G. (1988). "Full and Partial Minimax Estimation in Regression Analysis with Additional Linear Constraints." Linear Algebra and its Applications, 111, 279-292.

Steece, B.M. (1989). "Leverage in Bayesian Regression." Biometrics Journal, 7, 811-819.

Stein, C. (1956). "Inadmissibility of the Usual Estimator for the Mean of a Multivariate Normal Distribution." Proceedings of the Third Berkeley Symposium on Mathematics, Statistics and Probability. Berkeley: University of California Press, 197-206.

Stein, C. (1962). "Confidence Sets for the Mean of A Multivariate Normal Distribution." Journal of the Royal Statistical Society , Series B, 24, 265-296.

Stein, C. (1964). "Inadmissibility of the Usual Estimator for the Variance of a Normal Distribution with Unknown Mean." Annals Institute of Statistical Mathematics ,16, 155-160.

Stein, C. (1966). "An Approach to the Recovery of Inter-block Information on Balanced Incomplete Block Designs." Research Papers in Statistics edited by F.N. David. John Wiley and Sons, New York, 351-366.

Stein, C. (1981). "Estimation of the Mean of a Multivariate Normal Distribution." Annals of Statistics , 9, 1135-1151.

Stigler, S.M. (1990). "The 1988 Neyman Memorial Lecture: A Galtonian Perspective on Shrinkage Estimators." Statistical Science, 5(1), 147-155.

Strang, G. (1980). Linear Algebra and its Applications, 2nd edition, Academic Press. New York.

Strang, G. (1993). "The Fundamental Theorem of Linear Algebra." American Mathematical Monthly, 100(9), 848-859.

Strawderman, W.E. (1973). "Proper Bayes Minimax Estimators of the Multivariate Normal Mean Vector for the Case of Common Unknown Variances." Annals of Statistics, 1, 1189-1194.

Strawderman, W.E. (1978). "Minimax Adaptive Generalized Ridge Regression Estimators." Journal of the American Statistical Association, 73, 623-627.

Sugiura, N, and Takagi, Y. (1996). "Dominating James-Stein Positive Part Estimator for Normal Mean with Unknown Covariance Matrix." Communications in Statistics–Theory and Methods, 25(12), 2875-2900.

Sun, L. (1992). The Bayes Estimation Procedures for One and Two Way Hierarchical Models. Ph.D. Dissertation. Department of Statistics. University of Toronto.

Sun, L. (1995). "Risk Ratio and Minimaxity in Estimating the Multivariate Mean with Unknown Variance." Scandinavian Journal of Statistics, 22, 105-120.

Sun, L. (1996). "Shrinkage Estimation in the Two-Way Multivariate Normal Model." The Annals of Statistics, 24(2), 825-840.

Swamy, P.V.A.B., Mehta, J.S., Thurman, S.S., and Iyengar, N.S. (1985). "A Generalized Multicollinearity Index for Regression Analysis." Sankya: The Indian Journal of Statistics, B47(3), 401-431.

Swindel, B.F.(1976). "Good Ridge Estimators Based on Prior Information." Communications in Statistics-Theory and Methods, A5(11), 1065-1075.

Takada, H., Ullah, A. and Chen, Y.M. (1995) . "Estimation of the Seemingly Unrelated Regression Model when the Error Covariance Matrix is Singular." Journal of Applied Statistics , 22(4), 517-530.

Tamarkin, M. (1982). "A Simulation Study of the Stochastic Ridge k" Communications in Statistics -Simulation and Computation, 11(2), 159-173.

Tan, M. (1991). "Improved Estimators for the GMANOVA Problem with Application to Monte Carlo Simulation." Journal of Multivariate Analysis, 38, 262-274.

Terasvirta, T.(1981). "A Comparison of Mixed and Minimax Estimators of Linear Models." Communications in Statistics–Theory and Methods, 10(17), 1765-1778.

Terasvirta, T. (1986a). "Superiority Comparisons of Heterogeneous Linear Estimators." Communications in Statistics---Theory and Methods, 15(4), 1319-1336.

Terasvirta, T. (1986b). "Superiority Comparisons between Mixed Regression Estimators." Communications Statistics–Theory and Methods, 17(10), 3537-3546.

Theil, H.(1971). Principles of Econometrics. Wiley, New York.

Theil, H. and Goldberger, A.S. (1961). "On Pure and Mixed Estimation in Economics." International Economic Review 2, 65-78.

Theobald, C.M. (1974) . "Generalizations of Mean Square Error Applied to Ridge Regression." Journal of the Royal Statistical Society , B.36, 103-106.

Thisted, R.A. (1976). Ridge Regression, Minimax Estimation and Empirical Bayes Methods." Division of Biostatistics, Stanford University, Technical Report 28.

Tong, H.Q. (1996). "Convergence Rates for Empirical Bayes Estimators of Parameters in Multi-Parameter Exponential Families." Communications in Statistics 25(5).

Tong, H.Q. (1996). "Convergence Rates for Empirical Bayes Estimators of Parameters In Linear Regression Models." Communications in Statistics-Theory and Methods, 25(6), 1325-1334.

Toutenburg, H. (1982). Prior Information in Linear Models. John Wiley and Sons, New York.

Toutenburg, H. (1986). Weighted Mixed Regression with Application to Regressor's Nonresponse. 1. Theoretical Results. Preprint. Akademie der Wissenschaten der DDR Karl Weierstrass, Institut fur Mathematik, Berlin.

Toutenburg, H. (1988). "MSE Comparisons Between Restricted Least Squares, Mixed and Weighted Mixed Estimators with Special Emphasis to Nested Restrictions." Technical Report. Akademie der Wissenschaten der DDR Karl Weirstrasse Institut for Mathematik, Berlin.

Tracy, D.S. and Srivastava, A.K. (1992). "Selection of Biasing Parameters in Adaptive Ridge Regression Estimators." Econometric Reviews, 11(3), 367-377.

Trenkler, G. (1985). "Mean Square Error Comparisons of Estimators in Linear Regression." Communication in Statistics–Theory and Methods ,14, 2495-2509.

Trenkler, G. (1988). "Some Remarks on a Ridge Type Estimator and Good Prior Means." Communications in Statistics--Theory and Methods., 17(12), 4251-4256.

Trenkler, D. and Trenkler, G. (1981). "Ein Vergleich des Kleinste-Quadrate-Schatzers mit verzerrten Alternativen." In: Fleischman, B., Bloech, J., Fandel, G., Seifert, O., Weber, H. (eds.) Operations Research Proceedings, 1981, 218-227.

Tse, S.K. and Tso, G. (1996). "Shrinkage Estimation of Reliability for Exponentially Distributed Lifetimes." Communications in Statistics-Simulation , 25(2), 415-430.

Tukey, J.W. (1975). "Instead of Gauss Markov Least Squares, What?" Applied Statistics. R.P. Gupta (editor). North Holland Publishing Company, Amsterdam.

Ullah, A., Carter, R.A. L. and Srivastava, V.K. (1984). "The Sampling Distribution of Shrinkage Estimators and their F-Ratios in the Regression Model." Journal of Econometrics, 25, 109-122.

Ullah, A., Srivistava, V.K. and Chandra, R. (1983). "Properties of Shrinkage Estimators in Linear Regression when Disturbances are not Normal." Journal of Econometrics, 21, 389-402.

Ullah, A. and Ullah, S. (1978). "Double k-Class Estimators of Coefficients in Linear Regression." Econometrica, 46 (3), 705-722.

Ullah, A. and Vinod, H. D. (1984). "Improvement Ranges for Shrinkage Estimators with Stochastic Target." Communications In Statistics–Theory and Methods, 13(2), 207–215.

Van Loan, C.F. (1976). "Generalizing the Singular Value Decomposition." SIAM Journal of Numerical Analysis, 13, 76-83.

Varian, H.R. (1975). "A Bayesian Approach to Real Estate Assessment." in Studies in Bayesian Econometrics and Statistics in Honor of Leonard J. Savage, eds. Stephen E. Feinberg and Arnold Zellner. North Holland. Amsterdam.

Venter, J.H. and Steel, S.J. (1994). "Pre-Test Type Estimators for Selection of Simple Normal Models." Journal of Statistical Computation and Simulation, 51, 31-48.

Vinod, H.D. (1976). "Simulation and Extension of a Minimum Mean Squared Error Estimator in Comparison with Stein's." Technometrics, 18(4), 491-496.

Vinod, H.D. (1978). "A Survey for Ridge Regression and Related Techniques for Improvements Over Ordinary Least Squares." The Review of Economics and Statistics, 60, 121-131.

Vinod, H. D. (1980). "Improved Stein-Rule Estimator for Regression Problems." Journal of Econometrics, 12, 143-150.

Vinod, H.D. (1993). "Bootstrap, Jackknife Resampling and Simulation Methods: Applications in Econometrics." G.S. Maddala, C.R. Rao and H.D. Vinod editors. Handbook of Statistics :Econometrics Volume 11, North Holland Elsevier New York, 629-661.

Vinod, H.D. (1995). "Double Bootstrap for Shrinkage Estimators." Journal of Econometrics, 68, 287-302.

Vinod, H. D. and Raj, B. (1988). "Economic Issues in Bell System Divestiture: A Bootstrap Application." Applied Statistics , 37, 251-261.

Vinod, H.D. and Srivastava, V.K.(1995). "Large Sample Asymptotic Properties of the Double k-Class Estimators in Linear Regression Models." Econometric Reviews, 14(1), 75-100.

Vinod, H.D. and Ullah, A. (1981). Recent Advances in Regression Methods. Marcel Dekker, New York.

Vinod, H.D., Ullah, A. and Kadiyala, K. (1981). "Evaluation of Mean Squared Error of Certain Ridge Estimators Using Confluent Hypergeometric Functions." Sankya: The Indian Journal of Statistics: Series B, 43(3), 360-383.

Waikar, V.B. and Katti, S.K. (1971) "On a Two-Stage Estimate of the Mean." Journal of the American Statistical Association, 66, 75-81.

Wang, S., Tse, S., Chow, S. (1990). "On the Measures of Multicollinearity in Least Squares Regression." Statistics and Probability Letters, 9, 347-355.

Wax, Y. and Haitovsky, Y. (1980). "Generalized Ridge Regression, Least Square with Stochastic Prior Information and Bayesian Estimators." Applied Mathematics and Computation 7, 125-154.

Wei, L. and Trenkler, G. (1995). "Mean Square Error Matrix Superiority of Empirical Bayes Estimators under Misspecification." Test, 4(1), 187-205.

Weisberg, S. (1980). Applied Linear Regression. John Wiley and Son, New York.

Whittemore, A.S. (1989). "Errors in Variables Regression using Stein Estimates." The American Statistician., 43(4), 226-228.

Wichern, D.A. and Churchill, G.A. (1978). "A Comparison of Ridge Estimators." Technometrics, 20(3), 301-311.

Willan, A.R. and Watts, D.G. (1978). "Meaningful Multicollinearity Measures." Technometrics, 20, 407-412.

Wind, S.L. (1972). "Stein-James Estimators of a Multivariate Location Parameter." Annals of Mathematical Statistics, 43, 340-343.

Wind, S.L. (1973). "An Empirical Bayes Approach to Multiple Linear Regression." Annals of Statistics, 1, 93-103.

Withers, C.S. (1990). "Shrinkage Estimates based on Orthogonal Decomposition of the Sample Space." Communications in Statistics-Theory and Methods, 19(2), 505-526.

Wolfram, S. (1991). Mathematica, A System for Doing Mathematics by Computer, Second Edition. Addison Wesley, New York.

Yi, G. (1991). "Estimating the Variability of the Stein Estimator by Bootstrap." Economic Letters, 37, 293-298.

Zellner, A. (1962). "An Efficient Method of Estimating Seemingly Unrelated Regression Equations and Tests for Aggregation Bias." Journal of the American Statistical Association 57, 348-368.

Zellner, A. (1963). "Estimators for Seemingly Unrelated Regression Equations: Some Exact Finite Sample Results." Journal of the American Statistical Association, 58, 977-992.

Zellner, A. (1986a). "Bayesian Estimation and Prediction using Asymmetric Loss Functions." Journal of the American Statistical Association, 81, 446-451.

Zellner, A. (1986b). " On Assessing Prior Distributions and Bayesian Regression Analysis with g-Prior Distributions." in P Goel and A. Zellner (eds.), Bayesian Inference and Decision Techniques Essays in Honor of Bruno de Finetti, North Holland Amsterdam.

Zellner, A. (1994). "Bayesian and Non-Bayesian Estimation Using Balanced Loss Functions." in Statistical Decision Theory and Related Topics V, S.S. Gupta and J.O. Berger eds., 377-390.

Zellner, A. and Rossi, P.E. (1984). "Bayesian Analysis of Dichotomous Quantal Response Models," Journal of Econometrics, 25, 365-393.

Zellner, A. and Vandaele, W.(1974). "Bayes Stein Estimators for K Means, Regression and Simultaneous Equations Models." S.E. Feinberg and A. Zellner editors. Studies in Bayesian Econometrics and Statistics, Amsterdam, North Holland, 628-653.

Zontek, S. (1987). "On Characterization of Linear Admissible Estimators of a Result Due to C.R. Rao." Journal of Multivariate Analysis, 23, 1-12.

Author Index

Adkins, L. C., 41, 52, 54
Agnihotri, B.S., 24, 33
Ahmad, N., 51
Akdeniz, F., 48
Alam, K, 47
Allen, D.M., 153
Ali, M.A., 33, 39
Anderson, T.W., 424
Andrews, D. F., 44, 182

Baksalary, J.K., 33, 43, 47, 53, 239
Baldwin, K.F., 20, 49, 144,155,156,
    159, 160, 162, 164, 397
Banerjee, K.S. , 17
Baranchik, A.J., 16, 17, 18, 34, 42, 45
Barry, D., 50
Bar-Shalom, Y., 33
Basawa, I.V., 56
Baye,M.R., 58
Becker, R. A., 59
Belsey, D.A., 40
Beran, R., 52, 53, 112
Berger, J.O., 19, 23, 28, 32, 38, 48,
    168, 176, 203

Berliner, L.M., 28
Berndt, E., 55
Berry, J.C., 57
Bilodeau, M., 32, 34, 54
Birkes, D., 42
Blaylock, N.W., 35
Bock, M.E., 19, 22, 23, 25, 37, 43,
    48, 307, 338, 339
Boes, D.C., 83
Brandwein, A.C., 35, 38, 40, 43, 86
Breiman, L., 50
Brown, K.G, 22, 45
Brown, L., 45
Brown, P.J., 44
Brownstone, D., 37
Bucy, R.S., 556
Bunke, O., 29, 59

Campbell, S.L., 168
Carter, R.A.L., 18, 22, 27, 37, 46, 53,
    54
Casella, G., 27, 36, 45, 46
Cellier, D., 32, 48, 59
Chalton, D.0., 40

Chandra, R., 26
Chang,Y.T., 49, 54
Chano, C.H., 60
Chatterjee, S., 155, 165
Chaturvedi, A., 44
Chaubey, Y.P., 30, 298
Chawla, J.S, 30
Chen, J., 36, 46, 50
Chi, X.W., 42
Churchill, ,G.A., 145, 163
Copas, J.B., 29
Craig, A.T., 83, 85
Craven, P., 51
Cressie, N., 56
Crivelli, A., 49
Crouse, R.H., 49, 59
Czitrom, V., 31

Daniel, C., 32
Delaney, N.J. , 155, 165
Dempster, A.., 18, 20, 69, 168, 372,
        388
Dey, D., 45, 51, 57
Dhillon, U.S., 55
Dodge, Y., 42
Draper, N.R., 23, 166
Drygas, I.I., 53
Duncan, D.B., 18
Dwivedi, T.P., 30, 299

Edlund, P., 31
Efron, B., 16, 17, 18, 21, 22, 23, 24,
Efron, 34, 52, 103, 154, 155, 168,
        380, 432, 480 , 581
Efroymson, M.A., 166
Evans, S.N., 57

Farebrother, R.W., 20, 21, 24, 27, 61,
        116, 117, 123, 270, 273, 284
Fay, R.E., 23, 54
Feurguson, T.S., 216

Firinguetti, L., 33, 37
Flack, V.F., 32
Fomby, T.B., 27, 39, 41
Fortmann, T.E., 33
Fourdrinier, D., 47, 49, 59
Freedman, D.A., 39
Fule, E., 52

Galerneau, D.I., 156, 163
Gana, R., 49
George, E.I., 28, 35, 38, 40, 45, 59
Ghosh, J.K., 55
Ghosh, M., 31, 38, 42, 44, 45
Gibbons, D.G., 163
Giles, D.E.A., 38, 42, 58, 539, 540
Giles, J.A., 38, 42, 58
Girard, D.A., 39
Giri, N.C., 43, 56
Gnot, S., 49
Godambe, V.P., 39, 46, 585
Goldberger, A.S., 17, 27
Goldstein, M., 19
Golub, G.H., 51, 54, 155, 163
Gong, G., 154
Goutis, C., 57
Graybill, F., 83, 378
Greene, W., 55
Grewal, M.S., 44
Griffin, B.S., 15
Gross, J., 3, 10, 12, 53, 59, 67
Gruber, M.H.J., 3, 23, 25, 26, 27, 28,
        30, 34, 35, 41, 47, 56, 68, 123,
        168, 170, 172, 173, 196, 227,
        228, 229, 232, 284, 300, 374,
        556, 564, 573, 585
Guerrero, V.M., 46
Guo, Y.Y., 40

Hadzivukovic, S., 41, 156
Haff, L.R., 22, 36
Han, C., 34, 49

Harewood, S.L., 41
Harville, D.H., 18
Hasegawa, H., 48
Hawkins, D.L., 34
He'bel, P., 42
Heath, M., 51, 155, 156, 163
Heiligers, B., 54
Hemmerle, W.J., 32
Herriot, R.A., 23, 54
Hill, B.M., 45
Hill, R.C., 27, 37, 42, 52, 54
Hocking, R.R., 61
Hoerl, A. E., 2, 7, 9, 13, 16, 17, 18,
    20, 24, 25, 29, 32, 35, 37, 41,
    44, 47, 49, 53, 58, 68, 112,
    115, 118, 145, 148, 156, 157,
    160, 244, 372 386, 396, 397,
    580, 581, 582
Hoffmann, K., 47
Hogg, R.V., 83, 85
Holzworth, R.J., 59
Horn, S.D., 18
Hosmane, B.S., 33
Huang, H. C., 56
Hudson, H.M., 156
Humak, K.M.S., 29
Hwang, H.C., 36, 45, 46, 56

Ignaki, N., 48

James C., 1, 15, 17, 23, 69, 72, 79, 93,
    581
Jin, C., 49
Johnson, S.R., 27
Jones, M.C., 29
Judge, G.G., 22, 24, 29, 37, 42, 48,
    307, 338, 339

Kacinralar, S., 48
Kadane, J.B., 33, 45
Kadiyala, K., 25, 30, 37

Kagan, A.M., 174, 254
Kala, R., 53, 241
Kale, B.K., 39
Kalman, D., 3, 26, 33, 44, 46, 54, 56,
    69
Kalman, R.E., 556
Kariya, T., 34, 54
Karr, R.N, 17
Katti, S.J., 49
Keating, J.P., 31, 33
Kejian, L., 43, 48
Kempthorne, P.J., 30
Kennard, R.W., 2, 7, 9, 13, 16, 17, 18,
    20, 24, 25, 29, 32, 35, 37, 41,
    44, 47, 49, 53, 58, 68, 115,
    116, 118, 145, 148, 156, 157,
    160, 244, 372, 286, 396, 397,
    580, 581, 582
Ki, F., 36
Ki, Y.F., 40
Kim, M., 52
Kleffe, J., 40
Konno, Y., 51, 54
Krutchoff, R.G., 15
Kshirsagar, A.M., 500
Kubokawa, T., 38, 40, 51, 54, 57, 60
Kuks, J., 47, 59
Kullback, S., 37

Laird, N.M., 34
Lauter, H., 29, 59
Lauterbach, J., 29
Lawless, J. F., 145, 156, 161
Le Cessie, S., 41
Lehmann, E.L., 83
Lewis, F.L., 573
Liang, K.Y., 31, 39, 40
Liebler, R.A., 37
Lindley, D.V., 20, 22, 37, 40, 47, 83,
    156, 163, 378
Linnik, Y.V., 174, 254

Liski, E.P., 30, 33
Liu, X.H., 39, 40, 48
Louis, T.A., 34
Lowerre, J.M., 19
Lu, W., 43

Mackinnon, M.J., 32
Macullagh, P, 32
Mann, H.B., 526
Marchand, E., 43
Markiewicz, A., 47, 53, 54
Marquardt, D.W., 17, 20, 41, 156
Mason, R.L., 31.,33, 35
Mayer, L.S., 18, 115
Mc Donald, G.C., 37, 156
Md, A.K., 36
Meeden, G., 17
Mehta, J.S., 32, 51
Meinhold, R.J., 26, 556
Menjoge, S.S., 25, 53
Meyer, C.D., 168
Miller, A.J., 37
Mitra, S.K., 168
Mood, A., 83
Morrison, D.F., 495
Morris, C., 16, 17, 18, 21, 22, 23, 24,
    34, 45, 103, 168, 380, 432,
    436, 480
Mukerjee, R., 55
Mundlak, Y., 37

Nagata, Y., 48, 61
Nebebe, F., 36
Nelder, J.A., 32
Neter, J., 41
Neymann , J., 15
Nickerson, D.M, 30
Nieto, F.H, 46
Nomura, M., 30, 58
Noor, I., 32, 51
Nyquist, H., 30

Obenchain, R.L., 21, 32, 156, 163,
    164
Ohkubo, T., 58
Ohtani, K., 48, 51, 56, 58, 60, 61
Olman,V., 47, 59

Pal, N., 40, 60
Parker, D.F., 58
Peddada, S.D., 31
Peele,L., 26
Peffermann,,D., 18
Perron, F., 44
Phillips P.C.B, 60
Pitman, E.J.G., 31
Pliskin, J.L., 29
Pordzik, P.R., 43
Precht, M., 28
Press ,S.J., 3, 28
Price, M.J., 26
Puterman, M.L., 32

Quenouille, M., 30, 298

Raj, H.B., 52
Ralescu, S., 40
Rao, C.R. , 17, 20, 21, 22, 25, 27, 46,
    47, 50, 59, 63, 69, 111, 114,
    168, 173, 174, 207, 212, 235,
    251, 254, 257, 442, 581, 588
Rao, J.N.K. , 40
Rao, P.S.R.S., 25
Rao, P.S.S.N.V.P. , 28
Ravishanker, N., 51, 57
Rayner, R.K., 49
Reinsel, G., 27
Robbins, H., 5, 15, 17, 26, 62
Robert, C., 30, 36, 45, 47, 59
Roi, L.D., 52
Rolph, J.E., 18, 22, 28, 54, 156
Rubenfield, D., 55

Rubin, D.B., 23
Rukhin, A.L., 46
Rutherford, J.R., 15
Ryan, T.P., 26

Sajjan, S.G., 56
Saleh, A.K., 31
Samanta, S.K., 39
Sarkar, N., 58
Saxena, A., 27
Shao, P.Y.S., 308, 364, 365
Schatzoff, M., 20, 156,162, 164
Schipp, B., 29, 55
Schwing, R.C., 37, 59
Sclove, S.B., 15, 16, 17, 492
Searle, S.R., 95, 168, 170
Sen, P.K., 31, 33, 38, 55
Shannon, C.E., 37
Shao, P.Y.S., 45, 60, 69, 308, 364, 365 586
Sheafer, R.L., 52
Shiaishi, T., 51
Shieh, G., 38
Shinozaki, N., 22, 25, 34, 54, 207, 212, 442
Shulka, G., 44, 55
Silvapulle, M.J., 38
Silvey, S.D., 16
Singh, B., 302
Singh, R.K, 30, 45, 56
Singpurwalla, N.D., 26, 556
Sinha, B. K., 44
Smith, A.F.M., 19, 156, 163, 164
Snee, R.D., 20, 156
Sommers, R.W., 115
Soofi, E.S., 37
Srinivasan, R., 32
Srivatava, A.K., 42, 55
Srivastava, M.S., 32
Srivistava, V.K., 24, 26, 27, 33, 38, 47, 52, 53, 526, 539, 540

Stahlecker, P., 29
Stark, P.B., 57
Steece, B.M., 34
Steel, S.J., 45
Stein, C., 1, 2, 3, 4, 5, 6, 15, 16, 17, 18, 22, 23, 24, 26, 27, 28, 29, 30, 31, 33, 34, 35, 37, 38, 39, 40, 41, 42, 44, 45, 46, 47, 52, 53, 54, 55, 56, 57, 60, 61, 68, 69, 72, 79, 84, 86, 93, 581
Stigler, S.M., 35, 84, 88
Strang, G., 54
Strawderman, W.E., 25, 34, 35, 38, 40, 42, 43, 45, 54, 59, 60, 69, 86, 308, 364, 365, 581
Sugiura, N., 60
Sun, L., 47, 57
Swamy, P.V.A.B., 27
Swindel, B.F., 21, 29, 156

Takada, H., 50
Takagi, Y., 60
Tamarkin, M., 165
Terasvirta, T., 24, 28, 30
Theil, H., 17, 27, 61
Thisted,R., 156
Theobald, C.M., 128
Tong, H.Q. , 56
Toutenburg, H., 26, 28, 29, 50, 55, 59, 123, 581
Tracy, D.S., 42
Trenkler, D., 24
Trenkler, G., 24, 29, 51
Troskie, C.G., 40
Tse, S.K., 55
Tso, G., 55
Tukey, J..W., 19

Ullah, A., 22, 24, 25, 26, 27, 46, 47, 49, 53, 168, 178, 581
Ullah, S., 22, 53

Van Hoa, T., 44
Van Houwenlinging, J.C., 41
Van Loan, C.F., 34, 54
Van Nostrand, R.C., 23
Vandaele, W., 18, 19, 25, 174
Varian, H.R., 28
Venter, J.H., 45
Vinod, H.D., 21, 22, 24, 25, 27, 52,
　　113, 135, 156, 176, 178, 581

Wahba, G., 51, 155, 156, 163
Waikar, V.B., 49
Wald, A., 526
Wang, S., 25, 35, 145, 156, 161
Watts, D.G., 32
Wei, L., 51, 59
Wells, M.T., 49
Wermuth, N., 20, 156,162, 164
Whittemore, A.S., 40
Wichern, D.A., 145, 163
Willan, A.R., 32
Willke, T.A., 18, 114
Wind, S., 17, 18, 56, 69, 371, 372
Withers, C.S., 36
Wolfe, R.A., 52
Wood, F., 32
Wu, L., 51, 57

Zeger, K.Y., 31
Zellner, A., 18, 19, 25, 28, 34, 45, 51,
　　54, 58, 174, 533
Zmyslony, R., 49
Zontek, S., 20

Subject Index

Admissible estimator, 46
  definition of, 83
Admissible linear estimators, 20
Aiken estimator, 535
Alternative forms of the Bayes
  Estimator, 254 -256
Alternative Wind estimator, 504
Analysis of Variance, 61, 287, 374
Applications to
  aerospace tracking, 556
  air pollution, 59
  batting averages, 21
  breakup of Bell system
  Brownian motion, 57
  car price date, 59
  characteristics of rivets coming off
    an assembly line, 501
  DNA , 41
  ecological inference, 52
  evaluation of student performance,
    494
  fire alarm data, 19, 22
  gross national product, 34, 245

highway accidents, 59
import activity in the French
  economy, 43
insurance availability, 182
investments, 34, 533
labor force participation, 54
measurement of rivet diameters,
  406
mortgages, 55
navigation, 573
optimal control theory, 556
photography, 374
quality control, 556, 563
replicated industrial experiment,
  374
satellite tracking, 572
short term forecasting, 556
signal processing, 556
small area estimation, 23
standardized test scores, 23
superpopulation model, 20
survival of patients with ovarian
  cancer, 41

telephone companies, 52
Toxoplasmolosis, 21
underwater sonar, 556
voting decisions, 54
weather, 7, 56, 580
Approximate mean square error, 525,
    581
Approximate minimum mean square
    error estimators , 19, 186, 206,
    382,400, 505, 581, 583
Asymptotic approximation, 539
Asymptotic optimality (AO), 17
Asymptotic Unbiasedness, 71, 539
Augmented linear model, 242, 563
Average Mean Square Error, 190
    of empirical Bayes estimator of
        C.R. Rao, 377
    of empirical Bayes estimator of
        Dempster, 190, 409, 498,
        581
    of empirical Bayes estimator of
        Wind, 390
    of limited translation estimator,
        482
    of multivariate Bayes estimator,
        496, 498
    of positive part estimator, 312

Balanced loss function, 58
Basic estimable functions, 287
Bayes compromise problems, 30
Bayes estimator, 2, 61, 173, 496, 567,
    584
Bayesian point of view, 22, 173, 418
Best linear unbiased estimator
    (BLUE), 233, 235
Best   linear   unbiased   predictor
    (BLUP), 40, 56
Biased estimators, 24, 50
Bivariate linear model, 494
Bootstrap, 41, 49

Box Jenkins AR(1) model., 44

Classical confidence intervals, 34
Classical positive part, 307, 308, 310,
    316,320, 370
$(CMSE)_1$, 458, 587
    of Dempster estimator, 469
    of Wind Estimator, 462
$(CMSE)_r$, 458, 587
Colored noise, 561
Compact set, 436
Comparison of the conditional and
    average MSE, 455, 457
Comparison of the IMSE of the
    James-Stein estimator with that of
        the Least Square estimator,
        449, 453
Comparison of the MSE
    by computer simulation, 159-164
    of mixed estimators, 281
    of ridge and least square
        estimators, 125
    and the Relative Savings Loss, 330
    of two Bayes estimators averaging
        over their respective   priors,
        273, 283, 289
    of two ridge estimators, 274
Complete class theorems, 46
Complete Estimator, 83, 384
Compound risk, 442 ,485, 587
Computer intensive methods of
    choosing the perturbation factor k
    Allen's Press, 153
    bootstrap, 153, 164
    double bootstrap, 156
Condition number, 35, 37
Computer Programs
    Maple, 224, 324, 335, 349 ,362,
        429, 446, 452, 457, 467, 474,
        488, 543

Mathematica, 223, 324, 335, 349, 361, 428, 446, 451, 457, 467, 473, 482, 488, 543

TI 82 , 225, 429, 544

TI 92 , 226, 430,452 , 457, 468, 474, 489, 544

Computer Programs to obtain

Bayes risk of positive parts, 324, 335

frequentist risk of the BE and JS , 223-224, 428-430

frequentist risk and the relative loss of the positive parts, 349, 361

Computer Simulation, 158, 582

Conditional mean square error, 195, 509, 587

Contemporaneous correlation, 533, 589

Contraction estimator, 18, 114, 264, 270

Convex set, 436

Correlation coefficient, 495

Courant-Fisher minimax theorem, 205

Cross validation, 41, 50

Crossing points

of James-Stein and Bayes estimators, 426, 584

of positive parts, 348, 360

Definition of

estimable parametric function, 171

generalized inverse, 169

Moore Penrose inverse, 170

singular value decomposition, 170

Derivation

of alternative forms of the Bayes Estimators, 256

of linear Bayes estimator, 250

of ridge regression estimator, 112

as special cases of the Bayes estimator, 254

Degrees of freedom, 35

Dempster's prior, 18, 190, 583

Disadvantage of simulation, 158

Double bootstrap, 52, 156

Double f class estimators, 25

Double k class estimators, 22, 52

Edgeworth expansion, 49

Edgeworth type asymptotic expansions, 44

Eigenvalue, 35

Empirical Bayes, 61, 583,

confidence intervals, 29,34, 580

estimator (EBE), 2, 15, 54, 546, 583, 586

estimator of C.R. Rao, 373, 380, 414, 512

estimator of Dempster, 174, 403, 414, 420, 468, 587

estimator of Wind, 387, 414, 420, 459

predictor, 554

Entropy loss function, 47

Entry points to the literature, 23, 61

Estimable parametric function, 171, 195, 229, 496

Equivalent definitions of estimable parametric functions, 229

Estimating functions, 39, 46, 585

Estimator of

Aiken, 535

C.R. Rao, 373, 380, 414, 512

Dempster, 174,403,414, 420, 468, 587

Kubokawa, 60

Lindley and Smith, 162

Obenchain, 163

Wind, 387,414, 420, 459

Shao and Strawderman, 364, 586

Extended Gauss Markov Theorem,
    265
Extended unbiasedness, 252
Existence of k where ridge estimator
    has smaller MSE than LS, 120

Farebrother's 1976 Result, 123, 287
Fractional rank estimators, 40
Ferrar Glauber test, 139
Frequentist risk, 206, 265, 581

G correlated linear models, 545
"G prior", 18, 174
Γ- minimax, 42
Gauss Markov Theorem (GM), 229,
    265, 23 1, 232, 584
    proof of, 232
General forms of positive part
    estimator, 309, 311,312, 434
General Gauss-Markov model
    (GGM), 43, 53
Generalized cross validation, 125
Generalized inverse, 17, 169, 230, 584
Generalized James-Stein estimator
    (GJS), 176
Generalized linear model, 39
Generalized linear models, 32
Generalized ridge estimator
    of C.R.Rao, 121, 240, 256.264,
        270, 386, 497, 563
    of Hoerl and Kennard, 114, 129,
        386 , 556
Generalized singular value
    decomposition, 34
Geometrical motivation, 35, 233
GMANOVA model, 54
Half Kalman filter, 566
Helmert matrix, 400
Hemmerle's method of estimating
    perturbation matrix for ridge
    estimators, 147

Historical Survey, 15-61, 580
Hotellings T^2, 511

Ignaki's loss function, 48
$(IMSE)_1$, 458, 587
    comparison of estimator of
        Dempster with LS, 470
        Wind with LS, ,465
$(IMSE)_1$, 458, 587
    Dempster estimator, 469
    Inadmissible estimator, 45
Inadmissibility of the classical
    positive part, 308
Inadmissibility of the MLE, 45
    proof of , 94, 581
Inadmissibility of the James Stein
    estimator, 107, 307
Incomplete Beta and Gamma
    functions, 312
Individual Mean Square
    Error(IMSE), 117,129, 443, 445,
        551, 582
Individual Risk, 442
Inequality of C.R.Rao and Shinozaki,
    212

Jack-knife estimator, 30, 153
Jackknifed ridge estimators, 36, 298,
    585
James-Stein Estimator (JS), 1, 5-
    7, 580, 581, 583
    from Bayesian point of view, 173
    as empirical Bayes estimator, 99,
        255
    for estimable parametric functions,
        311
    from frequentist point of view, 186
    geometric motivation for, 86
    inadmissibility of, 107
    with Lindley correction (JSL), 176

motivation based on least square
  principle, 88
Stein's original argument for, 84

Kalman filter (KF), 32,33, 44, 46, 56,
  556, 589, 590
Kuks and Olman estimator, 59
Kullbach-Leibler information measure,
  48

Large sample approximations, 42
Limited translation estimators, 17,
  480, 588
  average individual risk of, 482
  average MSE of, 482
  form of, 481
  computer programs for risk of,
    488-489
Lindley correction, 40
Linear admissible estimators, 30
Linear Bayes Estimator, 40, 250, 254,
  265, 250, 373, 583, 588
Linear minimax estimator, 248, 293
Linear Model, 169, 229
Linear parametric functions, 46
LINEX loss function, 28, 42
LINEX loss function., 51
Logistic regression model, 41
L'Hospital's rule, 208

Mahalanobis loss function, 31, 51
MANOVA model, 54, 500, 533, 545,
  588
Maple programs, 224, 324, 335,
  349, 362, 429, 446, 452,
  457, 467, 474, 488, 543
Mathematica Programs, 223, 324,
  335, 349,361, 428, 446, 451,
  451, 457, 467, 473, 482,488,
  543
Matrices,

non negative definite, 239
positive definite, 121
positive semidefinite, 121
useful results, 238
Matrix Loss, 203
Maximum   likelihood   estimators
(MLE), 71,443
  inadmissibility of, 72
Mean Square Error (MSE)
  averaging over C.R. Rao's prior,
    377
  averaging over Dempsters's prior,
    190, 409, 498, 581
  averaging over a predictive loss
    function (MSEP), 211, 540
  averaging over Wind's prior, 390
  of Bayes estimator, 284
  of empirical Bayes estimators,
    377, 385, 398, 411
  frequentist of the least square and
  Bayes estimator, 266
  individual, 117,120
  of mixed estimator, 294
  of mixed estimators neglecting
    stochastic prior assumptions,
    95, 297
  of parametric functions, 117, 203,
    582
  of positive parts, 316, 326, 329,
    350
  of prediction, 196, 211, 397
  total, 117, 582
  without averaging over the prior
    distribution, 206, 385, 398,
    403
Metric space, 435
Minimax Bayes compromise
  problems, 30
Minimax estimators, 24, 55, 227, 563,
  584

Minimax linear estimator, 46, 248, 293

Minimum Mean Square Error Estimator (MMSE), 2, 66, 206, 383, 505, 583

Minimum Variance Unbiased Estimators (MVUE), 71

Minimum Variance Unbiased Linear Estimators (BLUE), 229
   in extended sense, 254, 265

Minitab, 374, 405, 561

Mixed estimator, 17, 41, 241, 242, 294, 563

Mixed model, 56

Modified relative savings loss, 482

Monte Carlo simulation, 36

Moore-Penrose generalized inverse, 17, 170

Multicollinearity problem, 24, 61, 167, 580

Multiple shrinkage estimator, 28, 40

Multivariate least square estimator, 493

Multivariate Normal Distribution, 73-75

Non-Central chi square distribution, 340

Non-full rank model, 22, 228, 229, 254

Nonnegative garrote, 50

Normal equation, 242

Normed linear spaces, 436

Numerical values of risk, RSL or RL of positive parts, 312, 319, 343

One way ANOVA model, 287

Operational ridge estimator, 38, 104

Optimization problem, 584
   for constrained least square estimator, 236

   for linear Bayes estimator, 251

Optimum c, 379, 385, 410, 417

Optimum G, 386

Optimum K, 386

Optimum ridge regression estimator, 385, 396

Order in probability, 526

Partial least squares, 50, 57

Perturbation factor k, 41
   sample estimators for, 144
   computer intensive methods for choice of, 153

Philosophy of mathematics, 45

Pitman closeness, 31, 33, 55, 580

Poisson distribution, 61

Positive part estimator, 25, 44, 45, 48, 52, 436, 432, 521, 585
   C.R. Rao's estimator, 438
   Dempster's estimator r models, 432
   PP_0–PP_4, 307, 310, 313, 433, 434, 585
   inadmissibility of classical positive part PP_0, 308
   relative savings loss of, 317, 347
   risk of general positive part, 355
   Wind's estimator, 434

Prediction
   of a future observation, 522
   of response variable in current linear model, 545

Predictive loss function, 196, 211, 550, 583

Preliminary test estimators, 24, 43, 52

Principal components regression, 24, 50

Prior Constraints, 235, 241

Prior distribution, 174, 497, 583
   multivariate normal, 254

Probit model, 54

Quadratic balanced loss functions, 45,
    586
Quadratic loss function, 203
    matrix of, 203
    predictive, 398
    total, 398

R estimable, 243
R linear models, 371, 373, 458, 544,
    586, 587
    multivariate, 512
Random effects linear model, 40
Recursive Bayes estimator, 556
Regression diagnostics, 35, 40
Relationships between different
    MSEs, 459
Relative loss, 206, 420, 509, 583
Relative weight of sample and prior
    information, 255
Relative savings loss (RSL), 17 ,103,
    381, 418 , 508, 583
Relevance function, 481
Reparametization of non-full rank
    linear model to one of full rank, 388
Ridge Estimator
    of C.R.Rao. 121, 536
    of Hoerl and Kennard, 114,
        125,129, 240, 243, 264, 270,
        566
    as special case of Bayes , minimax
        or mixed estimator, 244, 263
    as special case of mixed estimator,
        263
Ridge regression, 2, 7-15, 16, 112, 1
    20, 121,141, 144, 227, 536,
    556, 580, 584
Ridge Trace, 133, 582
Seemingly unrelated equation model
    (SURE), 19, 526, 533, 589

Shrinkage estimators, 19, 26, 43, 45,
    240
Signal to noise ratio, 274, 276
Simulated data, 405
Simulation study, 20 40, 42, 50, 54,
    55, 397, 582
Simulation studies of the performance
        of ridge type estimators by
    Delaney and Chatterjee, 164
    Dempster, Shatzoff and
        Wermuth,161, 163
    Gibbons, 162
    Hoerl, Kennard and Baldwin, 159
    Hoerl, Schuenemeyer and Hoerl.
        164
    Lawless and Wang, 160
    Mc Donald and Galerneau, 161
    Tamarkin, 165
    Wichern and Churchill, 162
Simultaneous estimation, 17, 372,
    544, 545, 586, 589
    for current individual, 547
    for different assumptions, 544, 590
.Simultaneous equations model., 19
Single linear model, 448
Singular value decomposition (SVD) ,
    54, 170 ,230
    definition, 170
Small s approximation , 27
Squared bias, 274
Standardized Regression Coefficients,
    134
Statistical decision theory, 46
Statistical Software packages , e.g.,
    SAS, Minitab, 244, 374, 485
Stein confidence sets, 53
Stein paradox, 2
Sufficient Statistic, 71, 83, 378, 384,
    392, 416
Sufficiency Principle, 83, 378,391,
    416, 507

Sum of squares decomposition, 35

Theobald's Result, 128, 224, 276
TI 82 programs, 225, 429, 544
TI 92 scripts, 226, 430,452 , 457,
        468, 474, 489, 544
Total
  MSE, 119, 397
  squared bias, 119
  variance, 119

Variance, 274
Variance components, 61, 581
Variance inflation factors, 136

W shaped, 364
Weighted bias, 276
Weighted least square estimator, 553
Weighted mean square error, 277
Weighted mixed regression, 28
Weighted signal to noise ratio, 276
White noise, 561
Wishart distribution, 379

Xestimable, 243
(X,R) estimable, 243

Printed in the United States
by Baker & Taylor Publisher Services